Essential Mathematics

Essential Mathematics

Margaret L. Lial
American River College

Stanley A. Salzman
American River College

Addison Wesley

Boston San Francisco New York
London Toronto Sydney Tokyo Singapore Madrid
Mexico City Munich Paris Cape Town Hong Kong Montreal

Publisher Jason A. Jordan
Executive Editor Maureen O'Connor
Editorial Project Management Ruth Berry and Suzanne Alley
Editorial Assistant Melissa Wright
Managing Editor/Production Supervisor Ron Hampton
Text and Cover Design Dennis Schaefer
Supplements Production Sheila C. Spinney
Production Services Elm Street Publishing Services, Inc.
Media Producer Lorie Reilly
Associate Producer Rebecca Martin
Marketing Manager Dona Kenly
Marketing Coordinator Heather Rosefsky
Prepress Services Buyer Caroline Fell
Technical Art Supervisor Joseph K. Vetere
First Print Buyer Hugh Crawford
Art Creation & Composition Services Pre-Press Company, Inc.
Cover Photo Credits Photos and Photos/Index Stock Imagery; Walter Bibikow/Index Stock Imagery; Tina Buckman/Index Stock Imagery
Photo Credits All photos from PhotoDisk except the following:

Bill Aron/PhotoEdit, p. 386 right; Bettmann/CORBIS, p. 264 bottom; Bongarts/A. Hassenstein/SportsChrome, pp. 465; Robert Brenner/PhotoEdit, p. 407; Christopher Briscoe/Photo Researchers, Inc., p. 464; John Cancalosi/Stock Boston, pp. 310 middle, 310 right; Cindy Charles/PhotoEdit, p. 28; CORBIS; Rob Crandall/The Image Works, p. 238; Bob Daemmrich/The Image Works, p. 302; © 2000 DaimlerChrysler. All rights reserved., pp. 91, 144, 456 top; Mary Kate Denny/PhotoEdit, p. 8 right; Duomo/CORBIS, p. 287; Kathy Ferguson/PhotoEdit, p. 107; Judy Gelles/Stock Boston, p. 443; Robert Ginn/PhotoEdit, p. 247; Francois Gohier/Photo Researchers, Inc., p. 216; Spencer Grant/Stock Boston, p. 256; John Griffin/The Image Works, p. 149; William Johnson/Stock Boston, p. 144; Earl & Nazima Kowall/CORBIS, p. 2; Craig Lovell/CORBIS, p. 373; Chris Marona/Photo Researchers, Inc., p. 288; Maslowski/Photo Researchers, Inc., p. 362; NASA, p. 295; Bill Nation/CORBIS-Sygma, p. 264 top; Michael Newman/PhotoEdit, pp. 1, 365, 395 left; Procter & Gamble Products, p. 395 right; Neal Preston/ CORBIS, p.111 top; Pascal Quittmelle/Stock Boston, p. 434; N. Richmond/The Image Works, p. 417 right; Rhoda Sidney/PhotoEdit, p. 61; WD-40, p. 417 left; David Young-Wolff/PhotoEdit, pp. 386.

Library of Congress Cataloging-in-Publication Data

Lial, Margaret L.
Essential mathematics./Margaret L. Lial, Stanley A. Salzman.
p. cm.
Includes index.
ISBN 0-201-75068-6 (Student Edition)
1. Mathematics. I. Salzman, Stanley A. II. Essential mathematics. III. Title.
QA37.2 .M55 2002
513'.1—dc21 00-068917

1 2 3 4 5 6 7 8 9 10 WC 04 03 02 01

Contents

List of Applications

List of Focus on Real-Data Applications

Preface

Essential Mathematics continues our ongoing commitment to provide the best possible text and supplements package that will help instructors teach and students succeed. To that end, we have addressed the diverse needs of today's students through an attractive design, updated applications and graphs, helpful features, careful explanation of concepts, and an expanded package of supplements and study aids. We have also responded to the suggestions of users and reviewers and have added many new examples and exercises based on their feedback.

The text is designed to help students achieve success in a developmental mathematics program. It provides the necessary review and coverage of whole numbers, fractions, decimals, ratio and proportion, percent, and measurement, as well as an introduction to algebra and geometry, and a preview of statistics. This text is part of a series that also includes the following books:

- *Prealgebra*, Second Edition, by Lial and Hestwood
- *Basic College Mathematics*, Sixth Edition, by Lial, Salzman, and Hestwood
- *Introductory Algebra*, Seventh Edition, by Lial, Hornsby, and McGinnis
- *Intermediate Algebra with Early Functions and Graphing*, Seventh Edition, by Lial, Hornsby, and McGinnis
- *Introductory and Intermediate Algebra*, Second Edition, by Lial, Hornsby, and McGinnis.

WHAT'S NEW IN THIS EDITION?

We believe students and instructors will welcome the following new features.

New Real-Life Applications We are always on the lookout for interesting data to use in real-life applications. As a result, we have included many new or updated examples and exercises throughout the text that focus on real-life applications of mathematics. Students are often asked to find data in a table, chart, graph, or advertisement. (See pp. 79, 143, and 269.) These applied problems provide an up-to-date flavor that will appeal to and motivate students. A comprehensive List of Applications appears at the beginning of the text.

New Figures and Photos Today's students are more visually oriented than ever. Thus, we have made a concerted effort to add mathematical figures, diagrams, tables, and graphs whenever possible. (See pp. 82, 208, and 319.) Many of the graphs use a style similar to that seen by students in today's print and electronic media. Photos have been incorporated to enhance applications in examples and exercises. (See pp. 104, 287, and 302.)

Increased Emphasis on Problem Solving Introduced at the end of Chapter 1, our six-step process for solving application problems has been refined and integrated throughout the text. The six steps, *Read*, *Plan*, *Estimate*, *Solve*, *State the Answer*, and *Check*, are emphasized in boldface type and repeated in specific problem-solving examples in Chapters 2, 3, 5, and 6. (See pp. 85, 145, and 167.)

Study Skills Component Poor study skills are a major reason why students do not succeed in math. A few generic tips sprinkled here and there are not enough to help students change their behavior. So, in this text, a desk-light icon at key points in the text directs students to a separate *Study Skills Workbook* containing carefully designed activities that correlate directly to the text. (See p. 241.) This unique workbook explains *how* the brain actually learns and remembers so students understand *why* the study skills activities will help them succeed in the course. Students are introduced to the workbook in a new To the Student section at the beginning of the text.

Focus on Real-Data Applications Each one-page activity presents a relevant and in-depth look at how mathematics is used in the real world. Designed to help instructors answer the often-asked question, "When will I ever use this stuff?," these activities ask students to read and interpret data from newspaper articles, the Internet, and other familiar, real-world sources. (See pp. 62, 168, and 354.) The activities are well-suited to collaborative work or they can be completed by individuals or used for open-ended class discussions. Instructor teaching notes and activity extensions are provided in the *Printed Test Bank and Instructor's Resource Guide*.

Diagnostic Pretest A diagnostic pretest is now included on p. xxiii of the text and covers all the material in the book, much like a sample final exam. This pretest can be used to facilitate student placement in the correct chapter according to skill level. The pretest also exposes students to the scope of the course content.

Chapter Openers New chapter openers feature real-world applications of mathematics that are relevant to students and tied to specific material within the chapters. Examples of topics include finding the best buy on cell phone service, home improvements, recipes, medical tests, fishing, and work/career applications. (See pp. 107, 185, and 311—Chapters 2, 3, and 5.)

Calculator Tips These optional tips, marked with a calculator icon, offer basic information and instruction for students using calculators in the course. (See pp. 252, 268, and 323.) In addition, An Introduction to Calculators is included as an appendix.

Test Your Word Power This new feature, incorporated into each chapter summary, helps students understand and master mathematical vocabulary. Key terms from the chapter are presented along with four possible definitions in a multiple-choice format. Answers and examples illustrating each term are provided. (See pp. 94, 231, and 349.)

What Familiar Features Have Been Retained?

We have retained the popular features of previous editions of the text, including the following:

Learning Objectives Each section begins with clearly stated, numbered objectives, and the material within sections is keyed to these objectives so that students know exactly what concepts are covered. (See pp. 129, 163, and 321.)

Cautions and Notes These color-coded and boxed comments, one of the most popular features of previous editions, warn students about common errors and emphasize important ideas throughout the exposition. (See pp. 66, 153, and 155.) There are more of these in the sixth edition than in the fifth, and the new text design makes them easier to spot; Cautions are highlighted in bright yellow and Notes are highlighted in green.

Margin Problems Margin problems, with answers immediately available on the bottom of the page, are found in every section of the text. (See pp. 79, 186, and 321.) This key feature allows students to immediately practice the material covered in the examples in preparation for the exercise sets. Based on reviewer feedback, we have added more margin exercises to the sixth edition.

Ample and Varied Exercise Sets The text contains a wealth of exercises to provide students with opportunities to practice, apply, connect, and extend the skills they are learning. Numerous illustrations, tables, graphs, and photos have been added to the exercise sets to help students visualize the problems they are solving. Problem types include skill building, writing, estimation, and calculator exercises as well as applications and correct-the-error problems. Exercises suitable for calculator work are marked in the student edition with a calculator icon. (See pp. 26, 70, and 253–256.)

Relating Concepts Exercises These sets of exercises help students tie concepts together and develop higher level problem-solving skills as they compare and contrast ideas, identify and describe patterns, and extend concepts to new situations. (See pp. 72, 162, and 256.) These exercises make great collaborative activities for pairs or small groups of students.

Summary Exercises There are four sets of in-chapter summary exercises: fractions, percent, geometry concepts, and operations with signed numbers. These exercises provide students with the all-important *mixed* practice they need at these critical points in their skill development. (See pp. 229 and 435.)

Ample Opportunity for Review Each chapter ends with a Chapter Summary featuring: Key Terms with definitions and helpful graphics, New Formulas, Test Your Word Power, and a Quick Review of each section's content with additional examples. Also included is a comprehensive set of Chapter Review Exercises keyed to individual sections, a set of Mixed Review Exercises, and a Chapter Test. Beginning with Chapter 2, each chapter concludes with a set of Cumulative Review Exercises. (See pp. 231–246, 297–310, and 445–464.)

What Content Changes Have Been Made?

We have worked hard to fine-tune and polish presentations of topics throughout the text based on user and reviewer feedback. Some of the content changes include the following:

- In Chapter 1, Whole Numbers, reading and understanding tables is now introduced in the first section, and a new section (1.9, Reading Pictographs, Bar Graphs, and Line Graphs) contributes additional application skills early in the course.
- In Chapter 4, Decimals, the sections on adding and subtracting decimals have been combined to provide more interesting options for application exercises.
- In Chapter 6, Percent, shortcuts for finding 200%, 300%, 10%, and 1% have been added to the shortcuts for finding 100% and 50%. The topics of percent proportion and identifying the parts in a percent problem are now treated in one section (6.3) rather than two sections so that students have an easier time relating these concepts.

What Supplements Are Available?

Our extensive supplements package includes testing materials, solutions manuals, tutorial software, videotapes, and a state-of-the-art Web site. For more information about any of the following supplement descriptions, please contact your Addison-Wesley sales consultant.

For the Student

***Student's Solutions Manual* (ISBN 0-321-09061-6)** The *Student's Solutions Manual* provides detailed solutions to the odd-numbered section exercises and to all margin,

Relating Concepts, Summary, Chapter Review, Chapter Test, and Cumulative Review exercises.

***Study Skills Workbook* (ISBN 0-321-09185-X)** A desk-light icon at key points in the text directs students to correlated activities in this unique workbook by Diana Hestwood and Linda Russell. The activities in the workbook teach students how to use the textbook effectively, plan their homework, take notes, make mind maps and study cards, manage study time, review a chapter, prepare for and take tests, evaluate test results, and prepare for a final exam. Students find out *how* their brains actually learn and remember, and what research tells us about ways to study effectively. A new To the Student section at the beginning of the text introduces students to the *Study Skills Workbook*.

Addison-Wesley Math Tutor Center The Addison-Wesley Math Tutor Center is staffed by qualified college mathematics instructors who tutor students on examples and exercises from their textbook. Tutoring is provided via toll-free telephone, toll-free fax, e-mail, and the Internet. White Board technology allows tutors and students to actually see problems being worked while they "talk" in real time over the Internet during tutoring sessions. The Math Tutor Center is accessed through a registration number that can be bundled free with a new textbook or purchased separately.

***Web Site:* www.MyMathLab.com** Ideal for lecture-based, lab-based, and on-line courses, MyMathLab.com provides students with a centralized point of access to the wide variety of on-line resources available with this text. The pages of the actual book are loaded into MyMathLab.com, and as students work through a section of the on-line text, they can link directly from the pages to supplementary resources (such as tutorial software, interactive animations, and audio and video clips) that provide instruction, exploration, and practice beyond what is offered in the printed book. MyMathLab.com generates personalized study plans for students and allows instructors to track all student work on tutorials, quizzes, and tests.

***InterAct Math® Tutorial Software* (ISBN 0-321-09056-X)** This interactive tutorial software provides algorithmically generated practice exercises that are correlated at the objective level to the content of the text. Every exercise in the program is accompanied by an example and a guided solution designed to involve students in the solution process. For Windows users, selected problems also include a video clip to help students visualize concepts. The software tracks student activity and scores and can generate printed summaries of students' progress. Instructors can use the InterAct Math® Plus course-management software to create, administer, and track tests and monitor student performance during practice sessions. (See For the Instructor.)

***InterAct MathXL:* www.mathxl.com** InterAct MathXL is a Web-based tutorial system that enables students to take practice tests and receive personalized study plans based on their results. Practice tests are correlated directly to the section objectives in the text, and once a student has taken a practice test, the software scores the test and generates a study plan that identifies strengths, pinpoints topics where more review is needed, and links directly to InterAct Math® tutorial software for additional practice and review. A course-management feature allows instructors to create and administer tests and view students' test results, study plans, and practice work. Students gain access to the InterAct MathXL Web site through a password-protected subscription, which can either be bundled free with a new copy of the text or purchased separately with a used book.

***Real-to-Reel Videotape Series* (ISBN 0-321-09088-8)** This series of videotapes, created specifically for *Essential Mathematics* and *Basic College Mathematics*, Sixth Edition, features an engaging team of math instructors who provide comprehensive lectures on every objective in the text. The videos include a stop-the-tape feature that encourages students to pause the video, work the presented example on their own, and then resume play to watch the video instructor go over the solution.

***Digital Video Tutor* (ISBN 0-321-09089-6)** This supplement provides the entire set of Real-to-Reel videotapes for the text in digital format on CD-ROM, making it easy and convenient for students to watch video segments from a computer, either at home or on campus. Available for purchase with the text at minimal cost, the Digital Video Tutor is ideal for distance learning and supplemental instruction.

FOR THE INSTRUCTOR

***Instructor's Solutions Manual* (ISBN 0-321-09060-8)** The *Instructor's Solutions Manual* provides complete solutions to all even-numbered section exercises.

***Answer Book* (ISBN 0-321-09058-6)** The *Answer Book* provides answers to all the exercises in the text.

***Printed Test Bank and Instructor's Resource Guide* (ISBN 0-321-09059-4)** The *Printed Test Bank* portion of this manual contains two diagnostic pretests, six free-response and two multiple-choice test forms per chapter, and two final exams. The *Instructor's Resource Guide* portion of the manual contains teaching suggestions for each chapter, additional practice exercises for every objective of every section, phonetic spellings for all key terms in the text, and teaching notes and extensions for the Focus on Real-Data Applications pages in the text.

***TestGen-EQ with QuizMaster EQ* (ISBN 0-321-09057-8)** This fully networkable software enables instructors to create, edit, and administer tests using a computerized test bank of questions organized according to the chapter content of the text. Six question formats are available, and a built-in question editor allows the user to create graphs, import graphics, and insert mathematical symbols and templates, variable numbers, or text. An "Export to HTML" feature allows practice tests to be posted to the Internet, and instructors can use QuizMaster-EQ to post quizzes to a local computer network so that students can take them on-line and have their results tracked automatically.

***InterAct Math® Plus* (ISBN 0-201-72140-6)** This networkable software provides course-management capabilities and network-based test administration for Addison-Wesley's InterAct Math® tutorial software. (See For the Student.) InterAct Math® Plus enables instructors to create and administer tests, summarize students' results, and monitor students' progress in the tutorial software, providing an invaluable teaching and tracking resource.

MathPass MathPass helps students succeed in their developmental mathematics courses by creating customized study plans based on diagnostic test results from ACT, Inc.'s Computer-Adaptive Placement Assessment and Support System (COMPASS®). MathPass pinpoints topics where the student needs in-depth study or targeted review and correlates these topics with the student's textbook and related supplements. The study plan can be saved as an HTML file that, when viewed on the Internet, links directly to text-specific, on-line resources. Instructors can add their own custom Web links to the HTML study plan. The MathPass learning system provides diagnostic assessment, focused instruction, and exit placement all in one package.

***Web Site:* www.MyMathLab.com** In addition to providing a wealth of resources for lecture-based courses, MyMathLab.com gives instructors a quick and easy way to create a complete on-line course. MyMathLab.com is hosted nationally at no cost to instructors, students, or schools, and it provides access to an interactive learning environment where all content is keyed directly to the text. Using a customized version of Blackboard™ as the course-management platform, MyMathLab.com lets instructors administer preexisting tests and quizzes or create their own, and it provides detailed tracking of all student work as well as a wide array of communication tools for course participants. Within MyMathLab.com, students link directly from on-line pages of their text to supplementary resources such as tutorial software, interactive animations, and audio and video clips.

Acknowledgments

The comments, criticisms, and suggestions of users, nonusers, instructors, and students have positively shaped this textbook over the years, and we are most grateful for the many responses we have received. The feedback gathered for this revision of the text was particularly helpful, and we especially wish to thank the following individuals who provided invaluable suggestions:

George Alexander, *University of Wisconsin Colleges*
Sonya Armstrong, *West Virginia State College*
Solveig R. Bender, *William Rainey Harper College*
Ernie Chavez, *Gateway Community College*
Terry Joe Collins, *Hinds Community College*
Martha Daniels, *Central Oregon Community College*
Donna Foster, *Piedmont Technical College*
Joe Howe, *St. Charles County Community College*
Rose Kaniper, *Burlington County College*
Douglas Lewis, *Yakima Valley Community College*
Wayne Miller, *Lee College*
Thea Philliou, *College of Santa Fe*
Richard D. Rupp, *Del Mar College*
Ellen Sawyer, *College of DuPage*
Lois Schuppig, *College of Mount St. Joseph*
Mary Lee Seitz, *Erie Community College—City Campus*
Kathryn Taylor, *Santa Ana College*
Sven Trenholm, *North Country Community College*
Bettie A. Truitt, *Black Hawk College*
Jackie Wing, *Angelina College*

Our sincere thanks go to these dedicated individuals at Addison-Wesley who worked long and hard to make this revision a success: Maureen O'Connor, Ruth Berry, Ron Hampton, Dennis Schaefer, Dona Kenly, Suzanne Alley, and Melissa Wright. Steven Pusztai of Elm Street Publishing Services provided his customary excellent production work. We are most grateful to Peg Crider for researching and writing the Focus on Real-Data Applications feature; Paul Van Erden for his accurate and useful index; Becky Troutman for preparing the comprehensive List of Applications; Abby Tanenbaum for writing the new Diagnostic Pretest; and Janis Cimperman, Luanne Galt, Mary Lee Seitz, Sam Tinsley, Thea Philliou, and Ellen Sawyer for accuracy checking the manuscript.

The ultimate measure of this textbook's success is whether it helps students master basic skills, develop problem-solving techniques, and increase their confidence in learning and using mathematics. In order for us, as authors, to know what to keep and what to improve for the next edition, we need to hear from you. Please tell us what you like and where you need additional help by sending an e-mail to math@awl.com. We appreciate your feedback.

In appreciation of your lasting support and never-ending enthusiasm: family, colleagues, and more than a generation of motivated students.

Stan Salzman

Feature Walk-Through

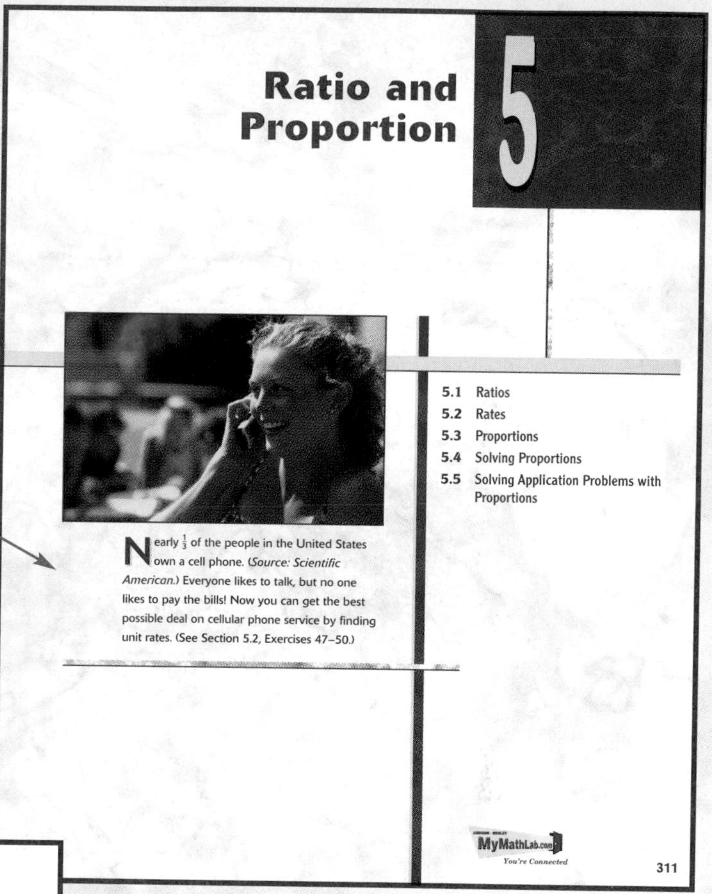
Ratio and Proportion

5

Nearly $\frac{1}{3}$ of the people in the United States own a cell phone. (*Source: Scientific American.*) Everyone likes to talk, but no one likes to pay the bills! Now you can get the best possible deal on cellular phone service by finding unit rates. (See Section 5.2, Exercises 47–50.)

MyMathLab.com You're Connected

311

New! Chapter Openers New chapter openers feature real-world applications of mathematics that are relevant to students and tied to specific material within the chapters.

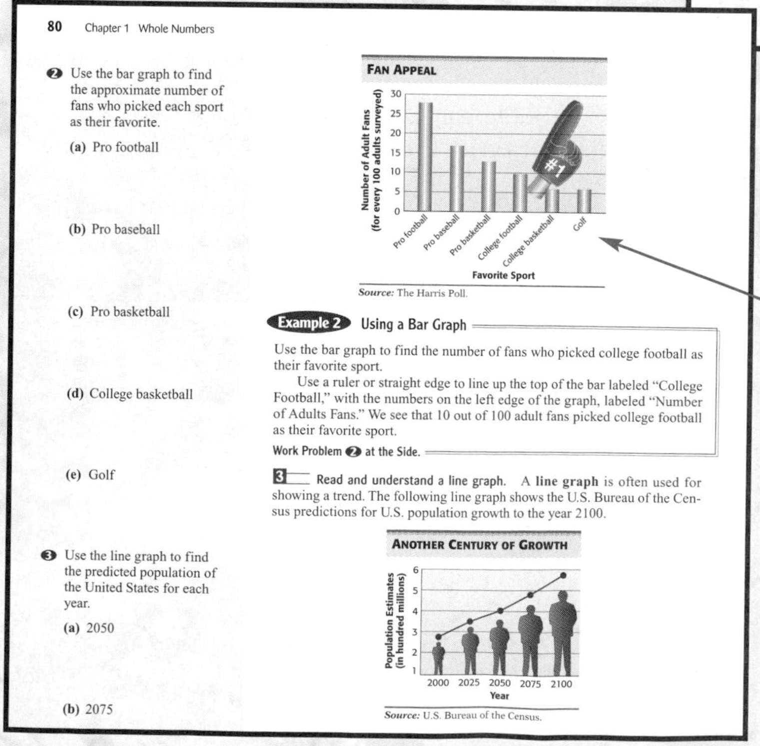
80 Chapter 1 Whole Numbers

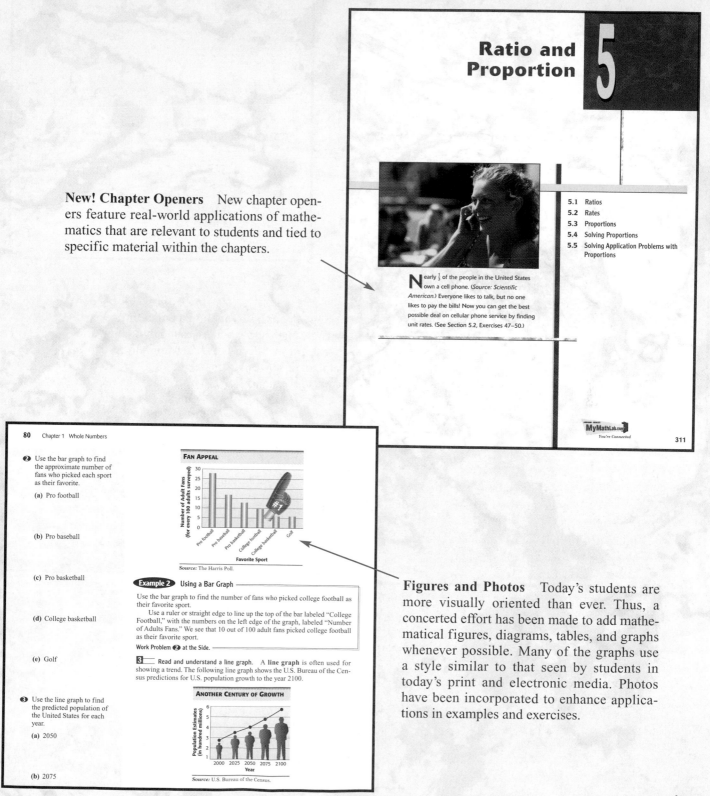

❷ Use the bar graph to find the approximate number of fans who picked each sport as their favorite.

(a) Pro football

(b) Pro baseball

(c) Pro basketball

(d) College basketball

(e) Golf

❸ Use the line graph to find the predicted population of the United States for each year.

(a) 2050

(b) 2075

Source: The Harris Poll.

Example 2 Using a Bar Graph

Use the bar graph to find the number of fans who picked college football as their favorite sport.

Use a ruler or straight edge to line up the top of the bar labeled "College Football," with the numbers on the left edge of the graph, labeled "Number of Adults Fans." We see that 10 out of 100 adult fans picked college football as their favorite sport.

Work Problem ❷ at the Side.

3 **Read and understand a line graph.** A **line graph** is often used for showing a trend. The following line graph shows the U.S. Bureau of the Census predictions for U.S. population growth to the year 2100.

Source: U.S. Bureau of the Census.

Figures and Photos Today's students are more visually oriented than ever. Thus, a concerted effort has been made to add mathematical figures, diagrams, tables, and graphs whenever possible. Many of the graphs use a style similar to that seen by students in today's print and electronic media. Photos have been incorporated to enhance applications in examples and exercises.

New! Relating Concepts These sets of exercises help students tie together topics and develop problem-solving skills as they compare and contrast ideas, identify and describe patterns, and extend concepts to new situations. These exercises make great collaborative activities for pairs or small groups of students.

72 Chapter 1 Whole Numbers

65. Mexico City Bank has total assets of 3,025,935,000 pesos. Round this amount to the nearest ten thousand, nearest million, and nearest billion pesos.

66. Round 621,999,652 to the nearest thousand, nearest ten thousand, and nearest hundred thousand.

67. In one year, the gross national product for the United States (sum of all goods and services sold) was $8,490,487,629,721. Round this amount to the nearest hundred thousand, nearest hundred million, and nearest billion. (*Source:* U.S. Department of Commerce.)

68. The U.S. Federal Food Assistance Program paid out $18,915,762,568 in food stamps. Round this amount to the nearest hundred thousand, nearest hundred million, and nearest ten billion. (*Source:* U.S. Department of Agriculture.)

RELATING CONCEPTS (Exercises 69–75) FOR INDIVIDUAL OR GROUP WORK

To see how both rounding and front end rounding are used in solving problems, ***work Exercises 69–75 in order.***

69. A number rounded to the nearest thousand is 72,000. What is the *smallest* whole number this could have been before rounding?

70. A number rounded to the nearest thousand is 72,000. What is the *largest* whole number this could have been before rounding?

71. When front end rounding is used, a whole number rounds to 8000. What is the *smallest* possible original number?

72. When front end rounding is used, a whole number rounds to 8000. What is the *largest* possible original number?

The table shows the average charitable contributions by amount of income.

CHARITABLE CONTRIBUTIONS
The average charitable contribution claimed on 1998 federal tax returns was $1075.

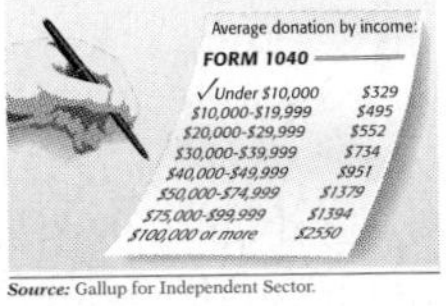
Average donation by income:
FORM 1040
✓Under $10,000 $329
$10,000-$19,999 $495
$20,000-$29,999 $552
$30,000-$39,999 $734
$40,000-$49,999 $951
$50,000-$74,999 $1379
$75,000-$99,999 $1394
$100,000 or more $2550

Source: Gallup for Independent Sector.

73. Round the amounts given for each income category to the nearest ten dollars.

74. Use front end rounding to round the amounts given for each income category.

75. **(a)** What is one advantage of using front end rounding instead of rounding to the nearest ten?

(b) What is one disadvantage?

Focus on

Real-Data Applications

Currency Exchange

When you travel between countries, you will exchange U.S. dollars for the local currency. The exchange rate between currencies changes daily, and you can easily find the updated rates using the Internet or any major newspaper. The table shown below has been extracted from the Bloomberg Currency Calculator Web page.

NORTH AMERICA/CARIBBEAN CURRENCY RATES

			Currency per 1 unit of USD	
Currency	Symbol	Value	Net Chg	Pct Chg
Canadian Dollar	CAD	1.4967	+0.0024	+0.1606
Cayman Islands	KYD	0.8282	—	—
Jamaica Dollar	JMD	45.1	+0.1	+0.2222
Mexican Peso	MXN	9.761	−0.014	−0.1432
United States Dollar	USD	1.00		

On February 4, 2001, the currency exchange rate from U.S. dollars to Mexican pesos was given as follows:

$1.00 U.S. was equivalent to 9.761 Mexican pesos

You can set up a proportion to convert dollars to pesos. For example, suppose you want to determine the number of pesos that is equivalent to $50.00.

$$\frac{\$1}{9.761 \text{ pesos}} = \frac{\$50}{x \text{ pesos}} \quad \text{or} \quad \frac{1}{9.761} = \frac{50}{x}$$

$$1 \cdot x = (9.761) \cdot 50$$

$$x = 488.05 \text{ pesos}$$

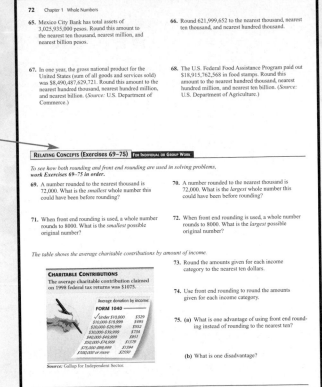

So $50 buys 488 pesos and 5 centavos.

1. Based on the currency exchange rates for February 4, 2001, find the amount of each local currency that is equivalent to $50 U.S. and find the number of U.S. dollars that is equivalent to 200 units of each local currency. Round your answers to the nearest hundredth.

(a) $50 = ______ Canadian dollars, and 200 Canadian dollars = ______ U.S. dollars.

(b) $50 = ______ Cayman Island dollars, and 200 Cayman Island dollars = ______ U.S. dollars.

(c) $50 = ______ Jamaican dollars, and 200 Jamaican dollars = ______ U.S. dollars.

2. Set up a proportion to find the number of U.S. dollars that was equivalent to 1 Mexican Peso. 1 Mexican peso was equivalent to $ ______(U.S.).

3. From Problem 2, you should recognize the conversion rate based on 1 Mexican peso as the expression $\frac{1}{9.761}$. What is the mathematical word that describes the relationship between the conversion rates 9.761 and $\frac{1}{9.761}$?

354

New! Focus on Real-Data Applications These one-page activities, found throughout the text, present even more relevant and in-depth looks at how mathematics is used in the real world. Designed to help instructors answer the often-asked question, "When will I ever use this stuff?," these activities ask students to read and interpret data from newspaper articles, the Internet, and other familiar, real sources. The activities are well suited to collaborative work and can also be completed by individuals or used for open-ended class discussions.

Calculator Tip When using a calculator to find unit prices, remember that division is *not* commutative. In Example 3 you wanted to find cost per ounce. Let the *order* of the *words* help you enter the numbers in the correct order.

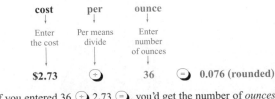

cost	per	ounce	
↓	↓	↓	
Enter the cost	Per means divide	Enter number of ounces	
↓	↓	↓	
$2.73	÷	36	= 0.076 (rounded)

If you entered 36 ÷ 2.73 =, you'd get the number of *ounces* per *dollar.* How could you use that information to find the best buy? (*Answer:* The best buy would be to get the greatest number of ounces per dollar.)

Calculator Tips These optional tips, marked with calculator icons, offer basic information and instruction for students using calculators in the course.

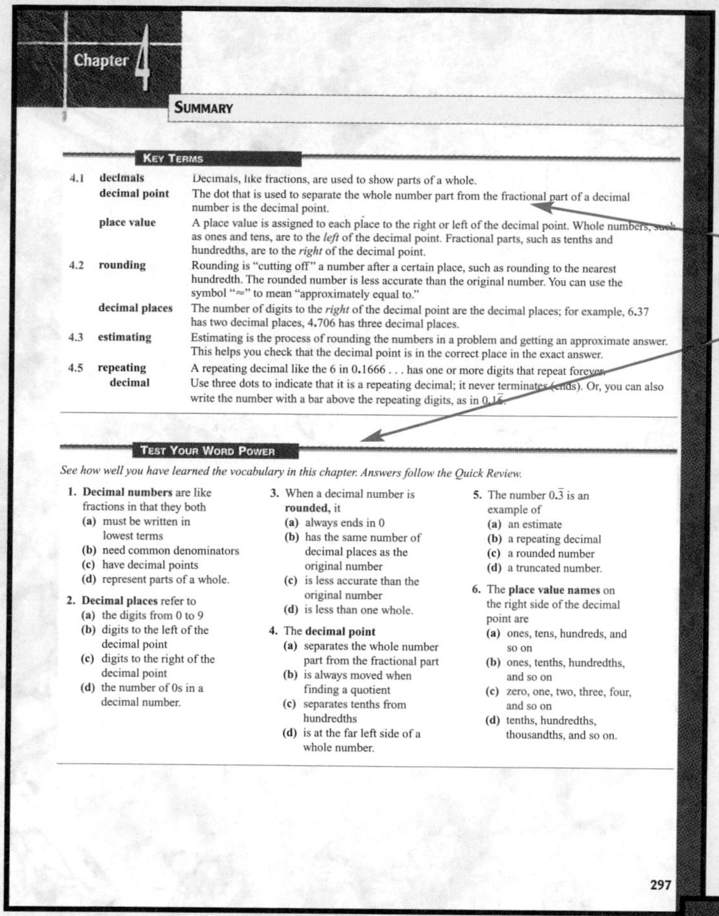
Chapter 4

SUMMARY

KEY TERMS

4.1 **decimals** Decimals, like fractions, are used to show parts of a whole.
decimal point The dot that is used to separate the whole number part from the fractional part of a decimal number is the decimal point.
place value A place value is assigned to each place to the right or left of the decimal point. Whole numbers, such as ones and tens, are to the *left* of the decimal point. Fractional parts, such as tenths and hundredths, are to the *right* of the decimal point.
4.2 **rounding** Rounding is "cutting off" a number after a certain place, such as rounding to the nearest hundredth. The rounded number is less accurate than the original number. You can use the symbol "≈" to mean "approximately equal to."
decimal places The number of digits to the *right* of the decimal point are the decimal places; for example, 6.37 has two decimal places, 4.706 has three decimal places.
4.3 **estimating** Estimating is the process of rounding the numbers in a problem and getting an approximate answer. This helps you check that the decimal point is in the correct place in the exact answer.
4.5 **repeating decimal** A repeating decimal like the 6 in 0.1666 . . . has one or more digits that repeat forever. Use three dots to indicate that it is a repeating decimal; it never terminates (ends). Or, you can also write the number with a bar above the repeating digits, as in $0.1\overline{6}$.

TEST YOUR WORD POWER

See how well you have learned the vocabulary in this chapter. Answers follow the Quick Review.

1. **Decimal numbers** are like fractions in that they both
(a) must be written in lowest terms
(b) need common denominators
(c) have decimal points
(d) represent parts of a whole.

2. **Decimal places** refer to
(a) the digits from 0 to 9
(b) digits to the left of the decimal point
(c) digits to the right of the decimal point
(d) the number of 0s in a decimal number.

3. When a decimal number is **rounded,** it
(a) always ends in 0
(b) has the same number of decimal places as the original number
(c) is less accurate than the original number
(d) is less than one whole.

4. The **decimal point**
(a) separates the whole number part from the fractional part
(b) is always moved when finding a quotient
(c) separates tenths from hundredths
(d) is at the far left side of a whole number.

5. The number $0.\overline{3}$ is an example of
(a) an estimate
(b) a repeating decimal
(c) a rounded number
(d) a truncated number.

6. The **place value names** on the right side of the decimal point are
(a) ones, tens, hundreds, and so on
(b) ones, tenths, hundredths, and so on
(c) zero, one, two, three, four, and so on
(d) tenths, hundredths, thousandths, and so on.

297

End-of-Chapter Material One of the most admired features of the Lial textbooks is the extensive and well-thought-out end of chapter material. At the end of each chapter, students will find:

Key Terms are listed, defined, and referenced back to the appropriate section number.

New! Test Your Word Power To help students understand and master mathematical vocabulary, Test Your Word Power has been incorporated in each Chapter Summary. Students are quizzed on Key Terms from the chapter in a multiple-choice format. Answers and examples illustrating each term are provided.

A Chapter Test helps students practice for the real thing.

New! Study Skills Component A desk-light icon at key points in the text directs students to a separate *Study Skills Workbook* containing activities correlated directly to the text. This unique workbook explains how the brain actually learns, so students understand *why* the study tips presented will help them succeed in the course.

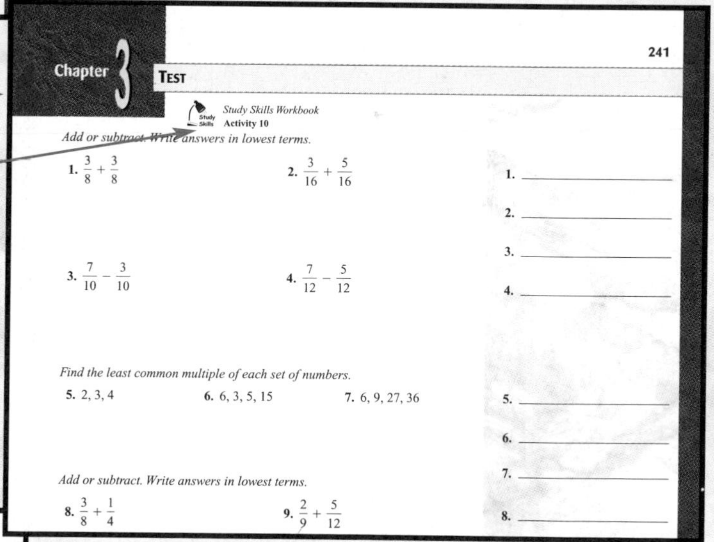
241

Chapter 3 TEST

Study Skills Workbook **Activity 10**

Add or subtract. Write answers in lowest terms.

1. $\frac{3}{8} + \frac{3}{8}$ 2. $\frac{3}{16} + \frac{5}{16}$

3. $\frac{7}{10} - \frac{3}{10}$ 4. $\frac{7}{12} - \frac{5}{12}$

Find the least common multiple of each set of numbers.

5. 2, 3, 4 6. 6, 3, 5, 15 7. 6, 9, 27, 36

Add or subtract. Write answers in lowest terms.

8. $\frac{3}{8} + \frac{1}{4}$ 9. $\frac{2}{9} + \frac{5}{12}$

1. ______
2. ______
3. ______
4. ______
5. ______
6. ______
7. ______
8. ______

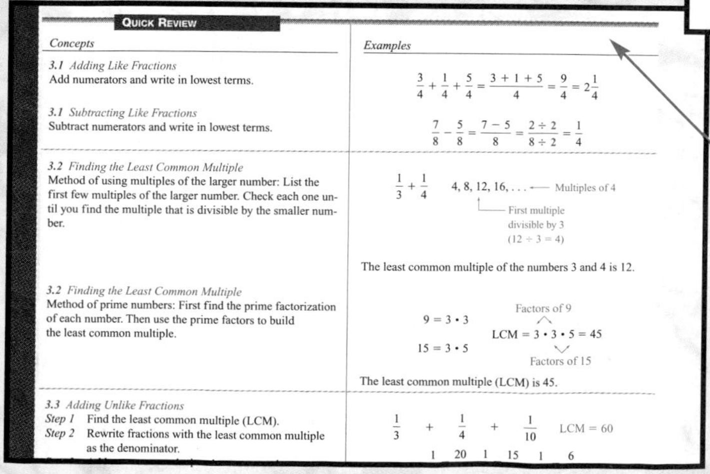
QUICK REVIEW

Concepts	Examples
3.1 Adding Like Fractions Add numerators and write in lowest terms.	$\frac{3}{4} + \frac{1}{4} + \frac{5}{4} = \frac{3+1+5}{4} = \frac{9}{4} = 2\frac{1}{4}$
3.1 Subtracting Like Fractions Subtract numerators and write in lowest terms.	$\frac{7}{8} - \frac{5}{8} = \frac{7-5}{8} = \frac{2 \div 2}{8 \div 2} = \frac{1}{4}$
3.2 Finding the Least Common Multiple Method of using multiples of the larger number: List the first few multiples of the larger number. Check each one until you find the multiple that is divisible by the smaller number.	$\frac{1}{3} + \frac{1}{4}$ 4, 8, 12, 16, . . . ← Multiples of 4; First multiple divisible by 3 (12 ÷ 3 = 4). The least common multiple of the numbers 3 and 4 is 12.
3.2 Finding the Least Common Multiple Method of prime numbers: First find the prime factorization of each number. Then use the prime factors to build the least common multiple.	9 = 3 • 3; 15 = 3 • 5; Factors of 9; LCM = 3 • 3 • 5 = 45; Factors of 15. The least common multiple (LCM) is 45.
3.3 Adding Unlike Fractions *Step 1* Find the least common multiple (LCM). *Step 2* Rewrite fractions with the least common multiple as the denominator.	$\frac{1}{3} + \frac{1}{4} + \frac{1}{10}$ LCM = 60

Quick Review sections give students not only the main concepts from the chapter (referenced back to the appropriate section), but also an adjacent example of each concept.

Review Exercises are keyed to the appropriate sections so that students can refer to examples of that type of problem if they need help.

Chapter 3 **REVIEW EXERCISES**

[3.1] *Add or subtract. Write answers in lowest terms.*

1. $\frac{5}{9}+\frac{2}{9}$
2. $\frac{2}{7}+\frac{3}{7}$
3. $\frac{1}{8}+\frac{3}{8}+\frac{2}{8}$
4. $\frac{5}{16}-\frac{3}{16}$
5. $\frac{3}{10}-\frac{1}{10}$
6. $\frac{5}{12}-\frac{3}{12}$
7. $\frac{36}{62}-\frac{10}{62}$
8. $\frac{68}{75}-\frac{43}{75}$

Solve each application problem. Write answers in lowest terms.

9. Nurse Suzie Brasher screened $\frac{7}{16}$ of her patients in her first hour on duty and $\frac{5}{16}$ of her patients in the second hour. What fraction of her patients did she screen in the two hours?
10. The Koats for Kids committee members completed $\frac{5}{8}$ of their Web-page design in the morning and $\frac{3}{8}$ in the afternoon. How much less did they complete in the afternoon than in the morning?

[3.2] *Find the least common multiple of each set of numbers.*

11. 4, 3
12. 5, 4
13. 10, 12, 20

Simplify by using the order of operations.

63. $6 \cdot \left(\frac{1}{3}\right)^2$
64. $\left(\frac{2}{3}\right)^2 \cdot 15$
65. $\left(\frac{3}{4}\right)^2 \cdot \left(\frac{8}{9}\right)^2$
66. $\frac{7}{8} \div \left(\frac{1}{8}+\frac{3}{4}\right)$
67. $\left(\frac{1}{2}\right)^2 \cdot \left(\frac{1}{4}+\frac{1}{2}\right)$
68. $\left(\frac{1}{4}\right)^3 + \left(\frac{5}{8}+\frac{3}{4}\right)$

MIXED REVIEW EXERCISES

Simplify by using the order of operations as necessary. Write answers in lowest terms and as whole or as mixed numbers when possible.

69. $\frac{5}{6}-\frac{1}{6}$
70. $\frac{7}{8}-\frac{3}{4}$
71. $\frac{29}{32}-\frac{5}{16}$
72. $\frac{1}{4}+\frac{1}{8}+\frac{5}{16}$
73. $6\frac{2}{3} - 4\frac{1}{2}$
74. $9\frac{1}{2} + 16\frac{3}{4}$
75. $7 - 1\frac{5}{8}$
76. $2\frac{3}{5} + 8\frac{5}{8} + \frac{5}{16}$
77. $32\frac{5}{12} - 17$
78. $\frac{7}{22}+\frac{3}{22}+\frac{3}{11}$
79. $\left(\frac{1}{4}\right)^2 \cdot \left(\frac{2}{5}\right)^3$
80. $\frac{3}{8} \div \left(\frac{1}{2}+\frac{1}{4}\right)$
81. $\left(\frac{2}{3}\right)^2 \cdot \left(\frac{1}{3}+\frac{1}{6}\right)$
82. $\left(\frac{2}{3}\right)^3 + \left(\frac{2}{3}-\frac{5}{9}\right)$

Mixed Review Exercises require students to solve problems without the help of section reference.

Cumulative Review Exercises **CHAPTERS 1–5**

Name the digit that has the given place value.

1. 216,475,038
   thousands
   tens
   millions
   hundred thousands
2. 340.6915
   hundredths
   ones
   ten-thousandths
   hundreds

Round each number as indicated.

3. 9903 to the nearest hundred
4. 617.0519 to the nearest tenth
5. $99.81 to the nearest dollar
6. $3.0555 to the nearest cent

First use front end rounding to round each number and estimate the answer. Then find the exact answer.

7. *Estimate:* + ___ *Exact:* 28 + 5206 + 351
8. *Estimate:* − ___ *Exact:* 63.1 − 5.692
9. *Estimate:* × ___ *Exact:* 4716 × 804
10. *Estimate:* × ___ *Exact:* 0.982 × 17.8
11. *Estimate:*) ___ *Exact:* $53\overline{)48{,}071}$
12. *Estimate:*) ___ *Exact:* $4.5\overline{)1638}$
13. *Estimate:* ___ • ___ = ___ *Exact:* $1\frac{5}{6} \cdot 3\frac{3}{5}$
14. *Estimate:* ___ ÷ ___ = ___ *Exact:* $5\frac{1}{4} \div \frac{7}{8}$
15. *Estimate:* ___ − ___ = ___ *Exact:* $2\frac{4}{5} - 1\frac{5}{6}$
16. *Estimate:* ___ + ___ = ___ *Exact:* $2\frac{9}{10} + 10\frac{1}{2}$

Cumulative Review Exercises gather various types of exercises from preceding chapters to help students remember and retain what they are learning throughout the course.

To the Student: Success in Mathematics

There are two main reasons why students have difficulty with mathematics:

- Students start in a course for which they do not have the necessary background knowledge.
- Students don't know how to study mathematics effectively.

Your instructor can help you decide whether this is the right course for you. We can give you some study tips.

Studying mathematics *is* different from studying subjects like English and history. The key to success is regular practice. This should not be surprising. After all, can you learn to play the piano or ski well without a lot of regular practice? The same is true for learning mathematics. Working problems nearly every day is the key to becoming successful. Here is a list of things that will help you succeed in studying mathematics.

1. *Attend class regularly.* Pay attention to what your instructor says and does in class, and take careful notes. In particular, note the problems the instructor works on the board and copy the complete solutions. Keep these notes separate from your homework to avoid confusion when you review them later.
2. Don't hesitate to *ask questions in class.* It is not a sign of weakness but of strength. There are always other students with the same question who are too shy to ask.
3. *Read your text carefully.* Many students read only enough to get by, usually only the examples. Reading the complete section will help you solve the homework problems. Most exercises are keyed to specific examples or objectives that will explain the procedures for working them.
4. Before you start on your homework assignment, *rework the problems the teacher worked in class.* This will reinforce what you have learned. Many students say, "I understand it perfectly when you do it, but I get stuck when I try to work the problem myself."
5. Do your homework assignment only *after reading the text* and reviewing your notes from class. Check your work against the answers in the back of the book. If you get a problem wrong and are unable to understand why, mark that problem and ask your instructor about it. Then practice working additional problems of the same type to reinforce what you have learned.
6. *Work as neatly as you can.* Write your symbols clearly, and make sure the problems are clearly separated from each other. Working neatly will help you to think clearly and also make it easier to review the homework before a test.
7. After you complete a homework assignment, *look over the text again.* Try to identify the main ideas that are in the lesson. Often they are clearly highlighted or boxed in the text.
8. *Use the chapter test at the end of each chapter as a practice test.* Work through the problems under test conditions, without referring to the text or the answers until you are finished. You may want to time yourself to see how long it takes you. When you finish, check your answers against those in the back of the book, and study the problems you missed.

9. *Keep all quizzes and tests that are returned to you,* and use them when you study for future tests and the final exam. These quizzes and tests indicate what concepts your instructor considers to be most important. Be sure to correct any problems on these tests that you missed, so you will have the corrected work to study.

10. *Don't worry if you do not understand a new topic right away.* As you read more about it and work through the problems, you will gain understanding. Each time you review a topic you will understand it a little better. Few people understand each topic completely right from the start.

Reading a list of study tips is a good start, but you may need some help actually *applying* the tips to your work in this math course.

Watch for this icon as you work in this textbook, particularly in the first few chapters. It will direct you to one of 12 activities in the *Study Skills Workbook* that comes with this text. Each activity helps you to actually *use* a study skills technique. These techniques will greatly improve your chances for success in this course.

- Find out *how your brain learns new material*. Then use that information to set up effective ways to learn math.
- Find out *why short-term memory is so short* and what you can do to help your brain remember new material weeks and months later.
- Find out *what happens when you "blank out" on a test* and simple ways to prevent it from happening.

All the activities in the *Study Skills Workbook* are brain-friendly ways to enjoy and succeed at math. Whether you need help with note taking, managing homework, taking tests, or preparing for a final exam, you'll find specific, clearly explained ideas that really work because they're based on research about how the brain learns and remembers.

Diagnostic Pretest

Study Skills — *Study Skills Workbook* **Activity 1**

[Chapter 1]

1. Use digits to write "eighty-nine million, twenty-three thousand, five hundred seven."

1. ________

2. Subtract. $\begin{array}{r} 7009 \\ -\ 2678 \\ \hline \end{array}$

2. ________

3. Divide. $20{,}213 \div 29$

3. ________

4. Round 88,658 to the nearest thousand.

4. ________

[Chapter 2]

5. Write $\frac{235}{8}$ as a mixed number.

5. ________

6. Write the prime factorization of 392 using exponents.

6. ________

7. A cake recipe calls for $2\frac{1}{4}$ cups of flour. How much flour is needed to make 5 cakes?

7. ________

8. First estimate the answer. Then multiply to find the exact answer. Write the exact answer as a mixed number.

$$5\frac{3}{4} \cdot 2\frac{1}{8}$$

8. *Estimate:* ________

Exact: ________

[Chapter 3]

9. Find the least common multiple of 5, 8, 12, and 30.

9. ________

10. Subtract. Write your answer in lowest terms.

$$\frac{9}{10} - \frac{4}{15}$$

10. ________

11. First estimate the answer. Then add to find the exact answer. Write the exact answer as a mixed number.

$$8\frac{2}{9} + 12\frac{5}{6}$$

11. *Estimate:* ________

Exact: ________

12. Use the order of operations to simplify.

$$\left(\frac{3}{4}\right)^2 + \left(\frac{5}{6} - \frac{2}{3}\right)$$

12. ________

[Chapter 4]

13. ______________

13. Round \$1.3852 to the nearest cent.

14. ______________

14. Write $6\frac{5}{9}$ as a decimal. Round to the nearest thousandth if necessary.

15. ______________

15. Use the order of operations to simplify $4.5^2 - 3.2 + 0.6 \cdot 12$.

16. ______________

16. Find the cost (to the nearest cent) of 5.3 pounds of chicken at \$1.59 per pound.

[Chapter 5]

17. ______________

17. Write the ratio "50 minutes to 4 hours" in lowest terms. Change to the same units if necessary.

18. ______________

18. Determine whether the proportion is true or false.

$$\frac{16.2}{23.6} = \frac{5.4}{8}$$

19. ______________

19. Find the unknown number in the proportion. Write your answer as a mixed number.

$$\frac{2\frac{1}{2}}{x} = \frac{\frac{3}{4}}{8}$$

20. ______________

20. Martha earns \$59.50 in 7 hours. How much will she earn in 40 hours?

[Chapter 6]

Write each number as a percent.

21. ______________

21. 0.582

22. ______________

22. $8\frac{3}{4}$

23. ______________

23. The price of a cell phone is \$128 plus 6% sales tax. Find the total cost of the phone including sales tax.

24. ______________

24. Roberto's annual salary was \$24,500. After his first year on the job, he received a raise to \$26,215. Find the percent of increase.

1 Whole Numbers

A typical Starbucks location has between 20 and 25 employees. Each of these employees is involved with work schedules, keeping track of their work hours, computing sales and sales taxes, and ordering inventory. An understanding of whole numbers is essential because each of these employees will be using mathematics as part of their jobs. (See Example 2 and Exercise 2 in Section 1.10.)

1.1 Reading and Writing Whole Numbers

Objectives

1. Identify whole numbers.
2. Give the place value of a digit.
3. Write a number in words or digits.
4. Read a table.

Study Skills Workbook
Activity 2

1 Identify the place value of the 4 in each whole number.

(a) 341

(b) 714

(c) 479

Knowing how to read and write numbers is an important step in learning mathematics.

1 Identify whole numbers. The **decimal system** of writing numbers uses the ten digits

0, 1, 2, 3, 4, 5, 6, 7, 8, 9

to write any number. For example, these digits can be used to write the **whole numbers:**

0, 1, 2, 3, 4, 5, 6, 7, 8, 9, 10, 11, 12, 13 . . .

The three dots indicate that the list goes on forever.

2 Give the place value of a digit. Each digit in a whole number has a **place value,** depending on its position in the whole number. The following place value chart shows the names of the different places used most often and has the whole number 29,022 entered.

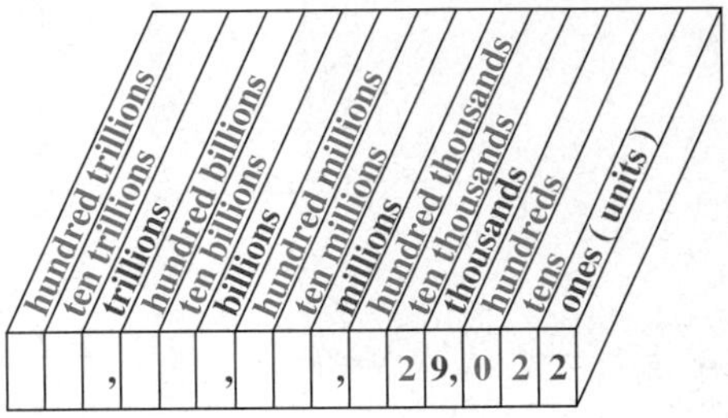

At 29,022 ft, Mt. Everest is the world's tallest mountain.

The tallest mountain in the world is Mt. Everest at 29,022 ft. Each of the 2s in 29,022 represents a different amount because of its position, or *place,* within the number. The *place value* of the 2 on the left is 2 ten thousands (20,000). The place value of the middle 2 is 2 tens (20), and the place value of the 2 on the right is 2 ones (2).

Example 1 Identifying Place Values

Identify the place value of 8 in each whole number.

(a) 28 — 8 ones **(b)** 85 — 8 tens **(c)** 869 — 8 hundreds

Notice that the value of 8 in each number is different, depending on its location (place) in the number.

Work Problem 1 at the Side.

Example 2 Identifying Place Values

Identify the place value of each digit in the number 725,283.

7 2 5 , 2 8 3

- 3 → 3 ones
- 8 → 8 tens
- 2 → 2 hundreds
- 5 → 5 thousands
- 2 → 2 ten thousands
- 7 → 7 hundred thousands

Answers

1. **(a)** tens **(b)** ones **(c)** hundreds

Notice the comma between the hundreds and thousands position in the number 725,283 in Example 2.

Work Problem ❷ at the Side.

Using Commas

Commas are used to separate each group of three digits, starting from the right. This makes numbers easier to read. (An exception: commas are frequently omitted in four-digit numbers such as 9748 or 1329.) Each three-digit group is called a **period.** Some instructors prefer to just call them **groups.**

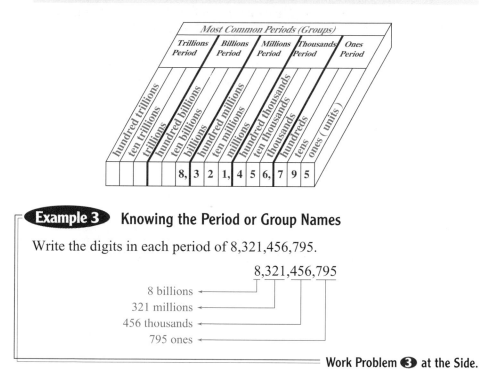

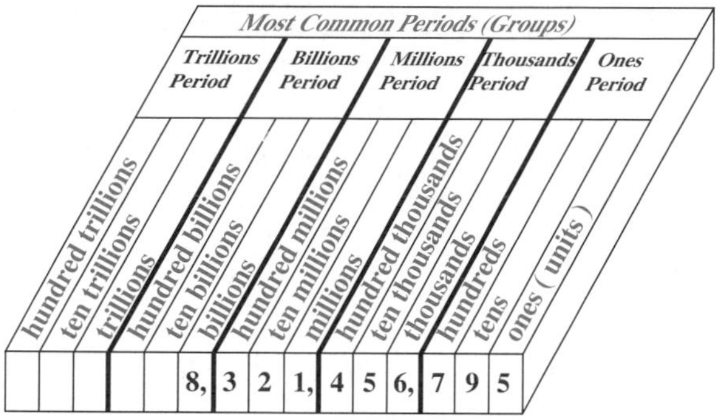

Example 3 **Knowing the Period or Group Names**

Write the digits in each period of 8,321,456,795.

8,321,456,795

8 billions ←
321 millions ←
456 thousands ←
795 ones ←

Work Problem ❸ at the Side.

Use the following rule to read a number with more than three digits.

Writing Numbers in Words

Start at the left when writing a number in words or saying it aloud. Write or say the digit names in each period (group), followed by the name of the period, except for the period name "ones," which is *not* used.

3 **Write a number in words or digits.** The following examples show how to write names for whole numbers.

Example 4 **Writing Numbers in Words**

Write each number in words.

(a) 57

This number means 5 tens and 7 ones, or 50 ones and 7 ones. Write the number as

fifty-seven.

Continued on Next Page

❷ Identify the place value of each digit.

(a) 24,386

(b) 371,942

❸ In the number 3,251,609,328 identify the digits in each period (group).

(a) billions period

(b) millions period

(c) thousands period

(d) ones period

ANSWERS

2. **(a)** 2: ten thousands
4: thousands
3: hundreds
8: tens
6: ones

(b) 3: hundred thousands
7: ten thousands
1: thousands
9: hundreds
4: tens
2: ones

3. **(a)** 3 **(b)** 251 **(c)** 609 **(d)** 328

4 Write each number in words.

(a) 15

(b) 68

(c) 293

(d) 902

5 Write each number in words.

(a) 7309

(b) 95,372

(c) 100,075,002

(d) 11,022,040,000

6 Rewrite each number using digits.

(a) one thousand, four hundred thirty-seven

(b) nine hundred seventy-one thousand, six

(c) eighty-two million, three hundred twenty-five

Answers

4. **(a)** fifteen **(b)** sixty-eight **(c)** two hundred ninety-three **(d)** nine hundred two

5. **(a)** seven thousand, three hundred nine **(b)** ninety-five thousand, three hundred seventy-two **(c)** one hundred million, seventy-five thousand, two **(d)** eleven billion, twenty-two million, forty thousand

6. **(a)** 1437 **(b)** 971,006 **(c)** 82,000,325

(b) 94

ninety-four

(c) 874

eight hundred seventy-four

(d) 601

six hundred one

CAUTION

The word *and* should never be used when writing whole numbers. You will often hear someone say "five hundred *and* twenty-two," but the use of "and" is not correct since "522" is a whole number. When you work with decimal numbers, the word *and* is used to show the position of the decimal point. For example, 98.6 is read as "ninety-eight *and* six tenths." Practice with decimal numbers is the topic of **Section 4.1.**

Work Problem **4** at the Side.

Example 5 Writing Numbers in Words by Using Period Names

Write each number in words.

(a) 725,283

seven hundred twenty-five **thousand,** two hundred eighty-three

(seven hundred twenty-five: Number in period; thousand: Name of period; two hundred eighty-three: Number in period (not necessary to write "ones"))

(b) 7835

seven **thousand,** eight hundred thirty-five

(thousand: Name of period; eight hundred thirty-five: No period name needed)

(c) 111,356,075

one hundred eleven **million,** three hundred fifty-six **thousand,** seventy-five

(d) 17,000,017,000

seventeen **billion,** seventeen **thousand**

Work Problem **5** at the Side.

Example 6 Writing Numbers in Digits

Rewrite each number using digits.

(a) seven **thousand,** eighty-five

7085

(b) two hundred fifty-six **thousand,** six hundred twelve

256,612

(c) nine **million,** five hundred fifty-nine

9,000,559

(Zeros indicate there are no thousands.)

Work Problem **6** at the Side.

Calculator Tip Does your calculator show a comma between each group of three digits? Probably not, but try entering a long number such as 34,629,075. Notice that there is no key with a comma on it, so you do not enter commas. A few calculators may show the position of the commas *above* the digits, like this

34'629'075

Most of the time you will have to write in the commas.

4 **Read a table.** A common way of showing number values is by using a **table.** Tables organize and display facts so that they are more easily understood. The following table shows some past facts and future predictions for the United States.

NUMBERS FOR THE NEW CENTURY

These estimated numbers give us a glimpse of what we can expect in the new century.

Year	1990s	2001	2020
U.S. population	261 million	281 million	338 million
Births	16 million	14 million	13 million
Household income	$42,936	$45,069	$53,375
Average salary	$21,129	$23,411	$28,050

Source: *Family Circle* magazine.

If you read from left to right along the row labeled "U.S. population," you find that the population in the 1990s was 261 million, then the population in 2001 was 281 million, and the estimated population for 2020 is 338 million.

Example 7 **Reading a Table**

Use the table to find each number, and write the number in words.

(a) The predicted household income in the year 2020

Read from left to right along the row labeled "Household income" until you reach the 2020 column and find $53,375.

Fifty-three thousand, three hundred seventy-five dollars

(b) The average salary in the 1990s

Read from left to right along the row labeled "Average salary." In the 1990s column you find $21,129.

Twenty-one thousand, one hundred twenty-nine dollars.

Work Problem 7 at the Side.

7 Use the table to find each number, and write the number in digits, or write the number in words.

(a) The number of births in 2000

(b) The predicted population in 2020

(c) Household income in the 1990s.

(d) The predicted average salary in 2020

ANSWERS

7. (a) 14,000,000
(b) 338,000,000
(c) forty-two thousand, nine hundred thirty-six dollars
(d) twenty-eight thousand, fifty dollars

Focus on

Real-Data Applications

'Til Debt Do You Part!

Bride's magazine recently released statistics comparing average wedding costs in the United States from 1990 and 1997. In 1990, the cost of the average wedding was $15,208. The average number of wedding guests had grown to 200 in 1997 compared to 170 in 1990.

Category	Average Cost in 1997
Misc. expenses (stationery, clergy, gifts, limousine)	$1260
Bouquets and other flowers	$ 756
Photography and videography	$1311
Music	$ 830
Engagement and wedding rings (bride and groom)	$4060
Rehearsal dinner	$ 698
Bride's wedding dress and headpiece	$ 989
Bridal attendants' dresses (average of 5 bridesmaids)	$ 790
Mother of the bride's apparel	$ 231
Men's formalwear (ushers, best man, groom)	$ 544
Wedding reception	$7635
Grand Total	

Source: Bride's 1997 Millennium Report: Wedding Love & Money.

1. What is the grand total of expenses shown in the chart?
2. How much more expensive was a wedding in 1997 compared to 1990?
3. The groom pays for the bouquets and flowers, the rehearsal dinner, the bride's engagement and wedding rings ($3500), the clergy ($232), and the groom's formalwear ($95). What is the total amount spent by the groom?
4. If you budgeted $38 per person for the wedding reception and you invited 200 guests to a wedding in 2001, how much money would you have spent compared to the 1997 wedding reception costs?
5. If you budget $5000 for the reception and the caterer charges $38 per person, how many guests can you invite? How much of your budget is left over?
6. If you budget $8500 for the reception and the caterer charges $38 per person, how many guests can you invite? How much of your budget is left over?
7. What type of arithmetic problem did you work to get the answers to Problems 6 and 7? What is the mathematical term for the "left over" budget?

1.1 EXERCISES

FOR EXTRA HELP: Student's Solutions Manual | MyMathLab.com | InterAct Math Tutorial Software | AW Math Tutor Center | www.mathxl.com | Digital Video Tutor CD 1 Videotape 1

Study Skills Workbook **Activity 3**

Write the digit for the given ***place value*** *in each whole number. See Examples 1 and 2.*

1. 2078
thousands
tens

2. 5139
thousands
ones

3. 18,015
ten thousands
hundreds

4. 75,229
ten thousands
ones

5. 7,628,592,183
millions
thousands

6. 1,700,225,016
billions
millions

Write the digits for the given ***period*** *(group) in each whole number. See Example 3.*

7. 2,768,543
millions
thousands
ones

8. 28,785,203
millions
thousands
ones

9. 60,000,502,109
billions
millions
thousands
ones

10. 100,258,100,006
billions
millions
thousands
ones

11. Do you think the fact that humans have four fingers and a thumb on each hand explains why we use a number system based on ten digits? Explain.

12. The decimal system uses ten digits. Fingers and toes are often referred to as digits. In your opinion, is there a relationship here? Explain.

Write each number in words. See Examples 4 and 5.

13. 23,115

14. 37,886

15. 346,009

16. 218,033

17. 25,756,665

18. 999,993,000

Write each number using digits. See Example 6.

19. thirty-two thousand, five hundred twenty-six

20. ninety-five thousand, one hundred eleven

21. ten million, two hundred twenty-three

22. one hundred million, two hundred

Write the numbers from each sentence using digits. See Example 6.

23. The Sunday newspaper says that it contains one hundred ninety-eight dollars in "money-saving" coupons. (*Source: Sacramento Bee.*)

24. The United States Postal Service set a record of two hundred eighty million, four hundred eighty-nine thousand postmarked pieces of mail on a single day. (*Source:* U.S. Postal Service.)

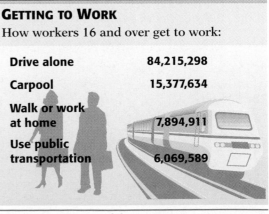

25. The Binney & Smith Company in Pennsylvania makes about two billion Crayola crayons each year. (*Source:* Binney & Smith Company.)

26. There will be four hundred seventy-three thousand new jobs for registered nurses in the year 2005. (*Source:* U.S. Department of Labor.)

27. In the year 2005, there will be five hundred thirty-two thousand new jobs for retail salespeople. (*Source:* U.S. Department of Labor.)

28. The total Temporary Assistance for Needy Families (TANF) in the United States in 1998 was twenty billion, five hundred twenty-eight million, four hundred ninety-one thousand dollars. (*Source:* U.S. Department of Health and Human Services.)

29. Rewrite eight hundred trillion, six hundred twenty-one million, twenty thousand, two hundred fifteen by using digits.

30. Rewrite 70,306,735,002,102 in words.

The table at the right shows various ways people get to work. Use the table to answer Exercises 31–34. See Example 7.

GETTING TO WORK

How workers 16 and over get to work:

Drive alone	**84,215,298**
Carpool	**15,377,634**
Walk or work at home	**7,894,911**
Use public transportation	**6,069,589**

Source: U.S. Bureau of the Census.

31. Which method of transportation is least used? Write the number in words.

32. Which method of transportation is most used? Write the number in words.

33. Find the number of people who walk to work or work at home, and write it in words.

34. Find the number of people who carpool, and write it in words.

1.2 ADDING WHOLE NUMBERS

There are four dollar bills at the left and two at the right. In all, there are six dollar bills.

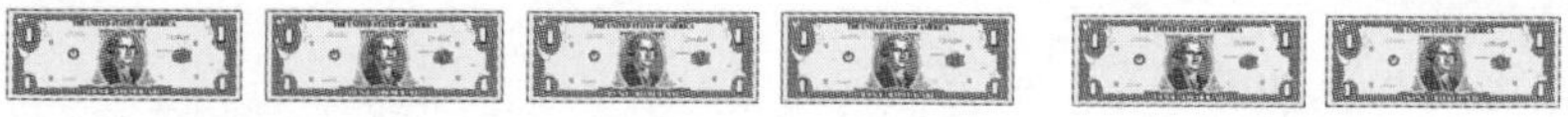

The process of finding the total is called **addition.** Here 4 and 2 were added to get 6. Addition is written with a + sign, so that

$$4 + 2 = 6.$$

1 Add two single-digit numbers. In addition, the numbers being added are called **addends,** and the resulting answer is called the **sum** or **total.**

$$\begin{array}{r l} 4 & \leftarrow \text{Addend} \\ +\ 2 & \leftarrow \text{Addend} \\ \hline 6 & \leftarrow \text{Sum (total)} \end{array}$$

Addition problems can also be written horizontally, as follows.

$$\underset{\text{Addend}}{4} \quad + \quad \underset{\text{Addend}}{2} \quad = \quad \underset{\text{Sum}}{6}$$

Commutative Property of Addition

By the **commutative property of addition,** changing the order of the addends in an addition problem does not change the sum.

For example, the sum of $4 + 2$ is the same as the sum of $2 + 4$. This allows the addition of the same numbers in a different order.

Example 1 **Adding Two Single-Digit Numbers**

Add, and then change the order of numbers to write another addition problem.

(a) $6 + 2 = 8$ and $2 + 6 = 8$

(b) $5 + 9 = 14$ and $9 + 5 = 14$

(c) $8 + 3 = 11$ and $3 + 8 = 11$

(d) $8 + 8 = 16$

Work Problem 1 at the Side.

Associative Property of Addition

By the **associative property of addition,** changing the grouping of the addends in an addition problem does not change the sum.

For example, the sum of $3 + 5 + 6$ may be found as follows.

$(3 + 5) + 6 = 8 + 6 = 14$ Parentheses tell what to do first.

Another way to add the same numbers is

$$3 + (5 + 6) = 3 + 11 = 14.$$

Either grouping gives the answer 14.

OBJECTIVES

1. Add two single-digit numbers.
2. Add more than two numbers.
3. Add when carrying is not required.
4. Add with carrying.
5. Solve application problems with carrying.
6. Check the answer in addition.

1 Add, and then change the order of numbers to write another addition problem.

(a) $6 + 3$

(b) $7 + 9$

(c) $7 + 8$

(d) $6 + 9$

ANSWERS

1. **(a)** 9; $3 + 6 = 9$ **(b)** 16; $9 + 7 = 16$ **(c)** 15; $8 + 7 = 15$ **(d)** 15; $9 + 6 = 15$

2 Add each column of numbers.

(a)
```
  6
  5
  3
  2
+ 9
```

(b)
```
  4
  6
  5
  3
+ 2
```

(c)
```
  8
  7
  9
  2
+ 1
```

(d)
```
  3
  8
  6
  4
+ 8
```

2 Add more than two numbers. To add several numbers, first write them in a column. Add the first number to the second. Add this sum to the third digit; continue until all the digits are used.

Example 2 Adding More Than Two Numbers

Add 2, 5, 6, 1, and 4.

```
   2   ⎤
       ⎬ 2 + 5 = 7
   5   ⎦       7 + 6 = 13
  (6) ─────────┘     13 + 1 = 14
  (1) ───────────────┘     14 + 4 = 18
+ (4) ─────────────────────┘
  18
```

NOTE

By the commutative and associative properties of addition, numbers may also be added starting at the bottom of a column. Adding from the top or adding from the bottom will give the same answer.

Work Problem 2 at the Side.

3 Add when carrying is not required. If numbers have two or more digits, you must arrange the numbers in columns so that the ones digits are in the same column, tens are in the same column, hundreds are in the same column, and so on. Next, you add column by column starting at the right.

Example 3 Adding without Carrying

Add 511 + 23 + 154 + 10.

First line up the numbers in columns, with the ones column at the right.

```
  Hundreds in a column
  Tens in a column
  Ones in a column

    5 1 1
      2 3
    1 5 4      Ones digits at
+     1 0      the right
```

Now start at the right and add the ones digits. Add the tens digits next, and finally, the hundreds digits.

```
    5 1 1
      2 3
    1 5 4
+     1 0
    6 9 8
```

Sum of ones
Sum of tens
Sum of hundreds

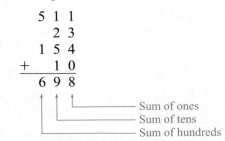

The sum of the four numbers is 698.

ANSWERS
2. **(a)** 25 **(b)** 20 **(c)** 27 **(d)** 29

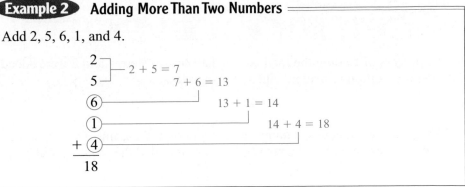

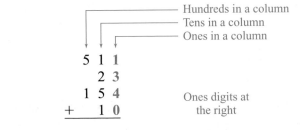

Work Problem ❸ at the Side.

4 Add with carrying. If the sum of the digits in any column is more than 9, use **carrying.**

Example 4 Adding with Carrying

Add 47 and 29.

Add ones.

$$\begin{array}{r} 47 \\ +\ 29 \\ \hline \end{array}$$

Sum of ones is 16.

Because 16 is 1 ten plus 6 ones, place 6 in the ones column and carry 1 to the tens column.

$$\begin{array}{r} 1 \\ 47 \\ +\ 29 \\ \hline 6 \end{array} \quad 7 + 9 = 16$$

Add the tens column, including the carried 1.

$$\begin{array}{r} 1 \\ 47 \\ +\ 29 \\ \hline 76 \end{array}$$

Sum of digits in tens column

Work Problem ❹ at the Side.

Example 5 Adding with Carrying

Add 324 + 7855 + 23 + 7 + 86.

Add the digits in the ones column.

$$\begin{array}{r} 2 \\ 324 \\ 7855 \\ 23 \\ 7 \\ +\quad 86 \\ \hline 5 \end{array}$$

Carry 2 to the tens column.

Sum of the ones column is 25.

Write 5 in the ones column.

In 25, the 5 represents 5 ones and is written in the ones column, while 2 represents 2 tens and is carried to the tens column.

Now add the digits in the tens column, including the carried 2.

$$\begin{array}{r} 12 \\ 324 \\ 7855 \\ 23 \\ 7 \\ +\quad 86 \\ \hline 95 \end{array}$$

Carry 1 to the hundreds column.

Sum of the tens column is 19.

Write 9 in the tens column.

Continued on Next Page

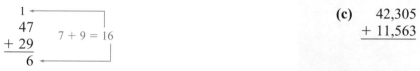

❸ Add.

(a) $\begin{array}{r} 34 \\ +\ 62 \\ \hline \end{array}$

(b) $\begin{array}{r} 478 \\ +\ 221 \\ \hline \end{array}$

(c) $\begin{array}{r} 42{,}305 \\ +\ 11{,}563 \\ \hline \end{array}$

❹ Add by carrying.

(a) $\begin{array}{r} 76 \\ +\ 19 \\ \hline \end{array}$

(b) $\begin{array}{r} 76 \\ +\ 18 \\ \hline \end{array}$

(c) $\begin{array}{r} 56 \\ +\ 37 \\ \hline \end{array}$

(d) $\begin{array}{r} 34 \\ +\ 49 \\ \hline \end{array}$

Answers

3. **(a)** 96 **(b)** 699 **(c)** 53,868

4. **(a)** 95 **(b)** 94 **(c)** 93 **(d)** 83

Add the hundreds column, including the carried 1.

```
  112
  324
 7855
   23
    7
+  86
  295
```

Carry 1 to the thousands column.

Sum of the hundreds column is 12.

Write 2 in the hundreds column.

Add the thousands column, including the carried 1.

```
  112
  324
 7855
   23
    7
+  86
 8295
```

Sum of the thousands column is 8.

Finally, 324 + 7855 + 23 + 7 + 86 = 8295.

Work Problem 5 at the Side.

NOTE

For additional speed, try to carry mentally. Do not write the carried number, but just remember it as you move to the top of the next column. Try this method. If it works for you, use it.

Work Problem 6 at the Side.

5 Solve application problems with carrying. In Section 1.10 we will describe how to solve application problems in more detail. The next two examples are application problems that require adding.

Example 6 Applying Addition Skills

On this map, the distance in miles from one location to another is written alongside the road. Find the shortest distance from Altamonte Springs to Clear Lake.

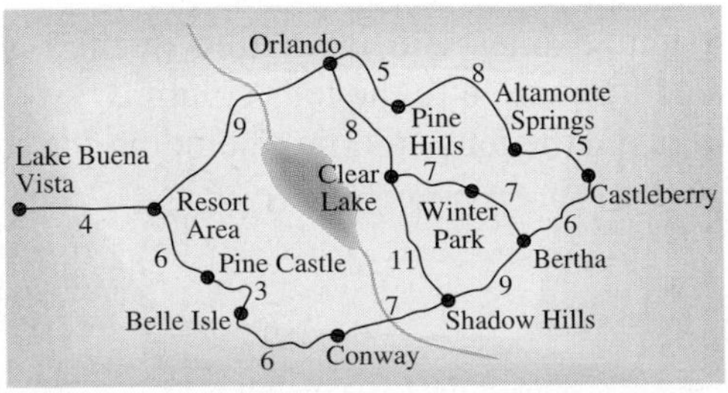

Continued on Next Page

5 Add by carrying as necessary.

(a)
```
  396
   87
   42
+ 651
```

(b)
```
  4271
   372
  8976
+  162
```

(c)
```
   57
    4
  392
  804
   51
+  27
```

(d)
```
  7821
   435
    72
   305
+ 1693
```

6 Add by carrying mentally.

(a)
```
   816
   363
    17
     2
     5
+ 7654
```

(b)
```
  3305
   650
   708
    29
    40
     6
+    3
```

(c)
```
  15,829
     765
      78
      15
       9
       7
+ 13,179
```

ANSWERS

5. **(a)** 1176 **(b)** 13,781 **(c)** 1335 **(d)** 10,326

6. **(a)** 8857 **(b)** 4741 **(c)** 29,882

Approach Add the mileage along various routes to determine the distances from Altamonte Springs to Clear Lake. Then select the shortest route.

Solution One way from Altamonte Springs to Clear Lake is through Orlando. Add the mileage numbers along this route.

8	Altamonte Springs to Pine Hills
5	Pine Hills to Orlando
+ 8	Orlando to Clear Lake
21 ⟶	miles from Altamonte Springs to Clear Lake, going through Orlando

Another way is through Bertha and Winter Park. Add the mileage numbers along this route.

5	Altamonte Springs to Castleberry
6	Castleberry to Bertha
7	Bertha to Winter Park
+ 7	Winter Park to Clear Lake
25 ⟶	miles from Altamonte Springs to Clear Lake through Bertha and Winter Park

The shortest way from Altamonte Springs to Clear Lake is through Orlando.

Work Problem 7 at the Side.

7 Use the map in Example 6 to find the shortest distance from Lake Buena Vista to Conway.

Example 7 Finding a Total Distance

Use the map in Example 6 to find the total distance from Shadow Hills to Castleberry to Orlando and back to Shadow Hills.

Approach Add the mileage from Shadow Hills to Castleberry to Orlando and back to Shadow Hills to find the total distance.

Solution Use the numbers from the map.

9	Shadow Hills to Bertha
6	Bertha to Castleberry
5	Castleberry to Altamonte Springs
8	Altamonte Springs to Pine Hills
5	Pine Hills to Orlando
8	Orlando to Clear Lake
+ 11	Clear Lake to Shadow Hills
52 ⟶	miles from Shadow Hills to Castleberry to Orlando and back to Shadow Hills

Work Problem 8 at the Side.

8 The road is closed between Orlando and Clear Lake, so this route cannot be used. Use the map in Example 6 to find the next shortest distance from Orlando to Clear Lake.

Example 8 Finding a Perimeter

Find the number of feet of fencing needed to enclose the lot shown in the figure.

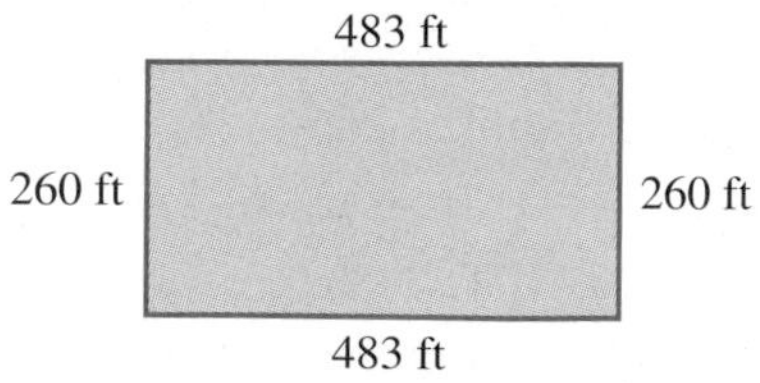

Approach Find the **perimeter,** or total distance around the lot, by adding the lengths of all the sides.

Continued on Next Page

Answers

7. 19 miles

8.

5	Orlando to Pine Hills
8	Pine Hills to Altamonte Springs
5	Altamonte Springs to Castleberry
6	Castleberry to Bertha
7	Bertha to Winter Park
+ 7	Winter Park to Clear Lake
38 miles	

9 Solve the problem.
Find the number of feet of fencing needed to enclose the lot shown.

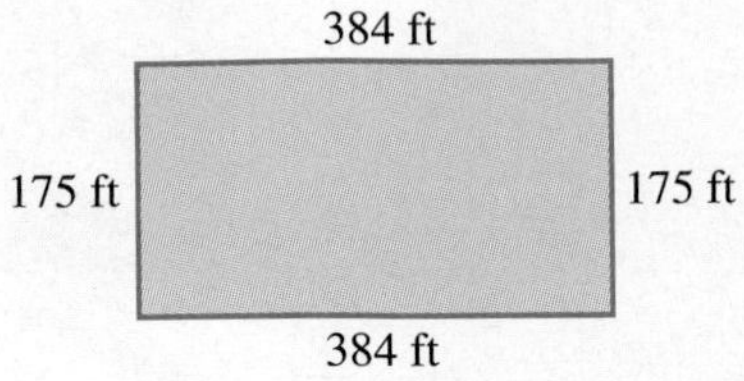

10 Check the following additions. If an answer is incorrect, find the correct answer.

(a)
```
   32
    8
    5
+  14
   59
```

(b)
```
  872
  539
   46
+ 152
 1609
```

(c)
```
   79
  218
    7
+ 639
  953
```

(d)
```
  21,892
  11,746
+ 43,925
  79,563
```

ANSWERS
9. 1118 ft
10. **(a)** correct **(b)** correct
(c) incorrect; should be 943
(d) incorrect; should be 77,563

Solution Use the lengths shown.

```
   483
   260
   483
+  260
  1486 ft
```

The amount of fencing needed is 1486 ft, which is the perimeter of (distance around) the lot.

Work Problem 9 at the Side.

6 **Check the answer in addition.** Checking the answer is an important part of problem solving. A common method for checking addition is to re-add from bottom to top. This is an application of the commutative and associative properties of addition.

Example 9 Checking Addition

Check the following addition.

```
             1428 ↑
Add down. |   738 |  Adding down and
          |    63 |  adding up should give
          |   125 |  the same answer.
          |    17 |  Add up.
          | + 485 └  Check.
          ↓  1428
```

Here the answers agree, so the sum is probably correct.

Example 10 Checking Addition

Check the following additions. Are they correct?

```
              1033 ↑  Correct, because both
(a)   785      785 |  answers are the same.
       63       63 |  Add up.
    + 185    + 185 └  Check.
     1033     1033

              2454 ↑  Error, because the
(b)   635      635 |  answers are different.
       73       73 |
      831      831 |  Add up.
    + 915    + 915 └  Check.
     2444     2444
```

Re-add to find that the correct sum is 2454.

Work Problem 10 at the Side.

1.2 Exercises

FOR EXTRA HELP: Student's Solutions Manual; MyMathLab.com; InterAct Math Tutorial Software; AW Math Tutor Center; www.mathxl.com; Digital Video Tutor CD 1 Videotape 1

Add. See Examples 1–3.

1. $\begin{array}{r} 34 \\ +\ 45 \\ \hline \end{array}$

2. $\begin{array}{r} 14 \\ +\ 13 \\ \hline \end{array}$

3. $\begin{array}{r} 56 \\ +\ 33 \\ \hline \end{array}$

4. $\begin{array}{r} 83 \\ +\ 15 \\ \hline \end{array}$

5. $\begin{array}{r} 317 \\ +\ 572 \\ \hline \end{array}$

6. $\begin{array}{r} 738 \\ +\ 261 \\ \hline \end{array}$

7. $\begin{array}{r} 318 \\ 151 \\ +\ 420 \\ \hline \end{array}$

8. $\begin{array}{r} 135 \\ 253 \\ +\ 410 \\ \hline \end{array}$

9. $\begin{array}{r} 6310 \\ 252 \\ +\ 1223 \\ \hline \end{array}$

10. $\begin{array}{r} 121 \\ 5705 \\ +\ 3163 \\ \hline \end{array}$

11. 721 + 23 + 834

12. 517 + 131 + 250

13. 1251 + 4311 + 2114

14. 3241 + 1513 + 2014

15. 12,142 + 43,201 + 23,103

16. 41,124 + 12,302 + 23,500

17. 3213 + 5715

18. 6344 + 1655

19. 38,204 + 21,020

20. 63,251 + 36,305

Add, carrying as necessary. See Examples 4 and 5.

21. $\begin{array}{r} 36 \\ +\ 78 \\ \hline \end{array}$

22. $\begin{array}{r} 91 \\ +\ 29 \\ \hline \end{array}$

23. $\begin{array}{r} 86 \\ +\ 69 \\ \hline \end{array}$

24. $\begin{array}{r} 37 \\ +\ 85 \\ \hline \end{array}$

25. $\begin{array}{r} 47 \\ +\ 74 \\ \hline \end{array}$

26. $\begin{array}{r} 79 \\ +\ 79 \\ \hline \end{array}$

27. $\begin{array}{r} 67 \\ +\ 78 \\ \hline \end{array}$

28. $\begin{array}{r} 96 \\ +\ 47 \\ \hline \end{array}$

29. $\begin{array}{r} 73 \\ +\ 29 \\ \hline \end{array}$

30. $\begin{array}{r} 68 \\ +\ 37 \\ \hline \end{array}$

31. $\begin{array}{r} 746 \\ +\ 905 \\ \hline \end{array}$

32. $\begin{array}{r} 621 \\ +\ 359 \\ \hline \end{array}$

33. $\begin{array}{r} 306 \\ +\ 848 \\ \hline \end{array}$

34. $\begin{array}{r} 798 \\ +\ 206 \\ \hline \end{array}$

35. $\begin{array}{r} 278 \\ +\ 135 \\ \hline \end{array}$

36. $\begin{array}{r} 172 \\ +\ 156 \\ \hline \end{array}$

37. $\begin{array}{r} 928 \\ +\ 843 \\ \hline \end{array}$

38. $\begin{array}{r} 686 \\ +\ 726 \\ \hline \end{array}$

39. $\begin{array}{r} 526 \\ +\ 884 \\ \hline \end{array}$

40. $\begin{array}{r} 116 \\ +\ 897 \\ \hline \end{array}$

41. $\begin{array}{r} 7968 \\ +\ 1285 \\ \hline \end{array}$

42. $\begin{array}{r} 1768 \\ +\ 8275 \\ \hline \end{array}$

43. $\begin{array}{r} 7896 \\ +\ 3728 \\ \hline \end{array}$

44. $\begin{array}{r} 9382 \\ +\ 7586 \\ \hline \end{array}$

45. $\begin{array}{r} 9625 \\ +\ 7986 \\ \hline \end{array}$

46. $\begin{array}{r} 5718 \\ 5623 \\ +\ 7436 \\ \hline \end{array}$

47. $\begin{array}{r} 9056 \\ 78 \\ 6089 \\ +\ 731 \\ \hline \end{array}$

48. $\begin{array}{r} 4022 \\ 709 \\ 8621 \\ +\ 37 \\ \hline \end{array}$

49. $\begin{array}{r} 18 \\ 708 \\ 9286 \\ +\ 636 \\ \hline \end{array}$

50. $\begin{array}{r} 1708 \\ 321 \\ 61 \\ +\ 8926 \\ \hline \end{array}$

51. $\begin{array}{r} 422 \\ 6074 \\ 435 \\ +\ 8663 \\ \hline \end{array}$

52. $\begin{array}{r} 6505 \\ 173 \\ 7044 \\ +\ 168 \\ \hline \end{array}$

53. $\begin{array}{r} 321 \\ 9603 \\ 8 \\ 21 \\ +\ 1604 \\ \hline \end{array}$

54. $\begin{array}{r} 7631 \\ 5983 \\ 7 \\ 36 \\ +\ 505 \\ \hline \end{array}$

55. $\begin{array}{r} 2109 \\ 63 \\ 16 \\ 3 \\ +\ 9887 \\ \hline \end{array}$

56. $\begin{array}{r} 322 \\ 6508 \\ 93 \\ 745 \\ 18 \\ +\ 2005 \\ \hline \end{array}$

57. $\begin{array}{r} 553 \\ 97 \\ 2772 \\ 437 \\ 63 \\ +\ 328 \\ \hline \end{array}$

58. $\begin{array}{r} 3187 \\ 810 \\ 527 \\ 76 \\ 2665 \\ +\ 317 \\ \hline \end{array}$

59. $\begin{array}{r} 413 \\ 85 \\ 9919 \\ 602 \\ 31 \\ +\ 1218 \\ \hline \end{array}$

60. $\begin{array}{r} 576 \\ 7934 \\ 60 \\ 781 \\ 5968 \\ +\ 371 \\ \hline \end{array}$

Check each addition. If an answer is incorrect, find the correct answer. See Examples 9 and 10.

61. ____ $\begin{array}{r} 621 \\ 486 \\ +\ 519 \\ \hline 1626 \end{array}$

62. ____ $\begin{array}{r} 852 \\ 679 \\ +\ 326 \\ \hline 1857 \end{array}$

63. ____ $\begin{array}{r} 179 \\ 214 \\ +\ 376 \\ \hline 759 \end{array}$

64. ____ $\begin{array}{r} 17 \\ 296 \\ 713 \\ +\ 94 \\ \hline 1220 \end{array}$

65. ____ $\begin{array}{r} 4713 \\ 28 \\ 615 \\ +\ 64 \\ \hline 5420 \end{array}$

66. ____ $\begin{array}{r} 3\,628 \\ 72 \\ 564 \\ +\ 7\,319 \\ \hline 11{,}583 \end{array}$

67. ____ $\begin{array}{r} 678 \\ 7\,952 \\ 56 \\ 718 \\ +\ 2\,173 \\ \hline 11{,}377 \end{array}$

68. ____ $\begin{array}{r} 516 \\ 8\,760 \\ 24 \\ 189 \\ +\ 1\,723 \\ \hline 11{,}212 \end{array}$

69. ____ $\begin{array}{r} 4\,714 \\ 27 \\ 77 \\ 8\,878 \\ +\ 636 \\ \hline 14{,}332 \end{array}$

70. ____ $\begin{array}{r} 6\,715 \\ 283 \\ 9\,617 \\ 13 \\ +\ 81 \\ \hline 16{,}719 \end{array}$

71. Explain the commutative property of addition in your own words. How is this used when checking an addition problem?

72. Explain the associative property of addition. How can this be used when adding columns of numbers?

For Exercises 73–76, use the map to find the shortest distance between each pair of cities. See Examples 6 and 7.

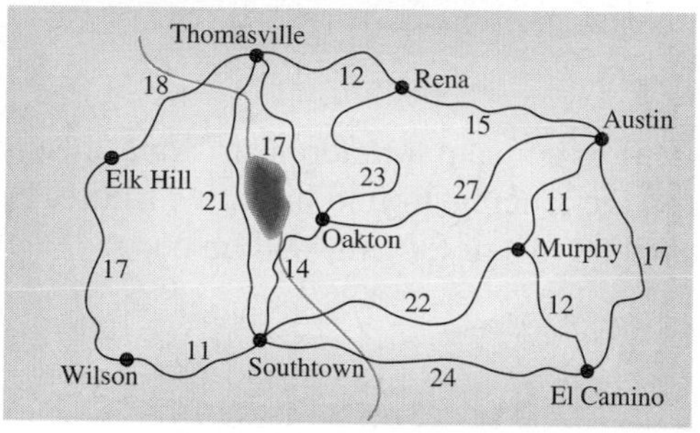

73. Southtown and Rena

74. Elk Hill and Oakton

75. Thomasville and Murphy

76. Murphy and Thomasville

Solve each application problem.

77. An oil change costs \$24 and a cooling system flush costs \$65. Find the total cost for both services.

78. Jean Hill ordered 35 gallons of interior paint and 73 gallons of exterior paint for her hardware store. How many gallons of paint did she order?

79. There are 413 women and 286 men on the sales staff. How many people are on the sales staff?

80. One department in an office building has 283 employees while another department has 218 employees. How many employees are in the two departments?

81. The Twin Lakes Food Bank raised \$3482 at a flea market and \$12,860 at their annual auction. Find the total amount raised at these two events.

82. A plane is flying at an altitude of 5924 ft. It then increases its altitude by 7284 ft. Find its new altitude.

Solve each problem involving perimeter. See Example 8.

83. A concrete curb is to be built around a parking lot. How many feet of curbing will be needed? (Disregard the driveways.)

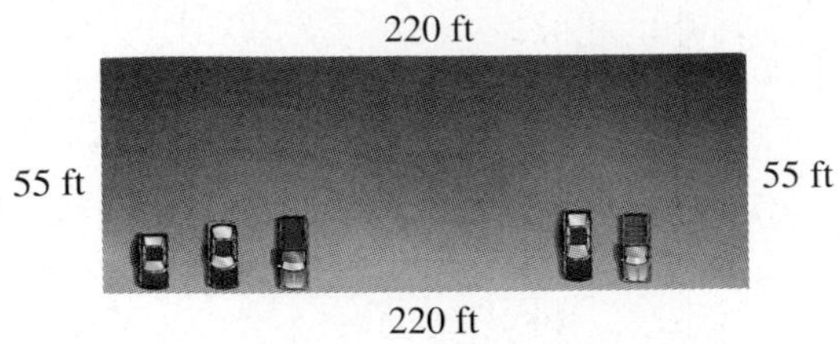

84. Because of heavy snowfall this winter, Maria needs to put new rain gutters around her entire roof. How many feet of gutters will she need?

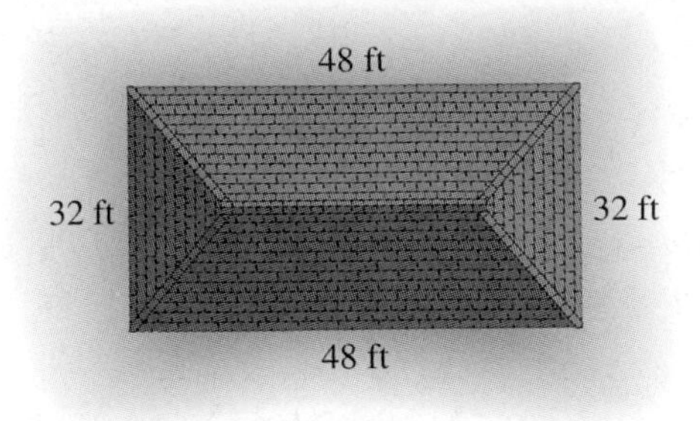

85. Martin plans to frame his back patio with redwood lumber. How many feet of lumber will he need?

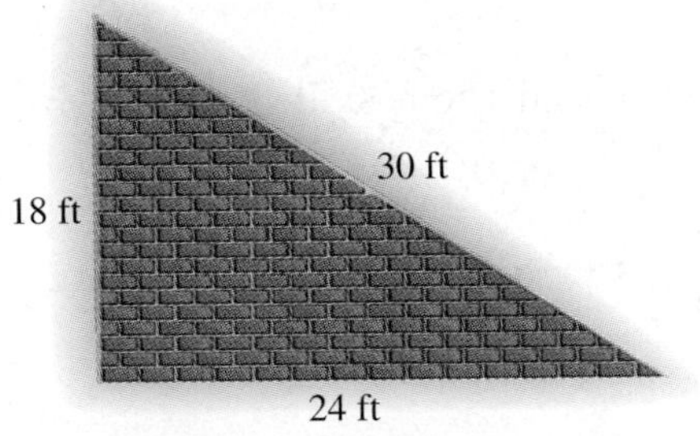

86. Due to a recent tornado, Carl Jones needs to replace all the fencing around his cow pasture. How many meters (m) of fencing will he need?

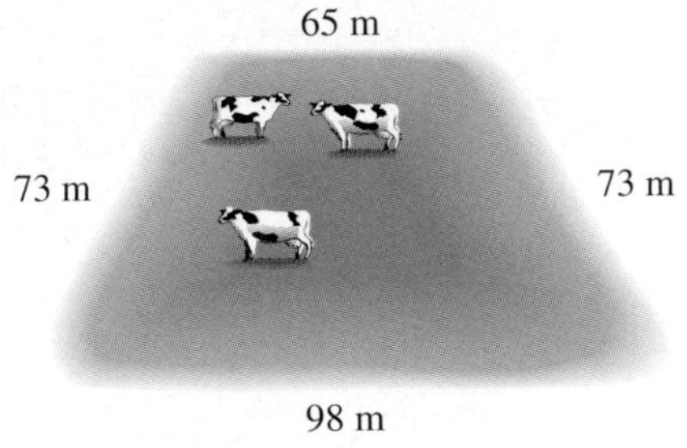

RELATING CONCEPTS (Exercises 87–94) FOR INDIVIDUAL OR GROUP WORK

Recall the place values of digits discussed in Section 1.1 and ***work Exercises 87–94 in order.***

87. Write the largest four-digit number possible using the digits 4, 1, 9, and 2. Use each digit once.

88. Using the digits 4, 1, 9, and 2, write the smallest four-digit number possible. Use each digit once.

89. Write the largest five-digit number possible using the digits 6, 2, and 7. Use each digit at least once.

90. Using the digits 6, 2, and 7, write the smallest five-digit number possible. Use each digit at least once.

91. Write the largest seven-digit number possible using the digits 4, 3, and 9. Use each digit at least twice.

92. Using the digits 4, 3, and 9, write the smallest seven-digit number possible. Use each digit at least twice.

93. Explain your rule or procedure for writing the largest number in Exercise 91.

94. Explain your rule or procedure for writing the smallest number in Exercise 92.

1.3 SUBTRACTING WHOLE NUMBERS

Suppose you have $8, and you spend $5 for gasoline. You then have $3 left. There are two different ways of looking at these numbers.

As an addition problem:

$5	+	$3	=	$8
↑		↑		↑
Amount spent		Amount left		Original amount

As a subtraction problem:

$8	−	$5	=	$3
↑	↑	↑		↑
Original amount	Subtraction symbol	Amount spent		Amount left

OBJECTIVES

1. Change addition problems to subtraction and subtraction problems to addition.
2. Identify the minuend, subtrahend, and difference.
3. Subtract when no borrowing is needed.
4. Check subtraction answers by adding.
5. Subtract by borrowing.
6. Solve application problems with subtraction.

1 Change addition problems to subtraction and subtraction problems to addition. As shown in the preceding box, an addition problem can be changed to a subtraction problem and a subtraction problem can be changed to an addition problem.

1 Write two subtraction problems for each addition problem.

(a) 6 + 1 = 7

(b) 7 + 4 = 11

(c) 15 + 22 = 37

(d) 23 + 55 = 78

Example 1 Changing Addition Problems to Subtraction

Change each addition problem to a subtraction problem.

(a) 4 + 1 = 5

Two subtraction problems are possible:

5 − 1 = 4 or 5 − 4 = 1

These figures show each subtraction problem.

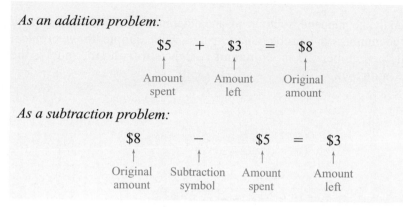

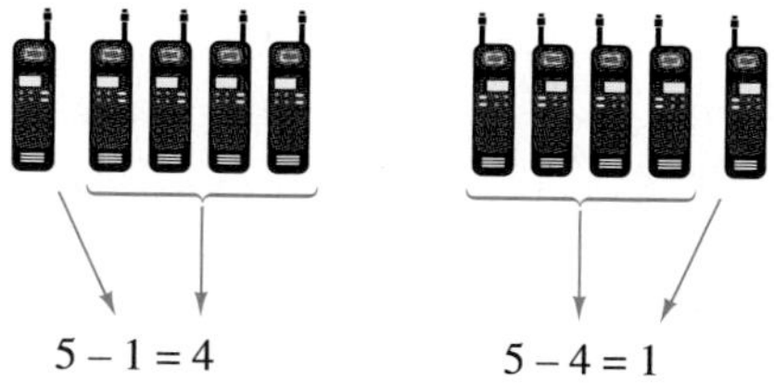

(b) 8 + 7 = 15

15 − 7 = 8 or 15 − 8 = 7

Work Problem 1 at the Side.

Example 2 Changing Subtraction Problems to Addition

Change each subtraction problem to an addition problem.

(a) 8 − 3 = 5
↓ ↓ ↓
8 = 3 + 5

It is also correct to write 8 = 5 + 3.

Continued on Next Page

ANSWERS

1. (a) 7 − 1 = 6 or 7 − 6 = 1
(b) 11 − 4 = 7 or 11 − 7 = 4
(c) 37 − 22 = 15 or 37 − 15 = 22
(d) 78 − 55 = 23 or 78 − 23 = 55

2 Write an addition problem for each subtraction problem.

(a) $6 - 4 = 2$

(b) $9 - 4 = 5$

(c) $21 - 15 = 6$

(d) $58 - 42 = 16$

3 Subtract.

(a) $\begin{array}{r} 63 \\ -\ 22 \\ \hline \end{array}$

(b) $\begin{array}{r} 47 \\ -\ 35 \\ \hline \end{array}$

(c) $\begin{array}{r} 429 \\ -\ 318 \\ \hline \end{array}$

(d) $\begin{array}{r} 3927 \\ -\ 2614 \\ \hline \end{array}$

(e) $\begin{array}{r} 5464 \\ -\ 324 \\ \hline \end{array}$

(b) $18 - 13 = 5$
$18 = 13 + 5$

(c) $29 - 13 = 16$
$29 = 13 + 16$

Work Problem 2 at the Side.

2 **Identify the minuend, subtrahend, and difference.** In subtraction, as in addition, the numbers in a problem have names. For example, in the problem $8 - 5 = 3$, the number 8 is the **minuend,** 5 is the **subtrahend,** and 3 is the **difference** or answer.

$$\underset{\text{Minuend}}{8} - \underset{\text{Subtrahend}}{5} = 3 \leftarrow \text{Difference (answer)}$$

$$\begin{array}{rl} 8 & \leftarrow \text{Minuend} \\ -\ 5 & \leftarrow \text{Subtrahend} \\ \hline 3 & \leftarrow \text{Difference} \end{array}$$

3 **Subtract when no borrowing is needed.** Subtract two numbers by lining up the numbers in columns so the digits in the ones place are in the same column. Next subtract by columns, starting at the right with the ones column.

Example 3 Subtracting Two Numbers

Subtract.

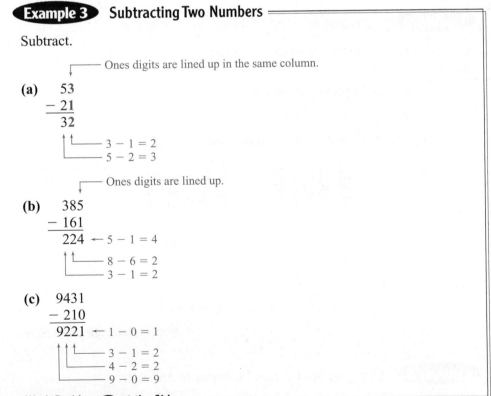

Ones digits are lined up in the same column.

(a) $\begin{array}{r} 53 \\ -\ 21 \\ \hline 32 \end{array}$

$3 - 1 = 2$
$5 - 2 = 3$

Ones digits are lined up.

(b) $\begin{array}{r} 385 \\ -\ 161 \\ \hline 224 \end{array}$ $\leftarrow 5 - 1 = 4$

$8 - 6 = 2$
$3 - 1 = 2$

(c) $\begin{array}{r} 9431 \\ -\ 210 \\ \hline 9221 \end{array}$ $\leftarrow 1 - 0 = 1$

$3 - 1 = 2$
$4 - 2 = 2$
$9 - 0 = 9$

Work Problem 3 at the Side.

4 **Check subtraction answers by adding.** Use addition to check your answer to a subtraction problem. For example, check $8 - 3 = 5$ by *adding* 3 and 5:

$$3 + 5 = 8, \quad \text{so} \quad 8 - 3 = 5 \quad \text{is correct.}$$

Answers

2. **(a)** $6 = 4 + 2$ **(b)** $9 = 4 + 5$ **(c)** $21 = 15 + 6$ **(d)** $58 = 42 + 16$

3. **(a)** 41 **(b)** 12 **(c)** 111 **(d)** 1313 **(e)** 5140

Example 4 Checking Subtraction by Using Addition

Check each answer.

(a)
$$\begin{array}{r} 89 \\ -\ 47 \\ \hline 42 \end{array}$$

Rewrite as an addition problem, as shown in Example 2.

Subtraction problem
$$\begin{array}{r} 89 \\ -\ 47 \\ \hline 42 \\ \hline \mathbf{89} \end{array}$$
Addition problem
$$\begin{array}{r} \mathbf{47} \\ \mathbf{+\ 42} \\ \hline \mathbf{89} \end{array}$$

Because 47 + 42 = 89, the subtraction was done correctly.

(b) 72 − 41 = 21

Rewrite as an addition problem.

$$72 = 41 + 21$$

But, 41 + 21 = 62, not 72, so the subtraction was done incorrectly. Rework the original subtraction to get the correct answer, 31.

(c)
$$\begin{array}{r} 374 \\ -\ 141 \\ \hline 233 \end{array}$$
Match: 141 + 233 = 374

The answer checks.

Work Problem 4 at the Side.

5 Subtract by borrowing. When a digit in the minuend is less than the one directly below it we subtract by using **borrowing.**

Example 5 Subtracting with Borrowing

Subtract 19 from 57.

Write the problem.

$$\begin{array}{r} 57 \\ -\ 19 \\ \hline \end{array}$$

In the ones column, 7 is **less** than 9, so to subtract, we must **borrow a 10** from the 5 (which represents 5 tens, or 50).

50 − 10 = 40 → 4 17 ← 10 + 7 = 17

$$\begin{array}{r} \not{5}\ \not{7} \\ -\ 1\ 9 \\ \hline \end{array}$$

Now we can subtract 17 − 9 in the ones column and then 4 − 1 in the tens column.

$$\begin{array}{r} 4\ 17 \\ \not{5}\ \not{7} \\ -\ 1\ 9 \\ \hline 3\ 8 \end{array}$$
Difference

Finally, 57 − 19 = 38. Check by adding 19 and 38; you should get 57.

Work Problem 5 at the Side.

4 Use addition to determine whether each answer is correct. If incorrect, what should it be?

(a) $\begin{array}{r} 76 \\ -\ 45 \\ \hline 31 \end{array}$

(b) $\begin{array}{r} 53 \\ -\ 22 \\ \hline 21 \end{array}$

(c) $\begin{array}{r} 374 \\ -\ 251 \\ \hline 113 \end{array}$

(d) $\begin{array}{r} 7531 \\ -\ 4301 \\ \hline 3230 \end{array}$

5 Subtract.

(a) $\begin{array}{r} 58 \\ -\ 19 \\ \hline \end{array}$

(b) $\begin{array}{r} 86 \\ -\ 38 \\ \hline \end{array}$

(c) $\begin{array}{r} 41 \\ -\ 27 \\ \hline \end{array}$

(d) $\begin{array}{r} 863 \\ -\ 47 \\ \hline \end{array}$

(e) $\begin{array}{r} 762 \\ -\ 157 \\ \hline \end{array}$

ANSWERS

4. **(a)** correct **(b)** incorrect, should be 31 **(c)** incorrect, should be 123 **(d)** correct

5. **(a)** 39 **(b)** 48 **(c)** 14 **(d)** 816 **(e)** 605

6 Subtract.

(a) $\begin{array}{r} 536 \\ -\ 75 \\ \hline \end{array}$

(b) $\begin{array}{r} 348 \\ -\ 79 \\ \hline \end{array}$

(c) $\begin{array}{r} 477 \\ -\ 389 \\ \hline \end{array}$

(d) $\begin{array}{r} 1437 \\ -\ 988 \\ \hline \end{array}$

(e) $\begin{array}{r} 8739 \\ -\ 3892 \\ \hline \end{array}$

Example 6 Subtracting with Borrowing

Subtract by borrowing as necessary.

(a) $\begin{array}{r} 7856 \\ -\ 137 \\ \hline \end{array}$

There is no need to borrow, as 4 is greater than 3.

10 + 6 = 16

$$\begin{array}{rcccc} & & & 4 & 16 \\ & 7 & 8 & \not{5} & \not{6} \\ - & & 1 & 3 & 7 \\ \hline & 7 & 7 & 1 & 9 \end{array} \quad \text{Difference}$$

(b) $\begin{array}{r} 635 \\ -\ 546 \\ \hline \end{array}$

600 − 100 = 500

100 + 20 = 120 (12 tens = 120)

10 + 5 = 15

$$\begin{array}{rccc} & 5 & 12 & 15 \\ & \not{6} & \not{3} & \not{5} \\ - & 5 & 4 & 6 \\ \hline & & 8 & 9 \end{array} \quad \text{Difference}$$

(c) $\begin{array}{r} 3648 \\ -\ 1769 \\ \hline \end{array}$

$$\begin{array}{rcccc} & 2 & 15 & 13 & 18 \\ & \not{3} & \not{6} & \not{4} & \not{8} \\ - & 1 & 7 & 6 & 9 \\ \hline & 1 & 8 & 7 & 9 \end{array}$$

Work Problem 6 at the Side.

Sometimes a minuend has 0s (zeros) in some of the positions. In such cases, borrowing may be a little more complicated than what we have shown so far.

Example 7 Borrowing with Zeros

Subtract.

$$\begin{array}{r} 4607 \\ -\ 3168 \\ \hline \end{array}$$

It is not possible to borrow from the tens position. Instead we must first borrow from the hundreds position.

600 − 100 = 500   100 + 0 = 100

$$\begin{array}{rcccc} & & 5 & 10 & \\ & 4 & \not{6} & \not{0} & 7 \\ - & 3 & 1 & 6 & 8 \\ \hline \end{array}$$

Now we may borrow from the tens position.

100 − 10 = 90

10 + 7 = 17

$$\begin{array}{rcccc} & & & 9 & \\ & & 5 & \not{10} & 17 \\ & 4 & \not{6} & \not{0} & \not{7} \\ - & 3 & 1 & 6 & 8 \\ \hline & & & & 9 \end{array}$$

Continued on Next Page

ANSWERS

6. (a) 461 (b) 269 (c) 88 (d) 449 (e) 4847

Complete the problem.

```
      9
    5 1̸0̸ 17
  4 6̸ 0̸ 7̸
– 3 1 6 8
  1 4 3 9   Difference
```

Check by adding 1439 and 3168; you should get 4607.

Work Problem 7 at the Side.

Example 8 Borrowing with Zeros

Subtract.

(a) $\begin{array}{r} 708 \\ -\ 149 \\ \hline \end{array}$

```
100 + 0 = 100        100 – 10 = 90 (9 tens = 90)
700 – 100 = 600      10 + 8 = 18

       9
    6 1̸0̸ 18
    7̸ 0̸ 8̸
  – 1 4 9
    5 5 9
```

(b) $\begin{array}{r} 380 \\ -\ 276 \\ \hline \end{array}$

```
80 – 10 = 70 (7 tens = 70)     10 + 0 = 10

      7 10
    3 8̸ 0̸
  – 2 7 6
    1 0 4
```

(c) $\begin{array}{r} 9000 \\ -\ 6999 \\ \hline \end{array}$

```
      9  9
    8 1̸0̸ 1̸0̸ 10
    9̸ 0̸ 0̸ 0̸
  – 6 9 9 9
    2 0 0 1
```

Work Problem 8 at the Side.

As we have seen, an answer to a subtraction problem can be checked by adding.

Example 9 Checking Subtraction by Using Addition

Using addition to check each answer.

(a)

	Check	
613	275	
– 275	+ 338	
338	613	Correct

(Match: 613 and 613)

Continued on Next Page

7 Subtract.

(a) $\begin{array}{r} 405 \\ -\ 363 \\ \hline \end{array}$

(b) $\begin{array}{r} 304 \\ -\ 237 \\ \hline \end{array}$

(c) $\begin{array}{r} 5073 \\ -\ 1632 \\ \hline \end{array}$

8 Subtract.

(a) $\begin{array}{r} 308 \\ -\ 159 \\ \hline \end{array}$

(b) $\begin{array}{r} 480 \\ -\ 275 \\ \hline \end{array}$

(c) $\begin{array}{r} 1570 \\ -\ 983 \\ \hline \end{array}$

(d) $\begin{array}{r} 7001 \\ -\ 5193 \\ \hline \end{array}$

(e) $\begin{array}{r} 4000 \\ -\ 1782 \\ \hline \end{array}$

ANSWERS

7. **(a)** 42 **(b)** 67 **(c)** 3441

8. **(a)** 149 **(b)** 205 **(c)** 587 **(d)** 1808 **(e)** 2218

9 Use addition to check each answer. If the answer is incorrect, find the correct answer.

(a)
$$
\begin{array}{r}
357 \\
- 168 \\
\hline
189
\end{array}
$$

(b)
$$
\begin{array}{r}
570 \\
- 328 \\
\hline
252
\end{array}
$$

(c)
$$
\begin{array}{r}
14{,}726 \\
- 8\ 839 \\
\hline
5\ 887
\end{array}
$$

10 Use the table from Example 10 to find, on average,

(a) how much more each year a person with a Bachelor's degree earns than a person with an Associate of Arts degree.

(b) how much more each year a person with an Associate of Arts degree earns than a person who is not a high school graduate.

(b)
$$
\begin{array}{r}
1915 \\
- 1635 \\
\hline
280
\end{array}
\quad\text{Match}\quad
\begin{array}{r}
\text{Check} \\
1635 \\
+ 280 \\
\hline
1915
\end{array}
\quad\text{Correct}
$$

(c)
$$
\begin{array}{r}
15{,}803 \\
- 7\ 325 \\
\hline
8\ 578
\end{array}
\quad\text{No match}\quad
\begin{array}{r}
\text{Check} \\
7\ 325 \\
+ 8\ 578 \\
\hline
15{,}903
\end{array}
\quad\text{Error}
$$

Rework the original problem to get the correct answer, 8478.

Work Problem 9 at the Side.

6 ▭ **Solve application problems with subtraction.** As shown in the next example, subtraction can be used to solve an application problem.

Example 10 Applying Subtraction Skills

Use the table to find how much more, on average, a person with an Associate of Arts degree earns each year than a high school graduate.

STUDENTS OF SUCCESS
The more education adults get, the higher their annual pay.

Education Level	Average Salary
Not a high-school graduate	$20,442
High-school graduate	$27,038
Some college	$31,128
Associate of Arts degree	$33,425
Bachelor's degree	$44,523
Master's degree	$55,384
Doctoral degree	$72,099
Professional degree	$98,197

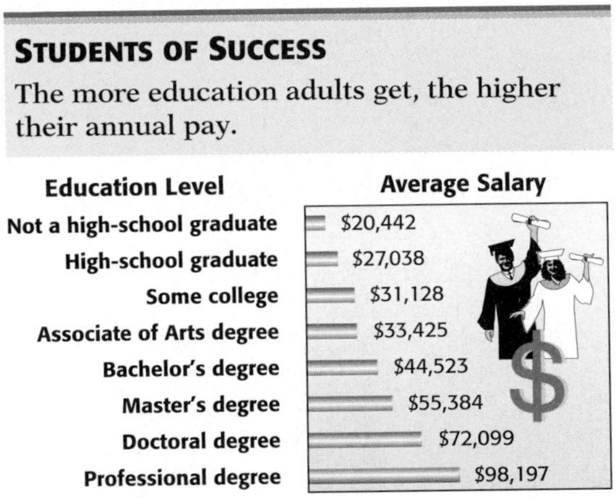

Source: U.S. Bureau of the Census.

Approach The average salary for a person with an Associate of Arts degree is $33,425 each year and the average for a high school graduate is $27,038. Find how much more a college graduate earns by subtracting $27,038 from $33,425.

Solution
$$
\begin{array}{r}
\$33{,}425 \\
- 27{,}038 \\
\hline
\$\ 6\ 387
\end{array}
\quad
\begin{array}{l}
\text{Associate of Arts degree} \\
\text{High school graduate} \\
\text{More earnings}
\end{array}
$$

On average, a person with an Associate of Arts degree earns $6387 more each year than a high school graduate.

Work Problem 10 at the Side.

1.3 Exercises

FOR EXTRA HELP — Student's Solutions Manual · MyMathLab.com · InterAct Math Tutorial Software · AW Math Tutor Center · www.mathxl.com · Digital Video Tutor CD 1 Videotape 1

Work each subtraction problem. Use addition to check each answer. See Examples 3 and 4.

1. 54 − 23	**2.** 19 − 12	**3.** 86 − 53	**4.** 78 − 35	**5.** 77 − 60
6. 95 − 71	**7.** 335 − 122	**8.** 602 − 301	**9.** 552 − 451	**10.** 888 − 215
11. 6821 − 610	**12.** 4420 − 310	**13.** 5546 − 2134	**14.** 1875 − 1362	**15.** 6259 − 4148
16. 9654 − 4323	**17.** 24,392 − 11,232	**18.** 57,921 − 34,801	**19.** 46,253 − 5 143	**20.** 75,904 − 3 702

Use addition to check each subtraction problem. If an answer is not correct, find the correct answer. See Example 4.

21. 43 − 31 12	**22.** 76 − 43 33	**23.** 89 − 27 63	**24.** 47 − 35 13	**25.** 382 − 261 131
26. 754 − 342 412	**27.** 4683 − 3542 1141	**28.** 5217 − 4105 1132	**29.** 8643 − 1421 7212	**30.** 9428 − 3124 6324

Subtract by borrowing as necessary. See Examples 5–8.

31. 45 − 27	**32.** 86 − 28	**33.** 94 − 49	**34.** 98 − 69	**35.** 57 − 38
36. 83 − 55	**37.** 828 − 547	**38.** 916 − 618	**39.** 771 − 252	**40.** 973 − 788

41. $\begin{array}{r} 9861 \\ -\ 684 \\ \hline \end{array}$ **42.** $\begin{array}{r} 6171 \\ -\ 1182 \\ \hline \end{array}$ **43.** $\begin{array}{r} 9988 \\ -\ 2399 \\ \hline \end{array}$ **44.** $\begin{array}{r} 3576 \\ -\ 1658 \\ \hline \end{array}$ **45.** $\begin{array}{r} 38{,}335 \\ -\ 29{,}476 \\ \hline \end{array}$

46. $\begin{array}{r} 61{,}278 \\ -\ 3\ 559 \\ \hline \end{array}$ **47.** $\begin{array}{r} 40 \\ -\ 37 \\ \hline \end{array}$ **48.** $\begin{array}{r} 80 \\ -\ 73 \\ \hline \end{array}$ **49.** $\begin{array}{r} 60 \\ -\ 37 \\ \hline \end{array}$ **50.** $\begin{array}{r} 70 \\ -\ 27 \\ \hline \end{array}$

51. $\begin{array}{r} 308 \\ -\ 289 \\ \hline \end{array}$ **52.** $\begin{array}{r} 600 \\ -\ 599 \\ \hline \end{array}$ **53.** $\begin{array}{r} 4041 \\ -\ 1208 \\ \hline \end{array}$ **54.** $\begin{array}{r} 4602 \\ -\ 2063 \\ \hline \end{array}$ **55.** $\begin{array}{r} 9305 \\ -\ 1530 \\ \hline \end{array}$

56. $\begin{array}{r} 7120 \\ -\ 6033 \\ \hline \end{array}$ **57.** $\begin{array}{r} 1580 \\ -\ 1077 \\ \hline \end{array}$ **58.** $\begin{array}{r} 3068 \\ -\ 2105 \\ \hline \end{array}$ **59.** $\begin{array}{r} 2006 \\ -\ 1850 \\ \hline \end{array}$ **60.** $\begin{array}{r} 8203 \\ -\ 5365 \\ \hline \end{array}$

61. $\begin{array}{r} 8240 \\ -\ 6056 \\ \hline \end{array}$ **62.** $\begin{array}{r} 7050 \\ -\ 6045 \\ \hline \end{array}$ **63.** $\begin{array}{r} 8503 \\ -\ 2816 \\ \hline \end{array}$ **64.** $\begin{array}{r} 16{,}004 \\ -\ 5\ 087 \\ \hline \end{array}$ **65.** $\begin{array}{r} 80{,}705 \\ -\ 61{,}667 \\ \hline \end{array}$

66. $\begin{array}{r} 81{,}000 \\ -\ 55{,}456 \\ \hline \end{array}$ **67.** $\begin{array}{r} 66{,}000 \\ -\ 34{,}444 \\ \hline \end{array}$ **68.** $\begin{array}{r} 77{,}000 \\ -\ 65{,}308 \\ \hline \end{array}$ **69.** $\begin{array}{r} 20{,}080 \\ -\ 13{,}496 \\ \hline \end{array}$ **70.** $\begin{array}{r} 80{,}056 \\ -\ 23{,}869 \\ \hline \end{array}$

Use addition to check each subtraction problem. If an answer is incorrect, find the correct answer. See Example 9.

71. $\begin{array}{r} 5382 \\ -\ 4634 \\ \hline 748 \end{array}$ **72.** $\begin{array}{r} 1671 \\ -\ 1325 \\ \hline 1346 \end{array}$ **73.** $\begin{array}{r} 2548 \\ -\ 2278 \\ \hline 270 \end{array}$ **74.** $\begin{array}{r} 5274 \\ -\ 1130 \\ \hline 4144 \end{array}$

75. $\begin{array}{r} 76{,}326 \\ -\ 44{,}539 \\ \hline 31{,}787 \end{array}$ **76.** $\begin{array}{r} 82{,}357 \\ -\ 14{,}396 \\ \hline 68{,}961 \end{array}$ **77.** $\begin{array}{r} 36{,}778 \\ -\ 17{,}405 \\ \hline 19{,}373 \end{array}$ **78.** $\begin{array}{r} 34{,}821 \\ -\ 17{,}735 \\ \hline 17{,}735 \end{array}$

79. An addition problem can be changed to a subtraction problem and a subtraction problem can be changed to an addition problem. Give two examples of each to demonstrate this.

80. Can you use the commutative and the associative properties in subtraction? Explain.

Solve each application problem. See Example 10.

81. A man burns 103 calories during 30 minutes of bowling while a woman burns 88 calories during 30 minutes of bowling. How many fewer calories did the woman burn than the man?

82. Lillian Kim has $729 in her checking account. After she writes a check to the bookstore for $249, how much is remaining in her account?

83. Toronto's skyline is dominated by the CN Tower, which rises 1821 ft. The Sears Tower in Chicago is 1454 ft high. Find the difference in height between the two structures.

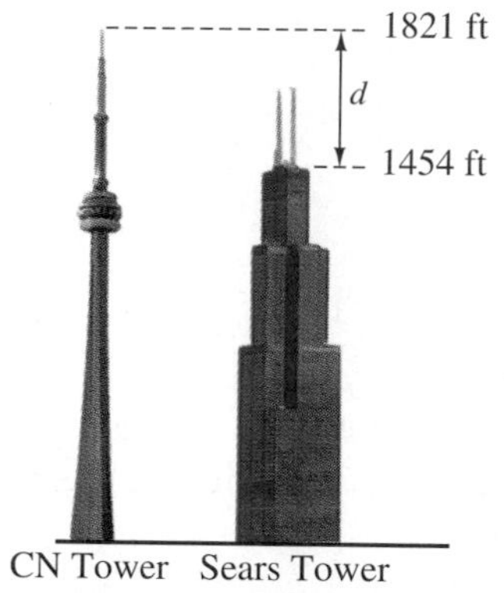

84. The fastest animal in the world, the peregrine falcon, dives at 217 miles per hour (mph). A Boeing 747 cruises at 580 mph. How much faster is the plane?

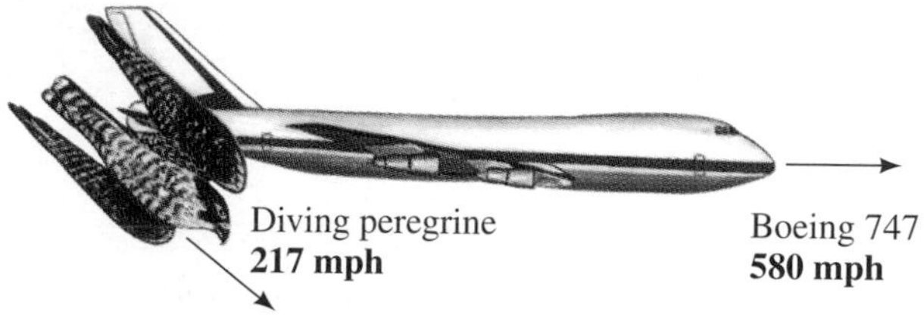

85. An airplane is carrying 254 passengers. When it lands in Atlanta, 133 passengers get off. How many passengers are left on the plane?

86. On Tuesday, 5822 people went to a soccer game, and on Friday, 7994 people went. How many more people went to the game on Friday?

87. Last fall 12,625 students enrolled in classes. In the spring semester, 11,296 students enrolled. How many more students enrolled in the fall semester than in the spring?

88. In 1964, its first year on the market, the Ford Mustang sold for $2500. In 2000, the Ford Mustang sold for $24,870. Find the increase in price. (*Source:* Ford Motor Company.)

89. Patriot Flag Company manufactured 14,608 flags and sold 5069. How many flags remain unsold?

90. Eye exams have been given to 14,679 children in the school district. If there are 23,156 students in the school district, how many have not received eye exams?

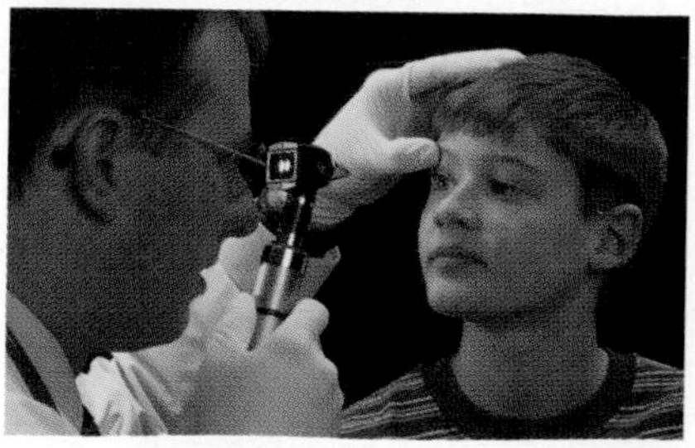

91. The Jordanos now pay rent of $650 per month. If they buy a house, their housing expense will be $913 per month. How much more will they pay per month if they buy a house?

92. A retired couple who used to receive a Social Security payment of $1479 per month now receive $1568 per month. Find the amount of the monthly increase.

93. On Monday, 11,594 people visited Arcade Amusement Park, and 12,352 people visited the park on Tuesday. How many more people visited the park on Tuesday?

94. Last month, Alice Blake earned $2382. This month she earned $2671. How much more did she earn this month than last month?

Solve each application problem. Add or subtract as necessary.

95. A survey of large hotels found that the average salary for a general manager of a deluxe spa and tennis resort is one hundred one thousand, five hundred dollars per year, while spa and tennis directors earn $44,000. How much more does a general manager earn than a spa and tennis director?

96. There are 24 million business enterprises in the United States. If only 7000 of these businesses are large businesses having 500 or more employees, how many are small and midsize businesses?

The table shows the deliveries made by Diana Lopez, a United Parcel Service driver. Use the table to answer Exercises 97–100.

PACKAGE DELIVERY (LOPEZ)

Day	Number of Deliveries
Monday	137
Tuesday	126
Wednesday	119
Thursday	89
Friday	147

97. How many more deliveries did Diana make on Monday than on Thursday?

98. How many more deliveries did Diana make on Friday than on Tuesday?

99. Find the total deliveries made on the two busiest days.

100. Find the total deliveries made on the two slowest days.

1.4 MULTIPLYING WHOLE NUMBERS

Suppose we want to know the total number of computers in a computer lab. The stations are arranged in four rows with three stations in each row. Adding the number 3 a total of 4 times gives 12.

$$3 + 3 + 3 + 3 = 12$$

This result can also be shown with a figure.

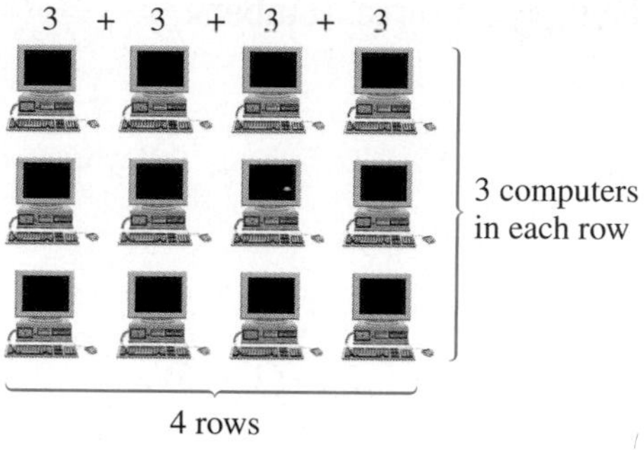

OBJECTIVES

1. Identify the parts of a multiplication problem.
2. Do chain multiplications.
3. Multiply by single-digit numbers.
4. Use multiplication shortcuts for numbers ending in zeros.
5. Multiply by numbers having more than one digit.
6. Solve application problems with multiplication.

1 Identify the parts of a multiplication problem. Multiplication is a shortcut for repeated addition. In the computer lab example, instead of *adding* 3 + 3 + 3 + 3 to get 12, we can *multiply* 3 by 4 to get 12. The numbers being multiplied are called **factors.** The answer is called the **product.** For example, the product of 3 and 4 can be written with the symbol ×, a raised dot, or parentheses, as follows.

$$\begin{array}{r} 3 \\ \times\ 4 \\ \hline 12 \end{array}$$

3 ← Factor (also called *multiplicand*)
× 4 ← Factor (also called *multiplier*)
12 ← Product (answer)

$3 \times 4 = 12$ ***or*** $3 \cdot 4 = 12$ ***or*** $(3)(4) = 12$

Work Problem 1 at the Side.

Commutative Property of Multiplication

By the **commutative property of multiplication,** the answer or product remains the same when the order of the factors is changed. For example,

$$3 \times 5 = 15 \quad \text{and} \quad 5 \times 3 = 15$$

CAUTION

Recall that addition also has a commutative property. Remember that 4 + 2 is the same as 2 + 4. Subtraction, however, is ***not*** commutative.

Example 1 Multiplying Two Numbers

Multiply. (Remember that a raised dot or parentheses means to multiply.)

(a) $3 \times 4 = 12$

(b) $6 \cdot 0 = 0$ (The product of any number and 0 is 0; if you give no money to each of 6 relatives, you give no money.)

(c) $(4)(8) = 32$

Work Problem 2 at the Side.

1 Identify the factors and the product in each multiplication problem.

(a) $2 \times 5 = 10$

(b) $6 \times 4 = 24$

(c) $7 \cdot 6 = 42$

(d) $(3)(9) = 27$

2 Multiply.

(a) 3×8

(b) 0×9

(c) $7 \cdot 5$

(d) $6 \cdot 5$

(e) $(3)(8)$

ANSWERS

1. (a) factors: 2, 5; product: 10
(b) factors: 6, 4; product: 24
(c) factors: 7, 6; product: 42
(d) factors: 3, 9; product: 27

2. (a) 24 **(b)** 0 **(c)** 35 **(d)** 30 **(e)** 24

3 Multiply.

(a) $2 \times 5 \times 3$

(b) $7 \cdot 1 \cdot 4$

(c) $(6)(4)(0)$

2 **Do chain multiplications.** Some multiplications contain more than two factors.

Associative Property of Multiplication

By the **associative property of multiplication,** grouping the factors differently does not change the product.

Example 2 Multiplying Three Numbers

Multiply $2 \times 3 \times 5$.

$$(2 \times 3) \times 5$$ Parentheses tell what to do first.

$$6 \times 5 = 30$$

Also,

$$2 \times (3 \times 5)$$

$$2 \times 15 = 30$$

Either grouping results in the same product.

Calculator Tip The calculator approach to Example 2 uses chain calculations.

2 (×) 3 (×) 5 (=) 30

A problem with more than two factors, such as the one in Example 2, is called a **chain multiplication.**

Work Problem 3 at the Side.

3 **Multiply by single-digit numbers.** Carrying may be needed in multiplication problems with larger factors.

Example 3 Multiplying with Carrying

Multiply.

(a) $\begin{array}{r} 53 \\ \times\ 4 \\ \hline \end{array}$

Start by multiplying in the ones column.

$$\begin{array}{r} ^{1} \\ 53 \\ \times\ 4 \\ \hline 2 \end{array} \qquad 4 \times 3 = \mathbf{12}$$

Carry the 1 to the tens column. Write 2 in the ones column.

Next, multiply 4 ones and 5 tens.

$$\begin{array}{r} ^{1} \\ 53 \\ \times\ 4 \\ \hline 2 \end{array} \qquad 4 \times 5 = \mathbf{20} \text{ tens}$$

Continued on Next Page

ANSWERS
3. (a) 30 (b) 28 (c) 0

Add the 1 that was carried to the tens column.

$$\begin{array}{r} 1 \\ 53 \\ \times\ 4 \\ \hline 212 \end{array} \quad 20 + 1 = \mathbf{21} \text{ tens}$$

(b) $\begin{array}{r} 724 \\ \times\ 5 \\ \hline \end{array}$

Work as shown.

$$\begin{array}{r} 12 \\ 724 \\ \times\ 5 \\ \hline 3620 \end{array}$$

← 5 × 4 = **20** ones; write 0 ones and carry 2 tens.

5 × 2 = **10** tens; add the 2 tens to get 12 tens; write 2 tens and carry 1 hundred.

5 × 7 = **35** hundreds; add the 1 hundred to get 36 hundreds.

Work Problem 4 at the Side.

4 **Use multiplication shortcuts for numbers ending in zeros.** The product of two whole number factors is also called a **multiple** of either factor. For example, since 4 • 2 = 8, the whole number 8 is a multiple of both 4 and 2. *Multiples of 10* are very useful when multiplying. A **multiple of 10** is a whole number that ends in 0, such as 10, 20, or 30; 100, 200, or 300; 1000, 2000, or 3000. There is a short way to multiply by these multiples of 10. Look at the following examples.

$$26 \times 1 = 26$$
$$26 \times \mathbf{10} = 26\mathbf{0}$$
$$26 \times \mathbf{100} = 26\mathbf{00}$$
$$26 \times \mathbf{1000} = 26{,}\mathbf{000}$$

Do you see a pattern? These examples suggest the following rule.

Multiplying by Multiples of 10

To multiply a whole number by 10, 100, or 1000, attach one, two, or three 0s to the right of the whole number.

Example 4 Using Multiples of 10 to Multiply

Multiply.

(a) 59 × 10 = 59**0** ← Attach 0.

(b) 74 × 100 = 74**00** ← Attach 00.

(c) 803 × 1000 = 803,**000** ← Attach 000.

Work Problem 5 at the Side.

You can also find the product of other multiples of 10 by attaching 0s.

4 Multiply.

(a) $\begin{array}{r} 62 \\ \times\ 4 \\ \hline \end{array}$

(b) $\begin{array}{r} 98 \\ \times\ 0 \\ \hline \end{array}$

(c) $\begin{array}{r} 758 \\ \times\ 8 \\ \hline \end{array}$

(d) $\begin{array}{r} 2831 \\ \times\ 7 \\ \hline \end{array}$

(e) $\begin{array}{r} 4714 \\ \times\ 8 \\ \hline \end{array}$

5 Multiply.

(a) 52 × 10

(b) 305 × 100

(c) 418 × 1000

ANSWERS

4. **(a)** 248 **(b)** 0 **(c)** 6064 **(d)** 19,817 **(e)** 37,712

5. **(a)** 520 **(b)** 30,500 **(c)** 418,000

6 Multiply.

(a) 17×40

(b) 58×300

(c) 180×30

(d) 4200×80

(e) 700×400

7 Complete each multiplication.

(a)
```
   35
 × 54
  140
  175
```

(b)
```
   76
 × 49
  684
  304
```

Answers

6. (a) 680 **(b)** 17,400 **(c)** 5400 **(d)** 336,000 **(e)** 280,000

7. (a) 1890 **(b)** 3724

Example 5 **Using Multiples of 10 to Multiply**

Multiply.

(a) 75×3000

Multiply 75 by 3, and then attach three 0s.

```
   75      75 × 3000 = 225,000
 ×  3
  225                 ← Attach 000.
```

(b) 150×70

Multiply 15 by 7, and then attach two 0s.

```
   15      150 × 70 = 10,500 ← Attach 00.
 ×  7
  105
```

Work Problem 6 at the Side.

5 **Multiply by numbers having more than one digit.** The next example shows multiplication when both factors have more than one digit.

Example 6 **Multiplying with More Than One Digit**

Multiply 46 and 23.

First multiply 46 by 3.

```
   1
   46
 ×  3
  138  ← 46 × 3 = 138
```

Now multiply 46 by 20.

```
   1
   46
 × 20
  920  ← 46 × 20 = 920
```

Add the results.

```
    46
  × 23
   138  ← 46 × 3
 + 920  ← 46 × 20
  1058
  ↑ Add.
```

Both 138 and 920 are called **partial products.** To save time, the 0 in 920 is usually not written.

```
   46
 × 23
  138
  92    ← 0 not written. Be very careful to
 1058     place the 2 in the tens column.
```

Work Problem 7 at the Side.

Example 7 Using Partial Products

Multiply.

(a)

$$\begin{array}{rl} 233 & \\ \times\ 132 & \\ \hline 466 & \\ 699 & \text{(Tens lined up)} \\ 233 & \text{(Hundreds lined up)} \\ \hline 30{,}756 & \text{Product} \end{array}$$

(b)

$$\begin{array}{r} 538 \\ \times\ 46 \\ \hline \end{array}$$

First multiply by 6.

$$\begin{array}{rl} {\scriptstyle 24} & \\ 538 & \\ \times\ 46 & \text{Carrying is needed here.} \\ \hline 3228 & \end{array}$$

Now multiply by 4, being careful to line up the tens.

$$\begin{array}{rl} {\scriptstyle 13} & \\ {\scriptstyle 24} & \\ 538 & \\ \times\ 46 & \\ \hline 3228 & \\ 2152 & \text{Finally, add the results.} \\ \hline 24{,}748 & \end{array}$$

Work Problem 8 at the Side.

When 0 appears in the multiplier, be sure to move the partial products to the left to account for the position held by the 0.

Example 8 Multiplying with Zeros

Multiply.

(a)

$$\begin{array}{rl} 137 & \\ \times\ 306 & \\ \hline 822 & \\ 000 & \text{(Tens lined up)} \\ 411 & \text{(Hundreds lined up)} \\ \hline 41{,}922 & \end{array}$$

(b)

$$\begin{array}{rl} 1406 & \\ \times\ 2001 & \\ \hline 1406 & \\ 0000 & \leftarrow \text{(0s to line up tens)} \\ 0000 & \leftarrow \text{(0s to line up hundreds)} \\ 2812 & \\ \hline 2{,}813{,}406 & \end{array}$$

$$\begin{array}{rl} 1406 & \\ \times\ 2001 & \\ \hline 1406 & \\ 281200 & \leftarrow \text{Zeros are written so this partial product starts in the thousands column.} \\ \hline 2{,}813{,}406 & \end{array}$$

8 Multiply.

(a) $\begin{array}{r} 46 \\ \times\ 14 \\ \hline \end{array}$

(b) $\begin{array}{r} 41 \\ \times\ 38 \\ \hline \end{array}$

(c) $\begin{array}{r} 75 \\ \times\ 63 \\ \hline \end{array}$

(d) $\begin{array}{r} 234 \\ \times\ 73 \\ \hline \end{array}$

(e) $\begin{array}{r} 835 \\ \times\ 189 \\ \hline \end{array}$

Answers

8. **(a)** 644 **(b)** 1558 **(c)** 4725 **(d)** 17,082 **(e)** 157,815

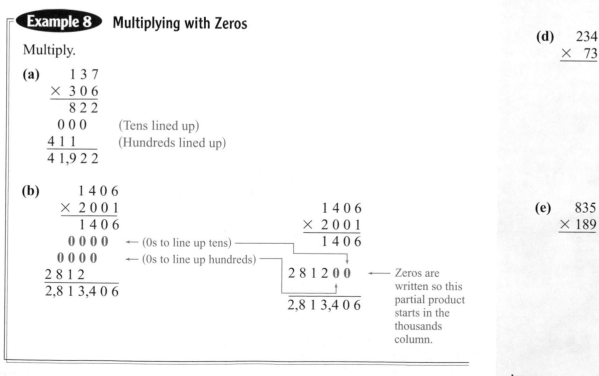

9 Multiply.

(a) 36×50

(b) 635×40

(c) 562×109

(d) 3526×6002

10 Find the total cost of the following items.

(a) 289 redwood planters at $12 per planter

(b) 106 Microsoft Intelli-mouse Explorers at $69 each

(c) 12 delivery vans at $24,300 per van

ANSWERS
9. **(a)** 1800 **(b)** 25,400 **(c)** 61,258 **(d)** 21,163,052
10. **(a)** $3468 **(b)** $7314 **(c)** $291,600

NOTE

In Example 8(b) in the alternative method on the right, 0s were inserted so that thousands were placed in the thousands column. This is a commonly used shortcut.

Work Problem 9 at the Side.

6 Solve application problems with multiplication. The next example shows how multiplication can be used to solve an application problem.

Example 9 Applying Multiplication Skills

Find the total cost of 24 cellular phones priced at $54 each.

Approach To find the cost of all the cellular phones, multiply the number of phones (24) by the cost of one phone ($54).

Solution Multiply 24 by 54.

$$\begin{array}{r} 24 \\ \times\ 54 \\ \hline 96 \\ 120 \\ \hline 1296 \end{array}$$

The total cost of the cellular phones is $1296.

Calculator Tip If you are using a calculator for Example 9, you will do this calculation.

24 (×) 54 (=) 1296

Work Problem 10 at the Side.

1.4 EXERCISES

FOR EXTRA HELP: Student's Solutions Manual | MyMathLab.com | InterAct Math Tutorial Software | AW Math Tutor Center | www.mathxl.com | Digital Video Tutor CD 1 Videotape 2

Work each chain multiplication. See Example 2.

1. $4 \times 1 \times 4$ **2.** $3 \times 5 \times 3$ **3.** $8 \times 6 \times 1$ **4.** $2 \times 4 \times 5$

5. $7 \cdot 8 \cdot 0$ **6.** $9 \cdot 0 \cdot 5$ **7.** $4 \cdot 1 \cdot 6$ **8.** $1 \cdot 5 \cdot 7$

9. (3)(2)(5) **10.** (4)(1)(9) **11.** (3)(0)(7) **12.** (0)(9)(4)

13. Explain in your own words the commutative property of multiplication. How do the commutative properties of addition and multiplication compare to each other?

14. Explain in your own words the associative property of multiplication. How do the associative properties of addition and multiplication compare to each other?

Multiply. See Example 3.

15. 25×6 **16.** 62×8 **17.** 34×7 **18.** 76×5

19. 642×5 **20.** 472×4 **21.** 624×3 **22.** 852×7

23. 2153×4 **24.** 1137×3 **25.** 2521×4 **26.** 2544×3

27. 2561×8 **28.** 7326×5 **29.** $36{,}921 \times 7$ **30.** $28{,}116 \times 4$

Multiply. See Examples 4 and 5.

31. 30×5 **32.** 20×7 **33.** 80×6 **34.** 70×5 **35.** 740×3

36. 200×7 **37.** 600×6 **38.** 860×7 **39.** 125×30 **40.** 246×50

41. $\begin{array}{r} 1485 \\ \times\ 30 \\ \hline \end{array}$

42. $\begin{array}{r} 8522 \\ \times\ 50 \\ \hline \end{array}$

43. $\begin{array}{r} 900 \\ \times\ 300 \\ \hline \end{array}$

44. $\begin{array}{r} 400 \\ \times\ 700 \\ \hline \end{array}$

45. $\begin{array}{r} 43{,}000 \\ \times\ 2\,000 \\ \hline \end{array}$

46. $\begin{array}{r} 11{,}000 \\ \times\ 9\,000 \\ \hline \end{array}$

47. 970 • 50

48. 730 • 40

49. 500 • 700

50. 850 • 700

51. 9700 • 200

52. 10,050 • 300

Multiply. See Examples 6–8.

53. $\begin{array}{r} 36 \\ \times\ 15 \\ \hline \end{array}$

54. $\begin{array}{r} 18 \\ \times\ 47 \\ \hline \end{array}$

55. $\begin{array}{r} 75 \\ \times\ 32 \\ \hline \end{array}$

56. $\begin{array}{r} 82 \\ \times\ 32 \\ \hline \end{array}$

57. $\begin{array}{r} 83 \\ \times\ 45 \\ \hline \end{array}$

58. (62)(31)

59. (58)(41)

60. (82)(67)

61. (67)(92)

62. (26)(33)

63. (28)(564)

64. (58)(312)

65. (619)(35)

66. (681)(47)

67. (55)(286)

68. $\begin{array}{r} 286 \\ \times\ 574 \\ \hline \end{array}$

69. $\begin{array}{r} 735 \\ \times\ 112 \\ \hline \end{array}$

70. $\begin{array}{r} 621 \\ \times\ 415 \\ \hline \end{array}$

71. $\begin{array}{r} 538 \\ \times\ 342 \\ \hline \end{array}$

72. $\begin{array}{r} 3228 \\ \times\ 751 \\ \hline \end{array}$

73. $\begin{array}{r} 9352 \\ \times\ 264 \\ \hline \end{array}$

74. $\begin{array}{r} 528 \\ \times\ 106 \\ \hline \end{array}$

75. $\begin{array}{r} 215 \\ \times\ 307 \\ \hline \end{array}$

76. $\begin{array}{r} 218 \\ \times\ 106 \\ \hline \end{array}$

77. $\begin{array}{r} 428 \\ \times\ 201 \\ \hline \end{array}$

78. $\begin{array}{r} 3706 \\ \times\ 208 \\ \hline \end{array}$

79. $\begin{array}{r} 6310 \\ \times\ 3078 \\ \hline \end{array}$

80. $\begin{array}{r} 3533 \\ \times\ 5001 \\ \hline \end{array}$

81. $\begin{array}{r} 2195 \\ \times\ 1038 \\ \hline \end{array}$

82. $\begin{array}{r} 1502 \\ \times\ 2009 \\ \hline \end{array}$

83. A classmate of yours is not clear on how to use a shortcut to multiply a whole number by 10, by 100, or by 1000. Write a short note explaining how this can be done.

84. Show two ways to multiply when a 0 is in the factor that is multiplying. Use the problem 291 × 307 to show this.

Solve each application problem. See Example 9.

85. There are 90 cartons of CDs loaded on a shipping pallet. If there are 200 loaded shipping pallets in the warehouse, how many cartons are in the warehouse?

86. A medical supply house has 30 bottles of vitamin C tablets, with each bottle containing 500 tablets. Find the total number of vitamin C tablets in the supply house.

87. There are 12 tomato plants to a flat. If a gardener buys 18 flats, find the total number of tomato plants he bought.

88. A hummingbird's wings beat about 65 times per second. How many times do the hummingbird's wings beat in 30 seconds?

89. A new Saturn automobile gets 38 miles per gallon on the highway. How many miles can it go on 11 gallons of gas?

90. Squid are being hauled out of the Santa Barbara Channel by the ton. They are then processed, renamed calamari, and exported. Last night 27 fishing boats each hauled out 40 tons of squid. What was the total catch for the night? (*Source: Santa Barbara News Press.*)

Find the total cost of the following items. See Examples 7–9.

91. 85 heater filters at $3 per filter

92. 38 employees at $64 per day

93. 65 rebuilt alternators at $24 per alternator

94. 76 flats of flowers at $22 per flat

95. 206 computers at $548 per computer

96. 520 printers at $219 per printer

Multiply.

97. 21 • 43 • 56

98. (600)(8)(75)(40)

Use addition, subtraction, or multiplication to solve each application problem.

99. In a forest-planting project, trees are planted 450 trees to an acre. Find the number of trees needed to plant 85 acres.

100. The largest living land mammal is the African elephant, and the largest mammal of all time is the blue whale. An African elephant weighs 15,225 pounds and a blue whale weighs 28 times that amount. Find the weight of the blue whale.

101. A large meal contains 1406 calories, while a small meal contains 348 calories. How many more calories are in the large meal than the small one?

102. As part of a Low Income Home Energy Assistance Program, the state of Florida will receive $1,100,000, while Vermont will receive $505,551. How much more will Florida receive than Vermont? (*Source:* U.S. Department of Energy.)

103. An insurance office purchased four computers at $680 each, four monitors at $295 each, and four printers at $230 each. Find the total cost of this equipment.

Relating Concepts (Exercises 104–113) For Individual or Group Work

Work Exercises 104–113 in order.

104. Add.

(a) $189 + 263$

(b) $263 + 189$

105. Your answers to Exercise 104(a) and (b) should be the same. This shows that the order of numbers in an addition problem does not change the sum. This is known as the ____________ property of addition.

106. Add. Recall that parentheses tell you what to do first.

(a) $(65 + 81) + 135$

(b) $65 + (81 + 135)$

107. Since the answers to Exercise 106(a) and (b) are the same, we see that grouping the addition of numbers in any order does not change the sum. This is known as the ____________ property of addition.

108. Multiply.

(a) 220×72

(b) 72×220

109. Since the answers to Exercise 108(a) and (b) are the same, we see that the product remains the same when the order of the factors is changed. This is known as the ____________ property of multiplication.

110. Multiply. Recall that parentheses tell you what to do first.

(a) $(26 \times 18) \times 14$

(b) $26(18 \times 14)$

111. Since the answers to Exercise 110(a) and (b) are the same, we see that the grouping of numbers in any order when multiplying gives the same product. This is known as the ____________ property of multiplication.

112. Do the commutative and associative properties apply to subtraction? Explain your answer using several examples.

113. Do the commutative and associative properties apply to division? Explain your answer using several examples.

1.5 DIVIDING WHOLE NUMBERS

Suppose the cost of a fast-food lunch is \$12 and is to be divided equally by three friends. Each person would pay \$4, as shown here.

\$12 total

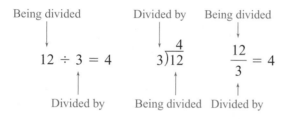

\$4 \$4 \$4

3 equal parts

OBJECTIVES

1. Write division problems in three ways.
2. Identify the parts of a division problem.
3. Divide 0 by a number.
4. Recognize that a number cannot be divided by 0.
5. Divide a number by itself.
6. Divide a number by 1.
7. Use short division.
8. Use multiplication to check the answer to a division problem.
9. Use tests for divisibility.

1 Write division problems in three ways. Just as $3 \cdot 4$, 3×4, and $(3)(4)$ are different ways of indicating the multiplication of 3 and 4, there are several ways to write 12 divided by 3.

Being divided ↓	Divided by ↓	Being divided ↓
$12 \div 3 = 4$	$3\overline{)12}$ with quotient 4	$\dfrac{12}{3} = 4$
↑ Divided by	↑ Being divided	↑ Divided by

We will use all three division symbols, $\div$, $\overline{)\ }$, and —. In courses such as algebra, a slash symbol, /, or a fraction bar, —, is most often used.

Example 1 Using Division Symbols

Write each division problem using two other symbols.

(a) $12 \div 4 = 3$

This division can also be written as

$$4\overline{)12} \text{ (quotient 3)} \quad \text{or} \quad \frac{12}{4} = 3.$$

(b) $\dfrac{15}{5} = 3$

$$15 \div 5 = 3 \quad \text{or} \quad 5\overline{)15} \text{ (quotient 3)}$$

(c) $5\overline{)20}$ (quotient 4)

$$20 \div 5 = 4 \quad \text{or} \quad \frac{20}{5} = 4$$

Work Problem 1 at the Side.

1 Write each division problem using two other symbols.

(a) $48 \div 6 = 8$

(b) $24 \div 6 = 4$

(c) $9\overline{)36}$ (quotient 4)

(d) $\dfrac{42}{6} = 7$

2 Identify the parts of a division problem. In division, the number being divided is the **dividend,** the number divided by is the **divisor,** and the answer is the **quotient.**

$$\text{dividend} \div \text{divisor} = \text{quotient}$$

$$\text{divisor}\overline{)\text{dividend}} \text{ (quotient on top)} \qquad \frac{\text{dividend}}{\text{divisor}} = \text{quotient}$$

ANSWERS

1. (a) $6\overline{)48}$ (quotient 8) and $\dfrac{48}{6} = 8$
(b) $6\overline{)24}$ (quotient 4) and $\dfrac{24}{6} = 4$
(c) $36 \div 9 = 4$ and $\dfrac{36}{9} = 4$
(d) $6\overline{)42}$ (quotient 7) and $42 \div 6 = 7$

2 Identify the dividend, divisor, and quotient.

(a) $20 \div 5 = 4$

(b) $18 \div 6 = 3$

(c) $\frac{28}{7} = 4$

(d) $\begin{array}{r}18\\2\overline{)36}\end{array}$

3 Divide.

(a) $0 \div 10$

(b) $\frac{0}{6}$

(c) $\frac{0}{24}$

(d) $37\overline{)0}$

Answers

2. **(a)** dividend: 20; divisor: 5; quotient: 4
   (b) dividend: 18; divisor: 6; quotient: 3
   (c) dividend: 28; divisor: 7; quotient: 4
   (d) dividend: 36; divisor: 2; quotient: 18
3. all 0

Example 2 Identifying the Parts of a Division Problem

Identify the dividend, divisor, and quotient.

(a) $35 \div 7 = 5$

35 ÷ **7** = **5** ← **Quotient**

↑ **Dividend** ↖ **Divisor**

(b) $\frac{100}{20} = 5$

Dividend ↓

$\frac{\mathbf{100}}{\mathbf{20}} = \mathbf{5}$ ← **Quotient**

↑ **Divisor**

(c) $\begin{array}{r}6\\12\overline{)72}\end{array}$

$\begin{array}{r}6\\12\overline{)72}\end{array}$ ← Quotient, ← Dividend

↑ Divisor

Work Problem 2 at the Side.

3 Divide 0 by a number. If no money, or \$0, is divided equally among five people, each person gets \$0. The general rule for dividing 0 follows.

Dividing 0 by a Number

The number **0** divided by any nonzero number is **0.**

Example 3 Dividing 0 by a Number

Divide.

(a) $0 \div 12 = 0$

(b) $0 \div 1728 = 0$

(c) $\frac{0}{375} = 0$

(d) $\begin{array}{r}0\\129\overline{)0}\end{array}$

Work Problem 3 at the Side.

Just as a subtraction such as $8 - 3 = 5$ can be written as the addition $8 = 3 + 5$, any division can be written as a multiplication. For example, $12 \div 3 = 4$ can be written as

$$3 \times 4 = 12 \quad \text{or} \quad 4 \times 3 = 12$$

Example 4 **Changing Division Problems to Multiplication**

Change each division problem to a multiplication problem.

(a) $\frac{20}{4} = 5$ becomes $4 \cdot 5 = 20$

(b) $8\overline{)48}$ with quotient 6 becomes $8 \cdot 6 = 48$

(c) $72 \div 9 = 8$ becomes $9 \cdot 8 = 72$

Work Problem 4 at the Side.

4 Write each division problem as a multiplication problem.

(a) $5\overline{)15}$ with quotient 3

(b) $\frac{32}{4} = 8$

(c) $48 \div 8 = 6$

4 **Recognize that a number cannot be divided by 0.** Division by 0 cannot be done. To see why, try to find

$$9 \div 0 = ?$$

As we have just seen, any division problem can be converted to a multiplication problem so that

$$\textbf{divisor} \cdot \textbf{quotient} = \textbf{dividend.}$$

If you convert the preceding problem to its multiplication counterpart, it reads

$$0 \cdot ? = 9.$$

You already know that 0 times any number must always be 0. Try any number you like to replace the "?" and you'll always get 0 instead of 9. Therefore, the division problem $9 \div 0$ cannot be done. Mathematicians say it is *undefined* and have agreed never to divide by 0. However, $0 \div 9$ *can* be done. Check by rewriting it as a multiplication problem.

$$0 \div 9 = 0 \quad \text{because} \quad 0 \cdot 9 = 0 \text{ is true.}$$

Dividing by 0

Since dividing by 0 cannot be done, we say that division by **0** is *undefined*. It is impossible to compute an answer.

Example 5 **Dividing Numbers by 0**

All the following are undefined.

(a) $\frac{6}{0}$ is undefined.

(b) $0\overline{)8}$ is undefined.

(c) $18 \div 0$ is undefined.

(d) $\frac{0}{0}$ is undefined.

ANSWERS

4. (a) $5 \cdot 3 = 15$ **(b)** $4 \cdot 8 = 32$ **(c)** $8 \cdot 6 = 48$

Division Involving 0

$0 \div \text{nonzero number} = 0$ and $\dfrac{0}{\text{nonzero number}} = 0$

but

$\dfrac{\text{nonzero number}}{0}$ and $\text{nonzero number} \div 0$ are undefined.

CAUTION

When 0 is the divisor in a problem, you write "undefined" as the answer. Never divide by 0.

Work Problem 5 at the Side.

Calculator Tip Try these two problems on your calculator. Jot down your answers.

9 (÷) 0 (=) ______ 0 (÷) 9 = ______

When you try to divide by 0, the calculator cannot do it, so it shows the word "Error" or the letter "E" (for error) in the display. But, when you divide 0 by 9 the calculator displays 0, which is the correct answer.

5 Divide a number by itself. What happens when a number is divided by itself? For example, what is $4 \div 4$ or $97 \div 97$?

Dividing a Number by Itself

Any nonzero number divided by itself is **1.**

Example 6 Dividing a Nonzero Number by Itself

Divide.

(a) $16 \div 16 = 1$

(b) $32\overline{)32}$ with quotient 1

(c) $\dfrac{57}{57} = 1$

Work Problem 6 at the Side.

6 Divide a number by 1. What happens when a number is divided by 1? For example, what is $5 \div 1$ or $86 \div 1$?

Dividing a Number by 1

Any number divided by 1 is itself.

---

5 Divide. If the division is not possible, write "undefined."

(a) $\dfrac{6}{0}$

(b) $\dfrac{0}{6}$

(c) $0\overline{)28}$

(d) $28\overline{)0}$

(e) $100 \div 0$

(f) $0 \div 100$

6 Divide.

(a) $6 \div 6$

(b) $15\overline{)15}$

(c) $\dfrac{37}{37}$

ANSWERS

5. **(a)** undefined **(b)** 0 **(c)** undefined **(d)** 0 **(e)** undefined **(f)** 0

6. all 1

Example 7 Dividing Numbers by 1

Divide.

(a) $8 \div 1 = 8$

(b) $\begin{array}{r} 26 \\ 1\overline{)26} \end{array}$

(c) $\dfrac{41}{1} = 41$

Work Problem 7 at the Side.

7 **Use short division.** **Short division** is a method of dividing a number by a one-digit divisor.

Example 8 Using Short Division

Divide: $3\overline{)96}$.

First, divide 9 by 3.

$$\begin{array}{r} 3 \\ 3\overline{)96} \end{array} \longleftarrow \frac{9}{3} = 3$$

Next, divide 6 by 3.

$$\begin{array}{r} 32 \\ 3\overline{)96} \end{array} \longleftarrow \frac{6}{3} = 2$$

Work Problem 8 at the Side.

When two numbers do not divide exactly, the leftover portion is called the **remainder.**

Example 9 Using Short Division with a Remainder

Divide 147 by 4.

Write the problem.

$$4\overline{)147}$$

Because 1 cannot be divided by 4, divide 14 by 4.

$$\begin{array}{r} 3 \\ 4\overline{)14^{2}7} \end{array} \qquad \frac{14}{4} = 3 \text{ with 2 left over}$$

Next, divide 27 by 4. The final number left over is the remainder. Use R to indicate the remainder, and write the remainder to the side.

$$\begin{array}{r} 3\;6\;\textbf{R3} \\ 4\overline{)14^{2}7} \end{array} \qquad \frac{27}{4} = 6 \text{ with 3 left over}$$

Work Problem 9 at the Side.

7 Divide.

(a) $6 \div 1$

(b) $1\overline{)18}$

(c) $\dfrac{36}{1}$

8 Divide.

(a) $2\overline{)18}$

(b) $3\overline{)39}$

(c) $4\overline{)88}$

(d) $2\overline{)462}$

9 Divide.

(a) $2\overline{)125}$

(b) $3\overline{)215}$

(c) $4\overline{)538}$

(d) $\dfrac{819}{5}$

ANSWERS

7. **(a)** 6 **(b)** 18 **(c)** 36

8. **(a)** 9 **(b)** 13 **(c)** 22 **(d)** 231

9. **(a)** 62 **R**1 **(b)** 71 **R**2 **(c)** 134 **R**2 **(d)** 163 **R**4

10 Divide.

(a) $5\overline{)937}$

(b) $\frac{675}{7}$

(c) $3\overline{)1885}$

(d) $8\overline{)1135}$

Example 10 Dividing with a Remainder

Divide 1809 by 7.

Divide 7 into 18.

$$\begin{array}{r} 2 \\ 7\overline{)18^{4}09} \end{array} \qquad \frac{18}{7} = 2 \text{ with 4 left over}$$

Divide 7 into 40.

$$\begin{array}{r} 2\ 5 \\ 7\overline{)18^{4}0^{5}9} \end{array} \qquad \frac{40}{7} = 5 \text{ with 5 left over}$$

Divide 7 into 59.

$$\begin{array}{r} 2\ 5\ 8\ \mathbf{R}3 \\ 7\overline{)18^{4}0^{5}9} \end{array} \qquad \frac{59}{7} = 8 \text{ with 3 left over}$$

Work Problem 10 at the Side.

NOTE

Short division takes practice but is useful in many situations.

8 **Use multiplication to check the answer to a division problem.** **Check** the answer to a division problem as follows.

Checking Division

(divisor × quotient) + remainder = dividend

Parentheses tell you what to do first: Multiply the divisor by the quotient, then add the remainder.

Example 11 Checking Division by Using Multiplication

Check each answer.

(a) $\begin{array}{r} 91\ \mathbf{R}3 \\ 5\overline{)458} \end{array}$

(divisor × quotient) + remainder = dividend

$$(5 \times 91) + 3$$

$$455 + 3 = 458$$

Matches original dividend, so the division was done correctly

Continued on Next Page

Answers

10. **(a)** 187 **R**2 **(b)** 96 **R**3
(c) 628 **R**1 **(d)** 141 **R**7

(b) $6\overline{)1437}$ = 239 **R4**

(divisor × quotient) + remainder = dividend

$(6 \times 239) + 4$

$1434 + 4 = 1438$ ← Does not match original dividend.

The answer does not check. Rework the original problem to get the correct answer, 239 **R**3.

CAUTION

A common error when checking division is to forget to add the remainder. Be sure to add any remainder when checking a division problem.

Work Problem 11 at the Side.

9 **Use tests for divisibility.** It is often important to know whether a number is *divisible* by another number. You will find this useful in Chapter 2 when writing fractions in lowest terms.

Divisibility

One whole number is **divisible** by another if the remainder is 0.

Use the following tests to decide whether one number is divisible by another number.

Tests for Divisibility

A number is divisible by

2 if it ends in 0, 2, 4, 6, or 8. These are the even numbers.
3 if the sum of its digits is divisible by 3.
4 if the last two digits make a number that is divisible by 4.
5 if it ends in 0 or 5.
6 if it is divisible by both 2 and 3.
7 has no simple test.
8 if the last three digits make a number that is divisible by 8.
9 if the sum of its digits is divisible by 9.
10 if it ends in 0.

The most commonly used tests are those for 2, 3, 5, and 10.

Divisibility by 2

A number is divisible by **2** if the number ends in 0, 2, 4, 6, or 8. All even numbers are divisible by 2.

11 Use multiplication to check each division. If an answer is incorrect, give the correct answer.

(a) $2\overline{)89}$ = 44 **R**1

(b) $8\overline{)739}$ = 92 **R**2

(c) $3\overline{)1223}$ = 407 **R**2

(d) $5\overline{)2383}$ = 476 **R**3

ANSWERS

11. **(a)** correct **(b)** incorrect, should be 92 **R**3 **(c)** correct **(d)** correct

12 Which numbers are divisible by 2?

a. 308

b. 219

c. 3626

d. 58,000

13 Which numbers are divisible by 3?

a. 751

b. 8625

c. 374,214

d. 205,633

14 Which numbers are divisible by 5?

a. 220

b. 445

c. 7251

d. 206,105

ANSWERS
12. all but b
13. b and c
14. all but c

Example 12 Testing for Divisibility by 2

Are the following numbers divisible by 2?

(a) 986
↑ Ends in 6

Because the number ends in 6, which is an even number, the number 986 is divisible by 2.

(b) 3255 is not divisible by 2.
↑ Ends in 5, and not in 0, 2, 4, 6, or 8

Work Problem 12 at the Side.

Divisibility by 3

A number is divisible by **3** if the sum of its digits is divisible by **3.**

Example 13 Testing for Divisibility by 3

Are the following numbers divisible by 3?

(a) 4251
Add the digits.

4 2 5 1

$$\mathbf{4} + \mathbf{2} + \mathbf{5} + \mathbf{1} = 12$$

Because 12 is divisible by 3, the number 4251 is divisible by 3.

(b) 29,806
Add the digits.

2 9 8 0 6

$$\mathbf{2} + \mathbf{9} + \mathbf{8} + \mathbf{0} + \mathbf{6} = 25$$

Because 25 is not divisible by 3, the number 29,806 is not divisible by 3.

CAUTION

Be careful when testing for divisibility by adding the digits. This method works only for the numbers 3 and 9.

Work Problem 13 at the Side.

Divisibility by 5 and by 10

A number is divisible by **5** if it ends in 0 or 5.
A number is divisible by **10** if it ends in 0.

Example 14 Testing for Divisibility by 5

Are the following numbers divisible by 5?

(a) 12,900 ends in 0 and is divisible by 5.

(b) 4325 ends in 5 and is divisible by 5.

(c) 392 ends in 2 and is not divisible by 5.

Work Problem 14 at the Side.

Example 15 **Testing for Divisibility by 10**

Are the following numbers divisible by 10?

(a) 700 and 9140 both end in 0 and are divisible by 10.

(b) 355 and 18,743 do not end in 0 and are not divisible by 10.

Work Problem 15 at the Side.

15 Which numbers are divisible by 10?

a. 350

b. 225

c. 4010

d. 12,560

ANSWERS

15. all but b

Focus on **Real-Data Applications**

Study Time

As a college student, you need to plan a schedule to accommodate many responsibilities: attending class, preparing for class, studying for exams, traveling to and from college, part-time work, family responsibilities, and personal time. First, you must calculate the amount of time dedicated to each of your obligations.

As an example, suppose that you are a full-time student enrolled in 12 credit hours of class: 4 credits of biology, 4 credits of computer science, 3 credits of mathematics, and 1 credit of physical education. Biology and computer science each have 3 hours of lecture and a 3-hour lab each week. Your biology and mathematics instructors recommend that you spend an additional 2 hours per week of study time for each hour of lecture time. Your physical education class is only 1 credit, but you are in class 3 hours each week.

Activity	Hours per Week
Class time	
Lab time	
Study time for mathematics and biology	
Travel time to and from college	5
Part-time work (including travel time)	25
Sleep (8 hours per day)	
Meals (3 hours per day)	
Hygiene (baths, dressing, etc.)	7
Other (housecleaning, laundry, etc.)	14

1. How many total hours are in one week?

2. Fill in the table entries for the number of hours in a week spent on class time, lab time, study time for biology and mathematics, sleep, and meals.

3. How many hours per week are spent on college-related activities?

4. How much more time is required for personal time (sleeping, eating, hygiene) than for college-related activities?

5. Based on the table data, how many hours per week are spent on the activities listed?

6. How many hours per week are available for other activities, such as dating, shopping, family responsibilities, and so on?

7. List two additional activities that are not listed in the table. Estimate the amount of time per week required for each activity.

8. Based on your analysis of the time constraints of this schedule, what changes would you make to your schedule and other obligations?

1.5 Exercises

FOR EXTRA HELP: Student's Solutions Manual; MyMathLab.com; InterAct Math Tutorial Software; AW Math Tutor Center; www.mathxl.com; Digital Video Tutor CD 1 Videotape 2

Write each division problem using two other symbols. See Example 1.

1. $15 \div 3 = 5$

2. $20 \div 5 = 4$

3. $\frac{45}{9} = 5$

4. $\frac{56}{8} = 7$

5. $2\overline{)16}$ with quotient 8

6. $8\overline{)48}$ with quotient 6

Divide. If the division is not possible, write "undefined." See Examples 3, 5, 6, and 7.

7. $7 \div 7$

8. $42 \div 6$

9. $\frac{14}{2}$

10. $\frac{8}{0}$

11. $36 \div 0$

12. $6 \div 6$

13. $\frac{24}{1}$

14. $\frac{12}{1}$

15. $12\overline{)0}$

16. $\frac{0}{7}$

17. $0\overline{)21}$

18. $\frac{2}{0}$

19. $\frac{15}{1}$

20. $\frac{0}{0}$

21. $\frac{8}{1}$

22. $\frac{0}{5}$

Divide by using short division. Use multiplication to check each answer. See Examples 8, 9, and 10.

23. $5\overline{)75}$

24. $4\overline{)84}$

25. $8\overline{)192}$

26. $6\overline{)168}$

27. $4\overline{)1216}$

28. $5\overline{)2305}$

29. $4\overline{)2509}$

30. $8\overline{)1335}$

31. $6\overline{)9137}$

32. $9\overline{)8371}$

33. $6\overline{)1854}$

34. $8\overline{)856}$

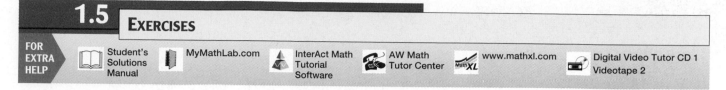

35. 24,040 ÷ 8

36. 8012 ÷ 4

37. 15,018 ÷ 3

38. 32,008 ÷ 8

39. 4867 ÷ 6

40. 5993 ÷ 7

41. 12,947 ÷ 5

42. 33,285 ÷ 9

43. 29,298 ÷ 4

44. 17,937 ÷ 6

45. 12,630 ÷ 4

46. 46,560 ÷ 7

47. $\dfrac{22{,}088}{4}$

48. $\dfrac{8199}{9}$

49. $\dfrac{74{,}751}{6}$

50. $\dfrac{72{,}543}{5}$

51. $\dfrac{71{,}776}{7}$

52. $\dfrac{77{,}621}{3}$

53. $\dfrac{128{,}645}{7}$

54. $\dfrac{172{,}255}{4}$

Use multiplication to check each answer. If an answer is incorrect, find the correct answer. See Example 11.

55. $\begin{array}{r} 375 \textbf{ R}2 \\ 5\overline{)1877} \end{array}$

56. $\begin{array}{r} 427 \textbf{ R}1 \\ 3\overline{)1282} \end{array}$

57. $\begin{array}{r} 1908 \textbf{ R}2 \\ 3\overline{)5725} \end{array}$

58. $\begin{array}{r} 432 \textbf{ R}3 \\ 5\overline{)2158} \end{array}$

59. $\begin{array}{r} 650 \textbf{ R}2 \\ 7\overline{)4692} \end{array}$

60. $\begin{array}{r} 663 \textbf{ R}5 \\ 9\overline{)5974} \end{array}$

61. $\begin{array}{r} 3\,568 \textbf{ R}2 \\ 6\overline{)21{,}409} \end{array}$

62. $\begin{array}{r} 532 \\ 6\overline{)3192} \end{array}$

63. $\begin{array}{r} 2\,002 \textbf{ R}3 \\ 8\overline{)16{,}019} \end{array}$

64. $\begin{array}{r} 4\,208 \\ 8\overline{)33{,}664} \end{array}$

65. $\begin{array}{r} 11{,}523 \textbf{ R}2 \\ 6\overline{)69{,}140} \end{array}$

66. $\begin{array}{r} 27{,}532 \textbf{ R}1 \\ 3\overline{)82{,}598} \end{array}$

67. $\begin{array}{r} 9\,628 \textbf{ R}7 \\ 9\overline{)86{,}655} \end{array}$

68. $\begin{array}{r} 7\,258 \textbf{ R}4 \\ 7\overline{)50{,}809} \end{array}$

69. $\begin{array}{r} 27{,}822 \\ 8\overline{)222{,}576} \end{array}$

70. $\begin{array}{r} 77{,}804 \\ 4\overline{)311{,}216} \end{array}$

71. Explain in your own words how to check a division problem using multiplication. Be sure to include what must be done if the quotient includes a remainder.

72. Describe the three divisibility rules that you feel might be most useful to you and tell why.

Solve each application problem.

73. A banquet caterer has 1165 dishes of the same pattern in inventory. If there are five dishes per place setting, how many place settings does the caterer have?

74. A school district will distribute 1620 new science books equally among 12 schools. How many books will each school receive?

75. Last year in Seattle, 56,000 people went to the circus over a 5-day period. If the same number of people attended each day, what was the daily attendance? (*Source:* Feld Entertainment, Vienna, Virginia.)

76. One gallon of orange juice will serve nine people. How many gallons are needed for 3483 people?

77. An estate of $127,400 is divided equally among seven family members. Find the amount received by each family member.

78. How many 5-lb bags of rice can be filled from 8750 lb of rice?

79. If 36 gallons of fertilizer are needed for each acre of land, find the number of acres that can be fertilized with 7380 gallons of fertilizer.

80. A roofing contractor has purchased 2268 squares (1 square measures 10 ft by 10 ft) of roofing material. If each home needs 21 squares of material, find the number of homes that can be roofed.

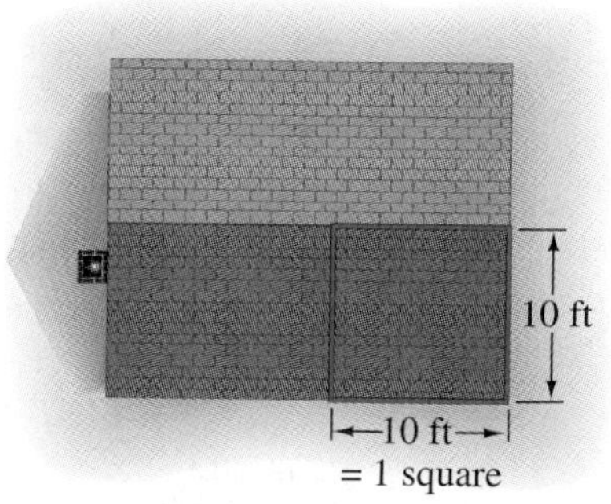

81. The Super Lotto payout of $8,100,000 will be divided equally by 36 people who purchased the winning ticket. Find the amount received by each person.

82. Ken Griffey Jr., signed a 9-year baseball contract for $117,000,000. Find his pay for each year. (*Source: USA Today.*)

83. Kaci Salmon, a supervisor at Albany Electric, earns $36,540 per year. Find the amount of her earnings in a 3-month period.

84. If Steven can assemble 168 light diffusers in an 8-hour shift, how many can he assemble in 3 hours?

Put a ✓ mark in the blank if the number at the left is divisible by the number at the top.
Put an X in the blank if the number is not divisible by the number at the top.
See Examples 12–15.

	2	3	5	10
85. 60	____	____	____	____
87. 92	____	____	____	____
89. 445	____	____	____	____
91. 903	____	____	____	____
93. 5166	____	____	____	____
95. 21,763	____	____	____	____

	2	3	5	10
86. 35	____	____	____	____
88. 96	____	____	____	____
90. 897	____	____	____	____
92. 500	____	____	____	____
94. 8302	____	____	____	____
96. 32,472	____	____	____	____

1.6 LONG DIVISION

If the total cost of 42 computer modems is \$3066, we can find the cost of each modem using **long division.** Long division is used to divide by a number with more than one digit.

OBJECTIVES

1. Do long division.
2. Divide numbers ending in 0 by numbers ending in 0.
3. Use multiplication to check division answers.

1 Do long division. In long division, estimate the various numbers by using a **trial divisor,** which is used to get a **trial quotient.**

Example 1 Using a Trial Divisor and a Trial Quotient

Divide: $42\overline{)3066}$.

Because 42 is closer to 40 than to 50, use the first digit of the divisor as a trial divisor.

$$\underset{\uparrow}{4}2 \quad \leftarrow \text{Trial divisor}$$

Try to divide the first digit of the dividend by 4. Since 3 cannot be divided by 4, use the first *two* digits, 30.

$$\frac{30}{4} = 7 \text{ with remainder } 2$$

$$\begin{array}{r} 7 \\ 42\overline{)3066} \end{array} \quad \leftarrow \text{Trial quotient}$$

7 goes over the 6, because $\frac{306}{42}$ is about 7.

Multiply 7 and 42 to get 294; next, subtract 294 from 306.

$$\begin{array}{rl} 7 & \\ 42\overline{)3066} & \\ \underline{294} & \leftarrow 7 \times 42 \\ 12 & \leftarrow 306 - 294 \end{array}$$

Bring down the 6 at the right.

$$\begin{array}{rl} 7 & \\ 42\overline{)3066} & \\ \underline{294\downarrow} & \\ 126 & \leftarrow \text{6 brought down} \end{array}$$

Use the trial divisor, 4.

First two digits of 126 $\rightarrow \frac{12}{4} = 3$

$$\begin{array}{rl} 73 & \\ 42\overline{)3066} & \\ \underline{294} & \\ 126 & \\ \underline{126} & \leftarrow 3 \times 42 = 126 \\ 0 & \end{array}$$

Check the answer by multiplying 42 and 73. The product should be 3066.

CAUTION

The ***first digit*** of the quotient in long division must be placed in the proper position over the dividend.

Work Problem 1 at the Side.

Example 2 Dividing to Find a Trial Quotient

Divide: $58\overline{)2730}$.

Use 6 as a trial divisor, since 58 is closer to 60 than to 50.

First two digits of dividend → $\frac{27}{6} = 4$ with 3 left over

$$\begin{array}{r} 4 \\ 58\overline{)2730} \\ \underline{232} \\ 41 \end{array}$$

4 ← Trial quotient

232 ← $4 \times 58 = 232$

41 ← $273 - 232 = 41$ (smaller than 58, the divisor)

Bring down the 0.

$$\begin{array}{r} 4 \\ 58\overline{)2730} \\ \underline{232}\downarrow \\ 410 \end{array}$$

410 ← 0 brought down

First two digits of 410 → $\frac{41}{6} = 6$ with 5 left over

$$\begin{array}{r} 46 \\ 58\overline{)2730} \\ \underline{232} \\ 410 \\ \underline{348} \\ 62 \end{array}$$

46 ← Trial quotient

348 ← $6 \times 58 = 348$

62 ← Greater than 58

The remainder, 62, is greater than the divisor, 58, so 7 should be used instead of 6.

$$\begin{array}{r} 47 \text{ R4} \\ 58\overline{)2730} \\ \underline{232} \\ 410 \\ \underline{406} \\ 4 \end{array}$$

406 ← $7 \times 58 = 406$

4 ← $410 - 406$

Work Problem 2 at the Side.

1 Divide.

(a) $14\overline{)1148}$

(b) $32\overline{)2048}$

(c) $61\overline{)4392}$

(d) $\frac{5394}{93}$

2 Divide.

(a) $48\overline{)2688}$

(b) $36\overline{)2236}$

(c) $65\overline{)5416}$

(d) $89\overline{)6649}$

ANSWERS

1. (a) 82 (b) 64 (c) 72 (d) 58
2. (a) 56 (b) 62 **R**4 (c) 83 **R**21 (d) 74 **R**63

Sometimes it is necessary to insert a 0 in the quotient.

Example 3 **Inserting 0s in the Quotient**

Divide: $42\overline{)8734}$.

Start as above.

$$\begin{array}{r} 2 \\ 42\overline{)8734} \\ \underline{84} \\ 3 \end{array} \quad \begin{array}{l} \\ \\ \leftarrow 2 \times 42 = 84 \\ \leftarrow 87 - 84 = 3 \end{array}$$

Bring down the 3.

$$\begin{array}{r} 2 \\ 42\overline{)8734} \\ \underline{84}{\downarrow} \\ 33 \end{array} \quad \begin{array}{l} \\ \\ \\ \leftarrow \text{3 brought down} \end{array}$$

Since 33 cannot be divided by 42, place a 0 in the quotient as a placeholder.

$$\begin{array}{r} 20 \\ 42\overline{)8734} \\ \underline{84} \\ 33 \end{array} \quad \begin{array}{l} \leftarrow \text{0 in quotient} \\ \\ \\ \\ \end{array}$$

Bring down the final digit, the 4.

$$\begin{array}{r} 20 \\ 42\overline{)8734} \\ \underline{84}{\downarrow} \\ 334 \end{array} \quad \begin{array}{l} \\ \\ \\ \leftarrow \text{4 brought down} \end{array}$$

Complete the problem.

$$\begin{array}{r} 207\ \mathbf{R}40 \\ 42\overline{)8734}\phantom{\ \mathbf{R}40} \\ \underline{84}\phantom{\ \mathbf{R}40} \\ 334\phantom{\ \mathbf{R}40} \\ \underline{294}\phantom{\ \mathbf{R}40} \\ 40\phantom{\ \mathbf{R}40} \end{array}$$

The answer is 207 **R**40.

CAUTION

There ***must be a digit*** in the quotient (answer) above every digit in the dividend once the answer has begun. Notice in Example 3 that a **0** was used to assure an answer digit above every digit in the dividend.

Work Problem 3 at the Side.

2 **Divide numbers ending in 0 by numbers ending in 0.** When the divisor and dividend both contain 0s at the far right, recall that these numbers are multiples of 10. As with multiplication, there is a short way to divide these multiples of 10. Look at the following examples.

$$26{,}000 \div 1 = 26{,}000$$
$$26{,}000 \div 10 = 2600$$
$$26{,}000 \div 100 = 260$$
$$26{,}000 \div 1000 = 26$$

Do you see a pattern? These examples suggest the following rule.

3 Divide.

(a) $34\overline{)3645}$

(b) $28\overline{)5768}$

(c) $39\overline{)15{,}933}$

(d) $78\overline{)23{,}462}$

ANSWERS

3. **(a)** 107 **R**7 **(b)** 206
(c) 408 **R**21 **(d)** 300 **R**62

4 Divide.

(a) $90 \div 10$

(b) $2400 \div 100$

(c) $206{,}000 \div 1000$

5 Divide.

(a) $50\overline{)6250}$

(b) $130\overline{)131{,}040}$

(c) $2600\overline{)195{,}000}$

ANSWERS
4. (a) 9 **(b)** 24 **(c)** 206
5. (a) 125 **(b)** 1008 **(c)** 75

Dividing a Whole Number by 10, 100, or 1000

Divide a whole number by 10, 100, or 1000 by dropping the appropriate number of 0s from the whole number.

Example 4 Dividing by Multiples of 10

Divide.

(a) $60 \div 10 = 6$ (One 0 in divisor; 0 dropped)

(b) $3500 \div 100 = 35$ (Two 0s in divisor; 00 dropped)

(c) $915{,}000 \div 1000 = 915$ (Three 0s in divisor; 000 dropped)

Work Problem 4 at the Side.

In the next example, we find the quotient for other multiples of 10 by dropping 0s.

Example 5 Dividing by Multiples of 10

Divide.

(a) $40\overline{)11{,}000}$ Drop one 0 from the divisor and the dividend.

```
   275
 4)1100
   8
   30
   28
    20
    20
     0
```

Since $1100 \div 4$ is 275, then $11{,}000 \div 40$ is also 275.

(b) $3500\overline{)31{,}500}$ Drop two 0s from the divisor and the dividend.

```
     9
 35)315
    315
      0
```

Since $315 \div 35$ is 9, then $31{,}500 \div 3500$ is also 9.

NOTE

Dropping 0s when dividing by multiples of 10 ***does not*** change the quotient (answer).

Work Problem 5 at the Side.

3 **Use multiplication to check division answers.** Answers in long division can be checked just as answers in short division were checked.

Example 6 **Checking Division by Using Multiplication**

Check each answer.

(a) $48\overline{)5324}$ with quotient 114 **R**43

$$\begin{array}{r} 114 \\ \times\ 48 \\ \hline 912 \\ 456 \\ \hline 5472 \\ +\ \ 43 \\ \hline 5515 \end{array}$$

Multiply the quotient and the divisor.

Add the remainder.

Result does *not* match dividend.

The answer does ***not*** check. Rework the original problem to get 110 **R**44.

(b) $716\overline{)26{,}492}$ with quotient 37

$$\begin{array}{r} 37 \\ 716\overline{)26{,}492} \\ 21\ 48 \\ \hline 5\ 012 \\ 5\ 012 \\ \hline 0 \end{array} \qquad \begin{array}{r} 716 \\ \times\ \ 37 \\ \hline 5\ 012 \\ 21\ 48 \\ \hline 26{,}492 \end{array}$$

Correct

Calculator Tip To check the answer to Example 6(a), don't forget to add the remainder.

48 (×) 110 (+) 44 (=) 5324

CAUTION

When checking a division problem, first multiply the quotient and the divisor. Then be sure to ***add any remainder*** before checking it against the original dividend.

Work Problem **6** at the Side.

6 Decide whether each answer is correct. If the answer is incorrect, find the correct answer.

(a)
$$\begin{array}{r} 38 \\ 16\overline{)608} \\ 48 \\ \hline 128 \\ 128 \\ \hline 0 \end{array}$$

(b)
$$\begin{array}{r} 42\ \text{R}178 \\ 426\overline{)19{,}170} \\ 17\ 04 \\ \hline 1\ 130 \\ 952 \\ \hline 178 \end{array}$$

(c)
$$\begin{array}{r} 57\ \text{R}18 \\ 514\overline{)29{,}316} \\ 25\ 70 \\ \hline 3\ 616 \\ 3\ 598 \\ \hline 18 \end{array}$$

ANSWERS

6. **(a)** correct **(b)** incorrect; should be 45 **(c)** correct

Focus on Real-Data Applications

Sharing Travel Expenses

Sharing costs for meals, entertainment, and travel can be a challenging mathematical problem, especially if the costs are incurred over several days and paid for by different people.

As an example, suppose that Donna and Jerry (couple), Peg and Richard (couple), and Jo (single) decide to spend a week touring Wales and Ireland. Donna has booked the accommodations, and each person will be responsible for their own meals and hotel expenses. However, to save on transportation costs, Richard (who lives in England) will use his personal car, and the group will share the costs for insurance, gasoline, parking, and the ferry. At the end of the trip, the costs were those listed in the table below. All the costs are rounded to the nearest U.S. dollar, although the travelers actually paid the amounts in British pounds and Irish punts.

Expense	Paid By	Costs (in U.S. dollars)
Supplemental car insurance	Richard	21
Ferry (prebooked)	Donna	220
Ferry (upgrade to fast boat)	Richard	40
Parking (Dublin, Ireland)	Richard	21
Gasoline (Wales)	Richard	46
Gasoline (Ireland)	Richard	13
Gasoline (Ireland)	Richard	22
Gasoline (Ireland)	Richard	24
Gasoline (Wales)	Richard	46
Gasoline (Wales)	Richard	21

1. What was the total amount spent on transportation?
2. To compute the shared costs, divide the total transportation costs by the number of travelers.
   (a) How many whole dollars does each traveler owe for transportation costs? (The division does not come out evenly; there is a remainder.)
   (b) How many dollars are left over (the remainder)? What should be done with the remainder costs to ensure a fair division of the costs?
   (c) If the total transportation costs are rounded up to the nearest multiple of 5, then how much would each traveler owe?
3. Using your answer from Problem 2(c), how much is Donna and Jerry's share of the costs as a couple? Do they owe money or are they owed money? If they owe money, how much and to whom should it be paid? If money is owed to them, who owes the money and how much is owed?
4. Using your answer from Problem 2(c), how much is Richard and Peg's share of the costs as a couple? Do they owe money or are they owed money? If they owe money, how much and to whom should it be paid? If money is owed to them, who owes the money and how much is owed?
5. Did each traveler pay the same amount toward transportation costs?

1.6 Exercises

FOR EXTRA HELP: Student's Solutions Manual | MyMathLab.com | InterAct Math Tutorial Software | AW Math Tutor Center | www.mathxl.com | Digital Video Tutor CD 1 Videotape 2

Decide where the first digit in the quotient would be located. Then without finishing the division, you can tell which of the three choices is the correct answer. Circle your choice. See Examples 1 and 2.

1. $25\overline{)550}$
2 22 220

2. $14\overline{)476}$
3 34 304

3. $18\overline{)4500}$
2 25 250

4. $42\overline{)7560}$
18 180 1800

5. $86\overline{)10,327}$
12 120 R7 1200

6. $46\overline{)24,026}$
5 52 522 R14

7. $52\overline{)68,025}$
13 130 R1 1308 R9

8. $12\overline{)116,953}$
974 R2 9746 R1 97,460

9. $21\overline{)149,826}$
71 713 7134 R12

10. $64\overline{)208,138}$
325 R2 3252 R10 32,521

11. $523\overline{)470,800}$
9 R100 90 R100 900 R100

12. $230\overline{)253,230}$
11 110 1101

Divide by using long division. Use multiplication to check each answer. See Examples 1–3.

13. $21\overline{)2272}$

14. $29\overline{)1827}$

15. $56\overline{)10,270}$

16. $83\overline{)39,692}$

17. $26\overline{)62,583}$

18. $28\overline{)84,249}$

19. $74\overline{)84,819}$

20. $238\overline{)186,948}$

21. $153\overline{)509,725}$

22. $308\overline{)26,796}$

23. $420\overline{)357,000}$

24. $900\overline{)153,000}$

Use multiplication to check each answer. If an answer in incorrect, find the correct answer. See Example 6.

25. $\begin{array}{r} 101 \textbf{ R4} \\ 35\overline{)3549} \end{array}$

26. $\begin{array}{r} 42 \textbf{ R26} \\ 64\overline{)2712} \end{array}$

27. $\begin{array}{r} 658 \textbf{ R9} \\ 28\overline{)18,424} \end{array}$

28. $\begin{array}{r} 239 \textbf{ R121} \\ 145\overline{)34,776} \end{array}$

29. $\begin{array}{r} 62 \textbf{ R3} \\ 614\overline{)38,068} \end{array}$

30. $\begin{array}{r} 174 \textbf{ R368} \\ 557\overline{)97,286} \end{array}$

31. Describe in your own words a shortcut you can use to divide multiples of 10 by 10, by 100, or by 1000. Write an example problem and solve it.

32. Suppose you have a division problem with a remainder in the answer. Explain how to check your answer by writing an example problem that has a remainder.

Solve each application problem by using addition, subtraction, multiplication, or division as needed. See Examples 3–5.

33. A private airplane flies 2304 miles to an air show in Oshkosh, Wisconsin, at 128 mph. How many hours did it take to get there?

34. The U.S. Government Printing Office uses 255,000 pounds of ink each year. If it does an equal amount of printing on each of 200 work days in a year, find the weight of the ink used each day. (*Source:* U.S. Government Printing Office.)

35. Don Gracey, the Mountain Timesmith, has serviced and repaired 636 clocks this year. He has worked on 272 wall clocks and 308 table clocks. The remainder were standing floor clocks. Find the number of floor clocks he worked on this year.

36. Two separated parents each share some of the $3718 education costs of their child. If one parent paid $1880, how much did the other pay?

37. Judy Martinez owes $3888 on a loan (including interest). Find her monthly payment if the loan is to be paid off in 36 months.

38. A consultant charged $13,050 for evaluating a school's compliance with the Americans with Disabilities Act. If the consultant worked 225 hours, find the rate charged per hour.

39. Clarence Hanks can assemble 42 circuit boards in 1 hour. How many circuit boards can he assemble in a 5-day workweek of 8 hours per day?

40. There are two conveyor lines in a factory, each of which packages 240 sacks of salt per hour. If the lines operate for 8 hours, find the total number of sacks of salt packaged by the two lines.

41. The average U.S. household of 2.5 people spent $2028 eating away from home last year. Find the average weekly cost of eating away from home. (*Hint:* 1 year equals 52 weeks.)(*Source:* U.S. Bureau of Labor Statistics consumer expenditure surveys.)

42. Former professional basketball player Junior Bridgeman now owns 120 Wendy's restaurants with 4080 employees. Find the average number of employees at each restaurant. (*Source:* National Basketball Retired Players Association.)

RELATING CONCEPTS (Exercises 43–50) FOR INDIVIDUAL OR GROUP WORK

Knowing and using the rules of divisibility is necessary in problem solving.
Work Exercises 43–50 in order.

43. If you have $0 and you divide this amount among three people, how much will each receive?

44. When 0 is divided by any nonzero number, the result is ________.

45. Divide.
$8 \div 0$

46. We say that division by 0 is *undefined* because it is ______________ (possible/impossible) to compute the answer. Give an example involving cookies that will support your answer.

47. Divide.

(a) $14 \div 1$

(b) $1\overline{)17}$

(c) $\frac{38}{1}$

48. Any number divided by 1 is the number itself. Is this also true when multiplying by 1? Give three examples that support your answer.

49. Divide.

(a) $32{,}000 \div 10$

(b) $32{,}000 \div 100$

(c) $32{,}000 \div 1000$

50. Write a rule that explains the shortcut for doing divisions like the ones in Exercise 49.

Focus on **Real-Data Applications**

Post Office Facts

The United States Postal Service posts a Web page on the Internet that gives a list of facts about their service. Some of those facts are given in the table below.

Resource/Service	Number
1. Mail collection boxes	312,000
2. Post offices	38,019
3. Delivery points	130 million
4. Pieces of First Class mail delivered each year	107 billion
5. Processing plants sorting and shipping the mail	331
6. Pounds of mail carried on commercial airline flights annually	2.7 billion
7. Commercial airline flights per day	15,000
8. Miles driven to move the mail annually	1.1 billion
9. Vehicles to pick up, transport, and deliver the mail	192,904
10. Customers a day who transact business at the post offices	7 million

Source: United States Postal Service, www.usps.com.

1. Write the number of delivery points in digits showing each period.
2. Write the number of miles driven annually to move the mail in digits showing each period.
3. Estimate the number of post offices, rounded to the nearest thousand.
4. Use the estimate found in Problem 3 to compute the average number of customers who transact business each day at post offices. Round the answer to the nearest person. (*Hint:* To find the average number of customers per day who transact business at post offices, divide the total number of customers per day by the number of post offices. Recall that you can use a short cut to simplify a division problem by crossing out the same number of 0s in the divisor and the dividend.)
5. Find the total number of commercial airline flights per year that carry postal freight. Write the result in words using period names. (*Hint:* Assume that planes fly 365 days per year.)
6. Find the average number of pounds of mail per commercial airline flight, rounded to the nearest 10 pounds. (*Hint:* Divide the total number of pounds of mail carried annually by the number of flights per year, calculated in Problem 5. Use the division shortcut described in Problem 4.)
7. Mail is sorted and shipped from processing plants to the post offices. Which of the following is a rough estimate of the number of post offices served by each processing plant: 10, 100, or 1000? Explain your choice.

1.7 Rounding Whole Numbers

One way to get a rough check on an answer is to *round* the numbers in the problem. **Rounding** a number means finding a number that is close to the original number, but easier to work with.

For example, the chancellor of a community college district might be discussing the need for more classrooms and lab facilities. In making her point, she probably would not need to say that the district has 61,832 students—she could probably just say there are 62,000 students, or even 60,000 students.

Objectives

1. Locate the place to which a number is to be rounded.
2. Round numbers.
3. Round numbers to estimate an answer.
4. Use front end rounding to estimate an answer.

1 Locate the place to which a number is to be rounded. The first step in rounding a number is to locate the *place to which the number is to be rounded.*

Example 1 Finding the Place to Which a Number Is to Be Rounded

Locate and draw a line under the place to which each number is to be rounded.

(a) Round 83 to the nearest ten. Is 83 closer to 80 or to 90?

83 is closer to 80. (Tens place: 8)

(b) Round 54,702 to the nearest thousand. Is it closer to 54,000 or to 55,000?

54,702 is closer to 55,000. (Thousands place: 4)

(c) Round 2,806,124 to the nearest hundred thousand. Is it closer to 2,800,000 or to 2,900,000?

2,806,124 is closer to 2,800,000. (Hundred thousands place: 8)

Work Problem 1 at the Side.

1 Locate and draw a line under the place to which each number is to be rounded. Then answer the question.

(a) 557 (nearest ten)

Is it closer to 550 or to 560? ______

(b) 1482 (nearest thousand)

Is it closer to 1000 or to 2000? ______

(c) 89,512 (nearest hundred)

Is it closer to 89,500 or 89,600? ______

(d) 546,325 (nearest ten thousand)

Is it closer to 540,000 or 550,000? ______

2 Round numbers. Use the following rules for rounding whole numbers.

Rounding Whole Numbers

Step 1 Locate the ***place*** to which the number is to be rounded. Draw a line under that place.

Step 2(a) Look only at the next digit to the right of the one you underlined. If it is ***5 or more,*** increase the underlined digit by 1.

Step 2(b) If the next digit to the right is ***4 or less,*** do not change the digit in the underlined place.

Step 3 ***Change*** all digits to the right of the underlined place to 0s.

Example 2 Using Rounding Rules for 4 or Less

Round 349 to the nearest hundred.

Step 1 Locate the place to which the number is being rounded. Draw a line under that place.

349 (Hundreds place: 3)

Continued on Next Page

Answers

1. (a) 557 is closer to 560
   (b) 1482 is closer to 1000
   (c) 89,512 is closer to 89,500
   (d) 546,325 is closer to 550,000

2 Round to the nearest ten.

(a) 43

(b) 92

(c) 164

(d) 6822

3 Round to the nearest thousand.

(a) 2635

(b) 5508

(c) 43,766

(d) 74,803

ANSWERS
2. **(a)** 40 **(b)** 90 **(c)** 160 **(d)** 6820
3. **(a)** 3000 **(b)** 6000 **(c)** 44,000 **(d)** 75,000

Step 2 Because the next digit to the right of the underlined place is 4, which is 4 or less, do *not* change the digit in the underlined place.

Next digit is 4 or less.
349
3 remains 3.

Step 3 Change all digits to the right of the underlined place to 0s.

349 rounded to the nearest hundred is 300.

In other words, 349 is closer to 300 than to 400.

Work Problem 2 at the Side.

Example 3 Using Rounding Rules for 5 or More

Round 36,833 to the nearest thousand.

Step 1 Find the place to which the number is to be rounded and draw a line under that place.

36,833
Thousands

Step 2 Because the next digit to the right of the underlined place is 8, which is 5 or more, add 1 to the underlined place.

Next digit is 5 or more.
36,833
Change 6 to 7.

Step 3 Change all digits to the right of the underlined place to 0s.

Change to 0.
36,833 rounded to the nearest thousand is 37,000.
Change 6 to 7.

In other words, 36,833 is closer to 37,000 than to 36,000.

Work Problem 3 at the Side.

Example 4 Using Rounding Rules

(a) Round 2382 to the nearest ten.

Step 1 2382
Tens place

Step 2 The next digit to the right is 2, which is 4 or less.

Next digit is 4 or less.
2382
Leave 8 as 8.

Step 3 2382 Change to 0.

2382 rounded to the nearest ten is 2380. In other words, 2382 is closer to 2380 than to 2390.

Continued on Next Page

(b) Round 13,961 to the nearest hundred.

Step 1 13,9̲61
Hundreds place

Step 2 The next digit to the right is 6.

Next digit is 5 or more.

13,9̲61

Change 9 to 10; write 0 and carry 1 into thousands place.
3 + carried 1 = 4

Change to 0.

Step 3 14,0̲61

13,961 rounded to the nearest hundred is 14,000.

NOTE

In Step 2 of Example 4(a), notice that the first three digits increased from 139 to 140 when we added 1 to the hundreds place.

(13,9)61 rounded to (14,0)00

Work Problem 4 at the Side.

Example 5 Rounding Large Numbers

(a) Round 37,892 to the nearest ten thousand.

Step 1 3̲7,892
Ten thousands place

Step 2 The next digit to the right is 7.

Next digit is 5 or more.

3̲7892
Change 3 to 4.

Change to 0.

Step 3 4̲7,892

37,892 rounded to the nearest ten thousand is 40,000.

(b) Round 528,498,675 to the nearest million.

Step 1 528̲,498,675
Millions place

Next digit is 4 or less.

Step 2 528̲,498,675
Leave 8 as 8.

Change to 0.

Step 3 528̲,498,675

528,498,675 rounded to the nearest million is 528,000,000.

Work Problem 5 at the Side.

4 Round each number as indicated.

(a) 7827 to the nearest ten

(b) 4342 to the nearest hundred

(c) 73,077 to the nearest hundred

(d) 56,961 to the nearest hundred

5 Round each number as indicated.

(a) 13,586 to the nearest ten thousand

(b) 724,518,715 to the nearest million

ANSWERS
4. **(a)** 7830 **(b)** 4300
(c) 73,100 **(d)** 57,000
5. **(a)** 10,000 **(b)** 725,000,000

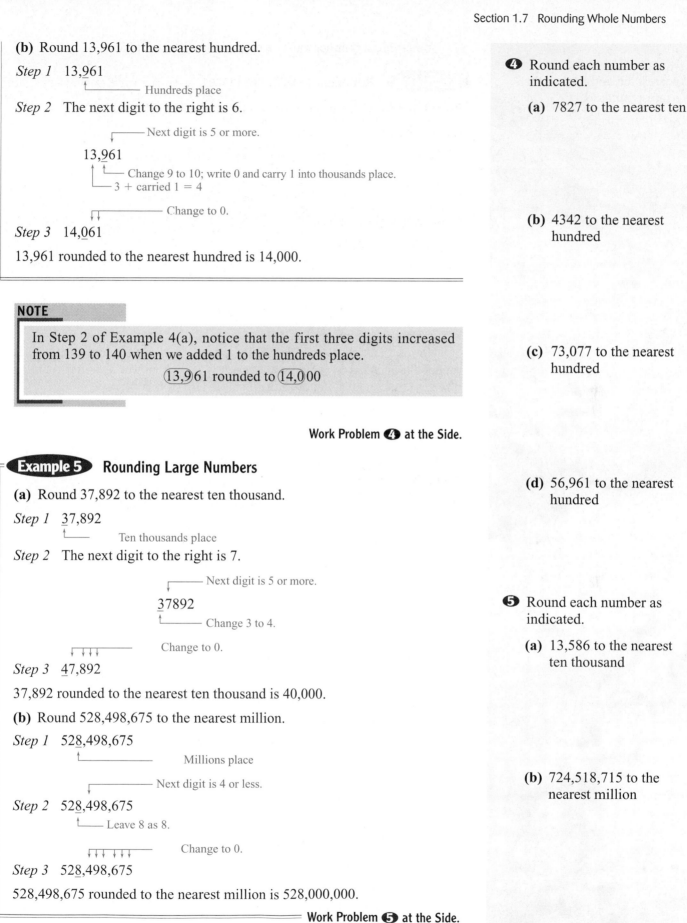

6 Round each number to the nearest ten and then to the nearest hundred.

(a) 267

(b) 549

(c) 8709

Sometimes a number must be rounded to different places.

Example 6 Rounding to Different Places

Round 648 **(a)** to the nearest ten and **(b)** to the nearest hundred.

(a) to the nearest ten

648 — Next digit is 5 or more. Tens place (4 + 1 = 5)

648 to the nearest ten is 650.

(b) to the nearest hundred

648 — Next digit is 4 or less. Hundreds place stays the same.

648 to the nearest hundred is 600.

Notice that 648 is rounded to the nearest ten is 650, and then 650 is rounded to the nearest hundred, the result is 700. If, however, 648 is rounded directly to the nearest hundred, the result is 600 (not 700).

CAUTION

Before rounding to a different place, always go back to the *original*, unrounded number.

Work Problem 6 at the Side.

Example 7 Applying Rounding Rules

Round each number to the nearest ten, nearest hundred, and nearest thousand.

(a) 4358

First round 4358 to the nearest ten.

4358 — Next digit is 5 or more. Tens place (5 + 1 = 6)

4358 rounded to the nearest ten is 4360.

Now go back to 4358, the *original* number, before rounding to the nearest hundred.

4358 — Next digit is 5 or more. Hundreds place (3 + 1 = 4)

4358 rounded to the nearest hundred is 4400.

Again, go back to the *original* number before rounding to the nearest thousand.

4358 — Next digit is 4 or less. Thousands place stays the same.

4358 rounded to the nearest thousand is 4000.

Continued on Next Page

ANSWERS
6. (a) 270; 300
(b) 550; 500
(c) 8710; 8700

(b) 680,914

First, round to the nearest ten.

Next digit is 4 or less.

680,91̲4

Tens place stays the same.

680,914 rounded to the nearest ten is 680,910.

Go back to 680,914, the *original* number, to round to the nearest hundred.

Next digit is 4 or less.

680,9̲14

Hundreds place stays the same.

680,914 rounded to the nearest hundred is 680,900.

Go back to the *original* number to round to the nearest thousand.

Next digit is 5 or more.

680̲,914

Thousands place $(0 + 1 = 1)$

680,914 rounded to the nearest thousand is 681,000.

Work Problem 7 at the Side.

7 Round each number to the nearest ten, nearest hundred, and nearest thousand.

(a) 2087

(b) 46,364

(c) 268,328

3 Round numbers to estimate an answer. Numbers may be rounded to estimate an answer. An estimated answer is one that is close to the exact answer and may be used as a check when the exact answer is found. The "≈" sign is often used to show that an answer has been rounded or estimated and is almost equal to the exact answer. ≈ means "approximately equal to."

8 Estimate each answer by rounding to the nearest ten.

(a)
```
  16
  74
  58
+ 31
```

(b)
```
  53
- 19
```

(c)
```
  37
× 84
```

Example 8 Using Rounding to Estimate an Answer

Estimate each answer by rounding to the nearest ten.

(a)
```
  76 ──→   80 ┐
  53 ──→   50 │ Rounded to the nearest ten
  38 ──→   40 │
+ 91 ──→ + 90 ┘
          260   Estimated answer
```

(b)
```
  27       30 ┐ Rounded to the nearest ten
- 14     - 10 ┘
           20   Estimated answer
```

(c)
```
  16       20 ┐ Rounded to the nearest ten
× 21     × 20 ┘
          400   Estimated answer
```

Work Problem 8 at the Side.

Example 9 Using Rounding to Estimate an Answer

Estimate each answer by rounding to the nearest hundred.

(a)
```
  152 ──→   200 ┐
  749 ──→   700 │ Rounded to the nearest hundred
  576 ──→   600 │
+ 819 ──→ + 800 ┘
           2300   Estimated answer
```

Continued on Next Page

Answers

7. (a) 2090; 2100; 2000
(b) 46,360; 46,400; 46,000
(c) 268,330; 268,300; 268,000

8. (a) $20 + 70 + 60 + 30 = 180$
(b) $50 - 20 = 30$
(c) $80 \times 40 = 3200$

9 Estimate each answer by rounding to the nearest hundred.

(a) 168
637
718
+ 883

(b) 842
− 475

(c) 723
× 478

10 Use front end rounding to estimate each answer.

(a) 36
3852
749
+ 5474

(b) 2583
− 765

(c) 441
× 76

ANSWERS
9. (a) $200 + 600 + 700 + 900 = 2400$
(b) $800 - 500 = 300$
(c) $500 \times 700 = 350{,}000$
10. (a) $40 + 4000 + 700 + 5000 = 9740$
(b) $3000 - 800 = 2200$
(c) $80 \times 400 = 32{,}000$

(b)

780	800	Rounded to the nearest hundred
− 536	− 500	
	300	Estimated answer

(c)

664	700	Rounded to the nearest hundred
× 843	× 800	
	560,000	Estimated answer

Work Problem 9 at the Side.

4 Use front end rounding to estimate an answer. A convenient way to estimate an answer is to use *front end rounding*. With **front end rounding,** we round to the highest possible place so that all the digits become 0 except the first one. For example, suppose you want to buy a big screen television for \$749, a home entertainment center for \$459, and a big comfortable chair for \$525. Using front end rounding, you can estimate the total cost of these purchases.

Television	\$749 →	\$700	
Entertainment center	\$459 →	500	
Big chair	\$525 →	+ 500	
		\$1700	← Estimated total cost

Example 10 Using Front End Rounding to Estimate an Answer

Estimate each answer using front end rounding.

(a)

3825	4000	All digits changed to 0 except first digit, which is rounded
72	70	
565	600	
+ 2389	+ 2000	
	6670	Estimated answer

(b)

6712	7000	First digit rounded and all others changed to 0
− 825	− 800	
	6200	Estimated answer

(c)

725	700	
× 86	× 90	
	63,000	Estimated answer

NOTE

When using front end rounding, all the digits become 0 except the highest-place digit (the first digit).

Work Problem 10 at the Side.

1.7 Exercises

FOR EXTRA HELP: Student's Solutions Manual · MyMathLab.com · InterAct Math Tutorial Software · AW Math Tutor Center · MathXL www.mathxl.com · Digital Video Tutor CD 1 Videotape 3

Round each number as indicated. See Examples 1–5.

1. 414 to the nearest ten

2. 307 to the nearest ten

3. 975 to the nearest ten

4. 826 to the nearest ten

5. 6771 to the nearest hundred

6. 5847 to the nearest hundred

7. 86,813 to the nearest hundred

8. 17,211 to the nearest hundred

9. 34,468 to the nearest hundred

10. 18,273 to the nearest hundred

11. 5996 to the nearest hundred

12. 4452 to the nearest hundred

13. 15,758 to the nearest thousand

14. 28,465 to the nearest thousand

15. 78,499 to the nearest thousand

16. 14,314 to the nearest thousand

17. 7760 to the nearest thousand

18. 49,706 to the nearest thousand

19. 14,988 to the nearest ten thousand

20. 6599 to the nearest ten thousand

21. 595,008 to the nearest ten thousand

22. 725,182 to the nearest ten thousand

23. 8,906,422 to the nearest million

24. 13,713,409 to the nearest million

Round each number to the nearest ten, nearest hundred, and nearest thousand. See Examples 6 and 7.

	Ten	Hundred	Thousand		Ten	Hundred	Thousand
25. 2368	______	______	______	**26.** 6483	______	______	______
27. 3374	______	______	______	**28.** 7632	______	______	______
29. 5049	______	______	______	**30.** 7065	______	______	______

	Ten	Hundred	Thousand
31. 4241	________	________	________
32. 7456	________	________	________
33. 19,539	________	________	________
34. 59,806	________	________	________
35. 26,292	________	________	________
36. 78,519	________	________	________
37. 64,503	________	________	________
38. 84,639	________	________	________

39. Write in your own words the three steps that you would use to round a number when the digit to the right of the place to which you are rounding is 5 or more.

40. Write in your own words the three steps that you would use to round a number when the digit to the right of the place to which you are rounding is 4 or less.

Estimate each answer by rounding to the nearest ten. Then find the exact answer. See Example 8.

41. *Estimate:* (Rounds to) *Exact:*

$$\begin{array}{r} 85 \\ 34 \\ 78 \\ +\ 93 \\ \hline \end{array}$$

42. *Estimate:* *Exact:*

$$\begin{array}{r} 56 \\ 24 \\ 85 \\ +\ 71 \\ \hline \end{array}$$

43. *Estimate:* *Exact:*

$$\begin{array}{r} 86 \\ -\ 34 \\ \hline \end{array}$$

44. *Estimate:* *Exact:*

$$\begin{array}{r} 57 \\ -\ 24 \\ \hline \end{array}$$

45. *Estimate:* *Exact:*

$$\begin{array}{r} 67 \\ \times\ 34 \\ \hline \end{array}$$

46. *Estimate:* *Exact:*

$$\begin{array}{r} 53 \\ \times\ 75 \\ \hline \end{array}$$

Estimate each answer by rounding to the nearest hundred. Then find the exact answer. See Example 9.

47. *Estimate:* (Rounds to) *Exact:*

$$\begin{array}{r} 863 \\ 735 \\ 438 \\ +\ 792 \\ \hline \end{array}$$

48. *Estimate:* *Exact:*

$$\begin{array}{r} 623 \\ 362 \\ 189 \\ +\ 736 \\ \hline \end{array}$$

49. *Estimate:* $-$ ____ *Exact:* $\begin{array}{r} 762 \\ -\ 375 \\ \hline \end{array}$

50. *Estimate:* $-$ ____ *Exact:* $\begin{array}{r} 614 \\ -\ 276 \\ \hline \end{array}$

51. *Estimate:* $\times$ ____ *Exact:* $\begin{array}{r} 368 \\ \times\ 436 \\ \hline \end{array}$

52. *Estimate:* $\times$ ____ *Exact:* $\begin{array}{r} 845 \\ \times\ 396 \\ \hline \end{array}$

Estimate each answer using front end rounding. Then find the exact answer. See Example 10.

53. *Estimate:* *Exact:*

Estimate	Rounds to	Exact
____	←	8215
____	←	56
____	←	729
+ ____	←	+ 3605

54. *Estimate:* + ____ *Exact:* $\begin{array}{r} 2685 \\ 73 \\ 592 \\ +\ 7183 \\ \hline \end{array}$

55. *Estimate:* $-$ ____ *Exact:* $\begin{array}{r} 796 \\ -\ 439 \\ \hline \end{array}$

56. *Estimate:* $-$ ____ *Exact:* $\begin{array}{r} 543 \\ -\ 174 \\ \hline \end{array}$

57. *Estimate:* $\times$ ____ *Exact:* $\begin{array}{r} 939 \\ \times\ 29 \\ \hline \end{array}$

58. *Estimate:* $\times$ ____ *Exact:* $\begin{array}{r} 864 \\ \times\ 74 \\ \hline \end{array}$

59. The number 3492 rounded to the nearest hundred is 3500, and 3500 rounded to the nearest thousand is 4000. But when 3492 is rounded to the nearest thousand it becomes 3000. Why is this true? Explain.

60. The use of rounding is helpful when estimating the answer to a problem. Why is this true? Give an example using either addition, subtraction, multiplication, or division to show how this works.

61. In 1900, the population of the United States was 76 million. Today it's 281 million. Round each of these numbers to the nearest ten million. (*Source:* Reiman Publications.)

62. In 1900, the average work week in the United States was 59 hours. Today it's 38 hours. Round each of these numbers to the nearest ten. (*Source:* Reiman Publications.)

63. In 1900, the life expectancy in the United States was 47 years. Today it's 76 years. Round these numbers to the nearest ten. (*Source:* Reiman Publications.)

64. In 1900, the average age of the population in the United States was 23. Today it's 35. Round these numbers to the nearest ten. (*Source:* Reiman Publications.)

65. Mexico City Bank has total assets of 3,025,935,000 pesos. Round this amount to the nearest ten thousand, nearest million, and nearest billion pesos.

66. Round 621,999,652 to the nearest thousand, nearest ten thousand, and nearest hundred thousand.

67. In one year, the gross national product for the United States (sum of all goods and services sold) was $8,490,487,629,721. Round this amount to the nearest hundred thousand, nearest hundred million, and nearest billion. (*Source:* U.S. Department of Commerce.)

68. The U.S. Federal Food Assistance Program paid out $18,915,762,568 in food stamps. Round this amount to the nearest hundred thousand, nearest hundred million, and nearest ten billion. (*Source:* U.S. Department of Agriculture.)

RELATING CONCEPTS (Exercises 69–75) FOR INDIVIDUAL OR GROUP WORK

To see how both rounding and front end rounding are used in solving problems, ***work Exercises 69–75 in order.***

69. A number rounded to the nearest thousand is 72,000. What is the *smallest* whole number this could have been before rounding?

70. A number rounded to the nearest thousand is 72,000. What is the *largest* whole number this could have been before rounding?

71. When front end rounding is used, a whole number rounds to 8000. What is the *smallest* possible original number?

72. When front end rounding is used, a whole number rounds to 8000. What is the *largest* possible original number?

The table shows the average charitable contributions by amount of income.

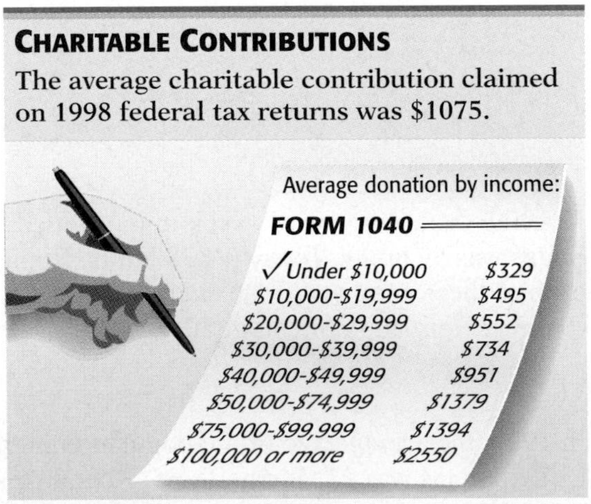

Source: Gallup for Independent Sector.

73. Round the amounts given for each income category to the nearest ten dollars.

74. Use front end rounding to round the amounts given for each income category.

75. **(a)** What is one advantage of using front end rounding instead of rounding to the nearest ten?

(b) What is one disadvantage?

1.8 Exponents, Roots, and Order of Operations

1 Identify an exponent and a base. The product 3 • 3 can be written as 3^2 (read as "3 squared"). The small raised number 2, called an **exponent,** says to use 2 factors of 3. The number 3 is called the **base.** Writing 3^2 as 9 is called *simplifying the expression.*

Example 1 Simplifying Expressions

Identify the exponent and the base, and then simplify each expression.

(a) 4^3

Base ⟶ 4^3 ⟵ Exponent $\quad 4^3 = 4 \times 4 \times 4 = 64$

(b) $2^5 = 2 \times 2 \times 2 \times 2 \times 2 = 32$
The base is 2 and the exponent is 5.

Work Problem 1 at the Side.

2 Find the square root of a number. Because $3^2 = 9$, the number 3 is called the **square root** of 9. The square root of a number is one of two identical factors of that number. Square roots of numbers are written with the symbol $\sqrt{\ }$.

$$\sqrt{9} = 3$$

Square Root

$$\sqrt{\textbf{number} \cdot \textbf{number}} = \sqrt{\textbf{number}^2} = \textbf{square root}$$

For example: $\sqrt{36} = \sqrt{6 \cdot 6} = \sqrt{6^2} = 6$ is the square root of 36.

To find the square root of 64 ask, "What number can be multiplied by itself (that is, *squared*) to give 64?" The answer is 8, so

$$\sqrt{64} = \sqrt{8 \cdot 8} = \sqrt{8^2} = 8.$$

A **perfect square** is a number that is the square of a ***whole number.*** The first few perfect squares are listed here.

Perfect Squares Table

$0 = 0^2$	$16 = 4^2$	$64 = 8^2$	$144 = 12^2$
$1 = 1^2$	$25 = 5^2$	$81 = 9^2$	$169 = 13^2$
$4 = 2^2$	$36 = 6^2$	$100 = 10^2$	$196 = 14^2$
$9 = 3^2$	$49 = 7^2$	$121 = 11^2$	$225 = 15^2$

Example 2 Using Perfect Squares

Find each square root.

(a) $\sqrt{16}$ Because $4^2 = 16$, $\sqrt{16} = 4$.

(b) $\sqrt{49} = 7$

(c) $\sqrt{0} = 0$

(d) $\sqrt{169} = 13$

Work Problem 2 at the Side.

3 Use the order of operations. Frequently problems may have parentheses, exponents, and square roots, and may involve more than one operation. Work these problems using the following **order of operations.**

Objectives

1. Identify an exponent and a base.
2. Find the square root of a number.
3. Use the order of operations.

Study Skills — *Study Skills Workbook* **Activity 4**

1 Identify the exponent and the base, and then simplify each expression.

(a) 5^2

(b) 6^3

(c) 2^4

(d) 3^4

2 Find each square root.

(a) $\sqrt{4}$

(b) $\sqrt{36}$

(c) $\sqrt{81}$

(d) $\sqrt{225}$

(e) $\sqrt{1}$

Answers

1. (a) 2; 5; 25 **(b)** 3; 6; 216 **(c)** 4; 2; 16 **(d)** 4; 3; 81
2. (a) 2 **(b)** 6 **(c)** 9 **(d)** 15 **(e)** 1

3 Simplify each expression.

(a) $2 + 6 + 3^2$

(b) $2^2 + 3^3$

(c) $4 \cdot 6 \div 12 - 2$

(d) $40 \div 5 \div 2$

(e) $8 + (14 \div 2) \cdot 6$

4 Simplify each expression.

(a) $10 - 4 + 3^2$

(b) $3^2 + 4^2 - (7 \cdot 2)$

(c) $2 \cdot \sqrt{64} - 5 \cdot 3$

(d) $20 \div 2 + (7 - 5)$

(e) $15 \cdot \sqrt{9} - 8 \cdot \sqrt{4}$

ANSWERS
3. **(a)** 17 **(b)** 31 **(c)** 0 **(d)** 4 **(e)** 50
4. **(a)** 15 **(b)** 11 **(c)** 1 **(d)** 12 **(e)** 29

Order of Operations

1. Do all operations inside ***parentheses*** or ***other grouping symbols.***
2. Simplify any expressions with ***exponents*** and find any ***square roots.***
3. ***Multiply*** or ***divide,*** proceeding from left to right.
4. ***Add*** or ***subtract,*** proceeding from left to right.

Example 3 Understanding the Order of Operations

Use the order of operations to simplify each expression.

(a) $8^2 + 5 + 2$

$\underbrace{8^2} + 5 + 2$

$\underbrace{8 \cdot 8} + 5 + 2$ Evaluate exponent first; 8^2 is $8 \cdot 8$.

$\underbrace{64 + 5} + 2$ Add from left to right.

$69 + 2 = 71$

(b) $\underbrace{35 \div 5} \cdot 6$ Divide first (start at left).

$7 \cdot 6 = 42$ Multiply.

(c) $9 + \underbrace{(20 - 4)} \cdot 3$ Work inside parentheses first.

$9 + \underbrace{16 \cdot 3}$ Multiply.

$9 + 48 = 57$ Add last.

(d) $12 \cdot \underbrace{\sqrt{16}} - 8 \cdot 4$ Find the square root first.

$\underbrace{12 \cdot 4} - \underbrace{8 \cdot 4}$ Multiply from left to right.

$48 - 32 = 16$ Subtract last.

Work Problem 3 at the Side.

Example 4 Using the Order of Operations

Use the order of operations to simplify each expression.

(a) $\underbrace{15 - 4} + 2$ Subtract first (start at left).

$11 + 2 = 13$ Add.

(b) $8 + \underbrace{(7 - 3)} \div 2$ Work inside parentheses first.

$8 + \underbrace{4 \div 2}$ Divide.

$8 + 2 = 10$ Add last.

(c) $4^2 \cdot 2^2 + \underbrace{(7 + 3)} \cdot 2$ Parentheses first.

$\underbrace{4^2} \cdot \underbrace{2^2} + 10 \cdot 2$ Evaluate exponents.

$\underbrace{16 \cdot 4} + \underbrace{10 \cdot 2}$ Multiply from left to right.

$64 + 20 = 84$ Add last.

(d) $4 \cdot \underbrace{\sqrt{25}} - 7 \cdot 2$ Find the square root first.

$\underbrace{4 \cdot 5} - \underbrace{7 \cdot 2}$ Multiply from left to right.

$20 - 14 = 6$ Subtract last.

NOTE

Getting a correct answer always depends on correctly using the order of operations.

Work Problem 4 at the Side.

1.8 EXERCISES

FOR EXTRA HELP — Student's Solutions Manual | MyMathLab.com | InterAct Math Tutorial Software | AW Math Tutor Center | www.mathxl.com | Digital Video Tutor CD 1 Videotape 3

Identify the exponent and the base, and then simplify each expression. See Example 1.

1. 4^2

2. 2^3

3. 5^2

4. 3^2

5. 12^2

6. 10^3

7. 15^2

8. 11^3

Use the Perfect Squares Table on page 73 to find each square root. See Example 2.

9. $\sqrt{9}$

10. $\sqrt{25}$

11. $\sqrt{64}$

12. $\sqrt{36}$

13. $\sqrt{100}$

14. $\sqrt{121}$

15. $\sqrt{144}$

16. $\sqrt{225}$

Fill in each blank. See Example 2.

17. $8^2 =$ ______ so $\sqrt{\quad} = 8$.

18. $9^2 =$ ______ so $\sqrt{\quad} = 9$.

19. $20^2 =$ ______ so $\sqrt{\quad} = 20$.

20. $30^2 =$ ______ so $\sqrt{\quad} = 30$.

21. $35^2 =$ ______ so $\sqrt{\quad} = 35$.

22. $38^2 =$ ______ so $\sqrt{\quad} = 38$.

23. $40^2 =$ ______ so $\sqrt{} = 40$.

24. $50^2 =$ ______ so $\sqrt{} = 50$.

25. $54^2 =$ ______ so $\sqrt{} = 54$.

26. $60^2 =$ ______ so $\sqrt{} = 60$.

27. Describe in your own words a perfect square. Of the two numbers 25 and 50, identify which is a perfect square and explain why.

28. Use the following list of words and phrases to write the four steps in the order of operations.

add	square root
exponents	subtract
multiply	divide
parentheses or other grouping symbols	

Simplify each expression by using the order of operations. See Examples 3 and 4.

29. $4^2 + 6 - 4$

30. $6^2 + 4 - 5$

31. $4 \cdot 8 - 7$

32. $5 \cdot 7 - 7$

33. $10 \cdot 4 \div 8$

34. $6 \cdot 8 \div 8$

35. $25 \div 5(8 - 4)$

36. $36 \div 18(7 - 3)$

37. $5 \cdot 3^2 + \frac{0}{8}$

38. $8 \cdot 3^2 - \frac{10}{2}$

39. $4 \cdot 1 + 8(9 - 2) + 3$

40. $3 \cdot 2 + 7(3 + 1) + 5$

41. $2^2 \cdot 3^3 + (20 - 15) \cdot 2$

42. $4^2 \cdot 5^2 + (20 - 9) \cdot 3$

43. $4 \cdot \sqrt{81} - 6 \cdot 3$

44. $2 \cdot \sqrt{100} - 3 \cdot 4$

45. $7 \cdot 3 + 4 \cdot 6 - 5$

46. $10 \cdot 3 + 6 \cdot 5 - 20$

47. $2^3 \cdot 3^2 + (14 - 4) \cdot 3$

48. $3^2 \cdot 4^2 + (15 - 6) \cdot 2$

49. $6 + 12 \div 3 + \frac{0}{5}$

50. $6 + 8 \div 2 + \frac{0}{8}$

51. $3^2 + 6^2 + (30 - 21) \cdot 2$

52. $4^2 + 5^2 + (25 - 9) \cdot 3$

53. $7 \cdot \sqrt{81} - 5 \cdot 6$

54. $6 \cdot \sqrt{64} - 6 \cdot 5$

55. $8 \cdot 2 + 5(3 \cdot 4) - 6$

56. $5 \cdot 2 + 3(5 + 3) - 6$

57. $4 \cdot \sqrt{49} - 7(5 - 2)$

58. $3 \cdot \sqrt{25} - 6(3 - 1)$

59. $7 \cdot (4 - 2) + \sqrt{9}$

60. $5 \cdot (4 - 3) + \sqrt{9}$

61. $7^2 + 3^2 - 8 + 5$

62. $3^2 - 2^2 + 3 - 2$

63. $5^2 \cdot 2^2 + (8 - 4) \cdot 2$

64. $5^2 \cdot 3^2 + (30 - 20) \cdot 2$

65. $8 + 8 \div 4 + 9 \cdot 2$

66. $8 + 3 \div 3 + 6 \cdot 3$

67. $5 \cdot \sqrt{36} - 7(7 - 4)$

68. $8 \cdot \sqrt{49} - 6(5 + 3)$

69. $3^2 - 2^3 + 4 \cdot 5$

70. $6^2 + 4^2 - 8 \cdot 4$

71. $9 + 4 \div 4 + 6 + \frac{0}{8}$

72. $2 + 12 \div 6 + 5 + \frac{0}{5}$

73. $5 \cdot \sqrt{36} - 3 \cdot 7$

74. $8 \cdot \sqrt{36} - 4 \cdot 6$

75. $7 \cdot \sqrt{16} - 5 \cdot \sqrt{25}$

76. $6 \cdot \sqrt{81} - 3 \cdot \sqrt{49}$

77. $7 \div 1 \cdot 8 \cdot 2 \div (21 - 5)$

78. $12 \div 4 \cdot 5 \cdot 4 \div (15 - 13)$

79. $15 \div 3 \cdot 2 \cdot 6 \div (14 - 11)$

80. $9 \div 1 \cdot 4 \cdot 2 \div (11 - 5)$

81. $6 \cdot \sqrt{25} - 4 \cdot \sqrt{16}$

82. $10 \cdot \sqrt{49} - 4 \cdot \sqrt{64}$

83. $5 \div 1 \cdot 10 \cdot 4 \div (17 - 9)$

84. $15 \div 3 \cdot 8 \cdot 9 \div (12 - 8)$

85. $8 \cdot 9 \div \sqrt{36} - 4 \div 2 + (14 - 8)$

86. $3 - 2 + 5 \cdot 4 \cdot \sqrt{144} \div \sqrt{36}$

87. $2 + 1 - 2 \cdot \sqrt{1} + 4 \cdot \sqrt{81} - 7 \cdot 2$

88. $6 - 4 + 2 \cdot 9 - 3 \cdot \sqrt{225} \div \sqrt{25}$

89. $5 \cdot \sqrt{36} \cdot \sqrt{100} \div 4 \cdot \sqrt{9} + 8$

90. $9 \cdot \sqrt{36} \cdot \sqrt{81} \div 2 + 6 - 3 - 5$

1.9 Reading Pictographs, Bar Graphs, and Line Graphs

We have all heard the saying "A picture is worth a 1000 words," and there may be some truth in this. Today, so much information and data is being presented in the form of pictographs, bar graphs, and line graphs that it is important to be able to read and understand these tools.

Objectives

1. Read and understand a pictograph.
2. Read and understand a bar graph.
3. Read and understand a line graph.

1 Read and understand a pictograph. A **pictograph** is a graph that uses pictures or symbols. It displays information that can be compared easily. However, since a symbol is used to represent a certain quantity, it can be difficult to determine the amount represented by a fraction of a symbol. The following pictograph compares the average amount that a household spends annually on public transportation in various regions of the country. In this pictograph, it is difficult to determine what fractional amount is represented by the partial subway car.

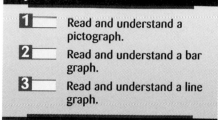

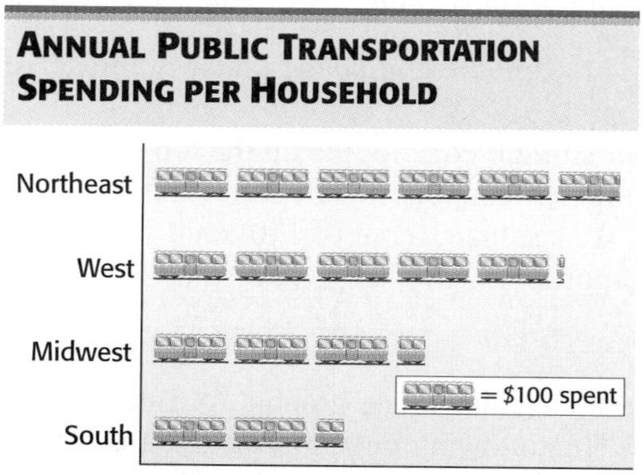

Source: Bureau of Labor Statistics *Consumer Expenditure Survey.*

Example 1 Using a Pictograph

Use the pictograph to answer each question.

(a) Which region of the country spends the least amount per household on public transportation?

The row representing the South has the fewest symbols. This means that the least amount is spent in the South.

(b) Approximately how much more is spent annually on public transportation by a household in the West than a household in the South?

The row representing the West has three more whole symbols than that for the South. This means that 3 • $100 or $300 more is spent on public transportation in the West than in the South.

Work Problem 1 at the Side.

2 Read and understand a bar graph. **Bar graphs** are useful for showing comparisons. For example, the following bar graph shows how many people out of every 100 fans surveyed chose each sport as their favorite.

1 Use the pictograph to answer each question.

(a) Which region of the country spends the most on public transportation?

(b) Approximately how much more is spent annually on public transportation by a household in the Northeast than a household in the Midwest?

Answers

1. (a) Northeast (b) ≈ $200

2 Use the bar graph to find the approximate number of fans who picked each sport as their favorite.

(a) Pro football

(b) Pro baseball

(c) Pro basketball

(d) College basketball

(e) Golf

3 Use the line graph to find the predicted population of the United States for each year.

(a) 2050

(b) 2075

(c) 2100

ANSWERS
2. **(a)** 28 out of 100 **(b)** 17 out of 100 **(c)** 13 out of 100 **(d)** 6 out of 100 **(e)** 6 out of 100
3. **(a)** 400,000,000 **(b)** 475,000,000 **(c)** 575,000,000

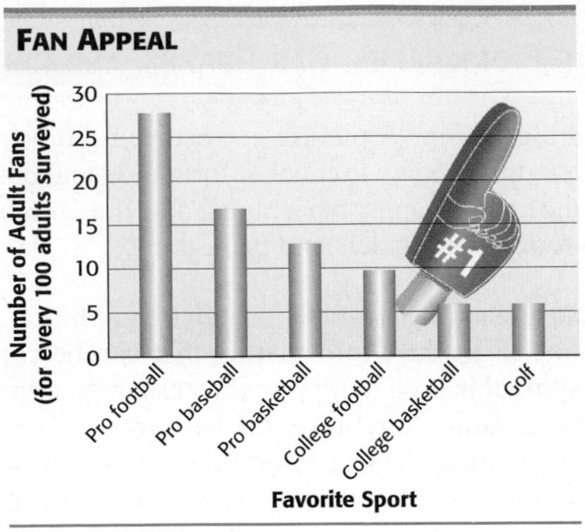

Source: The Harris Poll.

Example 2 **Using a Bar Graph**

Use the bar graph to find the number of fans who picked college football as their favorite sport.

Use a ruler or straight edge to line up the top of the bar labeled "College Football," with the numbers on the left edge of the graph, labeled "Number of Adults Fans." We see that 10 out of 100 adult fans picked college football as their favorite sport.

Work Problem 2 at the Side.

3 **Read and understand a line graph.** A **line graph** is often used for showing a trend. The following line graph shows the U.S. Bureau of the Census predictions for U.S. population growth to the year 2100.

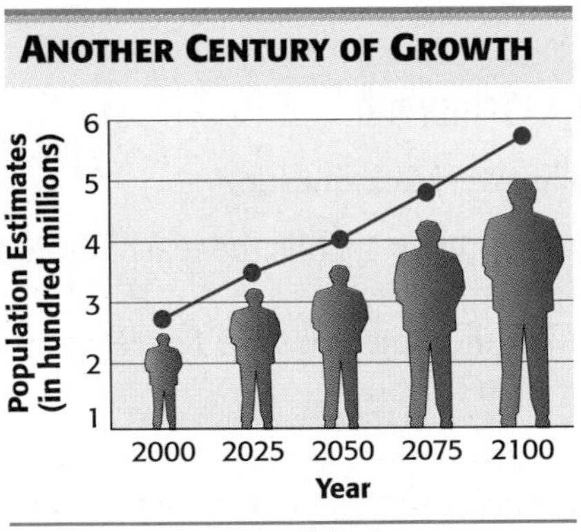

Source: U.S. Bureau of the Census.

Example 3 **Using a Line Graph**

Use the line graph to answer each question.

(a) What trend or pattern is shown in the graph?
The population will continue to increase.

(b) What is the estimated population for 2025?
Use a ruler or straight edge to line up the dot above the year labeled 2025 on the horizontal line with the numbers along the left edge of the graph. Notice that the label on the left side says "in hundred millions." Since the 2025 dot is halfway between 3 and 4, the population in 2025 is halfway between $3 \cdot 100{,}000{,}000$ and $4 \cdot 100{,}000{,}000$ or 300,000,000 and 400,000,000. That means that the predicted population in 2025 is about 350,000,000 people.

Work Problem 3 at the Side.

1.9 Exercises

FOR EXTRA HELP: Student's Solutions Manual | MyMathLab.com | InterAct Math Tutorial Software | AW Math Tutor Center | www.mathxl.com | Digital Video Tutor CD 1 Videotape 3

The following pictograph shows the approximate number of video rental outlets in the eight countries with the greatest number of outlets in the world. Use the pictograph to answer Exercises 1–6. See Example 1.

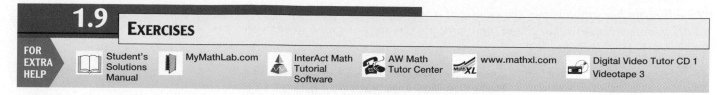

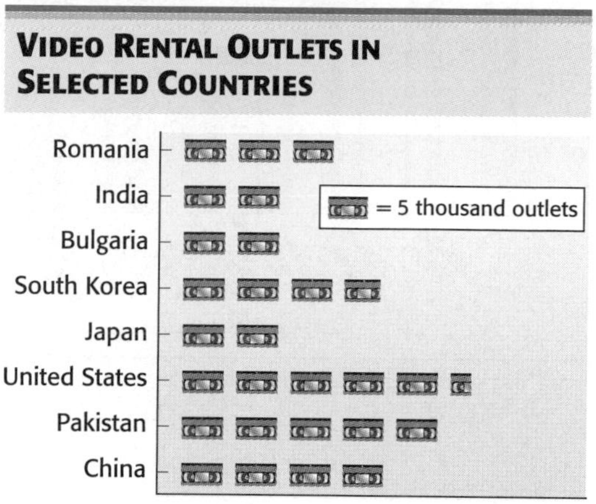

Source: *Screen Digest*—Top 8 Countries.

1. Find the number of video rental outlets in Romania.

2. How many video rental outlets are there in China?

3. Which country has the greatest number of outlets?

4. Which countries have the least number of outlets?

5. The United States has 28 thousand outlets. How many more is this than the second greatest number of outlets?

6. How many more outlets does China have than Bulgaria?

The following bar graph shows the results of a survey that was taken of 100 working adults to determine how they found their careers. Use the bar graph to answer Exercises 7–10. See Example 2.

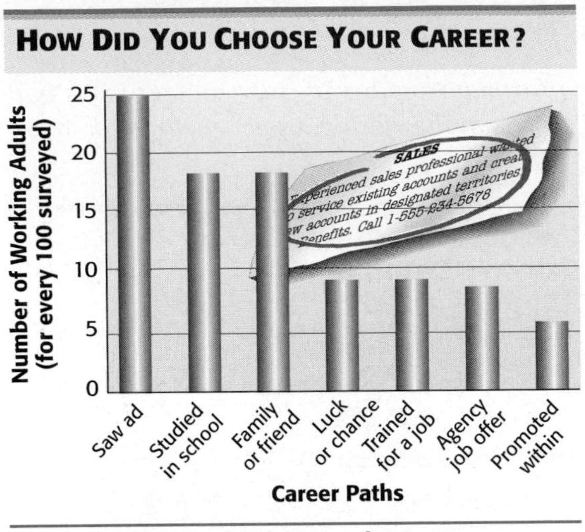

Source: Market Facts/TeleNation for Career Education Corporation.

7. How many people found their careers as a result of training for a job?

8. How many people found their careers because they studied for the career in school?

9. (a) Which career path was taken by the greatest number of people?

(b) How many people used this path?

10. (a) Which career path was taken by the least number of people?

(b) How many people used this path?

11. How many more people found their careers as a result of "Studied in school" than "Luck or chance"?

12. Find the total number of people who found their careers as a result of either "Studied in school" or "Trained for a job."

A local real estate association collected home sales data for their area and prepared the following line graph. Remembering that the sales data is shown in thousands, use the line graph to answer Exercises 13–16. See Example 3.

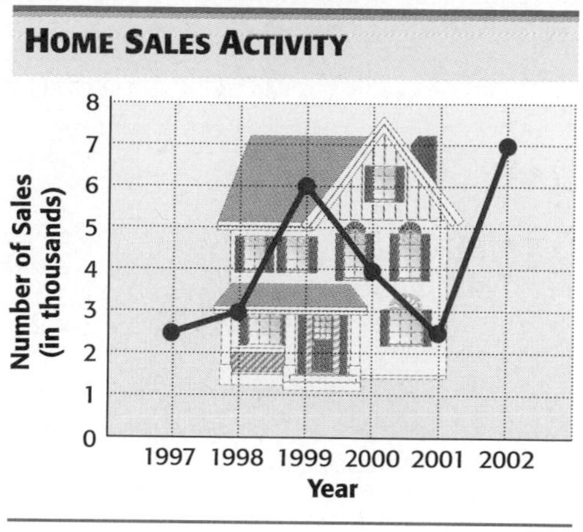

13. Which year had the greatest number of home sales? How many sales were there?

14. Which two years had the least number of home sales? How many sales were there in each of those years?

15. Find the increase in the number of home sales from 2001 to 2002.

16. Find the decrease in the number of home sales from 2000 to 2001.

17. Give three possible explanations for the decrease in home sales in 2000.

18. Give three possible explanations for the increase in home sales in 2002.

Relating Concepts (Exercises 19–23) For Individual or Group Work

Getting a correct answer in mathematics always depends on correctly using the order of operations. Insert grouping symbols (parentheses) so that each given expression evaluates to the given number. ***Work Exercises 19–23 in order.***

19. $5 + 1 \cdot 8 - 2$; evaluate to 46

20. $4 + 2 \cdot 5 + 1$; evaluate to 36

21. $36 \div 3 \cdot 3 \cdot 4$; evaluate to 16

22. $48 \div 2 \cdot 2 \cdot 2 + 2$; evaluate to 8

23. A fencing contractor is building 12 corrals like the one shown below for a dairy farmer.

(a) Use the order of operations to write an expression for the amount of fencing needed for all 12 corrals.

(b) How many feet of fencing are needed?

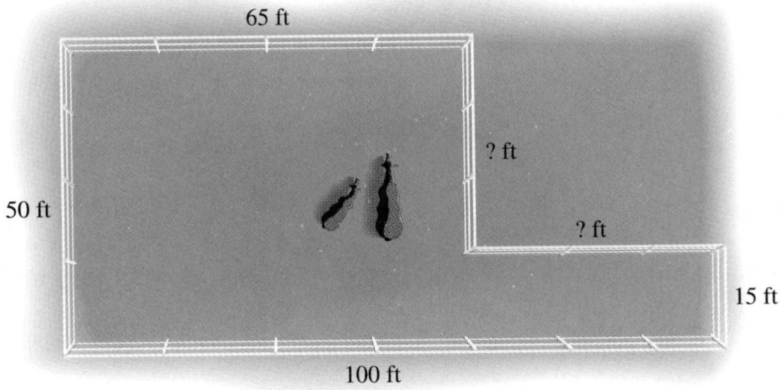

1.10 Solving Application Problems

Objectives

1. Find indicator words in application problems.
2. Solve application problems.
3. Estimate an answer.

Most problems involving applications of mathematics are written out in sentence form. You need to read the problem carefully to decide how to solve it.

1 Find indicator words in application problems. As you read an application problem, look for **indicator words** that help you determine the necessary operations—either addition, subtraction, multiplication, or division. Some of these indicator words are shown here.

Addition	Subtraction	Multiplication	Division	Equals
plus	less	product	divided by	is
more	subtract	double	divided into	the same as
more than	subtracted from	triple	quotient	equals
added to	difference	times	goes into	equal to
increased by	less than	of	divide	yields
sum	fewer	twice	divided equally	results in
total	decreased by	twice as much	per	are
sum of	loss of			
increase of	minus			
gain of	take away			

CAUTION

The word *and* does not always indicate addition, so it does not appear as an indicator word in the preceding table. Notice how the "and" shows the location of several different operation signs below.

The sum of 6 ***and*** 2 is $6 + 2$.
The difference of 6 ***and*** 2 is $6 - 2$.
The product of 6 ***and*** 2 is $6 \cdot 2$.
The quotient of 6 ***and*** 2 is $6 \div 2$.

2 Solve application problems. Solve application problems by using the following six steps.

Solving an Application Problem

Step 1 **Read** the problem carefully and be certain you *understand* what the problem is asking. It may be necessary to read the problem several times.

Step 2 Before doing any calculations, **work out a plan** and try to visualize the problem. Draw a sketch if possible. Know which facts are given and which must be found. Use *indicator words* to help decide on the *plan* (whether you will need to add, subtract, multiply, or divide).

Step 3 **Estimate** a *reasonable answer* by using rounding.

Step 4 **Solve** the problem by using the facts given and your plan.

Step 5 **State the answer.**

Step 6 **Check** your work. If the answer does not seem reasonable, begin again by rereading the problem.

1 Pick the most reasonable answer for each problem.

(a) A grocery clerk's hourly wage: $3; $11; $60

(b) The total length of five sports utility vehicles: 8 ft; 18 ft; 80 ft; 800 ft

(c) The cost of heart by-pass surgery: $500; $50,000; $5,000,000

2 Solve each problem.

(a) On a recent geology field trip, 84 fossils were collected. If the fossils are divided equally among John, Sean, Jenn, and Kara, how many fossils will each receive?

(b) There will be 384 people at an awards banquet. If each food server is assigned to 16 people, how many servers are needed?

CAUTION

Be careful not to begin solving a problem before you understand what the problem is asking. Be certain that you know what the problem is asking before you try to solve it.

3 ▦ **Estimate an answer.** The six problem-solving steps give a systematic approach for solving word problems. Each of the steps is important, but special emphasis should be placed on Step 3, estimating a *reasonable answer.* Many times an "answer" just does not fit the problem.

What is a reasonable answer? Read the problem and try to determine the approximate size of the answer. Should the answer be part of a dollar, a few dollars, hundreds, thousands, or even millions of dollars? For example, if a problem asks for the cost of a man's shirt, would an answer of $20 be reasonable? $1000? $0.65? $65?

CAUTION

Always estimate the answer, then look at your final result to be sure it fits your estimate and is reasonable. This step will give greater success in problem solving.

Work Problem ❶ at the Side.

Example 1 **Applying Division**

At a recent garage sale, the total sales were $584. If the money was divided equally among Tom, Rosetta, Maryann, and José, how much did each person get?

Step 1 **Read.** A reading of the problem shows that the four members in the group divided $584 equally.

Step 2 **Work out a plan.** The indicator words, *divided equally,* show that the amount each received can be found by dividing $584 by 4.

Step 3 **Estimate.** A reasonable answer would be a little less than $150 each, since $600 ÷ 4 = $150 ($584 rounded to $600).

Step 4 **Solve.** Find the actual answer by dividing $584 by 4.

$$\begin{array}{r} 146 \\ 4\overline{)584} \end{array}$$

Step 5 **State the answer.** Each person got $146.

Step 6 **Check.** The answer $146 is reasonable, as $146 is close to the estimated answer of $150. Is the answer $146 correct? Check by multiplying.

$$\begin{array}{r} \$146 \quad \text{Amount received by each person} \\ \times \quad 4 \quad \text{Number of people} \\ \hline \$584 \quad \text{Total sales; Matches number given in problem.} \end{array}$$

Work Problem ❷ at the Side.

ANSWERS
1. (a) $11 (b) 80 ft (c) $50,000
2. (a) 21 fossils (b) 24 servers

Example 2 Applying Addition

One week, Paula Crockett decided to total the sales at Starbucks. The daily sales figures were $2358 on Monday, $3056 on Tuesday, $2515 on Wednesday, $2475 on Thursday, $3378 on Friday, $3219 on Saturday, and $3008 on Sunday. Find the total sales for the week.

Step 1 **Read.** In this problem, the sales for each day are given and the total sales for the week must be found.

Step 2 **Work out a plan.** Add the daily sales to arrive at the weekly total.

Step 3 **Estimate.** Because the sales were about $3000 per day for a week of 7 days, a reasonable estimate would be around $21,000 ($7 \times \$3000 = \$21{,}000$).

Step 4 **Solve.** Find the actual answer by adding the sales for the 7 days.

$$\begin{array}{r l} \$\underline{20{,}009} & \text{Check by adding up.} \\ \$\ 2\,358 & \\ 3\,056 & \\ 2\,515 & \\ 2\,475 & \\ 3\,378 & \\ 3\,219 & \\ +\ \ 3\,008 & \\ \hline \$20{,}009 & \text{Sales for the week} \end{array}$$

Step 5 **State the answer.** Paula's total sales figure for the week was $20,009.

Step 6 **Check.** This answer of $20,009 is close to the estimate of $21,000, so it is reasonable. Add up the columns to check the exact answer.

Calculator Tip The calculator solution to Example 2 uses chain calculations to get

2358 (+) 3056 (+) 2515 (+) 2475 (+) 3378 (+) 3219 (+) 3008 (=) 20009

Work Problem 3 at the Side.

Step 4 **Solve.** Find the exact answer by subtracting 4084 from 21,382.

Example 3 Determining whether Subtraction Is Necessary

The number of students enrolled in Chabot College this year is 4084 fewer than the number enrolled last year. Enrollment last year was 21,382. Find the enrollment this year.

Step 1 **Read.** In this problem, the enrollment has decreased from last year to this year. The enrollment last year and the decrease in enrollment are given. This year's enrollment must be found.

Step 2 **Work out a plan.** The indicator word, ***fewer,*** shows that subtraction must be used to find the number of students enrolled this year.

Step 3 **Estimate.** Because the enrollment was about 21,000 students, and the decrease in enrollment is about 4000 students, a reasonable estimate would be 17,000 students ($21{,}000 - 4000 = 17{,}000$).

Continued on Next Page

3 Solve each problem.

(a) During the semester, Cindy received the following points on examinations and quizzes: 92, 81, 83, 98, 15, 14, 15, and 12. Find her total points for the semester.

(b) Stephanie Dixon works at the telephone order desk of a catalog sales company. One week she had the following number of customer contacts: Monday, 78; Tuesday, 64; Wednesday, 118; Thursday, 102; and Friday, 196. How many customer contacts did she have that week?

ANSWERS

3. (a) 410 points **(b)** 558 customer contacts

$$\begin{array}{r} 21{,}382 \\ -\ 4\,084 \\ \hline 17{,}298 \end{array}$$

Step 5 **State the answer.** The enrollment this year is 17,298 students.

Step 6 **Check.** The answer 17,298 is reasonable, as it is close to the estimate of 17,000. Check by adding.

$$\begin{array}{rl} 17{,}298 & \leftarrow \text{Enrollment this year} \\ +\ 4\,084 & \leftarrow \text{Decrease in enrollment} \\ \hline 21{,}382 & \leftarrow \text{Enrollment last year; Matches number given in problem.} \end{array}$$

Work Problem 4 at the Side.

Example 4 Solving a Two-Step Problem

In May, a landlord received $720 from each of eight tenants. After paying $2180 in expenses, how much rent money did the landlord have left?

Step 1 **Read.** The problem asks for the amount of rent remaining after expenses have been paid.

Step 2 **Work out a plan.** The wording *from each of eight tenants* indicates that the eight rents must be totaled. Since the rents are all the same, use multiplication to find the total rent received. Finally, subtract expenses.

Step 3 **Estimate.** The amount of rent is about $700, making the total rent received about $5600 ($700 × 8). The expenses are about $2000. A reasonable estimate of the amount remaining is $3600 ($5600 − $2000).

Step 4 **Solve.** Find the exact amount by first multiplying $720 by 8 (the number of tenants).

$$\begin{array}{r} \$\ 720 \\ \times\quad 8 \\ \hline \$5760 \end{array}$$

Finally, subtract the $2180 in expenses from $5760.

$$\begin{array}{r} \$5760 \\ -\ \$2180 \\ \hline \$3580 \end{array}$$

Step 5 **State the answer.** The amount remaining is $3580.

Step 6 **Check.** The answer of $3580 is reasonable, since it is close to the estimated answer of $3600. Check by adding the expenses to the amount remaining and then dividing by 8.

$$\$3580 + \$2180 = \$5760$$

$$8\overline{)5760} = \$720 \leftarrow \text{Matches the rent amount given in the problem.}$$

Work Problem 5 at the Side.

4 Solve each problem.

(a) A home occupies 1450 square feet, while an apartment occupies 980 square feet. Find the difference in the number of square feet of the two living units.

(b) Aim Electronics had 19,805 employees. After a layoff of 3980 employees, how many employees remain?

5 Solve each problem.

(a) Gwen is paid $315 for each car that she sells. If she sold five cars and had $280 in sales expense deducted, how much did she make?

(b) An Internet book company had sales of 12,628 books with a profit of $6 for each book sold. If 863 books are returned, how much profit remains?

ANSWERS
4. (a) 470 square feet **(b)** 15,825 employees
5. (a) $1295 **(b)** $70,590

1.10 EXERCISES

FOR EXTRA HELP | Student's Solutions Manual | MyMathLab.com | InterAct Math Tutorial Software | AW Math Tutor Center | www.mathxl.com | Digital Video Tutor CD 1 Videotape 3

Solve each application problem. First use front end rounding to estimate the answer. Then find the exact answer. See Examples 1–4.

1. To train for a triathlon, Beth Andrews rode her bike 80 miles on Monday, 75 miles on Tuesday, 135 miles on Wednesday, 40 miles on Thursday, and 52 miles on Friday. How many miles did she ride in this 5-day period?

Estimate:

Exact:

2. During a recent week, Starbucks Coffee sold 325 lb of Estate Java Coffee, 75 lb of Encanta Blend Coffee, 137 lb of Ethiopia Sidamo Coffee, 495 lb of Starbucks House Blend Decaf Coffee, and 105 lb of New Guinea Peaberry Coffee. Find the total number of pounds of coffee sold.

Estimate:

Exact:

3. In a certain country, there were 70 ATM burglaries and attempted burglaries in 1992 and 200 in 1997. How many more of these crimes were there in 1997 than in 1992? (*Source: ATM Crime and Security Newsletter*.)

Estimate:

Exact:

4. The amount of cash in ATMs ranges from $15,000 in small machines to $250,000 in large bank machines. How much more money is there in the large machines than in the small machines? (*Source:* Eric Sheppard of Security Corporation.)

Estimate:

Exact:

5. A packing machine can package 236 first-aid kits each hour. At this rate, find the number of first-aid kits packaged in 24 hours.

Estimate:

Exact:

6. If 450 admission tickets to a classic car show are sold each day, how many tickets are sold in a 12-day period?

Estimate:

Exact:

7. Ted Slauson, coordinator of Toys for Tots, has collected 2628 toys. If his group can give the same number of toys to each of 657 children, how many toys will each child receive?

Estimate:

Exact:

8. If profits of $680,000 are divided evenly among a firm's 1000 employees, how much money will each employee receive?

Estimate:

Exact:

9. The number of boaters and campers at the lake was 8392 on Friday. If this was 4218 more than the number of people at the lake on Wednesday, how many were there on Wednesday?

Estimate:

Exact:

10. The community has raised $52,882 for the homeless shelter. If the total amount needed for the shelter is $75,650, find the additional amount needed.

Estimate:

Exact:

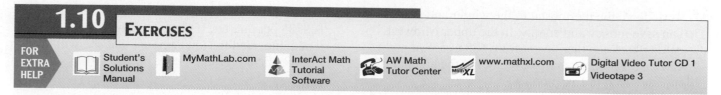

11. Turn down the thermostat in the winter and you can save money and energy. In the upper Midwest, setting back the thermostat from 68° to 55° at night can save $14 per month on fuel. Find the amount of money saved in five months.

Estimate:

Exact:

12. The cost of tuition and fees at a community college is $785 per quarter. If Gale Klein has five quarters remaining, find the total amount that she will need for tuition and fees.

Estimate:

Exact:

The table shows the average starting salaries for selected occupations. Refer to the table to answer Exercises 13–16. First use front end rounding to estimate the answer. Then find the exact answer.

ALL WALKS OF LIFE

Average starting salaries for selected occupations.

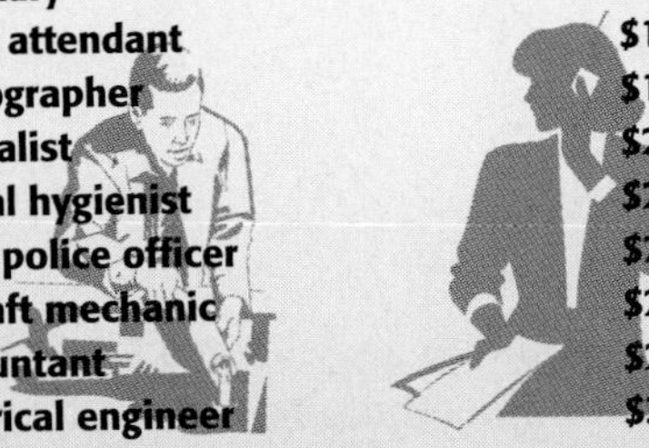

Secretary	$16,400
Flight attendant	$16,500
Photographer	$17,000
Journalist	$21,700*
Dental hygienist	$22,000
State police officer	$24,000
Aircraft mechanic	$29,000
Accountant	$29,800*
Electrical engineer	$38,700*

* requires a Bachelor's degree

Source: U.S. Department of Labor/U.S. Bureau of the Census.

13. How much more does an electrical engineer earn than an aircraft mechanic?

Estimate:

Exact:

14. How much more does an accountant earn than a journalist?

Estimate:

Exact:

15. Mr. Garrett is a journalist and Mrs. Garrett is an accountant. Mr. Harcos is a photographer and Mrs. Harcos is an electrical engineer.

(a) Which couple has higher earnings?

Estimate:

Exact:

(b) Find the difference in the earnings.

Estimate:

Exact:

16. Mr. Gonsalves is a dental hygienist and Mrs. Gonsalves is a secretary. Mr. Horton is a flight attendant and Mrs. Horton is a state police officer.

(a) Which couple has higher earnings?

Estimate:

Exact:

(b) Find the difference in the earnings.

Estimate:

Exact:

17. Dorene Cox decides to establish a budget. She will spend $450 for rent, $325 for food, $320 for child care, $182 for transportation, $150 for other expenses, and she will put the remainder in savings. If her monthly take-home pay is $1620, find her monthly savings.

Estimate:

Exact:

18. Jared Ueda had $2874 in his checking account. He wrote checks for $308 for auto repairs, $580 for child support, and $778 for an insurance payment. Find the amount remaining in his account.

Estimate:

Exact:

19. There are 43,560 square feet in one acre. How many square feet are there in 138 acres?

Estimate:

Exact:

20. The number of gallons of water polluted each day in an industrial area is 209,670. How many gallons of water are polluted each year? (Use a 365-day year.)

Estimate:

Exact:

The Internet was used to find the following minivan optional features and the price of each feature. Use this information to answer Exercises 21–24.

Safety and Security Options	
Option	**Cost**
Seven-passenger seating with two child seats	$375
Antilock brakes	$565
Rear window defroster	$195
Rear window wiper/washer	$115
Full-size spare tire	$125

Convenience and Comfort Options	
Option	**Cost**
Driver's side sliding door	$595
Roof rack	$215
Air conditioning	$860
Power windows	$340
Central locking system	$280

Source: www.edmunds.com

21. Find the total cost of all Safety and Security Options listed.

Estimate:

Exact:

22. Find the total cost of all Convenience and Comfort Options listed.

Estimate:

Exact:

23. A new-car dealer offers an option value package that includes seven-passenger seating with two child seats, antilock brakes, a central locking system, and air conditioning at a cost of $1550. If Jill buys the value package instead of paying for each option separately, how much will she save?

Estimate:

Exact:

24. A new-car dealer offers an option package that includes seven-passenger seating with two child seats, antilock brakes, full-size spare tire, power windows, air conditioning, and a roof rack for a total of $1750. How much will Samuel save if he buys the option package instead of paying for each option separately?

Estimate:

Exact:

25. The Enabling Supply House purchased 6 wheelchairs at $1256 each and 15 speech compression recorder-players at $895 each. Find the total cost.

Estimate:

Exact:

26. A college bookstore buys 17 computers at $506 each and 13 printers at $482 each. Find the total cost.

Estimate:

Exact:

27. Being able to identify indicator words is helpful in determining how to solve an application problem. Write three indicators words for each of these operations: add, subtract, multiply, and divide. Write two indicator words that mean equals.

28. Identify and explain the six steps used to solve an application problem. You may refer to the text if you need help, but use your own words.

29. Write in your own words why it is important to estimate a reasonable answer. Give three examples of what might be a reasonable answer to a math problem from your daily activities.

30. First estimate by rounding to thousands, then find the exact answer to the following problem.

$$7438 + 6493 + 2380$$

Do the two answers vary by more than 1000? Why? Will estimated answers always vary from exact answers?

Estimate:

Exact:

Solve each application problem. See Examples 1–4.

31. A package of 3 undershirts costs $12, and a package of 6 pairs of socks costs $15. Find the total cost of 30 undershirts and 18 pairs of socks.

32. Brian earned $8 per hour for 38 hours of work. Maria earned $9 per hour for 39 hours of work. Find their total combined income.

33. A car weighs 2425 lb. If its 582-lb engine is removed and replaced with a 634-lb engine, what will the car weigh?

34. Barbara has $2324 in her preschool operating account. She spends $734 from this account, and then the class parents raise $568 in a rummage sale. Find the balance in the account after she deposits the money from the rummage sale.

35. In a recent survey of high-priced hotels, the cost per night at Harrah's in Reno was $59, while the cost at Harrah's Lake Tahoe was $159 per night. Find the amount saved on a 5-night stay at Harrah's in Reno instead of staying at Harrah's Lake Tahoe. (*Source:* Harrah's Casinos and Hotels, weekday rates.)

36. The most expensive hotel room in a recent study was the Ritz-Carlton at $375 per night, while the least expensive was Motel 6 at $32 per night. Find the amount saved in a 4-night stay at Motel 6 instead of staying at the Ritz-Carlton. (*Source: USA Today*.)

37. A youth soccer association raised $7588 through fund-raising projects. After expenses of $838 were paid, the balance of the money was divided evenly among the 18 teams. How much did each team receive?

38. Feather Farms Egg Ranch collected 3545 eggs in the morning and 2575 eggs in the afternoon. If the eggs are packed in flats containing 30 eggs each, find the number of flats needed for packing.

39. A theater owner wants to provide enough seating for 1250 people. The main floor has 30 rows of 25 seats in each row. If the balcony has 25 rows, how many seats must be in each balcony row to satisfy the owner's seating requirements?

40. Jennie makes 24 grapevine wreaths per week to sell to gift shops. She works 40 weeks a year and packages six wreaths per box. If she ships equal quantities to each of five shops, find the number of boxes each store will receive.

Chapter 1

SUMMARY

Study Skills

Study Skills Workbook
Activity 5

KEY TERMS

1.1	**whole numbers**	The whole numbers are 0, 1, 2, 3, 4, 5, 6, 7, 8, and so on.
	place value	The place value of each digit in a whole number is determined by its position in the whole number.
	table	A table is a display of facts in rows and columns.
1.2	**addition**	The process of finding the total is addition.
	addends	The numbers being added in an addition problem are addends.
	sum (total)	The answer in an addition problem is called the sum.
	commutative property of addition	The commutative property of addition states that the order of numbers in an addition problem can be changed without changing the sum.
	associative property of addition	The associative property of addition states that grouping the addition of numbers differently does not change the sum.
	carrying	The process of carrying is used in an addition problem when the sum of the digits in a column is greater than 9.
	perimeter	The perimeter is the distance around the outside edges of a figure.
1.3	**minuend**	The number from which another number (the subtrahend) is being subtracted is the minuend.
	subtrahend	The subtrahend is the number being subtracted in a subtraction problem.
	difference	The answer in a subtraction problem is called the difference.
	borrowing	The method of borrowing is used in subtraction if a digit is less than the one directly below.
1.4	**factors**	The numbers being multiplied are called factors. For example, in $3 \times 4 = 12$, both 3 and 4 are factors.
	product	The answer in a multiplication problem is called the product.
	commutative property of multiplication	The commutative property of multiplication states that the product in a multiplication problem remains the same when the order of the factors is changed.
	associative property of multiplication	The associative property of multiplication states that grouping the numbers differently does not change the product.
	multiple	The product of two whole number factors is a multiple of those numbers.
1.5	**dividend**	The number being divided by another number in a division problem is the dividend.
	divisor	The divisor is the number doing the dividing in a division problem.
	quotient	The answer in a division problem is called the quotient.
	short division	A method of dividing a number by a one-digit divisor is short division.
	remainder	The remainder is the number left over when two numbers do not divide exactly.
1.6	**long division**	The process of long division is used to divide by a number with more than one digit.
1.7	**rounding**	Rounding is used to find a number that is close to the original number, but easier to work with. Use the $\approx$ sign, which means "approximately equal to."
	estimate	An estimated answer is one that is close to the exact answer.
	front end rounding	Rounding to the highest possible place so that all the digits become 0s except the first one.
1.8	**square root**	The square root of a whole number is the number that can be multiplied by itself to produce the given number.
	perfect square	A number that is the square of a whole number is a perfect square.
	order of operations	For problems or expressions with more than one operation, the order of operations tells what to do first, second, and so on to get the correct answer.

(continued)

1.9	**pictograph**	A graph that uses pictures or symbols to show data is a pictograph.
	bar graph	A graph that uses bars of various heights to show quantity is a bar graph.
	line graph	A graph that uses dots connected by lines to show trends is a line graph.
1.10	**indicator words**	Words in a problem that indicate the necessary operations—addition, subtraction, multiplication, or division—are indicator words.

Test Your Word Power

See how well you have learned the vocabulary in this chapter. Answers follow the Quick Review.

1. When using **addends** you are performing
   (a) division
   (b) subtraction
   (c) addition
   (d) multiplication.

2. The **subtrahend** is the
   (a) number being multiplied
   (b) number being subtracted
   (c) number being added
   (d) answer in division.

3. A **factor** is
   (a) the answer in an addition problem
   (b) one of two or more numbers being added
   (c) one of two or more numbers being multiplied
   (d) one of two or more numbers being divided.

4. The **divisor** is
   (a) the number being rounded
   (b) the number being multiplied
   (c) always the largest number
   (d) the number doing the dividing.

5. We use **rounding** to
   (a) avoid solving a problem
   (b) purely guess at the answer
   (c) help estimate a reasonable answer
   (d) find the remainder.

6. A **perfect square** is
   (a) the square of a whole number
   (b) the same as square root
   (c) similar to a perfect triangle
   (d) used when rounding.

Quick Review

Concepts	*Examples*
***1.1** Reading and Writing Whole Numbers* Do not use the word *and* when writing a whole number. Commas help divide the periods or groups for ones, thousands, millions, and billions. A comma is not needed when a number has four digits or fewer.	795 is written *seven hundred ninety-five.* 9,768,002 is written *nine million, seven hundred sixty-eight thousand, two.*
***1.2** Adding Whole Numbers* Add from top to bottom, starting with the ones column and working left. To check, add from bottom to top.	$\begin{array}{r} \underline{1\,1\,4\,0} \\ 6\,8\,7 \\ 2\,6 \\ 9 \\ +\ 4\,1\,8 \\ \hline 1\,1\,4\,0 \end{array}$ (Add up to check.) Addends: 687, 26, 9, 418; Sum: 1140
***1.2** Commutative Property of Addition* Changing the order of the addends in an addition problem does not change the sum.	$2 + 4 = 6$ $4 + 2 = 6$ By the commutative property, the sum is the same.
***1.2** Associative Property of Addition* Grouping the addends differently when adding does not change the sum.	$(2 + 3) + 4 = 9$ $2 + (3 + 4) = 9$ By the associative property, the sum is the same.
***1.3** Subtracting Whole Numbers* Subtract the subtrahend from the minuend to get the difference by borrowing when necessary. To check, add the difference to the subtrahend to get the minuend.	Problem: $\begin{array}{r} {\scriptstyle 6\,12\,18} \\ 4\,\not{7}\,\not{3}\,\not{8} \\ -\ 6\,4\,9 \\ \hline 4\,0\,8\,9 \end{array}$ Minuend, Subtrahend, Difference Check: $\begin{array}{r} 4\,0\,8\,9 \\ +\ 6\,4\,9 \\ \hline 4\,7\,3\,8 \end{array}$

Concepts	*Examples*
1.4 Multiplying Whole Numbers Use $\times$, • (a raised dot), or ( )( ) (two sets of parentheses) to indicate multiplication.	
The numbers being multiplied are called *factors.* The multiplicand is being multiplied by the multiplier, giving the product. When the multiplier has more than one digit, partial products must be used and added to find the product.	78 Multiplicand } Factors $\times$ 24 Multiplier } 312 Partial product 156 Partial product (move one position left) 1872 Product
1.4 Commutative Property of Multiplication The product in a multiplication problem remains the same when the order of the factors is changed.	$3 \times 4 = 12$ $4 \times 3 = 12$ By the commutative property, the product is the same.
1.4 Associative Property of Multiplication The grouping of factors differently when multiplying does not change the product.	$(2 \times 3) \times 4 = 24$ $2 \times (3 \times 4) = 24$ By the associative property, the product is the same.
1.5 Dividing Whole Numbers ÷ and $\overline{)}$ mean divide. Also a —, as in $\frac{25}{5}$, means to divide the top number (dividend) by the bottom number (divisor).	$4\overline{)88}$ = 22 (Divisor → 4; 88 ← Dividend; 22 ← Quotient) 88 0 $\frac{88}{4} = 22$ ← Quotient $88 \div 4 = 22$ (88: Dividend; 4: Divisor; 22: Quotient)
1.7 Rounding Whole Numbers **Rules for Rounding:** *Step 1* Locate the place to be rounded, and draw a line under it. *Step 2* If the next digit to the right is 5 or more, increase the underlined digit by 1. If the next digit is 4 or less, do not change the underlined digit. *Step 3* Change all digits to the right of the underlined place to 0s.	Round 726 to the nearest ten. Next digit is 5 or more. 7_2_6 Tens place increases by 1 (2 + 1 = 3). 7_2_6 rounds to 7_3_0. Round 1,498,586 to the nearest million. Next digit is 4 or less. _1_,498,586 Millions place does not change. 1,498,586 rounds to 1,000,000.
1.7 Front End Rounding Front end rounding is rounding to the highest possible place so that all the digits become 0 except the first digit.	Round each number using front end rounding. 76 rounds to 80 348 rounds to 300 6512 rounds to 7000 23,751 rounds to 20,000 652,179 rounds to 700,000

Concepts	*Examples*
***1.8** Order of Operations* Problems may have several operations. Work these problems using the order of operations. **1.** Do all operations inside parentheses or other grouping symbols. **2.** Simplify any expressions with exponents and find any square roots ($\sqrt{\ }$). **3.** Multiply or divide proceeding from left to right. **4.** Add or subtract proceeding from left to right.	Simplify, using the order of operations. $7 \cdot \sqrt{9} - 4 \cdot 5$ Find the square root. $7 \cdot 3 - 4 \cdot 5$ Multiply from left to right. $21 - 20 = 1$ Subtract.
***1.9** Reading Pictographs, Bar Graphs, and Line Graphs* A *pictograph* uses pictures or symbols to show data.	When reading a pictograph, be certain that you determine the quantity represented by each picture or symbol.
A *bar graph* uses bars of various heights to show quantity.	When reading a bar graph, use a straight edge to line up the top of the bar with the numbers along the left edge of the graph.
A *line graph* uses dots connected by lines to show trends.	When reading a line graph, use a straight edge to line up the dot with the numbers along the left edge of the graph.
***1.10** Application Problems* **Steps for Solving an Application Problem**	Manuel earns \$118 on Sunday, \$87 on Monday, and \$63 on Tuesday. Find his total earnings for the 3 days.
Step 1 **Read** the problem carefully, perhaps several times.	The earnings for each day are given, and the total for the 3 days must be found.
Step 2 **Work out a plan** before starting. Draw a sketch if possible.	Add the daily earnings to find the total.
Step 3 **Estimate** a reasonable answer.	Since the earnings were about \$100 + \$90 + \$60 = \$250, a reasonable estimate would be approximately \$250.
Step 4 **Solve** the problem.	\$268 Check by adding up. \$118 87 + 63 \$268 Total earnings
Step 5 **State the answer.**	Manuel's total earnings are \$268.
Step 6 **Check** your work. If the answer is not reasonable, start over.	The answer is reasonable, because it is close to the estimate of \$250.

ANSWERS TO TEST YOUR WORD POWER

1. (c) *Example:* In $2 + 3 = 5$, the 2 and the 3 are addends. **2. (b)** *Example:* In $5 - 4 = 1$, the 4 is the subtrahend. **3. (c)** *Example:* In $3 \times 5 = 15$, the numbers 3 and 5 are factors. **4. (d)** *Example:* In $8 \div 4 = 2$, $\frac{8}{4} = 2$, and $4\overline{)8}$ (quotient 2), the 4 is the divisor. **5. (c)** *Example:* We can use rounding to estimate our answer and then determine whether the exact answer is reasonable. **6. (a)** *Example:* 25 is the perfect square of 5 because $\sqrt{25} = 5$.

Chapter 1 Review Exercises

If you need help with any of these Review Exercises, look in the section indicated in brackets.

[1.1] *Write the digits for the given period or group in each number.*

1. 3582
thousands
ones

2. 64,234
thousands
ones

3. 105,724
thousands
ones

4. 1,768,710,618
billions
millions
thousands
ones

Rewrite each number in words.

5. 635

6. 15,310

7. 319,215

8. 62,500,005

Rewrite each number in digits.

9. ten thousand, eight

10. two hundred million, four hundred fifty-five

[1.2] *Add.*

11.
$$\begin{array}{r} 67 \\ +23 \\ \hline \end{array}$$

12.
$$\begin{array}{r} 48 \\ +\ 89 \\ \hline \end{array}$$

13.
$$\begin{array}{r} 807 \\ 4606 \\ +\quad 51 \\ \hline \end{array}$$

14.
$$\begin{array}{r} 8215 \\ 9 \\ +7433 \\ \hline \end{array}$$

15.
$$\begin{array}{r} 2130 \\ 453 \\ 8107 \\ +\quad 296 \\ \hline \end{array}$$

16.
$$\begin{array}{r} 5684 \\ 218 \\ 2960 \\ +\quad 983 \\ \hline \end{array}$$

17.
$$\begin{array}{r} 5\,732 \\ 11{,}069 \\ 37 \\ 1\,595 \\ +22{,}169 \\ \hline \end{array}$$

18.
$$\begin{array}{r} 3\,451 \\ 12{,}286 \\ 43 \\ 1\,291 \\ +32{,}784 \\ \hline \end{array}$$

[1.3] *Subtract.*

19. $\begin{array}{r} 52 \\ -18 \\ \hline \end{array}$

20. $\begin{array}{r} 38 \\ -15 \\ \hline \end{array}$

21. $\begin{array}{r} 375 \\ -186 \\ \hline \end{array}$

22. $\begin{array}{r} 573 \\ -389 \\ \hline \end{array}$

23. $\begin{array}{r} 6213 \\ -\ \ 458 \\ \hline \end{array}$

24. $\begin{array}{r} 5210 \\ -\ \ 883 \\ \hline \end{array}$

25. $\begin{array}{r} 2210 \\ -1986 \\ \hline \end{array}$

26. $\begin{array}{r} 99{,}704 \\ -73{,}838 \\ \hline \end{array}$

[1.4] *Multiply.*

27. $\begin{array}{r} 6 \\ \times 6 \\ \hline \end{array}$

28. $\begin{array}{r} 9 \\ \times 0 \\ \hline \end{array}$

29. 8×4

30. 8×8

31. (7)(5)

32. (8)(7)

33. $7 \cdot 8$

34. $9 \cdot 9$

Work each chain multiplication.

35. $2 \times 4 \times 6$

36. $9 \times 1 \times 5$

37. $4 \times 4 \times 3$

38. $2 \times 2 \times 2$

39. (7)(0)(5)

40. (7)(1)(6)

41. $6 \cdot 1 \cdot 8$

42. $7 \cdot 7 \cdot 0$

Multiply.

43. $\begin{array}{r} 32 \\ \times\ 5 \\ \hline \end{array}$

44. $\begin{array}{r} 46 \\ \times\ 8 \\ \hline \end{array}$

45. $\begin{array}{r} 58 \\ \times\ 9 \\ \hline \end{array}$

46. $\begin{array}{r} 98 \\ \times\ 1 \\ \hline \end{array}$

47. $\begin{array}{r} 589 \\ \times\ \ 7 \\ \hline \end{array}$

48. $\begin{array}{r} 781 \\ \times\ \ 7 \\ \hline \end{array}$

49. $\begin{array}{r} 1349 \\ \times\ \ \ 4 \\ \hline \end{array}$

50. $\begin{array}{r} 9163 \\ \times\ \ \ 5 \\ \hline \end{array}$

51. $\begin{array}{r} 8364 \\ \times\ \ \ 2 \\ \hline \end{array}$

52. $\begin{array}{r} 5440 \\ \times\ \ \ 6 \\ \hline \end{array}$

53. $\begin{array}{r} 93{,}105 \\ \times\ \ \ \ \ 5 \\ \hline \end{array}$

54. $\begin{array}{r} 21{,}873 \\ \times\ \ \ \ \ 8 \\ \hline \end{array}$

55. $\begin{array}{r} 45 \\ \times 15 \\ \hline \end{array}$

56. $\begin{array}{r} 63 \\ \times 28 \\ \hline \end{array}$

57. $\begin{array}{r} 98 \\ \times 12 \\ \hline \end{array}$

58. $\begin{array}{r} 68 \\ \times 75 \\ \hline \end{array}$

59. $\begin{array}{r} 472 \\ \times 33 \\ \hline \end{array}$

60. $\begin{array}{r} 392 \\ \times 77 \\ \hline \end{array}$

61. $\begin{array}{r} \times 4051 \\ \times 219 \\ \hline \end{array}$

62. $\begin{array}{r} 1527 \\ \times 328 \\ \hline \end{array}$

Find each total cost.

63. 20 CDs at \$15 per CD

64. 76 subscribers at \$14 per subscription

65. 318 drill bit sets at \$64 per set

66. 168 welder's masks at \$9 per mask

Multiply by using the shortcut for multiples of 10.

67. $\begin{array}{r} 280 \\ \times 50 \\ \hline \end{array}$

68. $\begin{array}{r} 340 \\ \times 70 \\ \hline \end{array}$

69. $\begin{array}{r} 517 \\ \times 400 \\ \hline \end{array}$

70. $\begin{array}{r} 637 \\ \times 500 \\ \hline \end{array}$

71. $\begin{array}{r} 16{,}000 \\ \times 8\,000 \\ \hline \end{array}$

72. $\begin{array}{r} 43{,}000 \\ \times 2\,100 \\ \hline \end{array}$

[1.5] *Divide. If the division is not possible, write "undefined."*

73. $15 \div 3$

74. $25 \div 5$

75. $42 \div 7$

76. $18 \div 9$

77. $\frac{54}{9}$

78. $\frac{36}{9}$

79. $\frac{49}{7}$

80. $\frac{0}{6}$

81. $\frac{148}{0}$

82. $\frac{0}{23}$

83. $\frac{64}{8}$

84. $\frac{81}{9}$

[1.5–1.6] *Divide.*

85. $3\overline{)276}$

86. $8\overline{)280}$

87. $6\overline{)26{,}532}$

88. $76\overline{)26{,}752}$

89. $2704 \div 18$

90. $15{,}525 \div 125$

[1.7] *Round as indicated.*

91. 476 to the nearest ten

92. 14,309 to the nearest hundred

93. 20,643 to the nearest thousand

94. 67,485 to the nearest ten thousand

Round each number to the nearest ten, nearest hundred, and nearest thousand. Remember to round from the original number.

	Ten	Hundred	Thousand
95. 3487	________	________	________
96. 20,065	________	________	________
97. 98,201	________	________	________
98. 352,118	________	________	________

[1.8] *Find each square root by using the Perfect Squares Table on page 73.*

99. $\sqrt{25}$

100. $\sqrt{64}$

101. $\sqrt{144}$

102. $\sqrt{196}$

Identify the exponent and the base, and then simplify each expression.

103. 7^3

104. 3^6

105. 5^3

106. 4^5

Simplify each expression by using the order of operations.

107. $8^2 - 10$

108. $4^2 - 8$

109. $2 \cdot 3^2 \div 2$

110. $9 \div 1 \cdot 2 \cdot 2 \div (11 - 2)$

111. $\sqrt{9} + 2 \cdot 3$

112. $6 \cdot \sqrt{16} - 6 \cdot \sqrt{9}$

[1.9] *The bar graph shows the number of parents out of 100 surveyed who nag their children about performing certain household chores.*

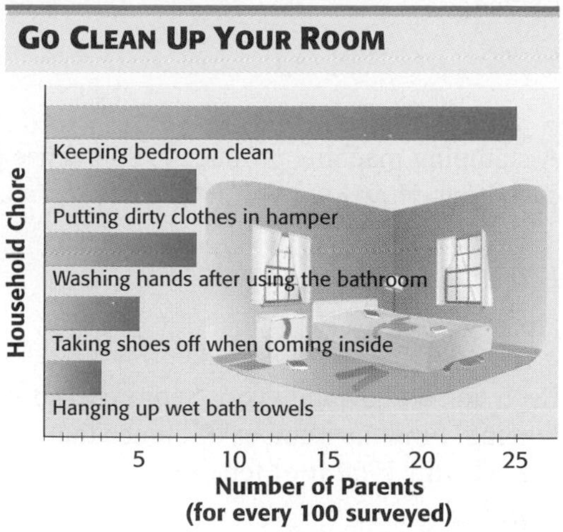

Source: Opinion Research Corporation for the Soap and Detergent Association.

113. How many parents nagged their children about washing hands after using the bathroom?

114. Find the number of parents who nagged their children about taking shoes off when coming inside.

115. Which household chore was nagged about by the greatest number of parents?

116. Which household chore was nagged about by the least number of parents.

[1.10] *Solve each application problem. First use front end rounding to estimate the answer. Then find the exact answer.*

117. Find the cost of 75 picnic coolers at \$10 per cooler.

Estimate:

Exact:

118. A pulley on an evaporative cooler turns 1400 revolutions per minute. How many revolutions will the pulley turn in 60 minutes?

Estimate:

Exact:

119. A McDonald's drink cooler contains 95 cups. Find the number of cups in four coolers.

Estimate:

Exact:

120. A drum contains 6000 brackets. How many brackets are in 30 drums?

Estimate:

Exact:

121. It takes 2000 hours of work to build one home. How many hours of work are needed to build 12 homes?

Estimate:

Exact:

122. A Japanese bullet train travels 80 miles in 1 hour. Find the number of miles traveled in 5 hours.

Estimate:

Exact:

123. The Houston Space Center charges \$15 for each adult admission and \$12 for each child. Find the total cost to admit a group of 18 adults and 26 children.

Estimate:

Exact:

124. A newspaper carrier has 62 customers who take the paper daily and 21 customers who take the paper on weekends only. A daily customer pays \$16 per month and a weekend-only customer pays \$7 per month. Find the total monthly collections.

Estimate:

Exact:

125. In a recent test of side-bagging lawn mowers, the most expensive model sold for \$350 and the least expensive model sold for \$170. Find the difference in price.

Estimate:

Exact:

126. Ruth Berry has \$876 in her checking account. She writes checks for \$408 and \$216. How much does she have left in her account?

Estimate:

Exact:

127. A food canner uses 1 lb of pork for every 175 cans of pork and beans. How many pounds of pork are needed for 8750 cans?

Estimate:

Exact:

128. A stamping machine produces 986 license plates each hour. How long will it take to produce 32,538 license plates?

Estimate:

Exact:

129. Nitrogen sulfate is used in farming to enrich nitrogen-poor soil. If 625 lb of nitrogen sulfate are spread per acre, how many acres can be spread with 32,500 lb of nitrogen sulfate?

Estimate:

Exact:

130. Each home in a subdivision requires 180 ft of fencing. Find the number of homes that can be fenced with 5760 ft of fencing material.

Estimate:

Exact:

Mixed Review Exercises*

Perform the indicated operations.

131. $\begin{array}{r} 56 \\ \times\ 5 \\ \hline \end{array}$

132. $\begin{array}{r} 83 \\ \times\ 8 \\ \hline \end{array}$

133. $\begin{array}{r} 207 \\ -\ 68 \\ \hline \end{array}$

134. $\begin{array}{r} 835 \\ -\ 247 \\ \hline \end{array}$

135. $\begin{array}{r} 662 \\ +\ 379 \\ \hline \end{array}$

136. $\begin{array}{r} 789 \\ +\ 872 \\ \hline \end{array}$

137. $\begin{array}{r} 38{,}140 \\ -\ 6\,078 \\ \hline \end{array}$

138. $\begin{array}{r} 29{,}156 \\ -\ 4\,209 \\ \hline \end{array}$

139. $21 \div 7$

140. $\dfrac{42}{6}$

141. $\begin{array}{r} 7\,218 \\ 3 \\ 18 \\ 1\,791 \\ 82{,}623 \\ +\ 1\,982 \\ \hline \end{array}$

142. $\begin{array}{r} 3\,812 \\ 5 \\ 22 \\ 1\,836 \\ 75{,}134 \\ +\ 2\,369 \\ \hline \end{array}$

143. $\dfrac{9}{0}$

144. $\dfrac{7}{1}$

145. $27{,}600 \div 4$

---

* The order of exercises in this final group does not correspond to the order in which topics occur in the chapter. This random ordering should help you prepare for the chapter test in yet another way.

146. $9220 \div 4$

147. $\begin{array}{r} 8430 \\ \times\ 128 \\ \hline \end{array}$

148. $\begin{array}{r} 25{,}817 \\ \times\ \ \ 4 \\ \hline \end{array}$

149. $34\overline{)3672}$

150. $68\overline{)14{,}076}$

151. Rewrite 376,853 in words.

152. Rewrite 408,610 in words.

153. Round 7549 to the nearest hundred.

154. Round 600,498 to the nearest thousand.

Find each square root.

155. $\sqrt{49}$

156. $\sqrt{81}$

Find each total cost.

157. 165 pairs of in-line skates at $36 per pair

158. 84 dishwashers at $370 per dishwasher

159. 208 baseball hats at $11 per hat

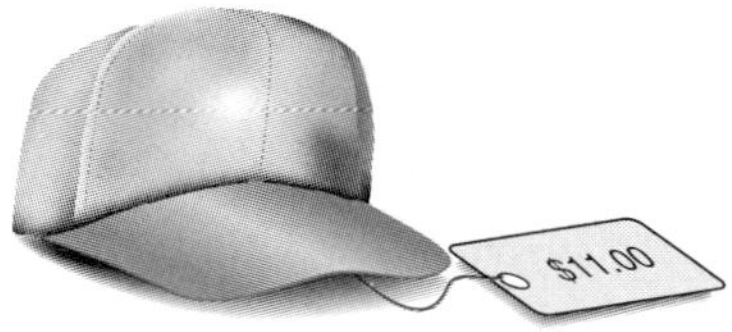

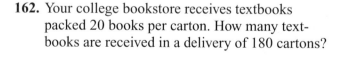

160. 607 boxes of avocados at $26 per box

Solve each application problem.

161. There are 52 playing cards in a deck. How many cards are there in nine decks?

162. Your college bookstore receives textbooks packed 20 books per carton. How many textbooks are received in a delivery of 180 cartons?

163. "Push type" gasoline-powered lawn mowers cost $100 less than self-propelled mowers that you walk behind. If a self-propelled mower costs $380, find the cost of a "push-type" mower.

164. The Country Day School wants to raise $218,450 to construct and equip a computer lab. If $103,815 has already been raised, how much more must be raised to reach the goal?

American River Raft Rentals lists the following daily raft rental fees. Notice that there is an additional $2 launch fee payable to the park system for each raft rented. Use this information to solve Exercises 165 and 166.

AMERICAN RIVER RAFT RENTALS

Size	Rental Fee	Launch Fee
4-person	$28	$2
6-person	$38	$2
10-person	$70	$2
12-person	$75	$2
16-person	$85	$2

Source: American River Raft Rentals.

165. On a recent Tuesday the following rafts were rented: 6 4-person; 15 6-person; 10 10-person; 3 12-person; and 2 16-person. Find the total receipts, including the $2 per raft launch fee.

166. On the 4th of July the following rafts were rented: 38 4-person; 73 6-person; 58 10-person; 34 12-person; and 18 16-person. Find the total receipts, including the $2 per raft launch fee.

Chapter 1 Test

Study Skills Workbook
Activity 6

Write each number in words.

1. 6106

2. 85,055

3. Use digits to write four hundred twenty-six thousand, five.

Add.

4.
$$\begin{array}{r} 762 \\ 72 \\ 5186 \\ +\ 2694 \\ \hline \end{array}$$

5.
$$\begin{array}{r} 17{,}063 \\ 7 \\ 12 \\ 1\ 505 \\ 93{,}710 \\ +\quad 333 \\ \hline \end{array}$$

Subtract.

6.
$$\begin{array}{r} 6006 \\ -\ 4953 \\ \hline \end{array}$$

7.
$$\begin{array}{r} 5062 \\ -\ 1978 \\ \hline \end{array}$$

Multiply.

8. $6 \times 8 \times 2$

9. $57 \cdot 3000$

10. $(95)(18)$

11.
$$\begin{array}{r} 7381 \\ \times\ 603 \\ \hline \end{array}$$

Divide. If the division is not possible, write "undefined."

12. $16\overline{)112{,}752}$

13. $\dfrac{835}{0}$

14. $19{,}241 \div 42$

15. $280\overline{)44{,}800}$

Round as indicated.

16. 5238 to the nearest ten

17. 67,509 to the nearest thousand.

1. ______________
2. ______________
3. ______________
4. ______________
5. ______________
6. ______________
7. ______________
8. ______________
9. ______________
10. ______________
11. ______________
12. ______________
13. ______________
14. ______________
15. ______________
16. ______________
17. ______________

Simplify each expression.

18. ______________

18. $6^2 + 3 \times 4$

19. $7 \cdot \sqrt{64} - 14 \cdot 2$

19. ______________

Solve each application problem. First use front end rounding to estimate the answer. Then find the exact answer.

20. *Estimate:* ______________
Exact: ______________

20. Amy collects the following monthly rents from the tenants in her four-plex: \$485, \$500, \$515, and \$425. After she pays expenses of \$785, how much does she have left?

21. *Estimate:* ______________
Exact: ______________

21. Hewlett-Packard assembles 542 computer printers each day. Find the number of work days it would take to assemble 67,750 computer printers.

22. *Estimate:* ______________
Exact: ______________

22. Kenée Shadbourne paid \$690 for tuition, \$185 for books, and \$68 for supplies. If this money was withdrawn from her checking account, which had a balance of \$1108, find her new balance.

23. *Estimate:* ______________
Exact: ______________

23. An electronics manufacturer assembles 208 digital cameras each hour for 4 hours and 238 camcorders each hour for the next 4 hours. Find the number of both types of camera assembled in the 8-hour period.

24. ______________

24. Explain in your own words the rules for rounding numbers. Give an example of rounding a number to the nearest ten thousand.

25. ______________

25. List the six steps for solving application problems.

2 Multiplying and Dividing Fractions

Most recipes include ingredients that use common fractions and mixed numbers in their measurements. The recipe shown here will make 4 dozen (48) cookies, but suppose you wanted to make 2 dozen, 8 dozen, or even $3\frac{1}{2}$ dozen cookies? You would have to multiply or divide each of the ingredients to arrive at the proper amounts needed. In this chapter we discuss multiplication and division of fractions, which you need to know when cooking or baking. (See Exercises 25–28 in Section 2.8.)

M & M COOKIES

- $\frac{1}{3}$ c. shortening
- $\frac{1}{3}$ c. butter, softened
- $\frac{1}{2}$ c. sugar
- $\frac{1}{2}$ c. packed brown sugar
- 1 egg
- 1 tsp. vanilla extract
- $1\frac{3}{4}$ c. flour
- $\frac{1}{2}$ tsp. soda
- $\frac{1}{4}$ tsp. salt
- $\frac{3}{4}$ c. chopped nuts
- 1 6oz. package M & M's

Cream shortening, butter, sugar, and brown sugar in mixer bowl until fluffy. Mix in egg and vanilla. Add sifted dry ingredients; mix well. Stir in nuts and candies. Drop by rounded teaspoonfuls 2 inches apart onto ungreased cookie sheet. Bake at 375 degrees for 8 to 10 minutes or until lightly browned. Cook slightly on cookie sheet. Remove to wire rack to cool completely. Yield: 4 dozen.

Source: *Recipes Are Naturally Good Eating.*

2.1 Basics of Fractions

Objectives

1. Use a fraction to show which part of a whole is shaded.
2. Identify the numerator and denominator.
3. Identify proper and improper fractions.

1 Write fractions for the shaded portions and the unshaded portions of each figure.

(a)

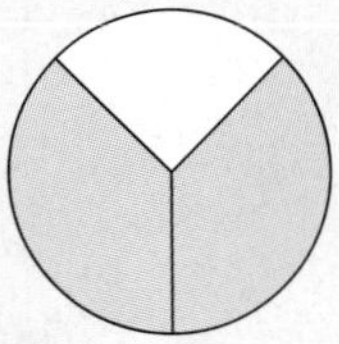

(b)

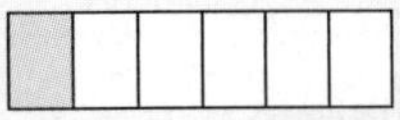

(c)

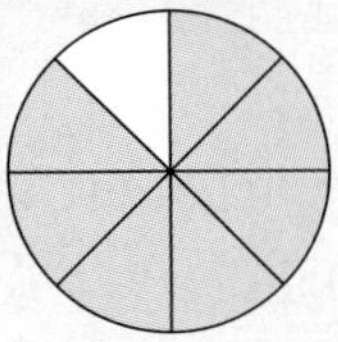

2 Write fractions for the shaded portions of each figure.

(a)

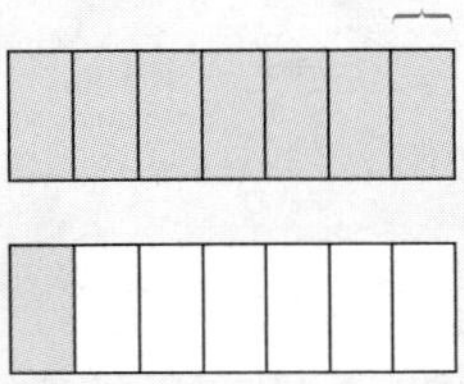

(b)

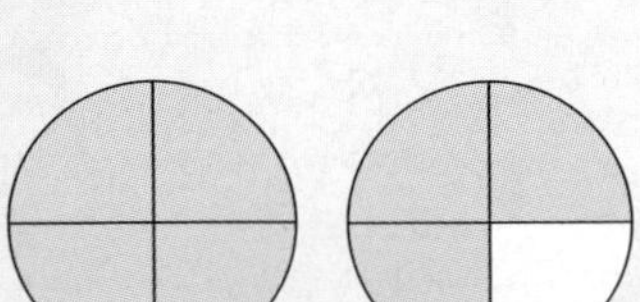

Answers

1. (a) $\frac{2}{3}$; $\frac{1}{3}$ (b) $\frac{1}{6}$; $\frac{5}{6}$ (c) $\frac{7}{8}$; $\frac{1}{8}$
2. (a) $\frac{8}{7}$ (b) $\frac{7}{4}$

In Chapter 1 we discussed whole numbers. Many times, however, we find that parts of whole numbers are considered. One way to write parts of a whole is with **fractions.** Another way is with decimals, which is discussed in Chapter 4.

1 Use a fraction to show which part of a whole is shaded. The number $\frac{1}{8}$ is a fraction that represents 1 of 8 equal parts. Read $\frac{1}{8}$ as "one eighth."

Example 1 Identifying Fractions

Use fractions to represent the shaded portions and the unshaded portions of each figure.

(a) The figure on the left has 4 equal parts. The 1 shaded part is represented by the fraction $\frac{1}{4}$. The *un*shaded part is $\frac{3}{4}$.

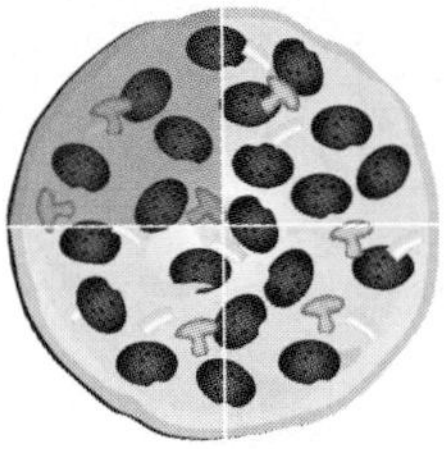

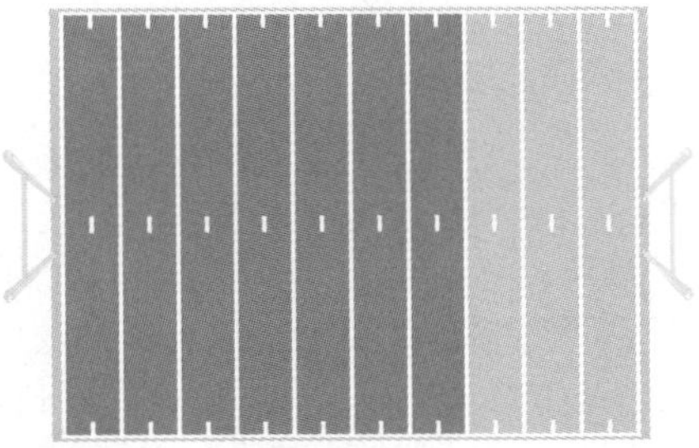

(b) The 3 shaded parts of the 10-part figure on the right are represented by the fraction $\frac{3}{10}$. The *un*shaded part is $\frac{7}{10}$.

Work Problem 1 at the Side.

Fractions can be used to show more than one whole object.

Example 2 Representing Fractions Greater Than 1

Use a fraction to represent the shaded part of each figure.

(a)

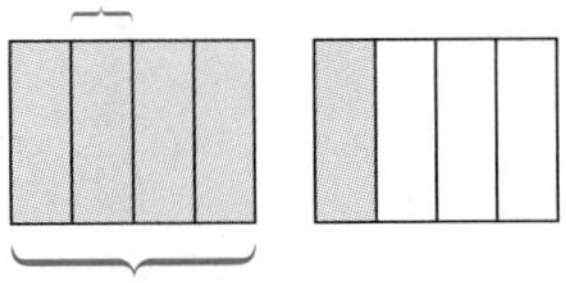

An area equal to 5 of the $\frac{1}{4}$ parts is shaded. Write this as $\frac{5}{4}$.

(b)

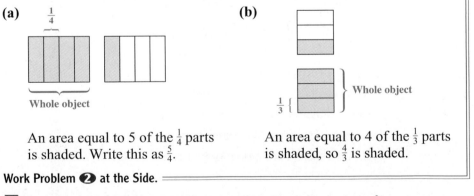

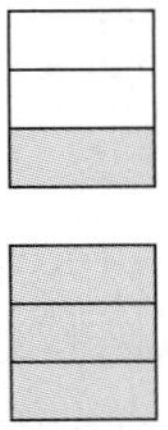

An area equal to 4 of the $\frac{1}{3}$ parts is shaded, so $\frac{4}{3}$ is shaded.

Work Problem 2 at the Side.

2 Identify the numerator and denominator. In the fraction $\frac{2}{3}$, the number 2 is the **numerator** and 3 is the **denominator**. The bar between the numerator and the denominator is the *fraction bar.*

$$\text{Fraction bar} \rightarrow \frac{2}{3} \begin{array}{l} \leftarrow \text{Numerator} \\ \leftarrow \text{Denominator} \end{array}$$

Numerators and Denominators

The **denominator** of a fraction shows the number of equivalent parts in the whole, and the **numerator** shows how many parts are being considered.

NOTE

Remember that a bar, —, is one of the division symbols, and that division by 0 is undefined. A fraction with a denominator of 0 is also undefined.

Example 3 **Identifying Numerators and Denominators**

Identify the numerator and denominator in each fraction.

(a) $\frac{5}{8}$ **(b)** $\frac{9}{4}$

$\frac{5}{8}$ ← Numerator (5) ← Denominator (8)

$\frac{9}{4}$ ← Numerator (9) ← Denominator (4)

Work Problem 3 at the Side.

3 **Identify proper and improper fractions.** Fractions are sometimes called *proper* or *improper* fractions.

Proper and Improper Fractions

If the numerator of a fraction is *smaller* than the denominator, the fraction is a **proper fraction.**

If the numerator is *greater than or equal to* the denominator, the fraction is an **improper fraction.**

Proper Fractions	Improper Fractions
$\frac{1}{2}, \frac{5}{11}, \frac{35}{36}$	$\frac{9}{7}, \frac{126}{125}, \frac{7}{7}$

Example 4 **Classifying Types of Fractions**

(a) Identify all proper fractions in this list.

$$\frac{3}{4}, \frac{5}{9}, \frac{17}{5}, \frac{9}{7}, \frac{3}{3}, \frac{12}{25}, \frac{1}{9}, \frac{5}{3}$$

Proper fractions have a numerator that is smaller than the denominator. The proper fractions are

$$\frac{3}{4} \leftarrow \text{3 is smaller than 4.} \quad \frac{5}{9}, \frac{12}{25}, \text{ and } \frac{1}{9}.$$

(b) Identify all improper fractions in the list in part (a).

Improper fractions have a numerator that is equal to or greater than the denominator. The improper fractions are

$$\frac{17}{5} \leftarrow \text{17 is greater than 5.} \quad \frac{9}{7}, \frac{3}{3}, \text{ and } \frac{5}{3}.$$

Work Problem 4 at the Side.

3 Identify the numerator and the denominator. Draw a picture with shaded parts to show each fraction. Your drawings may vary, but they should have the correct number of shaded parts.

(a) $\frac{2}{3}$

(b) $\frac{1}{4}$

(c) $\frac{8}{5}$

(d) $\frac{5}{2}$

4 From the following group of fractions:

$$\frac{2}{3}, \frac{4}{3}, \frac{3}{4}, \frac{8}{8}, \frac{3}{1}, \frac{1}{3}$$

(a) list all proper fractions;

(b) list all improper fractions.

Answers

3. **(a)** N: 2; D: 3

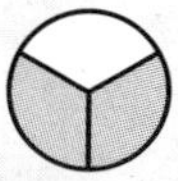

(b) N: 1; D: 4

(c) N: 8; D: 5

(d) N: 5; D: 2

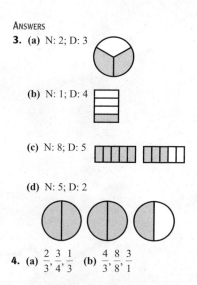

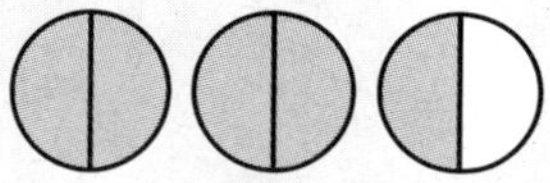

4. **(a)** $\frac{2}{3}, \frac{3}{4}, \frac{1}{3}$ **(b)** $\frac{4}{3}, \frac{8}{8}, \frac{3}{1}$

Focus on **Real-Data Applications**

Quilt Patterns

People who make quilts often base their designs on a block that is cut into a grid of 4, 9, 16, or 25 squares. The quilter chooses various colors for the pieces. Each quilt design shown was selected from an Archive of American Quilt Designs.

1. Identify the makeup of the block as 4, 9, 16, etc. Each color is what fractional part of the block?

Aunt Eliza's Star

Block size: ____

Tan: ____ Purple: ____

Green: ____

Barbara Fritchie Star

Block size: ____

Blue: ____ White: ____

Cobweb's

Block size: ____

Mauve: ____ Purple: ____

Yellow: ____

Handy Andy

Block size: ____

Blue: ____ White: ____

Peace and Plenty

Block size: ____

Aqua: ____ Yellow: ____

Whitish: ____

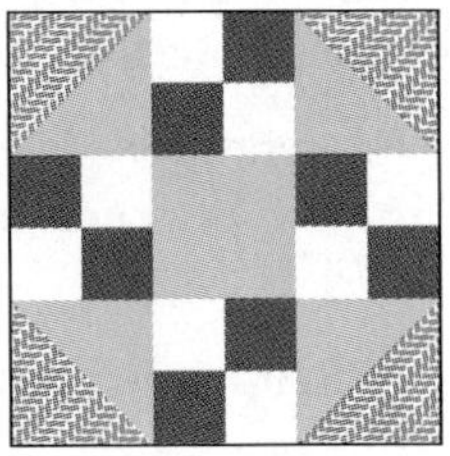

Prairie Queen

Block size: ____

Blue: ____ Grey: ____

Brown: ____ White: ____

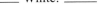

2. Use the blocks to design and color your own quilt patterns. Tell the fractional part of the block that is represented by each color.

3. Find the next two numbers in this pattern: 4, 9, 16, 25, ____, ____.

4. Explain how the pattern works.

2.1 EXERCISES

FOR EXTRA HELP: Student's Solutions Manual · MyMathLab.com · InterAct Math Tutorial Software · AW Math Tutor Center · www.mathxl.com · Digital Video Tutor CD 1 Videotape 4

Write fractions to represent the shaded and unshaded portions of each figure. See Examples 1 and 2.

Study Skills Workbook **Activity 3**

1.

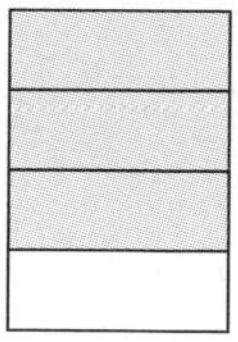

2.

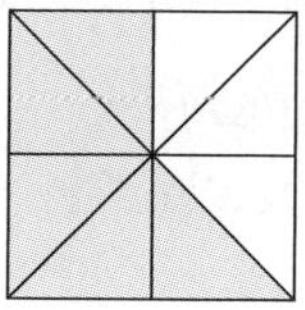

3.

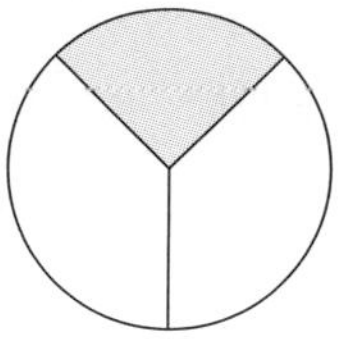

4.

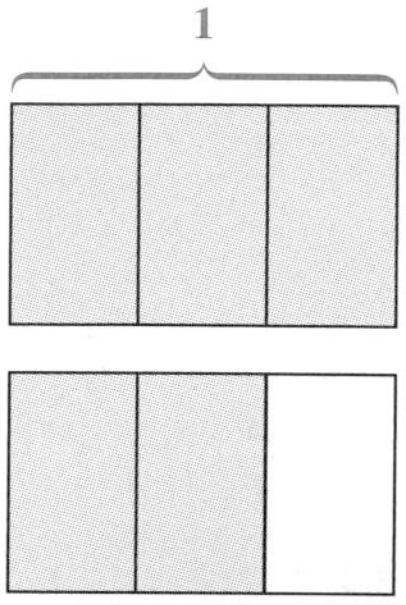

5.

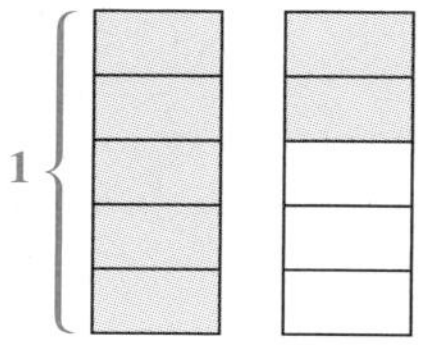

6.

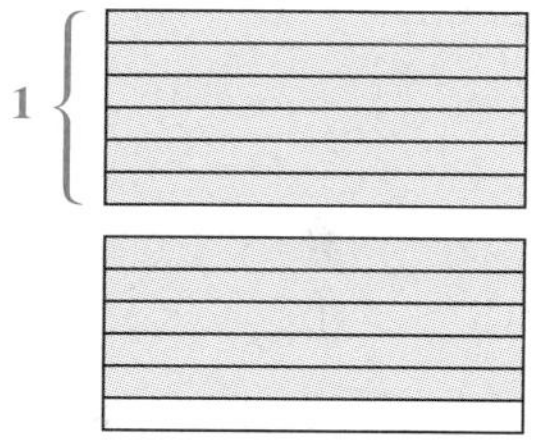

7. What fraction of these 11 coins are dimes?

8. What fraction of these 8 recording artists are men?

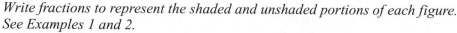

9. In an American Sign Language (ASL) class of 25 students, 8 are hearing impaired. What fraction of the students are hearing impaired?

10. Of 98 bicycles in a bike rack, 67 are mountain bikes. What fraction of the bicycles are not mountain bikes?

11. Of 134 employees at D. J. Diecast Company, 85 are part-time employees. What fraction of the employees are not part-time employees?

12. A college cheerleading squad has 12 members. If 5 of the cheerleaders are sophomores and the rest are freshmen, find the fraction of the members that are freshmen.

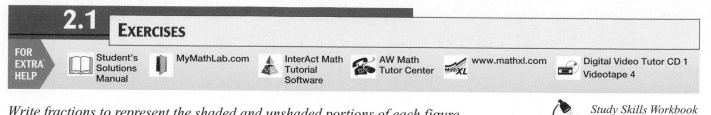

Identify the numerator and denominator. See Example 3.

	Numerator	Denominator
13. $\frac{3}{8}$	______	______
14. $\frac{5}{6}$	______	______
15. $\frac{11}{10}$	______	______
16. $\frac{7}{5}$	______	______

List the proper and improper fractions in each group. See Example 4.

	Proper	Improper
17. $\frac{8}{5}, \frac{1}{3}, \frac{5}{8}, \frac{6}{6}, \frac{12}{2}, \frac{7}{16}$	______	______
18. $\frac{1}{3}, \frac{3}{8}, \frac{16}{12}, \frac{10}{8}, \frac{6}{6}, \frac{3}{4}$	______	______
19. $\frac{3}{4}, \frac{3}{2}, \frac{5}{5}, \frac{9}{11}, \frac{7}{15}, \frac{19}{18}$	______	______
20. $\frac{12}{12}, \frac{15}{11}, \frac{13}{12}, \frac{11}{8}, \frac{17}{17}, \frac{19}{12}$	______	______

21. Write a fraction of your own choice. Label the parts of the fraction and write a sentence describing what each part represents. Draw a picture with shaded parts showing your fraction.

22. Give one example of a proper fraction and one example of an improper fraction. What determines whether a fraction is proper or improper? Draw pictures with shaded parts showing these fractions.

Fill in the blanks to complete each sentence.

23. The fraction $\frac{7}{8}$ represents ______ of the ______ equal parts into which a whole is divided.

24. The fraction $\frac{15}{16}$ represents ______ of the ______ equal parts into which a whole is divided.

25. The fraction $\frac{4}{32}$ represents ______ of the ______ equal parts into which a whole is divided.

26. The fraction $\frac{7}{64}$ represents ______ of the ______ equal parts into which a whole is divided.

2.2 Mixed Numbers

Objectives

1. Identify mixed numbers.
2. Write mixed numbers as improper fractions.
3. Write improper fractions as mixed numbers.

Suppose you had three whole cases of soft drink cans and half of another case. You would state this as a whole number and a fraction.

1 Identify mixed numbers. When a whole number and a fraction are written together, the result is a **mixed number.** For example, the mixed number

$$3\frac{1}{2} \quad \text{represents} \quad 3 + \frac{1}{2},$$

or 3 wholes and $\frac{1}{2}$ of a whole. Read $3\frac{1}{2}$ as "three and one half." As this figure shows, the mixed number $3\frac{1}{2}$ is equal to the improper fraction $\frac{7}{2}$.

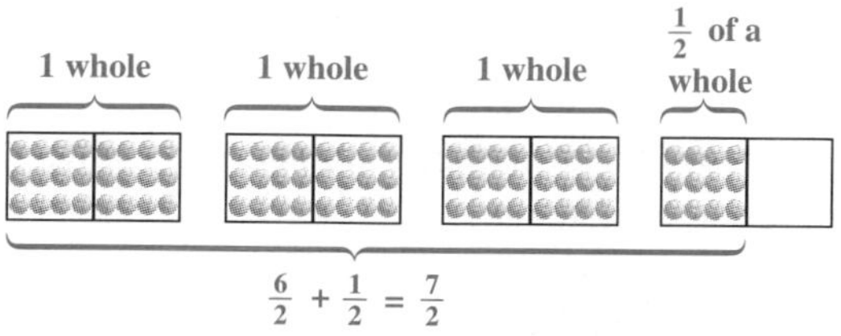

Work Problem 1 at the Side.

1 (a) Use these diagrams to write $1\frac{2}{3}$ as an improper fraction.

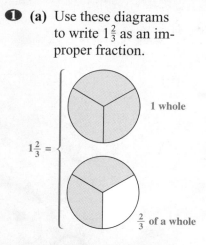

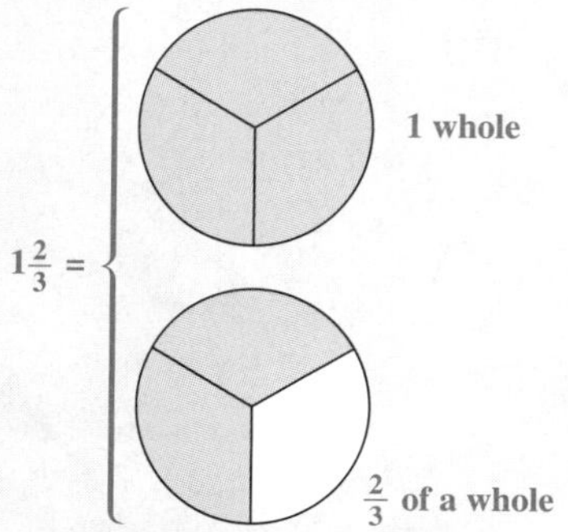

(b) Use these diagrams to write $2\frac{1}{4}$ as an improper fraction.

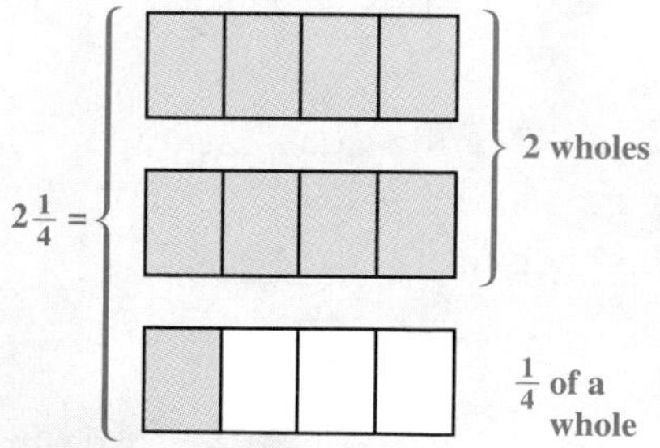

2 Write mixed numbers as improper fractions. Use the following steps to write $3\frac{1}{2}$ as an improper fraction without drawing a figure.

Step 1 Multiply 3 and 2.

$$3\frac{1}{2} \qquad 3 \cdot 2 = 6$$

Step 2 Add 1 to the product.

$$3\frac{1}{2} \qquad 6 + 1 = 7$$

Step 3 Use 7, from Step 2, as the numerator and 2 as the denominator.

$$3\frac{1}{2} = \frac{7}{2}$$

Same denominator

In summary, use the following steps to *write a mixed number as an improper fraction.*

Writing a Mixed Number as an Improper Fraction

Step 1 *Multiply* the denominator of the fraction and the whole number.

Step 2 *Add* to this product the numerator of the fraction.

Step 3 Write the result of Step 2 as the *numerator* and the original denominator as the *denominator.*

Answers

1. (a) $\frac{5}{3}$ (b) $\frac{9}{4}$

2 Write as improper fractions.

(a) $4\frac{1}{2}$

(b) $5\frac{3}{4}$

(c) $4\frac{7}{8}$

(d) $8\frac{5}{6}$

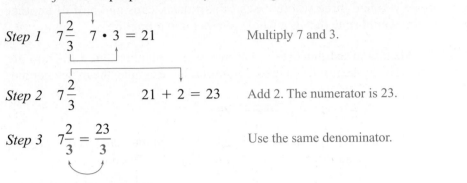

Example 1 **Writing a Mixed Number as an Improper Fraction**

Write $7\frac{2}{3}$ as an improper fraction (numerator greater than denominator).

Step 1 $7\frac{2}{3}$ $\quad 7 \cdot 3 = 21$ — Multiply 7 and 3.

Step 2 $7\frac{2}{3}$ $\quad 21 + 2 = 23$ — Add 2. The numerator is 23.

Step 3 $7\frac{2}{3} = \frac{23}{3}$ — Use the same denominator.

Work Problem 2 at the Side.

3 **Write improper fractions as mixed numbers.** Write an improper fraction as a mixed number as follows.

Writing an Improper Fraction as a Mixed Number

Write an **improper fraction** as a mixed number by dividing the numerator by the denominator. The quotient is the whole number (of the mixed number), the remainder is the numerator of the fraction part, and the denominator remains unchanged.

Example 2 **Writing Improper Fractions as Mixed Numbers**

Write each improper fraction as a mixed number.

(a) $\frac{17}{5}$

Divide 17 by 5.

$$\begin{array}{r} 3 \\ 5\overline{)17} \\ \underline{15} \\ 2 \end{array}$$

3 ← Whole number part

2 ← Remainder

The quotient **3** is the whole number part of the mixed number. The remainder **2** is the numerator of the fraction, and the denominator remains as **5.**

$$\frac{17}{5} = 3\frac{2}{5}$$

2 ← Remainder

5 Same denominator

We can verify this by using a diagram in which $\frac{17}{5}$ is shaded.

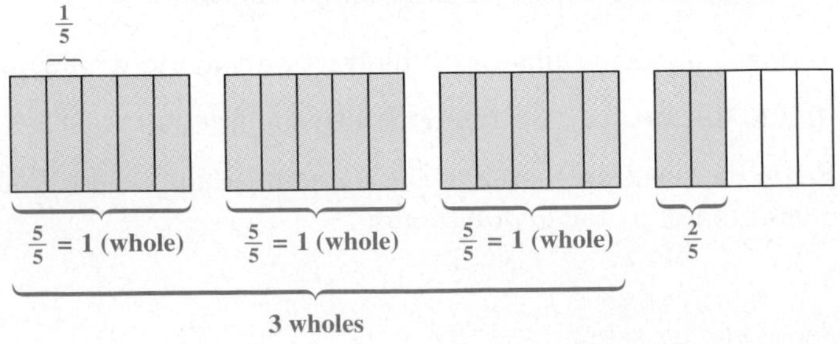

Continued on Next Page

ANSWERS

2. (a) $\frac{9}{2}$ (b) $\frac{23}{4}$ (c) $\frac{39}{8}$ (d) $\frac{53}{6}$

(b) $\frac{24}{4}$

Divide 24 by 4.

$$4\overline{)24} = 6 \quad \text{so} \quad \frac{24}{4} = 6$$

$$\begin{array}{r} 6 \\ 4\overline{)24} \\ \underline{24} \\ 0 \end{array} \leftarrow \text{No remainder}$$

NOTE

A proper fraction has a value that is less than 1, while an improper fraction has a value that is 1 or greater.

Work Problem 3 at the Side.

3 Write as whole or mixed numbers.

(a) $\frac{4}{3}$

(b) $\frac{9}{4}$

(c) $\frac{35}{5}$

(d) $\frac{85}{8}$

ANSWERS

3. **(a)** $1\frac{1}{3}$ **(b)** $2\frac{1}{4}$ **(c)** 7 **(d)** $10\frac{5}{8}$

Relating Concepts (Exercises 69–74) For Individual or Group Work

Knowing the basics of fractions is necessary in problem solving. ***Work Exercises 69–74 in order.***

69. Which of these fractions are proper fractions?

$\frac{2}{3}, \frac{4}{5}, \frac{8}{5}, \frac{3}{4}, \frac{6}{6}, \frac{7}{10}$

70. (a) The proper fractions in Exercise 69 are the ones where the ______________ is smaller than the ______________.

(b) Draw a picture with shaded parts to show each proper fraction in Exercise 69.

(c) The proper fractions in Exercise 69 are all ______________ than 1.
(less/greater)

71. Which of these fractions are improper fractions?

$\frac{5}{5}, \frac{3}{4}, \frac{10}{3}, \frac{2}{3}, \frac{5}{6}, \frac{6}{5}$

72. (a) The improper fractions in Exercise 71 are the ones where the ______________ is greater than or equal to the ______________.

(b) Draw a picture with shaded parts to show each mixed number in Exercise 71.

(c) The improper fractions in Exercise 71 are all ______________ than 1.
(less/greater)

73. Identify which of these fractions can be written as whole or mixed numbers and then write them as whole or mixed numbers.

$\frac{5}{3}, \frac{7}{8}, \frac{7}{7}, \frac{11}{6}, \frac{4}{5}, \frac{15}{16}$

74. (a) The fractions that can be written as whole or mixed numbers in Exercise 73 are ______________ fractions, and their value is
(proper/improper)
always ______________________________ 1.
(less than/greater than or equal to)

(b) Draw a picture with shaded parts to show each whole or mixed number in Exercise 73.

(c) Explain how to write an improper fraction as a whole or mixed number.

2.3 FACTORS

1 Find factors of a number. You will recall that numbers multiplied to give a product are called **factors.** Because $2 \cdot 5 = 10$, both 2 and 5 are factors of 10. The numbers 1 and 10 are also factors of 10, because

$$1 \cdot 10 = 10.$$

The various tests for divisibility show that 1, 2, 5, and 10 are the only whole number factors of 10. The products $2 \cdot 5$ and $1 \cdot 10$ are called **factorizations** of 10.

NOTE

You might want to review the tests for divisibility in Section 1.5. The ones that you will want to remember are those for 2, 3, 5, and 10.

Example 1 Using Factors

Find all possible two-number factorizations of each number.

(a) 12

$$1 \cdot 12 = 12 \qquad 2 \cdot 6 = 12 \qquad 3 \cdot 4 = 12$$

The factors of 12 are 1, 2, 3, 4, 6, and 12.

(b) 60

$$1 \cdot 60 = 60 \qquad 2 \cdot 30 = 60$$
$$3 \cdot 20 = 60 \qquad 4 \cdot 15 = 60$$
$$5 \cdot 12 = 60 \qquad 6 \cdot 10 = 60$$

The factors of 60 are 1, 2, 3, 4, 5, 6, 10, 12, 15, 20, 30, and 60.

Work Problem 1 at the Side.

Composite Numbers

A number with a factor other than itself or 1 is called a **composite number.**

Example 2 Identifying Composite Numbers

Which of the following numbers are composite?

(a) 6
Because 6 has factors of 2 and 3, numbers other than 6 or 1, the number 6 is composite.

(b) 17
The number 17 has only two factors, 17 and 1. It is not composite.

(c) 25
A factor of 25 is 5, so 25 is composite.

Work Problem 2 at the Side.

OBJECTIVES

1. Find factors of a number.
2. Identify prime numbers.
3. Find prime factorizations.

Study Skills Workbook
Activity 7

1 Find all the whole number factors of each number.

(a) 10

(b) 16

(c) 36

(d) 80

2 Which of these numbers are composite?

2, 4, 5, 6, 8, 10, 11, 13, 19, 21, 27, 28, 33, 36, 42

ANSWERS

1. **(a)** 1, 2, 5, 10 **(b)** 1, 2, 4, 8, 16
(c) 1, 2, 3, 4, 6, 9, 12, 18, 36
(d) 1, 2, 4, 5, 8, 10, 16, 20, 40, 80
2. 4, 6, 8, 10, 21, 27, 28, 33, 36, 42

3 Which of the following are prime?

4, 7, 9, 13, 17, 19, 29, 33

2 **Identify prime numbers.** Whole numbers that are not composite are called **prime numbers,** except 0 and 1, which are neither prime nor composite.

Prime Numbers

A **prime number** is a whole number that has exactly ***two different*** factors, ***itself*** and ***1.***

The number 3 is a prime number, since it can be divided evenly only by itself and 1. The number 8 is not a prime number (it is composite), since 8 can be divided evenly by 2 and 4, as well as by itself and 1.

CAUTION

A prime number has ***only two*** different factors, itself and 1. The number 1 is not a prime number because it does not have *two different* factors; the only factor of 1 is 1.

Example 3 Finding Prime Numbers

Which of the following numbers are prime?

2 5 11 15 27

The number 15 can be divided by 3 and 5, so it is not prime. Also, because 27 can be divided by 3 and 9, 27 is not prime. All the other numbers in the list are divisible by only themselves and 1, so they are prime.

Work Problem 3 at the Side.

3 **Find prime factorizations.** For reference, here are the prime numbers smaller than 100.

2,	3,	5,	7,	11,
13,	17,	19,	23,	29,
31,	37,	41,	43,	47,
53,	59,	61,	67,	71,
73,	79,	83,	89,	97

CAUTION

All prime numbers are odd numbers except the number 2. Be careful though, because *all odd numbers are not prime numbers.* For example, 9, 15, and 21 are odd numbers, but are *not* prime numbers.

The **prime factorization** of a number can be especially useful when we are adding or subtracting fractions and need to find a common denominator or write a fraction in lowest terms.

Prime Factorization

A **prime factorization** of a number is a factorization in which every factor is a *prime number.*

ANSWERS

3. 7, 13, 17, 19, 29

Example 4 **Determining the Prime Factorization**

Find the prime factorization of 12.

Try to divide 12 by the first prime, 2.

$$12 \div \underset{\uparrow \text{ First prime}}{2} = 6,$$

so

$$12 = 2 \cdot 6.$$

Try to divide 6 by the prime, 2.

$$6 \div 2 = 3,$$

so

$$12 = 2 \cdot \underbrace{2 \cdot 3}_{\text{Factorization of 6}}.$$

Because all factors are prime, the prime factorization of 12 is

$$2 \cdot 2 \cdot 3.$$

Work Problem 4 at the Side.

4 Find the prime factorization of each number.

(a) 8

(b) 42

(c) 18

(d) 40

Example 5 **Factoring by Using the Division Method**

Find the prime factorization of 48.

All prime factors	$2\overline{)48}$	Divide 48 by 2 (first prime).
	$2\overline{)24}$	Divide 24 by 2.
	$2\overline{)12}$	Divide 12 by 2.
	$2\overline{)6}$	Divide 6 by 2.
	$3\overline{)3}$	Divide 3 by 3.
	1	Continue to divide until the quotient is 1.

Because all factors (divisors) are prime, the prime factorization of 48 is

$$2 \cdot 2 \cdot 2 \cdot 2 \cdot 3.$$

In Chapter 1, we wrote $2 \cdot 2 \cdot 2 \cdot 2$ as 2^4, so the prime factorization of 48 can be written, using exponents, as

$$2 \cdot 2 \cdot 2 \cdot 2 \cdot 3 = 2^4 \cdot 3.$$

Work Problem 5 at the Side.

5 Find the prime factorization of each number. Write the factorization with exponents.

(a) 36

(b) 54

(c) 60

(d) 81

NOTE

When using the division method of factoring, the last quotient found is 1. The "1" is never used as a prime factor because 1 is neither prime nor composite. Besides, 1 times any number is the number itself.

ANSWERS

4. **(a)** $2 \cdot 2 \cdot 2$ **(b)** $2 \cdot 3 \cdot 7$ **(c)** $2 \cdot 3 \cdot 3$ **(d)** $2 \cdot 2 \cdot 2 \cdot 5$

5. **(a)** $2^2 \cdot 3^2$ **(b)** $2 \cdot 3^3$ **(c)** $2^2 \cdot 3 \cdot 5$ **(d)** 3^4

6 Write the prime factorization of each number by using exponents.

(a) 50

(b) 88

(c) 90

(d) 120

(e) 180

Example 6 Using Exponents with Prime Factorization

Find the prime factorization of 225.

All prime factors

$3\overline{)225}$	225 is not divisible by 2; use 3.
$3\overline{)75}$	Divide 75 by 3.
$5\overline{)25}$	25 is not divisible by 3; use 5.
$5\overline{)5}$	Divide by 5.
1	Continue to divide until the quotient is 1.

Write the prime factorization,

$$3 \cdot 3 \cdot 5 \cdot 5,$$

with exponents as

$$3^2 \cdot 5^2.$$

Work Problem 6 at the Side.

Another method of factoring is a *factor tree.*

Example 7 Factoring by Using a Factor Tree

Find the prime factorization of each number.

(a) 30

Try to divide by the first prime, 2. Write the factors under the 30. Circle the 2, since it is a prime.

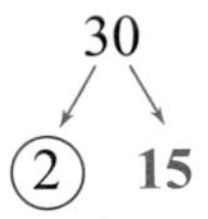

Since 15 cannot be divided evenly by 2, try the next prime, 3.

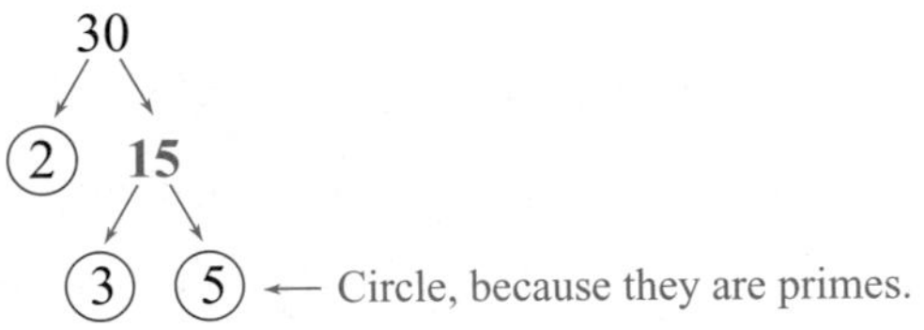

No uncircled factors remain, so the prime factorization (the circled factors) has been found.

$$30 = 2 \cdot 3 \cdot 5$$

(b) 24

Divide by 2.

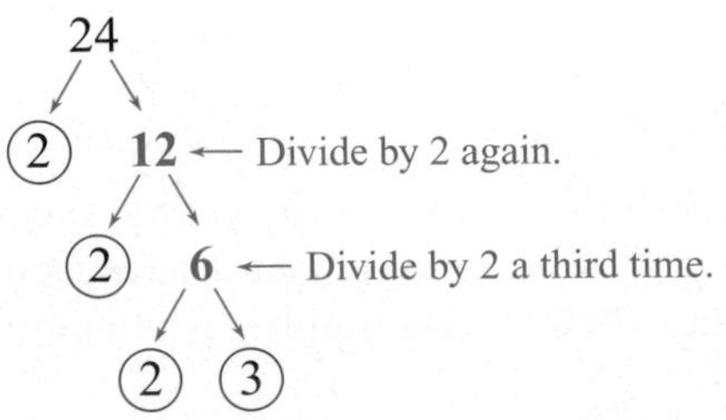

$24 = 2 \cdot 2 \cdot 2 \cdot 3$ or, using exponents, $24 = 2^3 \cdot 3$.

Continued on Next Page

ANSWERS

6. **(a)** $2 \cdot 5^2$ **(b)** $2^3 \cdot 11$ **(c)** $2 \cdot 3^2 \cdot 5$ **(d)** $2^3 \cdot 3 \cdot 5$ **(e)** $2^2 \cdot 3^2 \cdot 5$

(c) 45

Because 45 cannot be divided by 2, try 3.

45

③ **15** ◄— Divide by 3 again.

③ ⑤

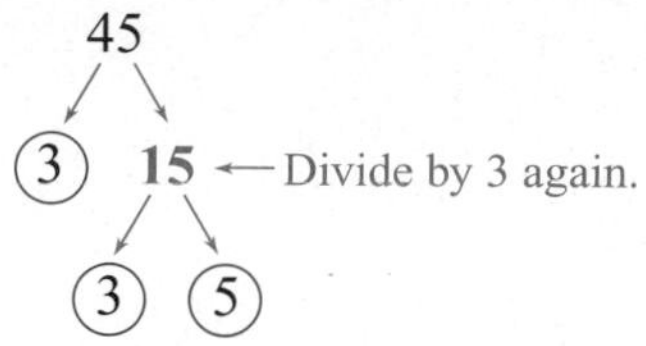

$45 = 3 \cdot 3 \cdot 5$ or, using exponents, $45 = 3^2 \cdot 5$.

NOTE

The diagrams used in Example 7 look like tree branches, and that is why this method is referred to as using a factor tree.

Work Problem ❼ at the Side.

❼ Complete each factor tree and give the prime factorization.

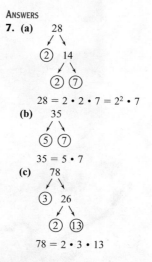

(a) 28

② **14**

○ ○

(b) 35

⑤

(c) 78

ANSWERS

7. (a) 28

② 14

② ⑦

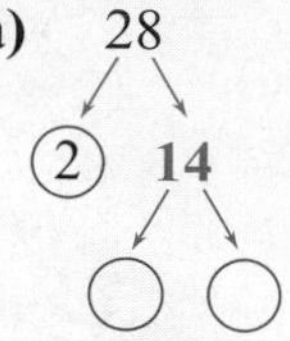

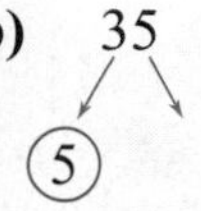

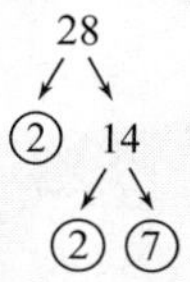

$28 = 2 \cdot 2 \cdot 7 = 2^2 \cdot 7$

(b) 35

⑤ ⑦

$35 = 5 \cdot 7$

(c) 78

③ 26

② ⑬

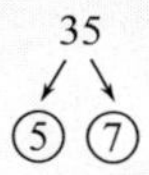

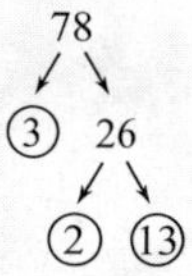

$78 = 2 \cdot 3 \cdot 13$

2.3 EXERCISES

FOR EXTRA HELP: Student's Solutions Manual | MyMathLab.com | InterAct Math Tutorial Software | AW Math Tutor Center | www.mathxl.com | Digital Video Tutor CD 1 Videotape 4

Find all the factors of each number. See Example 1.

1. 6

2. 15

3. 8

4. 28

5. 48

6. 30

7. 36

8. 20

9. 40

10. 60

11. 64

12. 84

Decide whether each number is prime *or* composite. *See Examples 2 and 3.*

13. 4

14. 6

15. 11

16. 7

17. 8

18. 9

19. 13

20. 65

21. 13

22. 17

23. 25

24. 26

25. 42

26. 47

27. 45

28. 53

Find the prime factorization of each number. Write answers with exponents when repeated factors appear. See Examples 4–7.

29. 10

30. 15

31. 20

32. 40

33. 25

34. 18

35. 36

36. 56

37. 68

38. 70

39. 72

40. 64

41. 44

42. 104

43. 100

44. 112

45. 125

46. 135

47. 180

48. 300

49. 320

50. 350

51. 360

52. 400

53. Give a definition in your own words of both a composite number and a prime number. Give three examples of each. Which whole numbers are neither prime nor composite?

54. With the exception of the number 2, all prime numbers are odd numbers. Nevertheless, all odd numbers are not prime numbers. Explain why these statements are true.

55. Explain the difference between finding all possible factors of 24 and finding the prime factorization of 24.

56. Use the division method to find the prime factorization of 36. Can you divide by 3s before you divide by 2s? Does the order of division change the answers?

Find the prime factorization of each number. Write answers using exponents.

57. 320

58. 640

59. 960

60. 1125

61. 1560

62. 2000

63. 1260

64. 2200

RELATING CONCEPTS (Exercises 65–70) FOR INDIVIDUAL OR GROUP WORK

An understanding of factors and factorization will be needed to solve fraction problems.
Work Exercises 65–70 in order.

65. A prime number is a whole number that has exactly two different factors, itself and 1. List all prime numbers smaller than 50.

66. Explain how to determine which of the numbers in Exercise 65 are prime.

67. With the exception of the number 2, all prime numbers are odd numbers. Why is this true?

68. Can a multiple of a prime number (for example, 6, 9, 12, and 15 are multiples of 3) be prime? Explain.

69. Find the prime factorization of 2100. Do not use exponents in your answer.

70. Write the answer to Exercise 69 using exponents for repeated factors.

2.4 Writing a Fraction in Lowest Terms

When working problems involving fractions, we must often compare two fractions to determine whether they represent the same portion of a whole. Consider the following diagrams.

$\frac{5}{6}$ is shaded.

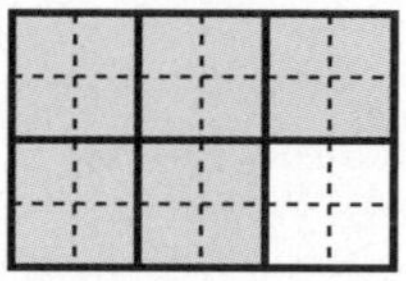

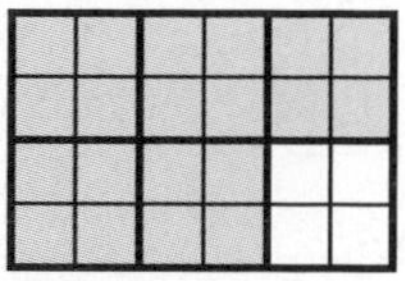

$\frac{20}{24}$ is shaded.

The figures show areas that are $\frac{5}{6}$ shaded and $\frac{20}{24}$ shaded. Because the shaded areas are equivalent (the same portion), the fractions $\frac{5}{6}$ and $\frac{20}{24}$ are **equivalent fractions.**

$$\frac{5}{6} = \frac{20}{24}$$

Because the numbers 20 and 24 both have 4 as a factor, 4 is called a **common factor** of the numbers. Other common factors of 20 and 24 are 1 and 2.

Work Problem 1 at the Side.

1 Tell whether a fraction is written in lowest terms. The fraction $\frac{5}{6}$ is written in *lowest terms* because the numerator and denominator have no common factor other than 1. However, the fraction $\frac{20}{24}$ is ***not*** in lowest terms because its numerator and denominator have common factors of 4, 2, and 1.

Writing a Fraction in Lowest Terms

A fraction is written in **lowest terms** when the numerator and denominator have no common factor other than 1.

Example 1 **Understanding Lowest Terms**

Are the following fractions in lowest terms?

(a) $\frac{3}{8}$

The numerator and denominator have no common factor other than 1, so the fraction is in lowest terms.

(b) $\frac{21}{36}$

The numerator and denominator have a common factor of 3, so the fraction is not in lowest terms.

Work Problem 2 at the Side.

2 Write a fraction in lowest terms using common factors. There are two common methods for writing a fraction in lowest terms. These methods are shown in the next examples. The first method works best when the numerator and denominator are small numbers.

Objectives

1. Tell whether a fraction is written in lowest terms.
2. Write a fraction in lowest terms using common factors.
3. Write a fraction in lowest terms using prime factors.
4. Tell whether two fractions are equivalent.

1 Decide whether the given factor is a common factor of both numbers.

(a) 8, 18; 2

(b) 32, 64; 8

(c) 16, 34; 16

(d) 56, 73; 1

2 Are the following fractions in lowest terms?

(a) $\frac{3}{4}$

(b) $\frac{5}{15}$

(c) $\frac{9}{15}$

(d) $\frac{17}{46}$

Answers

1. (a) yes **(b)** yes **(c)** no **(d)** yes
2. (a) yes **(b)** no **(c)** no **(d)** yes

3 Write in lowest terms.

(a) $\frac{6}{12}$

(b) $\frac{6}{8}$

(c) $\frac{48}{60}$

(d) $\frac{30}{80}$

(e) $\frac{16}{40}$

Example 2 Writing Fractions in Lowest Terms

Write each fraction in lowest terms.

(a) $\frac{20}{24}$

The largest common factor of 20 and 24 is 4. Divide both numerator and denominator by 4.

$$\frac{20}{24} = \frac{20 \div 4}{24 \div 4} = \frac{5}{6}$$

(b) $\frac{30}{50} = \frac{30 \div 10}{50 \div 10} = \frac{3}{5}$ Divide both numerator and denominator by 10.

(c) $\frac{24}{42} = \frac{24 \div 6}{42 \div 6} = \frac{4}{7}$ Divide both numerator and denominator by 6.

(d) $\frac{60}{72}$

Suppose we made an error and thought 4 was the largest common factor of 60 and 72. Dividing by 4 would give

$$\frac{60}{72} = \frac{60 \div 4}{72 \div 4} = \frac{15}{18}.$$

But $\frac{15}{18}$ is not in lowest terms, because 15 and 18 have a common factor of 3. So we divide by 3.

$$\frac{15}{18} = \frac{15 \div 3}{18 \div 3} = \frac{5}{6}$$

The fraction $\frac{60}{72}$ could have been written in lowest terms in one step by dividing by 12, the largest common factor of 60 and 72.

$$\frac{60}{72} = \frac{60 \div 12}{72 \div 12} = \frac{5}{6}$$

NOTE

Dividing the numerator and denominator by the same number results in an equivalent fraction.

In Example 2, we wrote fractions in lowest terms by dividing by a common factor. This method is summarized in the following steps.

The Method of Dividing by a Common Factor

Step 1 Find the largest number that will divide evenly into both the numerator and denominator. This number is a *common factor.*

Step 2 *Divide* both numerator and denominator by the common factor.

Step 3 *Check* to see whether the new fraction has any common factors (besides 1). If it does, repeat Steps 2 and 3. If the only common factor is 1, the fraction is in lowest terms.

Work Problem 3 at the Side.

ANSWERS

3. (a) $\frac{1}{2}$ (b) $\frac{3}{4}$ (c) $\frac{4}{5}$ (d) $\frac{3}{8}$ (e) $\frac{2}{5}$

3 Write a fraction in lowest terms using prime factors. The method of writing a fraction in lowest terms by division works well for fractions with small numerators and denominators. For larger numbers, it is common to use the method of *prime factors,* which is shown in the next example.

Example 3 Using Prime Factors

Write each fraction in lowest terms.

(a) $\frac{24}{42}$

Write the prime factorization of both numerator and denominator. See **Section 2.3** for help.

$$\frac{24}{42} = \frac{2 \cdot 2 \cdot 2 \cdot 3}{2 \cdot 3 \cdot 7}$$

Just as with the other method, divide both numerator and denominator by any common factors. Write a 1 by each factor that has been divided.

$$\frac{24}{42} = \frac{\overset{1}{\cancel{2}} \cdot 2 \cdot 2 \cdot \overset{1}{\cancel{3}}}{\underset{1}{\cancel{2}} \cdot \underset{1}{\cancel{3}} \cdot 7}$$

Multiply the remaining factors in both numerator and denominator.

$$\frac{24}{42} = \frac{1 \cdot 2 \cdot 2 \cdot 1}{1 \cdot 1 \cdot 7} = \frac{4}{7}$$

Finally, $\frac{24}{42}$ written in lowest terms is $\frac{4}{7}$.

(b) $\frac{162}{54}$

Write the prime factorization of both numerator and denominator.

$$\frac{162}{54} = \frac{2 \cdot 3 \cdot 3 \cdot 3 \cdot 3}{2 \cdot 3 \cdot 3 \cdot 3}$$

Now divide by the common factors. ***Do not forget to write the 1s.***

$$\frac{162}{54} = \frac{\overset{1}{\cancel{2}} \cdot \overset{1}{\cancel{3}} \cdot \overset{1}{\cancel{3}} \cdot \overset{1}{\cancel{3}} \cdot 3}{\underset{1}{\cancel{2}} \cdot \underset{1}{\cancel{3}} \cdot \underset{1}{\cancel{3}} \cdot \underset{1}{\cancel{3}}}$$

$$= \frac{1 \cdot 1 \cdot 1 \cdot 1 \cdot 3}{1 \cdot 1 \cdot 1 \cdot 1} = \frac{3}{1} = 3$$

(c) $\frac{18}{90}$

$$\frac{18}{90} = \frac{\overset{1}{\cancel{2}} \cdot \overset{1}{\cancel{3}} \cdot \overset{1}{\cancel{3}}}{\underset{1}{\cancel{2}} \cdot \underset{1}{\cancel{3}} \cdot \underset{1}{\cancel{3}} \cdot 5} = \frac{1 \cdot 1 \cdot 1}{1 \cdot 1 \cdot 1 \cdot 5} = \frac{1}{5}$$

CAUTION

In Example 3(c), all factors of the numerator were divided. But $1 \cdot 1 \cdot 1$ is still 1, so the final answer is $\frac{1}{5}$ (***not*** 5).

4 Use the method of prime factors to write each fraction in lowest terms.

(a) $\frac{12}{36}$

(b) $\frac{32}{56}$

(c) $\frac{74}{111}$

(d) $\frac{124}{340}$

5 Is each pair of fractions equivalent?

(a) $\frac{24}{48}$ and $\frac{36}{72}$

(b) $\frac{45}{60}$ and $\frac{50}{75}$

(c) $\frac{20}{4}$ and $\frac{110}{22}$

(d) $\frac{120}{220}$ and $\frac{180}{320}$

ANSWERS

4. (a) $\frac{1}{3}$ **(b)** $\frac{4}{7}$ **(c)** $\frac{2}{3}$ **(d)** $\frac{31}{85}$

5. (a) equivalent **(b)** not equivalent **(c)** equivalent **(d)** not equivalent

In Example 3, we wrote fractions in lowest terms using prime factors. This method is summarized as follows.

The Method of Prime Factors

Step 1 Write the *prime factorization* of both numerator and denominator.

Step 2 *Divide* both numerator and denominator by any common factors.

Step 3 *Multiply* the remaining factors in the numerator and denominator.

Work Problem 4 at the Side.

4 Tell whether two fractions are equivalent. The next example shows how to tell whether two fractions are equivalent.

Example 4 Determining whether Two Fractions Are Equivalent

Determine whether each pair of fractions is equivalent.

(a) $\frac{16}{48}$ and $\frac{24}{72}$

Use the method of prime factors to write each fraction in lowest terms.

$$\frac{16}{48} = \frac{\overset{1}{\cancel{2}} \cdot \overset{1}{\cancel{2}} \cdot \overset{1}{\cancel{2}} \cdot \overset{1}{\cancel{2}}}{\underset{1}{\cancel{2}} \cdot \underset{1}{\cancel{2}} \cdot \underset{1}{\cancel{2}} \cdot \underset{1}{\cancel{2}} \cdot 3} = \frac{1 \cdot 1 \cdot 1 \cdot 1}{1 \cdot 1 \cdot 1 \cdot 1 \cdot 3} = \frac{1}{3}$$

$$\frac{24}{72} = \frac{\overset{1}{\cancel{2}} \cdot \overset{1}{\cancel{2}} \cdot \overset{1}{\cancel{2}} \cdot \overset{1}{\cancel{3}}}{\underset{1}{\cancel{2}} \cdot \underset{1}{\cancel{2}} \cdot \underset{1}{\cancel{2}} \cdot \underset{1}{\cancel{3}} \cdot 3} = \frac{1 \cdot 1 \cdot 1 \cdot 1}{1 \cdot 1 \cdot 1 \cdot 1 \cdot 3} = \frac{1}{3}$$

Equivalent $\left(\frac{1}{3} = \frac{1}{3}\right)$

(b) $\frac{32}{52}$ and $\frac{64}{112}$

$$\frac{32}{52} = \frac{\overset{1}{\cancel{2}} \cdot \overset{1}{\cancel{2}} \cdot 2 \cdot 2 \cdot 2}{\underset{1}{\cancel{2}} \cdot \underset{1}{\cancel{2}} \cdot 13} = \frac{2 \cdot 2 \cdot 2}{1 \cdot 1 \cdot 13} = \frac{8}{13}$$

$$\frac{64}{112} = \frac{\overset{1}{\cancel{2}} \cdot \overset{1}{\cancel{2}} \cdot \overset{1}{\cancel{2}} \cdot \overset{1}{\cancel{2}} \cdot 2 \cdot 2}{\underset{1}{\cancel{2}} \cdot \underset{1}{\cancel{2}} \cdot \underset{1}{\cancel{2}} \cdot \underset{1}{\cancel{2}} \cdot 7} = \frac{1 \cdot 1 \cdot 1 \cdot 1 \cdot 2 \cdot 2}{1 \cdot 1 \cdot 1 \cdot 1 \cdot 7} = \frac{4}{7}$$

Not equivalent $\left(\frac{8}{13} \neq \frac{4}{7}\right)$

(c) $\frac{75}{15}$ and $\frac{60}{12}$

$$\frac{75}{15} = \frac{\overset{1}{\cancel{3}} \cdot \overset{1}{\cancel{5}} \cdot 5}{\underset{1}{\cancel{3}} \cdot \underset{1}{\cancel{5}}} = \frac{1 \cdot 1 \cdot 5}{1 \cdot 1} = 5$$

$$\frac{60}{12} = \frac{\overset{1}{\cancel{2}} \cdot \overset{1}{\cancel{2}} \cdot \overset{1}{\cancel{3}} \cdot 5}{\underset{1}{\cancel{2}} \cdot \underset{1}{\cancel{2}} \cdot 3} = \frac{1 \cdot 1 \cdot 1 \cdot 5}{1 \cdot 1 \cdot 1} = 5$$

Equivalent $(5 = 5)$

Work Problem 5 at the Side.

2.4 Exercises

Put a ✓ mark in the blank if the number at the left is divisible by the number at the top. Put an X in the blank if the number is not divisible by the number at the top. (For help, see Section 1.5.)

	2	3	5	10		2	3	5	10
1. 30	___	___	___	___	**2.** 60	___	___	___	___
3. 48	___	___	___	___	**4.** 36	___	___	___	___
5. 160	___	___	___	___	**6.** 175	___	___	___	___
7. 138	___	___	___	___	**8.** 120	___	___	___	___

Write each fraction in lowest terms. See Example 2.

9. $\frac{9}{12}$ **10.** $\frac{5}{10}$ **11.** $\frac{16}{24}$ **12.** $\frac{4}{12}$

13. $\frac{25}{40}$ **14.** $\frac{32}{48}$ **15.** $\frac{36}{42}$ **16.** $\frac{22}{33}$

17. $\frac{63}{70}$ **18.** $\frac{21}{35}$ **19.** $\frac{180}{210}$ **20.** $\frac{72}{80}$

21. $\frac{36}{63}$ **22.** $\frac{73}{146}$ **23.** $\frac{12}{600}$ **24.** $\frac{8}{400}$

25. $\frac{96}{132}$ **26.** $\frac{165}{180}$ **27.** $\frac{60}{108}$ **28.** $\frac{112}{128}$

Write the numerator and denominator of each fraction as a product of prime factors and divide by the common factors. Then write the fraction in lowest terms. See Example 3.

29. $\frac{18}{24}$ **30.** $\frac{16}{64}$ **31.** $\frac{35}{40}$

32. $\frac{20}{32}$ **33.** $\frac{90}{180}$ **34.** $\frac{36}{48}$

35. $\frac{36}{12}$

36. $\frac{192}{48}$

37. $\frac{72}{225}$

38. $\frac{65}{234}$

Tell whether each pair of fractions is equivalent or not equivalent. See Example 4.

39. $\frac{2}{4}$ and $\frac{16}{32}$

40. $\frac{4}{10}$ and $\frac{20}{50}$

41. $\frac{10}{24}$ and $\frac{12}{30}$

42. $\frac{22}{32}$ and $\frac{32}{48}$

43. $\frac{15}{24}$ and $\frac{35}{52}$

44. $\frac{21}{33}$ and $\frac{9}{12}$

45. $\frac{14}{16}$ and $\frac{35}{40}$

46. $\frac{27}{90}$ and $\frac{24}{80}$

47. $\frac{48}{6}$ and $\frac{72}{8}$

48. $\frac{45}{15}$ and $\frac{96}{32}$

49. $\frac{25}{30}$ and $\frac{65}{78}$

50. $\frac{24}{72}$ and $\frac{30}{90}$

51. What does it mean when a fraction is expressed in lowest terms? Give three examples.

52. Explain what equivalent fractions are and give an example of a pair of equivalent fractions. Show that they are equivalent.

Write each fraction in lowest terms.

53. $\frac{224}{256}$

54. $\frac{363}{528}$

55. $\frac{356}{178}$

56. $\frac{525}{105}$

2.5 MULTIPLYING FRACTIONS

1 Multiply fractions. Suppose that you give $\frac{1}{2}$ of your Energy Bar to your kickboxing partner Jennifer. Then Jennifer gives $\frac{1}{2}$ of her share to Tony. How much of the Energy Bar does Tony get to eat?

Start with a sketch showing the Energy Bar cut in half (2 equal pieces).

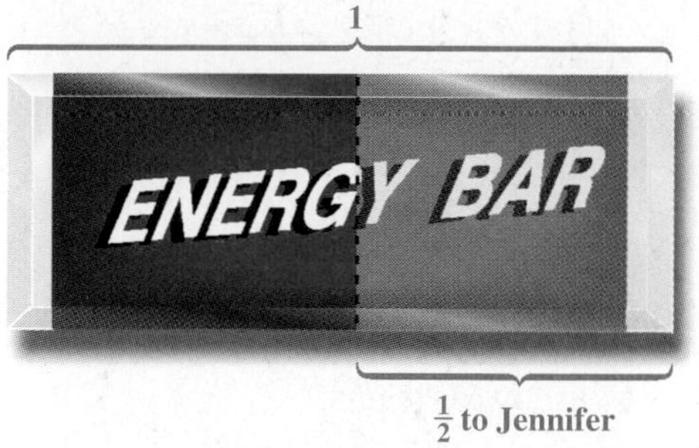

Next, take $\frac{1}{2}$ of the shaded area. (Here we are dividing $\frac{1}{2}$ into 2 equal parts and shading one darker than the other.)

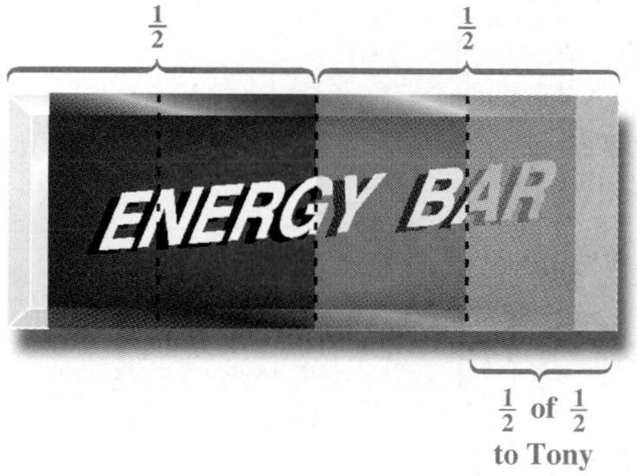

The sketch shows that Tony gets $\frac{1}{4}$ of the Energy Bar.

Tony gets $\frac{1}{2}$ of $\frac{1}{2}$ of the Energy Bar. When used between two fractions, the word **of** tells us to multiply.

$$\frac{1}{2} \text{ of } \frac{1}{2} \quad \text{means} \quad \frac{1}{2} \cdot \frac{1}{2}$$

Tony's share of the Energy Bar is

$$\frac{1}{2} \cdot \frac{1}{2} = \frac{1}{4}.$$

Work Problem 1 at the Side.

The rule for multiplying fractions follows.

Multiplying Fractions

Multiply two fractions by multiplying the numerators and multiplying the denominators.

OBJECTIVES

1. Multiply fractions.
2. Use a multiplication shortcut.
3. Multiply a fraction and a whole number.
4. Find the area of a rectangle.

1 Use these figures to find $\frac{1}{4}$ of $\frac{1}{2}$.

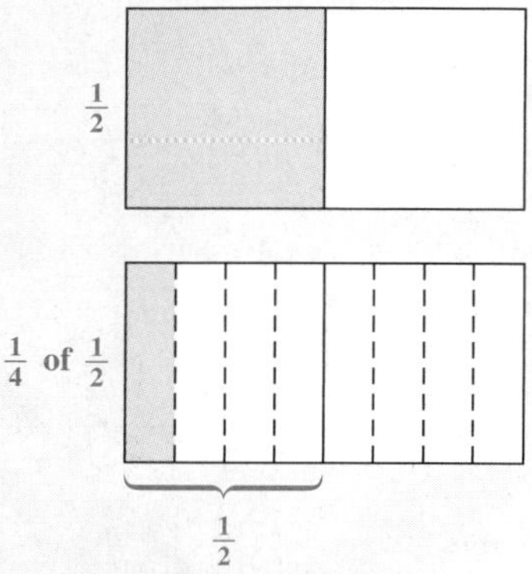

ANSWERS

1. $\frac{1}{8}$

Use this rule to find the product of $\frac{2}{3}$ and $\frac{1}{3}$ (multiply $\frac{2}{3}$ by $\frac{1}{3}$).

$$\frac{2}{3} \cdot \frac{1}{3} = \frac{2 \cdot 1}{3 \cdot 3} \begin{matrix} \leftarrow \text{Multiply numerators.} \\ \leftarrow \text{Multiply denominators.} \end{matrix}$$

$$= \frac{2}{9}$$

Finish multiplying.

$$\frac{2}{3} \cdot \frac{1}{3} = \frac{2 \cdot 1}{3 \cdot 3} = \frac{2}{9} \begin{matrix} \leftarrow 2 \cdot 1 = 2 \\ \leftarrow 3 \cdot 3 = 9 \end{matrix}$$

Check that the final result is in lowest terms. $\frac{2}{9}$ is in lowest terms because 2 and 9 have no common factor other than 1.

Example 1 Multiplying Fractions

Multiply. Write answers in lowest terms.

(a) $\frac{5}{8} \cdot \frac{3}{4}$

Multiply the numerators and multiply the denominators.

$$\frac{5}{8} \cdot \frac{3}{4} = \frac{5 \cdot 3}{8 \cdot 4} = \frac{15}{32}$$

Notice that 15 and 32 have no common factors other than 1, so the answer is in lowest terms.

(b) $\frac{4}{7} \cdot \frac{2}{5}$

$$\frac{4}{7} \cdot \frac{2}{5} = \frac{4 \cdot 2}{7 \cdot 5} = \frac{8}{35}$$

(c) $\frac{5}{8} \cdot \frac{3}{4} \cdot \frac{1}{2}$

$$\frac{5}{8} \cdot \frac{3}{4} \cdot \frac{1}{2} = \frac{5 \cdot 3 \cdot 1}{8 \cdot 4 \cdot 2} = \frac{15}{64}$$

Work Problem 2 at the Side.

2 Multiply. Write answers in lowest terms.

(a) $\frac{3}{4} \cdot \frac{1}{2}$

(b) $\frac{1}{3} \cdot \frac{3}{5}$

(c) $\frac{1}{8} \cdot \frac{5}{6} \cdot \frac{1}{2}$

(d) $\frac{1}{2} \cdot \frac{3}{4} \cdot \frac{3}{8}$

2 Use a multiplication shortcut. A multiplication shortcut that can be used with fractions is shown in Example 2.

Example 2 Using the Multiplication Shortcut

Multiply $\frac{5}{6}$ and $\frac{9}{10}$. Write the answer in lowest terms.

$$\frac{5}{6} \cdot \frac{9}{10} = \frac{5 \cdot 9}{6 \cdot 10} = \frac{45}{60} \quad \text{Not in lowest terms}$$

The numerator and denominator have a common factor other than 1, so write the prime factorization of each number.

$$\frac{5}{6} \cdot \frac{9}{10} = \frac{5 \cdot 9}{6 \cdot 10} = \frac{5 \cdot \overbrace{3 \cdot 3}}{\underbrace{2 \cdot 3} \cdot \underbrace{2 \cdot 5}}$$

Continued on Next Page

Answers

2. (a) $\frac{3}{8}$ (b) $\frac{1}{5}$ (c) $\frac{5}{96}$ (d) $\frac{9}{64}$

Next, divide by the common factors of 5 and 3.

$$\frac{5}{6} \cdot \frac{9}{10} = \frac{5 \cdot 9}{6 \cdot 10} = \frac{\overset{1}{\cancel{5}} \cdot \overset{1}{\cancel{3}} \cdot 3}{2 \cdot \underset{1}{\cancel{3}} \cdot 2 \cdot \underset{1}{\cancel{5}}}$$

Finally, multiply the remaining factors in the numerator and in the denominator.

$$\frac{5}{6} \cdot \frac{9}{10} = \frac{\mathbf{1} \cdot \mathbf{1} \cdot 3}{2 \cdot \mathbf{1} \cdot 2 \cdot \mathbf{1}} = \frac{3}{4} \quad \text{Lowest terms}$$

As a shortcut, instead of writing the prime factorization of each number, find the product of $\frac{5}{6}$ and $\frac{9}{10}$ as follows.

First, divide both 5 and 10 by 5. $\frac{\overset{1}{\cancel{5}}}{6} \cdot \frac{9}{\underset{2}{\cancel{10}}}$

Next, divide both 6 and 9 by 3. $\frac{\overset{1}{\cancel{5}}}{\underset{2}{\cancel{6}}} \cdot \frac{\overset{3}{\cancel{9}}}{\underset{2}{\cancel{10}}}$

Finally, multiply. $\frac{1 \cdot 3}{2 \cdot 2} = \frac{3}{4}$

CAUTION

When using the multiplication shortcut, you are dividing a numerator and a denominator. Be certain that you divide a numerator and a denominator ***by the same number.*** If you do all possible divisions, your answer will be in lowest terms.

Example 3 Using the Multiplication Shortcut

Use the multiplication shortcut to find each product. Write the answers in lowest terms and as mixed numbers where possible.

(a) $\frac{6}{11} \cdot \frac{7}{8}$

Divide both 6 and 8 by 2. Next, multiply.

$$\frac{\overset{3}{\cancel{6}}}{11} \cdot \frac{7}{\underset{4}{\cancel{8}}} = \frac{3 \cdot 7}{11 \cdot 4} = \frac{21}{44} \quad \text{Lowest terms}$$

(b) $\frac{7}{10} \cdot \frac{20}{21}$

Divide 7 and 21 by 7, and divide 10 and 20 by 10.

$$\frac{\overset{1}{\cancel{7}}}{\underset{1}{\cancel{10}}} \cdot \frac{\overset{2}{\cancel{20}}}{\underset{3}{\cancel{21}}} = \frac{1 \cdot 2}{1 \cdot 3} = \frac{2}{3} \quad \text{Lowest terms}$$

Continued on Next Page

3 Use the multiplication shortcut to find each product.

(a) $\frac{3}{4} \cdot \frac{2}{3}$

(b) $\frac{6}{11} \cdot \frac{33}{21}$

(c) $\frac{20}{4} \cdot \frac{3}{40} \cdot \frac{1}{3}$

(d) $\frac{18}{17} \cdot \frac{1}{36} \cdot \frac{2}{3}$

(c) $\frac{35}{12} \cdot \frac{32}{25}$

$$\frac{\overset{7}{\cancel{35}}}{\underset{3}{\cancel{12}}} \cdot \frac{\overset{8}{\cancel{32}}}{\underset{5}{\cancel{25}}} = \frac{7 \cdot 8}{3 \cdot 5} = \frac{56}{15} \text{ or } 3\frac{11}{15} \quad \text{Mixed number}$$

(d) $\frac{2}{3} \cdot \frac{8}{15} \cdot \frac{3}{4}$

$$\frac{\overset{1}{\cancel{2}}}{\underset{1}{\cancel{3}}} \cdot \frac{\overset{4}{\cancel{8}}}{15} \cdot \frac{\overset{1}{\cancel{3}}}{\underset{\underset{1}{\cancel{2}}}{\cancel{4}}} = \frac{1 \cdot 4 \cdot 1}{1 \cdot 15 \cdot 1} = \frac{4}{15} \quad \text{Lowest terms}$$

This shortcut is especially helpful when the fractions involve large numbers.

NOTE

There is no specific order that must be used when dividing numerators and denominators as long as both numerator and denominator are divided by the same number.

Work Problem 3 at the Side.

3 Multiply a fraction and a whole number. The rule for multiplying a fraction and a whole number follows.

Multiplying a Whole Number and a Fraction

Multiply a whole number and a fraction by writing the whole number as a fraction with a denominator of 1.

For example, write the whole numbers 8, 10, and 25 as follows.

$$8 = \frac{8}{1}, \quad 10 = \frac{10}{1}, \quad \text{and} \quad 25 = \frac{25}{1}$$

Example 4 Multiplying by Whole Numbers

Multiply. Write answers in lowest terms and as whole numbers where possible.

(a) $8 \cdot \frac{3}{4}$

Write 8 as $\frac{8}{1}$ and multiply.

$$8 \cdot \frac{3}{4} = \frac{\overset{2}{\cancel{8}}}{1} \cdot \frac{3}{\underset{1}{\cancel{4}}} = \frac{2 \cdot 3}{1 \cdot 1} = \frac{6}{1} = 6$$

Continued on Next Page

ANSWERS

3. (a) $\frac{\overset{1}{\cancel{3}}}{\underset{2}{\cancel{4}}} \cdot \frac{\overset{1}{\cancel{2}}}{\underset{1}{\cancel{3}}} = \frac{1}{2}$ (b) $\frac{\overset{2}{\cancel{6}}}{\underset{1}{\cancel{11}}} \cdot \frac{\overset{3}{\cancel{33}}}{\underset{7}{\cancel{21}}} = \frac{6}{7}$

(c) $\frac{\overset{1}{\cancel{20}}}{4} \cdot \frac{\overset{1}{\cancel{3}}}{\underset{2}{\cancel{40}}} \cdot \frac{1}{\underset{1}{\cancel{3}}} = \frac{1}{8}$

(d) $\frac{\overset{1}{\cancel{18}}}{17} \cdot \frac{1}{\underset{\underset{1}{\cancel{2}}}{\cancel{36}}} \cdot \frac{\overset{1}{\cancel{2}}}{3} = \frac{1}{51}$

(b) $12 \cdot \frac{5}{6}$

$$12 \cdot \frac{5}{6} = \frac{\overset{2}{\cancel{12}}}{1} \cdot \frac{5}{\underset{1}{\cancel{6}}} = \frac{2 \cdot 5}{1 \cdot 1} = \frac{10}{1} = 10$$

Work Problem 4 at the Side.

4 Find the area of a rectangle. To find the area of a rectangle (the amount of surface inside the rectangle), use the following formula.

Area of a Rectangle

The area of a rectangle is equal to the length multiplied by the width.

area = length • width

For example, the rectangle shown here has an area of 12 square feet (ft^2).

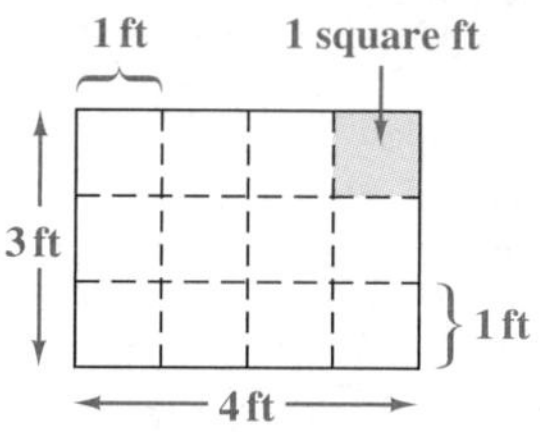

Area = length • width
Area = 4 ft • 3 ft
Area = 12 ft^2

(See Section 8.3 for more information on area.)

Example 5 **Applying Fraction Skills**

Find the area of each rectangle.

(a) Find the area of this floor tile.

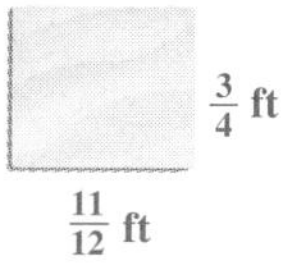

$$\text{Area} = \text{length} \cdot \text{width}$$

$$\text{Area} = \frac{11}{12} \cdot \frac{3}{4}$$

$$= \frac{11}{\underset{4}{\cancel{12}}} \cdot \frac{\overset{1}{\cancel{3}}}{4} \quad \text{Divide numerator and denominator.}$$

$$= \frac{11}{16}\ \text{ft}^2$$

Continued on Next Page

4 Multiply. Write answers in lowest terms.

(a) $5 \cdot \frac{1}{5}$

(b) $\frac{5}{3} \cdot 12 \cdot \frac{3}{4}$

(c) $40 \cdot \frac{3}{10}$

(d) $\frac{3}{25} \cdot \frac{5}{11} \cdot 99$

Answers

4. (a) 1 **(b)** 15 **(c)** 12 **(d)** $\frac{27}{5}$ or $5\frac{2}{5}$

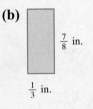

(b) Find the area of this wallpaper sample.

Multiply the length and width.

$$\text{area} = \frac{7}{9} \bullet \frac{3}{14}$$

$$= \frac{\overset{1}{\cancel{7}}}{\underset{3}{\cancel{9}}} \bullet \frac{\overset{1}{\cancel{3}}}{\underset{2}{\cancel{14}}} \qquad \text{Divide numerator and denominator.}$$

$$= \frac{1}{6} \text{ square yard } (\text{yd}^2)$$

Work Problem 5 at the Side.

5 Find the area of each rectangle.

(a)

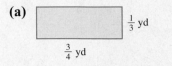

(b)

(c) a nature area that is $\frac{7}{5}$ miles by $\frac{5}{8}$ miles

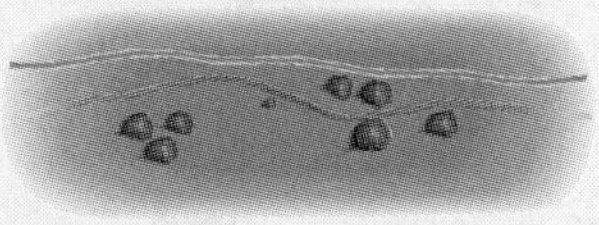

ANSWERS

5. **(a)** $\frac{1}{4}$ yd^2

(b) $\frac{7}{24}$ in.^{2}

(c) $\frac{7}{8}$ mi^2

2.5 EXERCISES

FOR EXTRA HELP: Student's Solutions Manual | MyMathLab.com | InterAct Math Tutorial Software | AW Math Tutor Center | www.mathxl.com | Digital Video Tutor CD 2 Videotape 5

Multiply. Write answers in lowest terms. See Examples 1–3.

1. $\frac{3}{4} \cdot \frac{1}{2}$

2. $\frac{1}{2} \cdot \frac{2}{3}$

3. $\frac{2}{3} \cdot \frac{5}{8}$

4. $\frac{2}{5} \cdot \frac{3}{4}$

5. $\frac{8}{5} \cdot \frac{15}{32}$

6. $\frac{4}{9} \cdot \frac{12}{7}$

7. $\frac{2}{3} \cdot \frac{7}{12} \cdot \frac{9}{14}$

8. $\frac{7}{8} \cdot \frac{16}{21} \cdot \frac{1}{2}$

9. $\frac{3}{4} \cdot \frac{5}{6} \cdot \frac{2}{3}$

10. $\frac{2}{5} \cdot \frac{3}{8} \cdot \frac{2}{3}$

11. $\frac{9}{22} \cdot \frac{11}{16}$

12. $\frac{5}{12} \cdot \frac{7}{10}$

13. $\frac{5}{8} \cdot \frac{16}{25}$

14. $\frac{6}{11} \cdot \frac{22}{15}$

15. $\frac{14}{25} \cdot \frac{65}{48} \cdot \frac{15}{28}$

16. $\frac{35}{64} \cdot \frac{32}{15} \cdot \frac{27}{72}$

17. $\frac{16}{25} \cdot \frac{35}{32} \cdot \frac{15}{64}$

18. $\frac{39}{42} \cdot \frac{7}{13} \cdot \frac{7}{24}$

Multiply. Write answers in lowest terms and as whole or mixed numbers where possible. See Example 4.

19. $6 \cdot \frac{5}{6}$

20. $40 \cdot \frac{3}{4}$

21. $\frac{5}{8} \cdot 64$

22. $\frac{3}{5} \cdot 45$

23. $28 \cdot \frac{3}{4}$

24. $30 \cdot \frac{3}{10}$

25. $36 \cdot \frac{5}{8} \cdot \frac{9}{15}$

26. $35 \cdot \frac{3}{5} \cdot \frac{1}{2}$

27. $100 \cdot \frac{21}{50} \cdot \frac{3}{4}$

28. $200 \cdot \frac{7}{8}$

29. $\frac{3}{5} \cdot 400$

30. $\frac{3}{7} \cdot 490$

31. $\frac{2}{3} \cdot 284$

32. $\frac{12}{25} \cdot 430$

33. $\frac{28}{21} \cdot 640 \cdot \frac{15}{32}$

34. $\frac{21}{13} \cdot 520 \cdot \frac{7}{20}$

35. $\frac{54}{38} \cdot 684 \cdot \frac{5}{6}$

36. $\frac{76}{43} \cdot 473 \cdot \frac{5}{19}$

Find the area of each rectangle. See Example 5.

37.

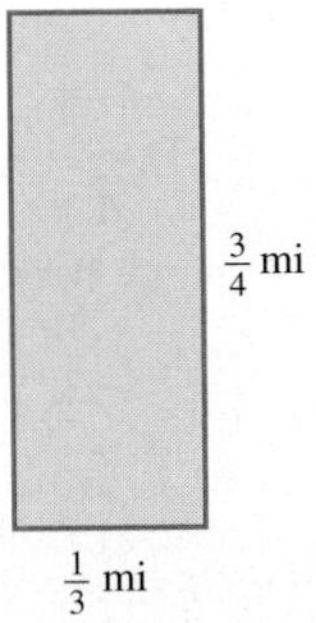

38.

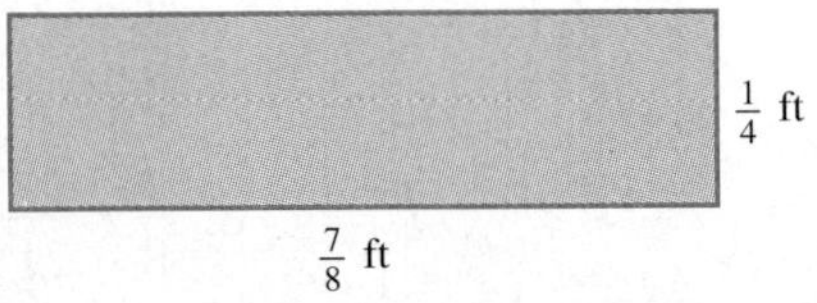

39.

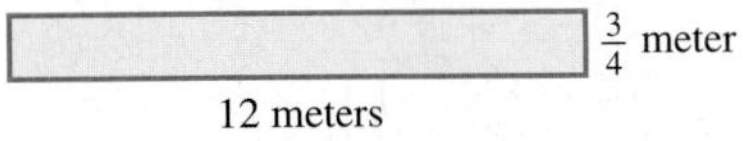

40.

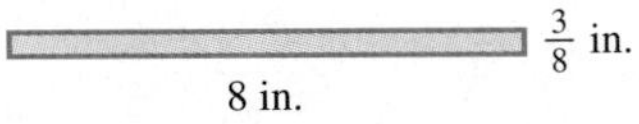

41.

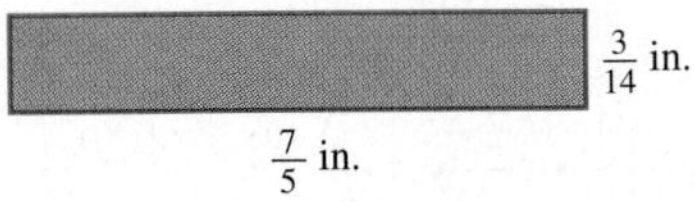

42.

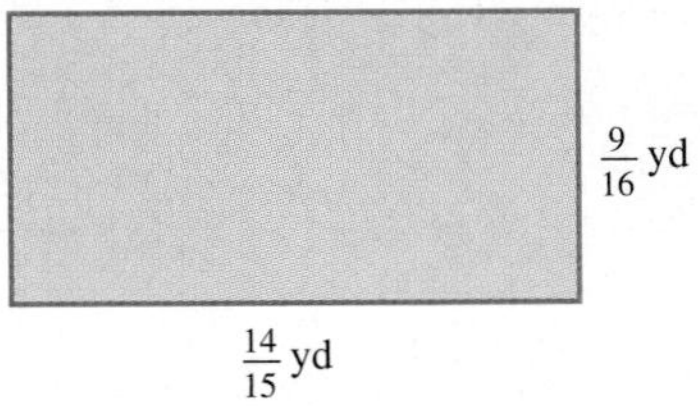

43. Write in your own words the rule for multiplying fractions. Make up an example problem to show how this works.

44. A useful shortcut when multiplying fractions is to divide a numerator and a denominator by the same number. Describe how this works and give an example.

Solve each application problem. Write answers in lowest terms and as whole or mixed numbers where possible. See Example 5.

45. Find the area of a heating-duct grill having a length of 2 yd and a width of $\frac{3}{4}$ yd.

46. Find the area of the top of a computer desk having a length of 2 yd and a width of $\frac{7}{8}$ yd.

47. A parcel of land measures $\frac{1}{2}$ mile by 2 miles. Find the total area of the parcel.

48. A motorcycle race course is $\frac{3}{4}$ mile wide by 6 miles long. Find the area of the race course.

49. Parking lot A is $\frac{1}{4}$ mile long and $\frac{3}{16}$ mile wide, while parking lot B is $\frac{3}{8}$ mile long and $\frac{1}{8}$ mile wide. Which parking lot has the larger area?

50. The Rocking Horse Ranch is $\frac{3}{4}$ mile long and $\frac{1}{2}$ mile wide. The Silver Spur Ranch is $\frac{5}{8}$ mile long and $\frac{4}{5}$ mile wide. Which ranch has the larger area?

Relating Concepts (Exercises 51–56) For Individual or Group Work

Front end rounding can be used to estimate an answer when multiplying fractions. ***Work Exercises 51–56 in order.***

The table shows the 10 best-selling vehicle models in the nation for the month of March.

MARCH 2000 TOP SELLERS
THE 10 BEST-SELLING VEHICLE MODELS

Vehicle	March Sales
Ford F-series pickup	90,522
Chevrolet C/K pickup	61,391
Ford Explorer	46,684
Ford Ranger	43,026
Ford Taurus	41,335
Dodge Caravan	37,457
Toyota Camry	36,076
Honda Accord	35,696
Dodge Ram	35,444
Ford Windstar	30,264

Source: *Autodata* April 4, 2000.

51. Use front end rounding to estimate the total sales for the 10 best-selling vehicles in the month of March.

52. Find the exact answer for total sales of the 10 best-selling vehicles sold in March using the exact numbers.

53. If $\frac{3}{4}$ of the Ford Explorers sold had 4-wheel drive, use front end rounding to estimate, and then find the exact number of Ford Explorers sold having 4-wheel drive.

54. If $\frac{3}{7}$ of the Dodge Rams sold had V-8 engines, use front end rounding to estimate, and then find the exact number of Dodge Rams sold having V-8 engines. Round to the nearest whole number.

Compare the estimated answers and the exact answers in Exercises 53 and 54. How can you get an estimated answer that is closer to the exact answer? Try rounding the number of vehicles to some multiple of the denominator in the fraction.

55. Refer to Exercise 53. Round the number of Ford Explorers to some multiple of the denominator in $\frac{3}{4}$. Now estimate the answer, showing your work.

56. Refer to Exercise 54. Round the number of Dodge Rams to some multiple of the denominator in $\frac{3}{7}$. Now estimate the answer, showing your work.

2.6 APPLICATIONS OF MULTIPLICATION

1 Solve fraction application problems using multiplication. Many application problems are solved by multiplying fractions. Use the following indicator words for multiplication.

product
double
triple
times
of
twice
twice as much

Look for these indicator words in the following examples.

Example 1 Applying Indicator Words

Lois Stevens gives $\frac{1}{10}$ of her income to her church. One month she earned \$1980. How much did she give to the church that month?

Step 1 **Read** the problem. The problem asks us to find the amount of money given to the church.

Step 2 **Work out a plan.** The indicator word is ***of:*** Stevens gave $\frac{1}{10}$ ***of*** her income. The word ***of*** indicates multiplication, so find the amount given to the church by multiplying $\frac{1}{10}$ and \$1980.

Step 3 **Estimate** a reasonable answer. Round the income of \$1980 to \$2000. Then divide \$2000 by 10 to find $\frac{1}{10}$ of the income (one of 10 equal parts). Our estimate is \$2000 ÷ 10 = \$200. (Recall the shortcut for dividing by 10; drop one 0 from the dividend.)

Step 4 **Solve** the problem.

$$\text{amount} = \frac{1}{\underset{1}{\cancel{10}}} \cdot \frac{\overset{198}{\cancel{1980}}}{1} = \frac{198}{1} = 198$$

Step 5 **State the answer.** Stevens gave \$198 to her church that month.

Step 6 **Check.** The answer, \$198, is close to our estimate of \$198.

Work Problem 1 at the Side.

Example 2 Solving a Fraction Application Problem

Of the 42 students in a biology class, $\frac{2}{3}$ went on a field trip. How many went on the trip?

Step 1 **Read** the problem. The problem asks us to find the number of students who went on the field trip.

Step 2 **Work out a plan.** Reword the problem to read

$\frac{2}{3}$ **of** the students went on a field trip.

↑ Indicator word for multiplication

Continued on Next Page

OBJECTIVE

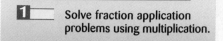

1 Solve fraction application problems using multiplication.

1 Solve each problem.

(a) Shafali Patel pays $\frac{1}{4}$ of her monthly salary as a house payment. If her monthly salary is \$4320, find her monthly house payment.

(b) A retiring police officer will receive $\frac{5}{8}$ of her highest annual salary as retirement income. If her highest annual salary is \$48,000, how much will she receive as retirement income?

ANSWERS

1. (a) \$1080 **(b)** \$30,000

2 At one pharmacy, $\frac{3}{16}$ of the prescriptions are paid by a third party (insurance company). If 2816 prescriptions are filled, find the number paid by a third party.
Use the six problem-solving steps.

Step 3 **Estimate** a reasonable answer. Round the number of students in the class from 42 to 40. Then, $\frac{1}{2}$ of 40 is 20. Since $\frac{2}{3}$ is more than $\frac{1}{2}$, our estimate is that "more than 20 students" went on the trip.

Step 4 **Solve** the problem. Find the number who went on the trip by multiplying $\frac{2}{3}$ and 42.

$$\text{number who went} = \frac{2}{3} \cdot 42$$

$$= \frac{2}{\cancel{3}_{1}} \cdot \frac{\cancel{42}^{14}}{1} = \frac{28}{1} = 28$$

Step 5 **State the answer.** 28 students went on the trip.

Step 6 **Check.** The answer, 28, fits our estimate of "more than 20."

Work Problem 2 at the Side.

Example 3 Finding a Fraction of a Fraction

In her will, a woman divides her estate into 6 equal parts. Five of the 6 parts are given to relatives. Of the sixth part, $\frac{1}{3}$ goes to the Salvation Army. What fraction of her total estate goes to the Salvation Army?

Step 1 **Read** the problem. The problem asks for the fraction of an estate that goes to the Salvation Army.

Step 2 **Work out a plan.** Reword the problem to read the Salvation Army gets $\frac{1}{3}$ **of** $\frac{1}{6}$.

↑ Indicator word for multiplication

3 In a certain community, $\frac{1}{3}$ of the residents speak a foreign language. Of those speaking a foreign language, $\frac{3}{4}$ speak Spanish. What fraction of the residents speak Spanish?
Use the six problem-solving steps.

Step 3 **Estimate** a reasonable answer. If the estate is divided into 6 equal parts and each of these parts was divided into 3 equal parts, we would have $6 \cdot 3 = 18$ equal parts. Our estimate is $\frac{1}{18}$.

Step 4 **Solve** the problem. The Salvation Army gets $\frac{1}{3}$ **of** $\frac{1}{6}$.

↑ Indicator word

To find the fraction that the Salvation Army is to receive, multiply $\frac{1}{3}$ and $\frac{1}{6}$.

$$\text{fraction to Salvation Army} = \frac{1}{3} \cdot \frac{1}{6}$$

$$= \frac{1}{18}$$

Step 5 **State the answer.** The Salvation Army gets $\frac{1}{18}$ of the total estate.

Step 6 **Check.** The answer, $\frac{1}{18}$, matches our estimate.

Work Problem 3 at the Side.

ANSWERS
2. 528 prescriptions
3. $\frac{1}{4}$ speak Spanish

Example 4 Using Fractions with a Circle Graph

The circle graph, or pie chart, shows how people in a survey answered the question "Are unidentified flying objects (UFOs) real or imaginary?" If 1500 people were in the survey, find the number who answered "not sure."

OUT OF THIS WORLD?
When asked in a survey if UFOs actually exist, respondents answered:

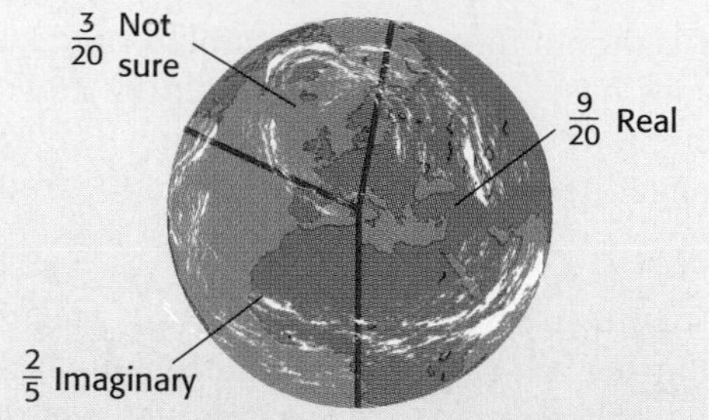

***Source*:** Yankelovich Partners for *Life* magazine.

Step 1 **Read** the problem. The problem asks for the number of people who answered "not sure" in the survey.

Step 2 **Work out a plan.** Reword the problem to read

$\frac{3}{20}$ **of** 1500 people answered "not sure."
↑
Indicator word for multiplication

Step 3 **Estimate** a reasonable answer. $\frac{1}{2}$ of 1500 people is 750 people. $\frac{3}{20}$ is less than $\frac{1}{2}$, so our estimate is "less than 750 people."

Step 4 **Solve** the problem. Find the number who were "not sure" by multiplying $\frac{3}{20}$ and 1500.

$$\text{number "not sure"} = \frac{3}{20} \cdot 1500$$

$$= \frac{3}{20} \cdot \frac{1500}{1}$$

$$= \frac{3}{\underset{1}{\cancel{\underset{\cancel{2}}{20}}}} \cdot \frac{\overset{75}{\cancel{\overset{\cancel{150}}{1500}}}}{1} \quad \text{Divide both numerator and denominator.}$$

$$= \frac{225}{1} = 225$$

Step 5 **State the answer.** 225 people were "not sure."

Step 6 **Check.** The answer, 225 people, fits our estimate of "less than 750 people."

Work Problem 4 at the Side.

4 Solve each problem using the six problem-solving steps. Use the circle graph in Example 4.

(a) What fraction of the people answered that UFOs are real?

(b) What number of people answered that UFOs are real?

(c) What fraction of the people answered that UFOs are imaginary?

(d) What number of people answered that UFOs are imaginary?

ANSWERS
4. (a) $\frac{9}{20}$ **(b)** 675 people **(c)** $\frac{2}{5}$
(d) 600 people

Focus on **Real-Data Applications**

Heart-Rate Training Zone

Performing aerobic exercise is beneficial both for improving aerobic fitness and for burning fat. For best results, you should keep your heart rate within the training zone for a minimum of 12 minutes. If you train at the higher end of the training zone, you will burn glycogen and improve aerobic fitness. Training for longer periods at the lower end of the training zone results in your body using fat reserves for energy.

Example: The training zone (TZ) is based on your heart rate (HR) for one minute. To see if you are in the training zone, measure your heart rate for 15 seconds. Compare it to the 15-second training zone. Find the exact answer, and then round to the nearest whole number.

Instruction	Calculation	Example (age 22)
Calculate maximum heart rate (MHR)	220 − your age	220 − 22 = 198
Calculate lower limit of training zone (TZ)	$\frac{3}{5} \times (\text{MHR})$	$\frac{3}{5} \times (198) = \frac{594}{5} = 118\frac{4}{5}$
Calculate upper limit of training zone (TZ)	$\frac{4}{5} \times (\text{MHR})$	$\frac{4}{5} \times (198) = \frac{792}{5} = 158\frac{2}{5}$
Calculate the exact 15-second training zone. Round the results to the nearest whole number.	$\left(\frac{1}{4} \times \text{lower TZ}, \frac{1}{4} \times \text{Upper TZ}\right)$	$\frac{1}{4} \times \frac{594}{5} = 29\frac{7}{10}; \frac{1}{4} \times \frac{792}{5} = 39\frac{3}{5}$ $29\frac{7}{10} < \text{HR} < 39\frac{3}{5}$ $30 < \text{HR} < 40$

Age	MHR	Lower Limit of TZ	Upper Limit of TZ	15-Second TZ (exact)	15-Second TZ (rounded)
18					
25					
30					
40					
50					
60					

1. Suppose you work in a physical fitness center and decide to design a poster to remind the clients of the training zone for their age. Compute the exact 15-second training zone for people of each of the following ages. Write fractions in lowest terms. Then round the answers to the nearest whole.
2. Explain why the lower and upper training zones (TZ) are multiplied by $\frac{1}{4}$.
3. Explain why the 15-second training zone is lower for a person aged 50 in comparison to a person aged 20.
4. Based on the chart, what would you tell a 45-year-old person about their 15-second training zone?

2.6 Exercises

FOR EXTRA HELP Student's Solutions Manual | MyMathLab.com | InterAct Math Tutorial Software | AW Math Tutor Center | www.mathxl.com | Digital Video Tutor CD 2 Videotape 5

Solve each application problem. Look for indicator words. See Examples 1–4.

1. A file cabinet top is $\frac{3}{4}$ yd by $\frac{2}{3}$ yd. Find its area.

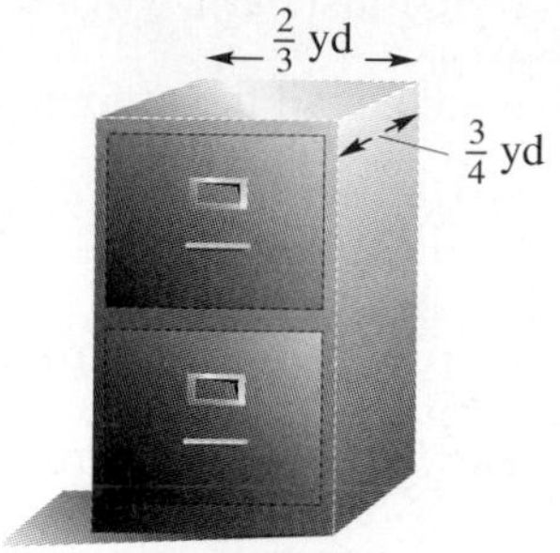

2. A dog bed is $\frac{7}{8}$ yd by $\frac{10}{9}$ yd. Find its area.

3. A cookie sheet is $\frac{4}{3}$ ft by $\frac{2}{3}$ ft. Find its area.

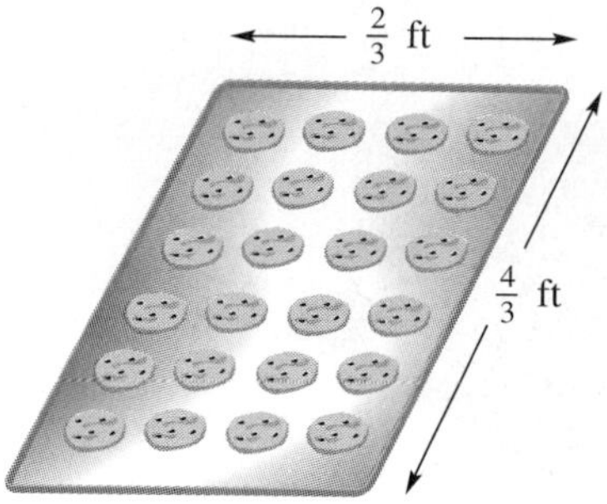

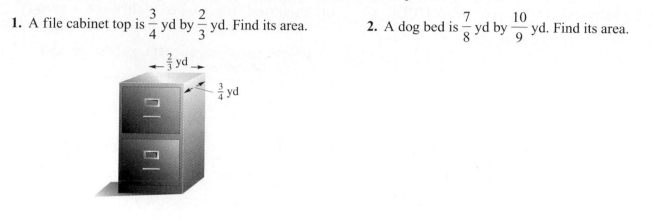

4. Al is helping Tim make a mahogany lamp table for Jill's birthday. Find the area of the top of the table if it is $\frac{4}{5}$ yd long by $\frac{3}{8}$ yd wide.

5. One-third of the players elected to the Baseball Hall of Fame were pitchers. If 183 players are in the Hall of Fame, how many were pitchers? (*Source:* National Baseball Hall of Fame.)

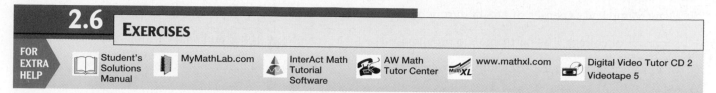

6. A minimarket sells 2500 items, of which $\frac{3}{25}$ are classified as junk food. How many of the items are junk food?

7. Dan Crump had expenses of \$6848 during one semester of college. His part-time job provided $\frac{3}{8}$ of the amount he needed. How much did he earn on his job?

8. Erin Hernandez produces \$5680 in profits for her employer. If her personal earnings are $\frac{2}{5}$ of these profits, find the amount of her earnings.

9. A school gives scholarships to $\frac{5}{24}$ of its 1800 freshmen. How many freshman students received scholarships?

10. Jason Todd estimates that it will cost him \$8400, including living expenses, to attend college full time for one year. If he must earn $\frac{3}{4}$ of the cost and borrow the balance, find the amount that he must earn.

11. At the Garlic Festival Fun Run, $\frac{5}{12}$ of the runners are women. If there are 780 runners, how many are women?

12. A hotel has 408 rooms. Of these rooms, $\frac{9}{17}$ are for nonsmokers. How many rooms are for nonsmokers?

In a recent year, Americans purchased one billion books. The circle graph shows the purchase of these books by age group over a one-year period. Use this information to work Exercises 13–18.

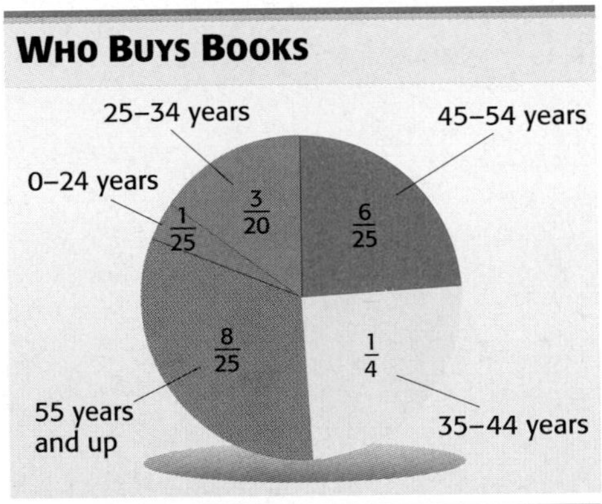

Source: Book Industry Study Group.

13. Which age group purchased the least number of books? How many did this group purchase?

14. Which age group purchased the greatest number of books? How many did this group purchase?

15. Find the number of books purchased by those in the 35–44 year age group.

16. Find the number of books purchased by those in the 45–54 age group.

17. Without actually adding the fractions given for all the age groups, explain why their sum has to be 1.

18. Refer to Exercise 17. Suppose you added all the fractions for the age groups and did not get 1 as an answer. List some possible explanations.

The table shows the earnings for the Gomes family last year and the circle graph shows how they spent their earnings. Use this information to answer Exercises 19–24.

Month	Earnings	Month	Earnings
January	\$3050	July	\$3160
February	\$2875	August	\$2355
March	\$3325	September	\$2780
April	\$3020	October	\$3675
May	\$2880	November	\$3310
June	\$3265	December	\$4305

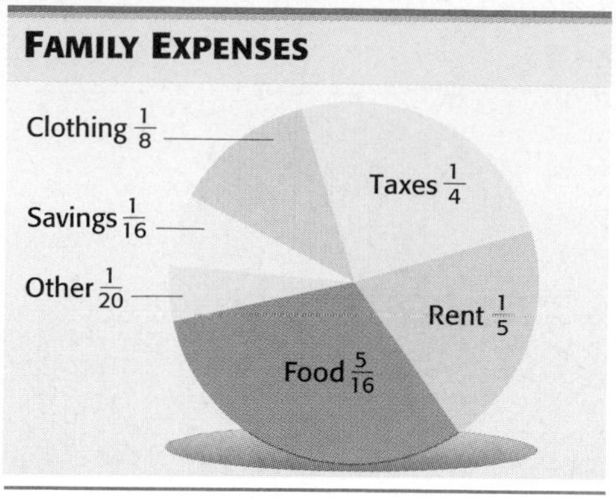

19. Find the Gomes family's total income for the year.

20. How much of their annual earnings went to taxes?

21. Find the amount of their rent for the year.

22. How much did they spend for food during the year?

23. Find their annual savings.

24. How much of their annual income was spent on clothing?

25. Here is how one student solved a multiplication problem. Find the error and solve the problem correctly.

$$\frac{9}{10} \times \frac{20}{21} = \frac{\overset{3}{\cancel{9}}}{\underset{1}{\cancel{10}}} \times \frac{\overset{2}{\cancel{20}}}{\underset{3}{\cancel{21}}} = \frac{6}{3} = 2$$

26. When two whole numbers are multiplied, the product is always larger than the numbers being multiplied. When two proper fractions are multiplied, the product is always smaller than the numbers being multiplied. Are these statements true? Why or why not?

Solve each application problem.

27. Pamela Denny is a waitress and earned \$112 in one 8-hour day. How much money did she earn in 3 hours?

28. Ruth Berry jogged 40 miles in 8 hours. How far did she jog in 5 hours?

29. LaDonna Washington is running for city council. She needs to get $\frac{2}{3}$ of her votes from senior citizens and 27,000 votes in all to win. How many votes does she need from voters other than the senior citizens?

30. The start-up cost of a Subs and Sandwich Shop is \$32,000. If the bank will loan you $\frac{9}{16}$ of the start up and you must pay the balance, how much more will you need to open a shop?

31. A will states that $\frac{7}{8}$ of an estate is to be divided among relatives. Of the remaining estate, $\frac{1}{4}$ goes to the American Cancer Society. What fraction of the estate goes to the American Cancer Society?

32. A couple has $\frac{1}{5}$ of their total investments in stocks. Of the remaining investment, $\frac{1}{8}$ is invested in bonds. What fraction of the total investment is invested in bonds?

2.7 Dividing Fractions

Objectives

1. Find the reciprocal of a fraction.
2. Divide fractions.
3. Solve application problems in which fractions are divided.

1 Find the reciprocal of a fraction. To divide fractions, we need to know how to find the **reciprocal** of a fraction.

Reciprocal of a Fraction

Two numbers are reciprocals of each other if their product is 1. To find the reciprocal of a fraction, interchange the numerator and denominator.

For example, the reciprocal of $\frac{3}{4}$ is $\frac{4}{3}$.

$$\text{Fraction} \quad \frac{3}{4} \times \frac{4}{3} \quad \text{Reciprocal}$$

NOTE

Notice that you invert, or "flip" a fraction to find its reciprocal.

Example 1 Finding Reciprocals

Find the reciprocal of each fraction.

(a) The reciprocal of $\frac{1}{4}$ is $\mathbf{\frac{4}{1}}$ because $\frac{1}{4} \cdot \frac{4}{1} = \frac{4}{4} = 1$.

(b) The reciprocal of $\frac{2}{3}$ is $\mathbf{\frac{3}{2}}$ because $\frac{2}{3} \cdot \frac{3}{2} = \frac{6}{6} = 1$.

(c) The reciprocal of $\frac{3}{5}$ is $\mathbf{\frac{5}{3}}$ because $\frac{3}{5} \cdot \frac{5}{3} = \frac{15}{15} = 1$.

(d) The reciprocal of 8 is $\mathbf{\frac{1}{8}}$ because $\frac{8}{1} \cdot \frac{1}{8} = \frac{8}{8} = 1$. Think of 8 as $\frac{8}{1}$.

Work Problem 1 at the Side.

NOTE

Every number has a reciprocal except 0. The number 0 has no reciprocal because there is no number that can be multiplied by 0 to get 1.

$$0 \cdot (\text{reciprocal}) = 1$$

↑ There is no number to use here that will give an answer of 1. When you multiply by 0, you always get 0.

1 Find the reciprocal of each fraction.

(a) $\frac{5}{8}$

(b) $\frac{2}{5}$

(c) $\frac{9}{4}$

(d) 5

Answers

1. (a) $\frac{8}{5}$ **(b)** $\frac{5}{2}$ **(c)** $\frac{4}{9}$ **(d)** $\frac{1}{5}$

In Chapter 1, we saw that the division problem $12 \div 3$ asks how many 3s are in 12. In the same way, the division problem $\frac{2}{3} \div \frac{1}{6}$ asks how many $\frac{1}{6}$s are in $\frac{2}{3}$. The figure illustrates $\frac{2}{3} \div \frac{1}{6}$.

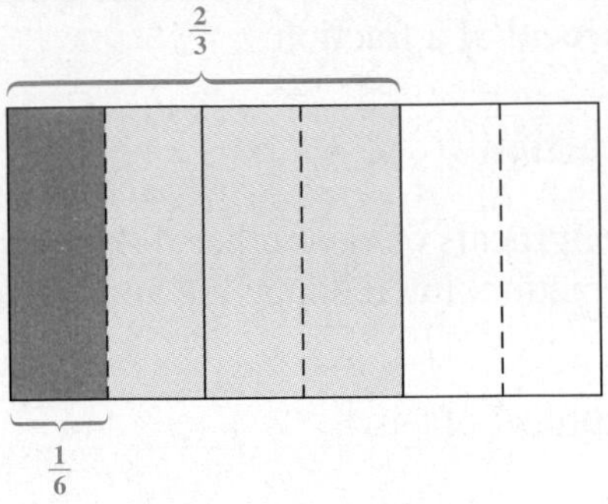

The figure shows that there are 4 of the $\frac{1}{6}$s in $\frac{2}{3}$, or

$$\frac{2}{3} \div \frac{1}{6} = 4.$$

2 **Divide fractions.** We will use reciprocals to divide fractions.

Dividing Fractions

To divide two fractions, multiply the first fraction by the reciprocal of the second fraction.

Example 2 Dividing One Fraction by Another

Divide. Write answers in lowest terms.

(a) $\frac{7}{8} \div \frac{15}{16}$

The reciprocal of $\frac{15}{16}$ is $\frac{16}{15}$.

Reciprocals

$$\frac{7}{8} \div \frac{15}{16} = \frac{7}{8} \cdot \frac{16}{15}$$

Change division to multiplication.

$$= \frac{7}{\cancel{8}_{1}} \cdot \frac{\cancel{16}^{2}}{15}$$ Divide the numerator and denominator by 8.

$$= \frac{7 \cdot 2}{1 \cdot 15}$$ Multiply.

$$= \frac{14}{15}$$

Continued on Next Page

(b) $\dfrac{\frac{4}{5}}{\frac{3}{10}}$

$$\frac{\frac{4}{5}}{\frac{3}{10}} = \frac{4}{5} \div \frac{3}{10}$$ Rewrite by using the ÷ symbol for division.

$$= \frac{\overset{1}{\cancel{4}}}{\cancel{5}_{1}} \cdot \frac{\overset{2}{\cancel{10}}}{3}$$ The reciprocal of $\frac{3}{10}$ is $\frac{10}{3}$. Change "÷" to "•", and divide the numerator and denominator by 5.

$$= \frac{4 \cdot 2}{1 \cdot 3}$$ Multiply.

$$= \frac{8}{3} = 2\frac{2}{3}$$ Mixed number

CAUTION

Be certain that the divisor fraction is changed to its reciprocal *before* you divide numerators and denominators by common factors.

Work Problem 2 at the Side.

Example 3 Dividing with a Whole Number

Divide. Write all answers in lowest terms and as whole or mixed numbers where possible.

(a) $5 \div \frac{1}{4}$

Write 5 as $\frac{5}{1}$. Next, use the reciprocal of $\frac{1}{4}$, which is $\frac{4}{1}$.

$$5 \div \frac{1}{4} = \frac{5}{1} \cdot \frac{4}{1}$$ Reciprocal of $\frac{1}{4}$ is $\frac{4}{1}$.

Reciprocals

$$= \frac{5 \cdot 4}{1 \cdot 1}$$ Multiply.

$$= \frac{20}{1} = 20$$ Whole number

Continued on Next Page

2 Divide. Write answers in lowest terms.

(a) $\frac{1}{2} \div \frac{3}{4}$

(b) $\frac{5}{8} \div \frac{7}{8}$

(c) $\dfrac{\frac{2}{3}}{\frac{4}{5}}$

(d) $\dfrac{\frac{5}{6}}{\frac{7}{12}}$

Answers

2. (a) $\frac{2}{3}$ (b) $\frac{5}{7}$ (c) $\frac{5}{6}$ (d) $1\frac{3}{7}$

3 Divide. Write answers in lowest terms and as whole or mixed numbers where possible.

(a) $12 \div \frac{3}{4}$

(b) $8 \div \frac{8}{9}$

(c) $\frac{4}{5} \div 6$

(d) $\frac{3}{8} \div 4$

4 Solve each problem using the six problem-solving steps.

(a) How many $\frac{3}{4}$-quart leaf-blower fuel tanks can be filled from 15 quarts of fuel?

(b) How many $\frac{2}{3}$-gallon garden sprayers can be filled from 36 gallons of insect spray?

ANSWERS

3. **(a)** 16 **(b)** 9 **(c)** $\frac{2}{15}$ **(d)** $\frac{3}{32}$

4. **(a)** 20 fuel tanks **(b)** 54 sprayers

(b) $\frac{2}{3} \div 6$

Write 6 as $\frac{6}{1}$. The reciprocal of $\frac{6}{1}$ is $\frac{1}{6}$.

$$\frac{2}{3} \div \frac{6}{1} = \frac{2}{3} \cdot \frac{1}{6}$$

Reciprocals

$$= \frac{\overset{1}{\cancel{2}}}{3} \cdot \frac{1}{\underset{3}{\cancel{6}}}$$ Divide the numerator and denominator by 2, then multiply.

$$= \frac{1 \cdot 1}{3 \cdot 3} = \frac{1}{9}$$

Work Problem 3 at the Side.

3 Solve application problems in which fractions are divided. Many application problems require division of fractions. Recall that typical indicator words for division are ***goes into, per, divide, divided by, divided equally,*** and ***divided into.***

Example 4 Applying Fraction Skills

Sophia, the manager of the Village Deli, must fill a 10-gallon dill pickle crock with salt brine. She has only a $\frac{2}{3}$-gallon container to use. How many times must she fill the $\frac{2}{3}$-gallon container and empty it into the 10-gallon crock?

Step 1 **Read** the problem. We need to find the number of times Sophia needs to use a $\frac{2}{3}$-gallon container in order to fill a 10-gallon crock.

Step 2 **Work out a plan.** We can solve the problem by finding the number of times 10 can be divided by $\frac{2}{3}$.

Step 3 **Estimate** a reasonable answer. Round $\frac{2}{3}$-gallon to 1-gallon. In order to fill the 10-gallon container, she would have to use the 1-gallon container 10 times, so our estimate is 10.

Step 4 **Solve** the problem.

Reciprocals

$$10 \div \frac{2}{3} = \frac{10}{1} \cdot \frac{3}{2}$$ Use the reciprocal, $\frac{3}{2}$, and change "÷" to "•".

$$= \frac{\overset{5}{\cancel{10}}}{1} \cdot \frac{3}{\underset{1}{\cancel{2}}}$$ Divide the numerator and denominator, and then multiply.

$$= \frac{15}{1} = 15$$

Step 5 **State the answer.** Sofia must fill the container 15 times.

Step 6 **Check.** The answer, 15 times, is close to our estimate of 10 times.

Work Problem 4 at the Side.

Example 5 Applying Fraction Skills

At the Happi-Time Day Care Center, $\frac{6}{7}$ of the total operating fund goes to classroom operation. If there are 18 classrooms, what fraction of the classroom operating amount does each classroom receive?

Step 1 **Read** the problem. Since $\frac{6}{7}$ of the total operating fund must be split into 18 parts, we must find the fraction of the classroom operating amount received by each classroom.

Step 2 **Work out a plan.** We must divide the fraction of the total operating fund going to classroom operation $\left(\frac{6}{7}\right)$ by the number of classrooms (18).

Step 3 **Estimate** a reasonable answer. Round $\frac{6}{7}$ to 1. If all of the operating expenses (1 whole) were divided between 18 classrooms, each classroom would receive $\frac{1}{18}$ of the operating expenses, our estimate.

Step 4 **Solve** the problem. We solve by dividing $\frac{6}{7}$ by 18.

$$\frac{6}{7} \div 18 = \frac{6}{7} \div \frac{18}{1}$$ The reciprocal of $\frac{18}{1}$ is $\frac{1}{18}$.

$$= \frac{\overset{1}{\cancel{6}}}{7} \cdot \frac{1}{\underset{3}{\cancel{18}}}$$ Use the reciprocal, change "÷" to "•", and divide the numerator and denominator by 6.

$$= \frac{1}{21}$$ Multiply.

Step 5 **State the answer.** Each classroom receives $\frac{1}{21}$ of the total operating fund.

Step 6 **Check.** The answer, $\frac{1}{21}$, is close to our estimate of $\frac{1}{18}$.

Work Problem 5 at the Side.

5 Solve each problem using the six problem-solving steps.

(a) The top 12 employees at Mayfield Manufacturing will divide $\frac{3}{4}$ of the annual bonus money. What fraction of the bonus money will each employee receive?

(b) The four top-performing students at Tulsa Community College will divide $\frac{1}{3}$ of the scholarship money awarded to students. What fraction of the scholarship money will each of these top students receive?

ANSWERS

5. (a) $\frac{1}{16}$ of the bonus money

(b) $\frac{1}{12}$ of the scholarship money

Focus on Real-Data Applications

Hotel Expenses

Mathematics teachers attending conferences in New Orleans, Louisiana, and San Jose, California, found the following information about hotel rates on the organizations' Internet Web sites.

Hotel	Single	Double	Triple	Quad	Suites
New Orleans (January 2001)					
Marriott	\$116	\$121	\$121	\$121	\$603
Sheraton (Club)	\$157	\$169	\$194	\$219	\$598
San Jose (February 2001)					
Crowne Plaza	\$157	\$157	\$167	\$177	—
Hilton Towers	\$167	\$187	\$207	\$227	—
Hyatt Sainte Claire	\$156	\$176	\$196	\$216	—

1. The double rate is for two people sharing a room. What fractional part does each person pay?
2. (a) Multiply the double rate at the Hilton Towers in San Jose by $\frac{1}{2}$. What is the result?
   (b) Divide the double rate at the Hilton Towers in San Jose by 2. What is the result?
   (c) Explain what happened. How much money would one person owe if he or she shared a double room at the Hilton Towers in San Jose?
3. The triple rate is for three people sharing a room. What fractional part does each person pay?
4. How much money would one person owe if he or she shared a triple room at the Hyatt Sainte Claire in San Jose? How much money would each person save if they could book a triple room at the Crowne Plaza instead of the Hyatt Sainte Claire? Find your answer using two different methods, based on your observations in Problem 2.
5. The quad rate is for four people sharing a room. What fractional part does each person pay?
6. How much money would one person owe if he or she shared a quad room at the Sheraton in New Orleans? Find your answer using two different methods, based on your observations in Problem 2.
7. How many people would have to share a suite at the Sheraton in New Orleans for the cost per person to be less than sharing a quad room at the same hotel? Round the answer to the nearest whole number. (*Hint:* Estimate the cost per person for a quad room first. Then estimate the number of people needed to share the cost of the suite. Check your work using actual values.)
8. Suppose you have a travel allotment of \$500 that can be spent on transportation, hotel, food, and registration fees. You and a colleague decide to attend the 3-day New Orleans conference and plan to share a room at the Marriott. Registration costs \$150; the flight costs \$129 round-trip; the taxi ride to the hotel costs \$20 per person each way; and you budget \$35 per day for meals. How much out-of-pocket expense will you have to pay? How much would you save if you could recruit a third person to share the room?

2.7 EXERCISES

FOR EXTRA HELP: Student's Solutions Manual | MyMathLab.com | InterAct Math Tutorial Software | AW Math Tutor Center | www.mathxl.com | Digital Video Tutor CD 2 Videotape 5

Find the reciprocal of each number. See Example 1.

1. $\frac{2}{3}$

2. $\frac{3}{4}$

3. $\frac{8}{5}$

4. $\frac{12}{7}$

5. $\frac{5}{6}$

6. $\frac{13}{20}$

7. 4

8. 10

Divide. Write answers in lowest terms and as whole or mixed numbers where possible. See Examples 2 and 3.

9. $\frac{1}{4} \div \frac{3}{4}$

10. $\frac{3}{8} \div \frac{5}{8}$

11. $\frac{7}{8} \div \frac{1}{3}$

12. $\frac{7}{8} \div \frac{3}{4}$

13. $\frac{3}{4} \div \frac{5}{3}$

14. $\frac{4}{5} \div \frac{9}{4}$

15. $\frac{7}{9} \div \frac{7}{36}$

16. $\frac{5}{8} \div \frac{5}{16}$

17. $\frac{15}{32} \div \frac{5}{64}$

18. $\frac{7}{12} \div \frac{14}{15}$

19. $\dfrac{\frac{13}{20}}{\frac{4}{5}}$

20. $\dfrac{\frac{9}{10}}{\frac{3}{5}}$

21. $\dfrac{\frac{5}{6}}{\frac{25}{24}}$

22. $\dfrac{\frac{28}{15}}{\frac{21}{5}}$

23. $12 \div \frac{2}{3}$

24. $7 \div \frac{1}{4}$

25. $\dfrac{15}{\frac{2}{3}}$

26. $\dfrac{9}{\frac{3}{4}}$

27. $\dfrac{\frac{4}{7}}{8}$

28. $\dfrac{\frac{7}{10}}{3}$

Solve each application problem by using division. See Examples 4 and 5.

29. Ms. Shaffer has a piece of property with an area that is $\frac{8}{9}$ acre. She wishes to divide it into 4 equal parts for her children. How many acres of land will each child get?

30. The Sweepstakes Lottery pays out $\frac{7}{8}$ of the total revenue to 14 top winners. What fraction of the total revenue does each winner receive?

31. Some college roommates want to make pancakes for their neighbors. They need 5 cups of flour, but have only a $\frac{1}{3}$-cup measuring cup. How many times will they need to fill their measuring cup?

32. Robert Cockrill has 10 quarts of lubricating oil. If each lubricating reservoir in a lathe holds $\frac{1}{3}$ quart of oil, how many reservoirs can be filled?

33. How many $\frac{1}{8}$-ounce eye drop dispensers can be filled with 11 ounces of eye drops?

34. It is estimated that each guest at a party will eat $\frac{5}{16}$ lb of peanuts. How many guests may be served with 10 lb of peanuts?

35. Pam Trizlia had a small pickup truck that could carry $\frac{2}{3}$-cord of firewood. Find the number of trips needed to deliver 40 cords of wood.

36. Manuel Servin has a 200-yd roll of weather stripping material. Find the number of pieces of weather stripping $\frac{5}{8}$ yd in length that may be cut from the roll.

37. A batch of double chocolate chip cookies requires $\frac{3}{4}$ lb of chocolate chips. If you have 9 lb of chocolate chips, how many batches of cookies can you make?

38. An upholsterer uses $\frac{7}{8}$ lb of brass tacks to reupholster one bar stool. How many bar stools can she reupholster with 21 lb of brass tacks?

39. Your classmate is confused on how to divide by a fraction. Write a short note telling him how this should be done.

40. If you multiply positive proper fractions, the product is smaller than the fractions multiplied. When you divide by a proper fraction, is the quotient smaller than the numbers in the problem? Prove your answer with examples.

41. Mike and Charlie have completed $\frac{4}{5}$ of their river rafting excursion down the Colorado River. If they have gone 304 miles so far, find the number of miles remaining in the excursion.

42. Sheila has been working on a job for 63 hours. The job is $\frac{7}{9}$ finished. How many *more* hours must she work to finish the job?

43. The Bridge Lighting Committee has raised $\frac{7}{8}$ of the funds necessary for their lighting project. If this amounts to \$840,000, how much additional money must be raised?

44. The school bond committee has raised $\frac{9}{16}$ of the funds needed to promote the bond measure. If the amount raised so far is \$45,000, how much additional money must be raised?

RELATING CONCEPTS (Exercises 45–50) FOR INDIVIDUAL OR GROUP WORK

Many application problems are solved using multiplication and division of fractions.
Work Exercises 45–50 in order.

45. Perhaps the most common indicator word for multiplication is the word *of.* Circle the words in the list below that are also indicator words for multiplication.

more than	per
double	twice
times	product
less than	difference
equals	twice as much

46. Circle the words in the list below that are indicator words for division.

fewer	sum of
goes into	divide
per	quotient
equals	double
loss of	divided by

47. To divide two fractions, multiply the first fraction by the ____________ of the second fraction.

48. Find the reciprocals for each number.

$\frac{3}{4}$; $\frac{7}{8}$; 5; $\frac{12}{19}$

The size of a square postage stamp is shown here. Use this to answer Exercises 49 and 50.

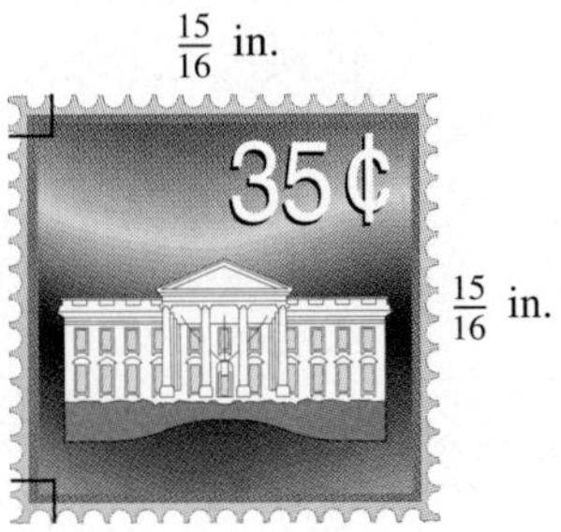

49. Find the perimeter (distance around the outside edges) of the postage stamp. Explain how to find the perimeter of any 3-, 4-, 5-, or 6-sided figure.

50. Find the area of the postage stamp. Explain how to find the area of any square.

2.8 Multiplying and Dividing Mixed Numbers

In Section 2.2 we worked with mixed numbers—a whole number and a fraction written together. Many of the fraction problems you encounter in everyday life involve mixed numbers.

Objectives

1. Estimate the answer and multiply mixed numbers.
2. Estimate the answer and divide mixed numbers.
3. Solve application problems with mixed numbers.

1 Estimate the answer and multiply mixed numbers. When multiplying mixed numbers, it is a good idea to estimate the answer first. Then multiply the mixed numbers by using the following steps.

Multiplying Mixed Numbers

Step 1 *Change* each mixed number to an improper fraction.

Step 2 *Multiply* as fractions.

Step 3 Simplify the answer, which means to write it in *lowest terms*, and change it to a mixed number or whole number where possible.

To estimate the answer, round each mixed number to the nearest whole number. If the numerator is *half* of the denominator or *more,* round up the whole number part. If the numerator is *less* than half the denominator, leave the whole number as it is.

$$1\frac{5}{8} \quad \begin{matrix} \leftarrow \text{5 is more than 4.} \\ \leftarrow \text{Half of 8 is 4.} \end{matrix} \quad 1\frac{5}{8} \text{ rounds up to 2}$$

$$3\frac{2}{5} \quad \begin{matrix} \leftarrow \text{2 is less than } 2\frac{1}{2}. \\ \leftarrow \text{Half of 5 is } 2\frac{1}{2}. \end{matrix} \quad 3\frac{2}{5} \text{ rounds to 3}$$

Work Problem 1 at the Side.

1 Round each mixed number to the nearest whole number.

(a) $3\frac{2}{3}$

(b) $5\frac{2}{5}$

(c) $2\frac{3}{4}$

(d) $4\frac{7}{12}$

(e) $6\frac{1}{2}$

(f) $1\frac{4}{9}$

Example 1 Multiplying Mixed Numbers

First estimate the answer. Then multiply to get an exact answer. Simplify your answers.

(a) $2\frac{1}{2} \cdot 3\frac{1}{5}$

Estimate the answer by rounding the mixed numbers.

$$2\frac{1}{2} \text{ rounds to 3} \quad \text{and} \quad 3\frac{1}{5} \text{ rounds to 3}$$

$$3 \cdot 3 = 9 \quad \text{Estimated answer}$$

To find the exact answer, change each mixed number to an improper fraction.

Step 1 $2\frac{1}{2} = \frac{5}{2}$ and $3\frac{1}{5} = \frac{16}{5}$

Continued on Next Page

Answers

1. **(a)** 4 **(b)** 5 **(c)** 3 **(d)** 5 **(e)** 7 **(f)** 1

2 First estimate the answer. Then multiply to find the exact answer. Write answers in lowest terms. Simplify your answers.

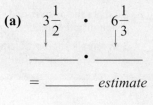

(a) $3\frac{1}{2} \cdot 6\frac{1}{3}$

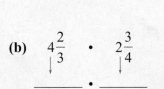

$\underline{\quad\quad} \cdot \underline{\quad\quad}$

$= \underline{\quad\quad}$ *estimate*

(b) $4\frac{2}{3} \cdot 2\frac{3}{4}$

$\underline{\quad\quad} \cdot \underline{\quad\quad}$

$= \underline{\quad\quad}$ *estimate*

(c) $3\frac{3}{5} \cdot 4\frac{4}{9}$

$\underline{\quad\quad} \cdot \underline{\quad\quad}$

$= \underline{\quad\quad}$ *estimate*

(d) $5\frac{1}{4} \cdot 3\frac{2}{5}$

$\underline{\quad\quad} \cdot \underline{\quad\quad}$

$= \underline{\quad\quad}$ *estimate*

Next, multiply.

$$2\frac{1}{2} \cdot 3\frac{1}{5} = \frac{5}{2} \cdot \frac{16}{5} = \frac{\overset{1}{\cancel{5}}}{\underset{1}{\cancel{2}}} \cdot \frac{\overset{8}{\cancel{16}}}{\underset{1}{\cancel{5}}} = \underbrace{\frac{1 \cdot 8}{1 \cdot 1}}_{\text{Step 2}} = \underbrace{\frac{8}{1} = 8}_{\text{Step 3}}$$

The estimated answer is 9 and the exact answer is 8. The exact answer is reasonable.

(b) $3\frac{5}{8} \cdot 4\frac{4}{5}$

$3\frac{5}{8}$ rounds to 4 and $4\frac{4}{5}$ rounds to 5

$4 \cdot 5 = 20$ Estimated answer

Now find the exact answer.

$$3\frac{5}{8} \cdot 4\frac{4}{5} = \underbrace{\frac{29}{8} \cdot \frac{24}{5}}_{\text{Step 1}} = \frac{29}{\underset{1}{\cancel{8}}} \cdot \frac{\overset{3}{\cancel{24}}}{5} = \underbrace{\frac{29 \cdot 3}{1 \cdot 5} = \frac{87}{5}}_{\text{Step 2}}$$

As a mixed number,

$$\underbrace{\frac{87}{5} = 17\frac{2}{5}}_{\text{Step 3}}. \quad \text{Simplified answer}$$

The estimate was 20, so the exact answer is reasonable.

(c) $1\frac{3}{5} \cdot 3\frac{1}{3}$

$1\frac{3}{5}$ rounds to 2 and $3\frac{1}{3}$ rounds to 3

$2 \cdot 3 = 6$ Estimated answer

The exact answer is

$$1\frac{3}{5} \cdot 3\frac{1}{3} = \frac{8}{\underset{1}{\cancel{5}}} \cdot \frac{\overset{2}{\cancel{10}}}{3} = \frac{8 \cdot 2}{1 \cdot 3} = \frac{16}{3} = 5\frac{1}{3}.$$

The estimate was 6, so the exact answer is reasonable.

Work Problem 2 at the Side.

2 Estimate the answer and divide mixed numbers. Just as you did when multiplying mixed numbers, it is also a good idea to estimate the answer when dividing mixed numbers. To divide mixed numbers, use the following steps.

ANSWERS

2. (a) *Estimate:* $4 \cdot 6 = 24$; *Exact:* $22\frac{1}{6}$

(b) *Estimate:* $5 \cdot 3 = 15$; *Exact:* $12\frac{5}{6}$

(c) *Estimate:* $4 \cdot 4 = 16$; *Exact:* 16

(d) *Estimate:* $5 \cdot 3 = 15$; *Exact:* $17\frac{17}{20}$

Dividing Mixed Numbers

Step 1 *Change* each mixed number to an improper fraction.

Step 2 Use the *reciprocal* of the second fraction (divisor).

Step 3 *Multiply.*

Step 4 Simplify the answer, which means to write it in *lowest terms*, and change it to a mixed number or whole number where possible.

NOTE

Recall that the reciprocal of a fraction is found by interchanging the numerator and the denominator.

Example 2 Dividing Mixed Numbers

First estimate the answer. Then divide to find the exact answer. Simplify your answers.

(a) $2\frac{2}{5} \div 1\frac{1}{2}$

First estimate the answer by rounding each mixed number to the nearest whole number.

$$2\frac{2}{5} \div 1\frac{1}{2}$$

Rounded

$$2 \div 2 = 1 \quad \text{Estimated answer}$$

To find the exact answer, first change each mixed number to an improper fraction.

$$2\frac{2}{5} \div 1\frac{1}{2} = \underbrace{\frac{12}{5} \div \frac{3}{2}}_{\textit{Step 1}}$$

Next, use the reciprocal of the second fraction and multiply.

$$\frac{12}{5} \div \frac{3}{2} = \underbrace{\frac{\overset{4}{\cancel{12}}}{5} \cdot \frac{2}{\underset{1}{\cancel{3}}}}_{\textit{Step 2}} = \underbrace{\frac{4 \cdot 2}{5 \cdot 1}}_{\textit{Step 3}} = \underbrace{\frac{8}{5} = 1\frac{3}{5}}_{\textit{Step 4}} \quad \text{Simplified answer}$$

Reciprocals ($\frac{3}{2}$ and $\frac{2}{3}$)

The estimate was 1, so the exact answer is reasonable.

Continued on Next Page

3 First estimate the answer. Then divide to find the exact answer. Simplify all answers.

(a) $3\frac{1}{8} \div 6\frac{1}{4}$

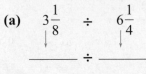

$\_\_\_\_ \div \_\_\_\_$

$= \_\_\_\_$ *estimate*

(b) $5\frac{1}{3} \div 1\frac{1}{4}$

$\_\_\_\_ \div \_\_\_\_$

$= \_\_\_\_$ *estimate*

(c) $8 \div 5\frac{1}{3}$

$\_\_\_\_ \div \_\_\_\_$

$= \_\_\_\_$ *estimate*

(d) $13\frac{1}{2} \div 18$

$\_\_\_\_ \div \_\_\_\_$

$= \_\_\_\_$ *estimate*

(b) $8 \div 3\frac{3}{5}$

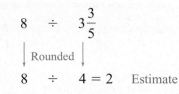

$8 \div 3\frac{3}{5}$ — Rounded — $8 \div 4 = 2$ Estimate

Now find the exact answer.

$$8 \div 3\frac{3}{5} = \frac{8}{1} \div \frac{18}{5} = \frac{\overset{4}{\cancel{8}}}{1} \cdot \frac{5}{\underset{9}{\cancel{18}}} = \frac{20}{9} = 2\frac{2}{9}$$

Write 8 as $\frac{8}{1}$. Reciprocals

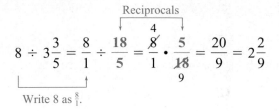

The estimate was 2, so the exact answer is reasonable.

(c) $4\frac{3}{8} \div 5$

$4\frac{3}{8} \div 5$ — Rounded — $4 \div 5 = \frac{4}{1} \div \frac{5}{1} = \frac{4}{1} \cdot \frac{1}{5} = \frac{4}{5}$ Estimate (Reciprocals)

The exact answer is

$$4\frac{3}{8} \div 5 = \frac{35}{8} \div \frac{5}{1} = \frac{\overset{7}{\cancel{35}}}{8} \cdot \frac{1}{\underset{1}{\cancel{5}}} = \frac{7}{8}.$$

Reciprocals. Write 5 as $\frac{5}{1}$.

The estimate was $\frac{4}{5}$, so the exact answer is reasonable.

Work Problem 3 at the Side.

3 Solve application problems with mixed numbers. The next two examples show how to solve application problems involving mixed numbers.

Example 3 Applying Multiplication Skills

The local Habitat for Humanity chapter is looking for 11 contractors who will each donate $3\frac{1}{4}$ days of labor to a community building project. How many days of labor will be donated in all?

Step 1 **Read** the problem. The problem asks for the total days of labor donated by the 11 contractors.

Step 2 **Work out a plan.** Multiply the number of contractors (11) and the amount of labor that each donates ($3\frac{1}{4}$ days).

Continued on Next Page

Answers

3. (a) *Estimate:* $3 \div 6 = \frac{1}{2}$; *Exact:* $\frac{1}{2}$

(b) *Estimate:* $5 \div 1 = 5$; *Exact:* $4\frac{4}{15}$

(c) *Estimate:* $8 \div 5 = 1\frac{3}{5}$; *Exact:* $1\frac{1}{2}$

(d) *Estimate:* $14 \div 18 = \frac{7}{9}$; *Exact:* $\frac{3}{4}$

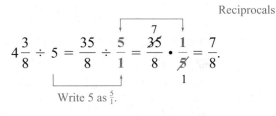

Step 3 **Estimate** a reasonable answer. Round $3\frac{1}{4}$ days to 3 days. Multiply 3 days by 11 contractors $(3 \cdot 11)$ to get an estimate of 33 days.

Step 4 **Solve** the problem. Find the exact answer.

$$11 \cdot 3\frac{1}{4} = 11 \cdot \frac{13}{4}$$

$$= \frac{11}{1} \cdot \frac{13}{4} = \frac{143}{4} = 35\frac{3}{4}$$

Step 5 **State the answer.** The community building project will receive $35\frac{3}{4}$ days of donated labor.

Step 6 **Check.** The answer, $35\frac{3}{4}$ days, is close to our estimate of 33 days.

Work Problem 4 at the Side.

Example 4 **Applying Division Skills**

A dome tent for backpacking requires $7\frac{1}{4}$ yd of nylon material. How many tents can be made from $65\frac{1}{4}$ yd of material?

Step 1 **Read** the problem. The problem asks how many tents can be made from $65\frac{1}{4}$ yd of material.

Step 2 **Work out a plan.** Divide the number of yards of cloth $\left(65\frac{1}{4}\text{ yd}\right)$ by the number of yards needed for one tent $\left(7\frac{1}{4}\text{ yd}\right)$.

Step 3 **Estimate** a reasonable answer.

$$65\frac{1}{4} \div 7\frac{1}{4}$$

Rounded

$$65 \div 7 \approx 9 \text{ tents} \quad \text{Estimate}$$

Step 4 **Solve** the problem. The exact answer is

$$65\frac{1}{4} \div 7\frac{1}{4} = \frac{261}{4} \div \frac{29}{4}$$

$$= \frac{\overset{9}{\cancel{261}}}{\underset{1}{\cancel{4}}} \cdot \frac{\overset{1}{\cancel{4}}}{\underset{1}{\cancel{29}}} = \frac{9}{1} = 9. \quad \text{Matches estimate}$$

Step 5 **State the answer.** 9 tents can be made from $65\frac{1}{4}$ yd of cloth.

Step 6 **Check.** The answer, 9, is close to our estimate.

Work Problem 5 at the Side.

NOTE

When rounding mixed numbers to estimate the answer to a problem, the estimated answer usually varies somewhat from the exact answer. However, the importance of the estimated answer is that it will show you whether your exact answer is reasonable or not.

4 Use the six problem-solving steps. Simplify all answers.

(a) If one hotel room requires $1\frac{7}{8}$ gallons of paint, find the number of gallons needed for 9 hotel rooms.

(b) Clare earns $\$9\frac{1}{4}$ per hour. How much would she earn in $6\frac{1}{2}$ hours? Write the answer as a mixed number.

5 Use the six problem-solving steps. Simplify all answers.

(a) The manufacture of one outboard engine propeller requires $4\frac{3}{4}$ lb of brass. How many propellers can be manufactured from 57 lb of brass?

(b) Student help is paid $\$6\frac{1}{4}$ per hour. Find the number of hours of student help that can be paid for with $150.

ANSWERS

4. (a) *Estimate:* $2 \cdot 9 = 18$; *Exact:* $16\frac{7}{8}$ gal

(b) *Estimate:* $9 \cdot 7 = 63$; *Exact:* $\$60\frac{1}{8}$

5. (a) *Estimate:* $57 \div 5 \approx 11$; *Exact:* 12 propellers

(b) *Estimate:* $150 \div 6 = 25$; *Exact:* 24 hr

Focus on Real-Data Applications

Recipes

The side of the corn starch box shown has useful recipes for Fun-Time Dough and for Great Gravy. Suppose you have made the recipe before and know from experience that 1 pound makes enough Fun-Time Dough for three children.

ARGO

CORN STARCH

FAVORITE RECIPES

Fun-Time Dough

$1\frac{1}{2}$ cups Argo Corn Starch
$\frac{1}{2}$ cup flour
2 cups water
2 tsp cream of tartar
1 cup salt
1 T. vegetable oil

Mix all ingredients together in saucepan. Cook over medium heat, stirring constantly, until mixture gathers on the stirring spoon and forms dough. This will take about 6 minutes. Dump onto waxed paper until cool enough to handle and knead to form a pliable mass. Store in covered container or plastic bag. Food coloring may be added to make different colors.
Makes about 2 lbs. of Fun-Time Dough.

Great Gravy

3 T. bacon fat or meat drippings
2 T. Argo Corn Starch
$1\frac{1}{2}$ cups water
$\frac{3}{4}$ tsp. salt
$\frac{1}{8}$ tsp. pepper

Blend fat and Argo Corn Starch over low heat until it is a rich brown color, stirring constantly. Gradually add water, salt, and pepper. Heat to boiling over direct heat and then boil gently 2 minutes, stirring constantly.
Makes $1\frac{1}{2}$ cups.

Satisfaction Guaranteed

1. If you make the recipe as written, you will have enough Fun-Time Dough for how many children?

2. Suppose you work in a day care center. How many pounds of Fun-Time Dough will you need for nine children?

3. If you double the recipe, you would have 4 pounds of dough. By what fraction should you multiply each ingredient amount to make 3 pounds of dough?

4. Fill in the blanks with the ingredient amounts needed to make 3 pounds of Fun-Time Dough. Show your work.

Corn starch: _______ Flour: _______

Water: _______ Cream of tartar: _______

Salt: _______ Vegetable oil: _______

You decide to make Great Gravy for Thanksgiving dinner.

5. How much gravy does the recipe make?

6. Suppose you decide to double the recipe. By what factor will you change the ingredient amounts?

7. You did not make enough! Suppose you decide to halve the recipe to make more gravy. Now, by what factor will you change the ingredient amounts?

8. If you need 4 cups of gravy, by what factor must you multiply each ingredient amount? Explain how you determined your answer.

2.8 EXERCISES

First estimate the answer. Then multiply to find the exact answer. Simplify all answers. *See Example 1.*

1. *Exact:* $3\frac{1}{4} \cdot 2\frac{1}{2}$

Estimate: ____ • ____ = ____

2. *Exact:* $3\frac{1}{2} \cdot 1\frac{1}{4}$

Estimate: ____ • ____ = ____

3. *Exact:* $1\frac{2}{3} \cdot 2\frac{7}{10}$

Estimate: ____ • ____ = ____

4. *Exact:* $4\frac{1}{2} \cdot 2\frac{1}{4}$

Estimate: ____ • ____ = ____

5. *Exact:* $3\frac{1}{9} \cdot 1\frac{2}{7}$

Estimate: ____ • ____ = ____

6. *Exact:* $6\frac{1}{4} \cdot 3\frac{1}{5}$

Estimate: ____ • ____ = ____

7. *Exact:* $8 \cdot 6\frac{1}{4}$

Estimate: ____ • ____ = ____

8. *Exact:* $6 \cdot 2\frac{1}{3}$

Estimate: ____ • ____ = ____

9. *Exact:* $4\frac{1}{2} \cdot 2\frac{1}{5} \cdot 5$

Estimate: ____ • ____ • ____ = ____

10. *Exact:* $5\frac{1}{2} \cdot 1\frac{1}{3} \cdot 2\frac{1}{4}$

Estimate: ____ • ____ • ____ = ____

11. *Exact:* $3 \cdot 1\frac{1}{2} \cdot 2\frac{2}{3}$

Estimate: ____ • ____ • ____ = ____

12. *Exact:* $\frac{2}{3} \cdot 3\frac{2}{3} \cdot \frac{6}{11}$

Estimate: ____ • ____ • ____ = ____

First estimate the answer. Then divide to find the exact answer. Simplify all answers. *See Example 2.*

13. *Exact:* $2\frac{1}{2} \div 7\frac{1}{2}$

Estimate: ____ ÷ ____ = ____

14. *Exact:* $1\frac{1}{8} \div 2\frac{1}{4}$

Estimate: ____ ÷ ____ = ____

15. *Exact:* $2\frac{1}{2} \div 3$

Estimate: ____ ÷ ____ = ____

16. *Exact:* $2\frac{3}{4} \div 2$

Estimate: ____ ÷ ____ = ____

17. *Exact:* $9 \div 2\frac{1}{2}$

Estimate: ____ ÷ ____ = ____

18. *Exact:* $5 \div 1\frac{7}{8}$

Estimate: ____ ÷ ____ = ____

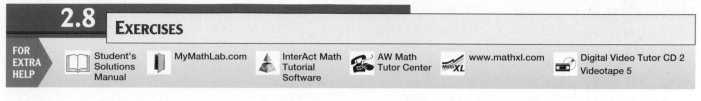

19. *Exact:*

$\frac{3}{4} \div 1\frac{3}{4}$

Estimate:

____ ÷ ____ = ____

20. *Exact:*

$\frac{3}{4} \div 2\frac{1}{2}$

Estimate:

____ ÷ ____ = ____

21. *Exact:*

$1\frac{7}{8} \div 6\frac{1}{4}$

Estimate:

____ ÷ ____ = ____

22. *Exact:*

$8\frac{2}{5} \div 3\frac{1}{2}$

Estimate:

____ ÷ ____ = ____

23. *Exact:*

$5\frac{2}{3} \div 6$

Estimate:

____ ÷ ____ = ____

24. *Exact:*

$5\frac{3}{4} \div 2$

Estimate:

____ ÷ ____ = ____

For Exercises 25–40, first estimate the answer. Then solve each application problem by using the six problem-solving steps. Simplify all answers. See Examples 3 and 4.

Use the recipe for Quaker Choc Oat-Chip Cookies to work Exercises 25–28.

Quaker Choc Oat-Chip Cookies

1 cup (2 sticks) **margarine or butter, softened**
$1\frac{1}{4}$ cups firmly packed brown sugar
$\frac{1}{2}$ cup granulated sugar
2 eggs
2 tablespoons milk
2 teaspoons vanilla
$1\frac{3}{4}$ cups all purpose flour
1 teaspoon baking soda
$\frac{1}{2}$ teaspoon salt (optional)
$2\frac{1}{2}$ cups Quaker® Oats (quick or old fashioned, uncooked)
One 12-ounce package (2 cups) Nestle® Toll House® semi-sweet chocolate morsels
1 cup coarsely chopped nuts (optional)

Heat oven to 375°F. **Beat** margarine and sugars until creamy.
Add eggs, milk and vanilla; beat well.
Add combined flour, baking soda and salt; mix well. **Stir** in oats, chocolate morsels and nuts; mix well.
Drop by rounded measuring tablespoonfuls onto ungreased cookie sheet.
Bake 9 to 10 minutes for a chewy cookie or 12 to 13 minutes for a crisp cookie.
Cool 1 minute on a cookie sheet; remove to wire rack. Cool completely.
ABOUT 5 DOZEN

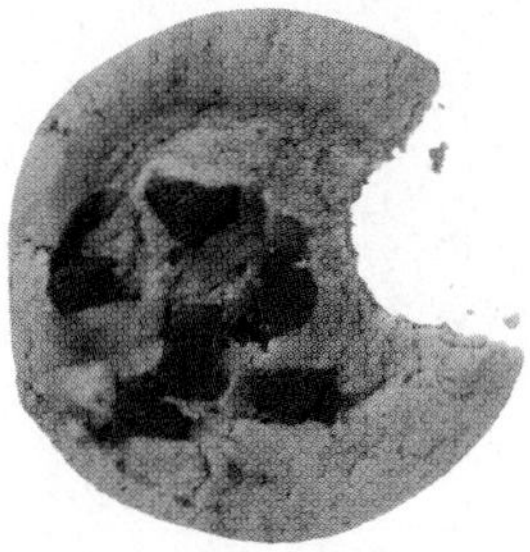

Source: Reprinted by permission of the Quaker Oats Company.

25. If the recipe is doubled, find the amount needed of each ingredient.

(a) Quaker Oats

Estimate:
Exact:

(b) Brown sugar

Estimate:
Exact:

(c) Flour

Estimate:
Exact:

26. If the recipe is tripled, find the amount needed of each ingredient.

(a) Flour

Estimate:
Exact:

(b) Quaker Oats

Estimate:
Exact:

(c) Granulated sugar

Estimate:
Exact:

27. How much of each ingredient is needed if you bake one-half of the recipe?

(a) Brown sugar

Estimate:
Exact:

(b) Granulated sugar

Estimate:
Exact:

(c) Flour

Estimate:
Exact:

28. How much of each ingredient is needed if you bake one-third of the recipe?

(a) Quaker Oats

Estimate:
Exact:

(b) Salt

Estimate:
Exact:

(c) Brown sugar?

Estimate:
Exact:

29. Each home in a housing development needs $109\frac{1}{2}$ yd of prefinished baseboard. How many homes can be fitted if there are 1314 yd of baseboard available?

Estimate:

Exact:

30. How many $1\frac{1}{8}$-ounce vials can be filled from 99 ounces of normal saline solution?

Estimate:

Exact:

31. An insect spray manufactured by Dutch Chemicals Incorporated is a mixture of $1\frac{3}{4}$ ounces of chemical per 1 gallon of water. How many ounces of chemical are needed for $12\frac{1}{2}$ gallons of water?

Estimate:

Exact:

32. To put roofing on the new homes in the Happy Trails subdivision, Duncan needs $37\frac{3}{4}$ lb of roofing nails per home. How many pounds of roofing nails will he need for 36 homes?

Estimate:

Exact:

33. Write the three steps for multiplying mixed numbers. Use your own words.

34. Refer to Exercise 33. In your own words, write the additional step that must be added to the rule for multiplying mixed numbers to make it the rule for dividing mixed numbers.

35. Manufacturing a fishing boat anchor requires $10\frac{3}{8}$ lb of steel. Find the number of anchors that can be manufactured with 25,730 lb of steel.

Estimate:

Exact:

36. To remodel the apartments in her complex, Joleen needs $62\frac{1}{2}$ square yards of carpet for each apartment unit. How many apartment units can she remodel with 6750 square yards of carpet?

Estimate:

Exact:

37. A manufacturer of bird feeders cuts spacers from a tube that is $9\frac{3}{4}$ in. long. How many spacers can be cut from the tube if each spacer must be $\frac{3}{4}$-in. thick?

Estimate:

Exact:

38. A building contractor must move 12 tons of sand. If his truck can carry $\frac{3}{4}$ ton of sand, how many trips must he make to move the sand?

Estimate:

Exact:

39. A fishing guide's boat motor uses $12\frac{3}{4}$ gallons of fuel on a full-day fishing trip and $7\frac{1}{8}$ gallons of fuel on a half-day fishing trip. Find the total number of gallons of fuel used in 28 full-day trips and 16 half-day trips.

Estimate:

Exact:

40. One necklace can be completed in $6\frac{1}{2}$ minutes, while a bracelet takes $3\frac{1}{8}$ minutes. Find the total time that it takes to complete 36 necklaces and 22 bracelets.

Estimate:

Exact:

Chapter 2

SUMMARY

Study Skills Workbook
Activity 5

KEY TERMS

2.1	**numerator**	The number above the fraction bar in a fraction is called the numerator. It shows how many of the equivalent parts are being considered.
	denominator	The number below the fraction bar in a fraction is called the denominator. It shows the number of equal parts in a whole.
	proper fraction	In a proper fraction, the numerator is smaller than the denominator. The fraction is less than 1.
	improper fraction	In an improper fraction, the numerator is greater than or equal to the denominator. The fraction is equal to or greater than 1.
2.2	**mixed number**	A mixed number includes a fraction and a whole number written together.
2.3	**factors**	Numbers that are multiplied to give a product are factors.
	composite number	A composite number has at least one factor other than itself and 1.
	prime number	A prime number is a whole number other than 0 and 1 that has exactly two factors, itself and 1.
	factorizations	The numbers that can be multiplied to give a specific number (product) are factorizations of that number.
	prime factorization	In a prime factorization, every factor is a prime number.
2.4	**equivalent fractions**	Two fractions are equivalent when they represent the same portion of a whole.
	common factor	A common factor is a number that can be divided into two or more whole numbers.
	lowest terms	A fraction is written in lowest terms when its numerator and denominator have no common factor other than 1.
2.5	**multiplication shortcut**	When multiplying or dividing fractions, the process of dividing a numerator and denominator by a common factor can be used as a shortcut.
2.7	**reciprocal**	Two numbers are reciprocals of each other if their product is 1. To find the reciprocal of a fraction, interchange the numerator and the denominator.

NEW FORMULA

Area of a rectangle: Area = length • width

TEST YOUR WORD POWER

See how well you have learned the vocabulary in this chapter. Answers follow the Quick Review.

1. A **numerator** is
- **(a)** a number greater than 5
- **(b)** the number above the fraction bar in a fraction
- **(c)** any number
- **(d)** the number below the fraction bar in a fraction.

2. A **proper fraction**
- **(a)** has a value less than 1
- **(b)** has a whole number and a fraction
- **(c)** has a value greater than 1
- **(d)** is equal to 1.

3. A **mixed number** is
- **(a)** equal to 1
- **(b)** less than 1
- **(c)** a whole number and a fraction written together
- **(d)** a number multiplied by another number.

4. A **factor** is
- **(a)** one of two or more numbers that are added to get another number
- **(b)** the answer in division
- **(c)** one of two or more numbers that are multiplied to get another number
- **(d)** the answer in multiplication.

5. A whole number greater than 1 is **prime** if
- **(a)** it cannot be factored
- **(b)** it has just one factor
- **(c)** it has only itself and 1 as a factor
- **(d)** it has at least two different factors.

6. A **common factor** can
- **(a)** only be divided by itself and 1
- **(b)** be divided into two or more whole numbers
- **(c)** never be divided by 2
- **(d)** only be divided by the numbers 5 and 10.

7. A fraction is in **lowest terms** when
- **(a)** it cannot be divided
- **(b)** it is a common fraction
- **(c)** its numerator and denominator have no common factor other than 1
- **(d)** it has a value less than 1.

8. To find the **reciprocal** of a fraction
- **(a)** multiply it by itself
- **(b)** interchange the numerator and the denominator
- **(c)** change it to an improper fraction
- **(d)** change it to lowest terms.

QUICK REVIEW

Concepts	*Examples*
2.1 *Types of Fractions* **Proper** Numerator smaller than denominator.	$\frac{2}{3}, \frac{3}{4}, \frac{15}{16}, \frac{1}{8}$
Improper Numerator equal to or greater than denominator.	$\frac{17}{8}, \frac{19}{12}, \frac{11}{2}, \frac{5}{3}, \frac{7}{7}$
2.2 *Converting Fractions* **Mixed to Improper** Multiply denominator by whole number, add numerator, and place over denominator.	$7\frac{2}{3} = \frac{23}{3}$ ← $3 \times 7 + 2$ Same denominator
Improper to Mixed Divide numerator by denominator and place remainder over denominator.	$\frac{17}{5} = 3\frac{2}{5}$ Same denominator

Concepts	*Examples*
2.3 Prime Numbers Determine whether a whole number is evenly divisible only by itself and 1. (By definition, 0 and 1 are not prime.)	The prime numbers less than 100 are 2, 3, 5, 7, 11, 13, 17, 19, 23, 29, 31, 37, 41, 43, 47, 53, 59, 61, 67, 71, 73, 79, 83, 89, and 97.
2.3 Finding the Prime Factorization of a Number Divide each factor by a prime number by using a diagram that forms the shape of tree branches.	Find the prime factorization of 30. Use a factor tree. 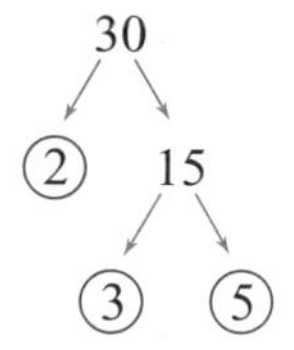 Prime factors are circled. $30 = 2 \cdot 3 \cdot 5$
2.4 Writing Fractions in Lowest Terms Divide the numerator and denominator by the greatest common factor.	Write $\frac{30}{42}$ in lowest terms. $\frac{30}{42} = \frac{30 \div 6}{42 \div 6} = \frac{5}{7}$
2.5 Multiplying Fractions **1.** Multiply numerators by numerators and denominators by denominators. **2.** Write answers in lowest terms if the multiplication shortcut was not used.	Multiply. $\frac{6}{11} \cdot \frac{7}{8} = \frac{\overset{3}{\cancel{6}}}{11} \cdot \frac{7}{\underset{4}{\cancel{8}}} = \frac{3 \cdot 7}{11 \cdot 4} = \frac{21}{44}$
2.7 Finding the Reciprocal To find the reciprocal of a fraction, interchange the numerator and denominator.	Find the reciprocal of each fraction. $\frac{3}{4}$ The reciprocal of $\frac{3}{4}$ is $\frac{4}{3}$. $\frac{8}{5}$ The reciprocal of $\frac{8}{5}$ is $\frac{5}{8}$. 9 The reciprocal of 9 is $\frac{1}{9}$.
2.7 Dividing Fractions Use the reciprocal of the second fraction (divisor) and multiply as fractions.	Divide. $\frac{25}{36} \div \frac{15}{18} = \frac{\overset{5}{\cancel{25}}}{\underset{2}{\cancel{36}}} \cdot \frac{\overset{1}{\cancel{18}}}{\underset{3}{\cancel{15}}} = \frac{5 \cdot 1}{2 \cdot 3} = \frac{5}{6}$ Reciprocals

Concepts	*Examples*
2.8 Multiplying Mixed Numbers First estimate the answers. Then follow these steps. *Step 1* *Change* each mixed number to an improper fraction. *Step 2* *Multiply.* *Step 3* Simplify the answer which means to write it in *lowest terms*, and change it to a mixed number or whole number where possible.	First estimate the answer. Then multiply to get the exact answer. *Estimate:* $1\frac{3}{5} \cdot 3\frac{1}{3}$ Rounded $\downarrow$ $2 \cdot 3 = 6$ *Exact:* $1\frac{3}{5} \cdot 3\frac{1}{3} = \frac{8}{\cancel{5}_1} \cdot \frac{\cancel{10}^2}{3} = \frac{8 \cdot 2}{1 \cdot 3} = \frac{16}{3} = 5\frac{1}{3}$ Close to estimate
2.8 Dividing Mixed Numbers First estimate the answer. Then follow these steps. *Step 1* *Change* each mixed number to an improper fraction. *Step 2* Use the *reciprocal* of the second fraction (divisor). *Step 3* *Multiply.* *Step 4* Simplify the answer which means to write it in *lowest terms*, and change it to a mixed number or whole number where possible.	First estimate the answer. Then divide to get the exact answer. *Estimate:* $3\frac{5}{9} \div 2\frac{2}{5}$ Rounded $\downarrow$ $4 \div 2 = 2$ *Exact:* $3\frac{5}{9} \div 2\frac{2}{5} = \frac{32}{9} \div \frac{12}{5} = \frac{\cancel{32}^8}{9} \cdot \frac{5}{\cancel{12}_3} = \frac{40}{27} = 1\frac{13}{27}$ Close to estimate

ANSWERS TO TEST YOUR WORD POWER

1. (b) *Example:* In $\frac{3}{8}$, the numerator is 3. **2. (a)** *Example:* $\frac{1}{2}$, $\frac{3}{4}$, and $\frac{7}{8}$ are all proper fractions with a value less than 1. **3. (c)** *Example:* $2\frac{3}{8}$ and $5\frac{3}{4}$ are mixed numbers. **4. (c)** *Example:* Since $3 \cdot 5 = 15$, the numbers 3 and 5 are factors of 15. **5. (c)** *Example:* 3, 5, and 11 are prime numbers; 4, 8, and 12 are composite numbers. **6. (b)** *Example:* 3 is a common factor of both 6 and 9 because it can be evenly divided into each of them. **7. (c)** *Example:* $\frac{3}{8}$, $\frac{4}{5}$, and $\frac{5}{6}$ are in lowest terms but $\frac{6}{8}$, $\frac{3}{6}$, and $\frac{2}{4}$ are not. **8. (b)** *Example:* The reciprocal of $\frac{3}{8}$ is $\frac{8}{3}$, the reciprocal of $\frac{25}{4}$ is $\frac{4}{25}$, and the reciprocal of 6 or $\frac{6}{1}$ is $\frac{1}{6}$.

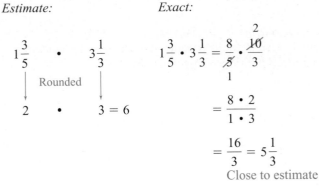

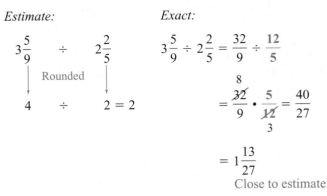

Chapter 2 Review Exercises

[2.1] *Write the fraction that represents each shaded portion.*

1. 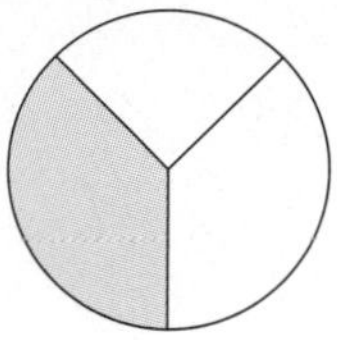

2. 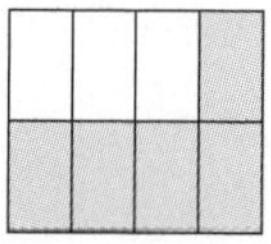

3. 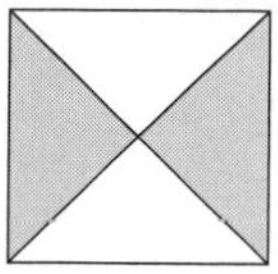

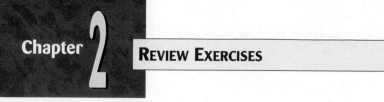

List the proper and improper fractions in each group.

	Proper	Improper
4. $\frac{1}{8}, \frac{4}{3}, \frac{5}{5}, \frac{3}{4}, \frac{2}{3}$	______	______
5. $\frac{6}{5}, \frac{15}{16}, \frac{16}{13}, \frac{1}{8}, \frac{5}{3}$	______	______

[2.2] *Write each mixed number as an improper fraction. Write each improper fraction as a mixed number.*

6. $5\frac{2}{3}$ **7.** $10\frac{4}{5}$ **8.** $\frac{17}{8}$ **9.** $\frac{63}{5}$

[2.3] *Find all factors of each number.*

10. 6 **11.** 24 **12.** 55 **13.** 90

Write the prime factorization of each number by using exponents.

14. 27 **15.** 150 **16.** 168

Simplify each expression.

17. 6^2 **18.** $5^2 \cdot 2^3$ **19.** $8^2 \cdot 3^3$ **20.** $4^3 \cdot 2^5$

[2.4] *Write each fraction in lowest terms.*

21. $\frac{9}{12}$ **22.** $\frac{30}{36}$ **23.** $\frac{75}{80}$

Write the numerator and denominator of each fraction as a product of prime factors. Then, write the fraction in lowest terms.

24. $\frac{25}{60}$ **25.** $\frac{384}{96}$

Decide whether each pair of fractions is equivalent or not equivalent, using the method of prime factors.

26. $\frac{3}{4}$ and $\frac{48}{64}$

27. $\frac{3}{4}$ and $\frac{42}{58}$

[2.5–2.8] *Multiply. Write answers in lowest terms, and as mixed numbers or whole numbers where possible.*

28. $\frac{2}{3} \cdot \frac{3}{4}$

29. $\frac{3}{10} \cdot \frac{5}{8}$

30. $\frac{70}{175} \cdot \frac{5}{14}$

31. $\frac{44}{63} \cdot \frac{3}{11}$

32. $\frac{5}{16} \cdot 48$

33. $\frac{5}{8} \cdot 1000$

Divide. Write answers in lowest terms, and as mixed numbers or whole numbers where possible.

34. $\frac{1}{2} \div \frac{3}{4}$

35. $\frac{5}{6} \div \frac{1}{2}$

36. $\dfrac{\frac{15}{18}}{\frac{10}{30}}$

37. $\dfrac{\frac{7}{8}}{\frac{7}{16}}$

38. $7 \div \frac{7}{8}$

39. $18 \div \frac{3}{4}$

40. $\frac{4}{5} \div 6$

41. $\frac{2}{3} \div 5$

42. $\dfrac{\frac{12}{13}}{3}$

Find the area of each rectangle.

43.

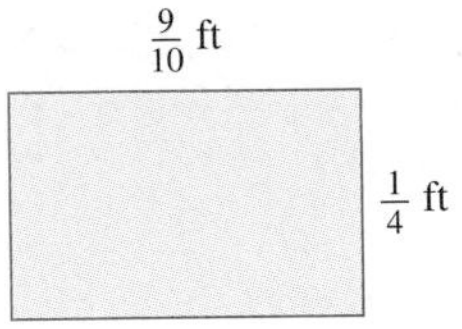

44.

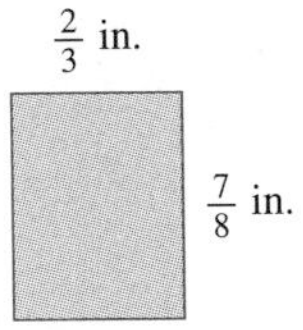

45. Find the area of a display shelf having a length of 10 ft and a width of $\frac{3}{4}$ ft.

46. Find the area of a rectangle having a length of 48 yd and a width of $\frac{3}{4}$ yd.

First estimate the answer. Then multiply or divide to find the exact answer. Simplify all answers.

47. *Exact:*

$5\frac{1}{2} \cdot 1\frac{1}{4}$

Estimate:

____ • ____ = ____

48. *Exact:*

$2\frac{1}{4} \cdot 7\frac{1}{8} \cdot 1\frac{1}{3}$

Estimate:

____ • ____ • ____ = ____

49. *Exact:*

$15\frac{1}{2} \div 3$

Estimate:

_____ ÷ _____ = _____

50. *Exact:*

$4\frac{3}{4} \div 6\frac{1}{3}$

Estimate:

_____ ÷ _____ = _____

Solve each application problem by using the six problem-solving steps.

51. How many $\frac{7}{8}$-lb boxes of sheetrock nails can be filled from 294 lb of sheetrock nails.

52. An estate is divided so that each of 5 children receives equal shares of $\frac{2}{3}$ of the estate. What fraction of the total estate will each receive?

53. How many window-blind pull cords can be made from $157\frac{1}{2}$ yd of cord if $4\frac{3}{8}$ yd of cord are needed for each blind? First estimate, and then find the exact answer.

Estimate:

Exact:

54. Working as a bookkeeper, Neta Fitzgerald is paid $\$8\frac{1}{2}$ per hour. Find her earnings for a week in which she works 38 hours. First estimate, and then find the exact answer.

Estimate:

Exact:

55. Ebony Wilson purchased 100 lb of rice at the food co-op. After selling $\frac{1}{4}$ of this to her neighbor, she gives $\frac{2}{3}$ of the remaining rice to her parents. How many pounds of rice does she have left?

56. Tiffany Crowder received her Social Security check for \$1275. After paying $\frac{1}{3}$ of this amount for rent, she paid $\frac{2}{5}$ of the remaining amount for food, utilities, and transportation. How much money does she have left?

57. The Springvale Parish will divide $\frac{5}{8}$ of its library budget evenly among the 4 largest parish libraries. What fraction of the total library budget will each of these libraries receive?

58. In a morning of deep-sea fishing, 8 fishermen catch $\frac{2}{3}$ ton of halibut. If they divide the fish evenly, how much will each receive?

Mixed Review Exercises

Multiply or divide as indicated. Simplify all answers.

59. $\frac{3}{8} \cdot \frac{2}{3}$

60. $\frac{2}{3} \cdot \frac{3}{5}$

61. $12\frac{1}{2} \cdot 2\frac{1}{2}$

62. $12\frac{1}{2} \cdot 2\frac{1}{4}$

63. $\frac{\frac{4}{5}}{8}$

64. $\frac{\frac{5}{8}}{4}$

65. $\frac{15}{31} \cdot 62$

66. $3\frac{1}{4} \div 1\frac{1}{2}$

Write each mixed number as an improper fraction. Write each improper fraction as a mixed number.

67. $\frac{7}{3}$

68. $\frac{143}{5}$

69. $5\frac{2}{3}$

70. $38\frac{3}{8}$

Write the numerator and denominator of each fraction as a product of prime factors, then write the fraction in lowest terms.

71. $\frac{8}{12}$

72. $\frac{108}{210}$

Write each fraction in lowest terms.

73. $\frac{46}{69}$

74. $\frac{24}{72}$

75. $\frac{44}{110}$

76. $\frac{87}{261}$

Solve each application problem.

77. The directions on a can of fabric glue say to apply $3\frac{1}{2}$ ounces of glue to each square yard of fabric. How many ounces are needed for $43\frac{5}{9}$ square yards? First estimate, and then find the exact answer.

Estimate:

Exact:

78. Jack Anderson Trucking purchased some diesel fuel additive. The instructions say to use $7\frac{1}{4}$ quarts of additive with each in-ground tank of fuel. How many quarts are needed for $25\frac{1}{2}$ in-ground tanks? First estimate, and then find the exact answer.

Estimate:

Exact:

79. A postage stamp is $\frac{7}{8}$ in. by $\frac{7}{8}$ in. Find its area.

80. A section of marble countertop is $\frac{1}{2}$ yard by $\frac{5}{8}$ yard. What is its area?

Chapter 2 Test

Study Skills Workbook
Activity 8

Write a fraction to represent each shaded portion.

1. 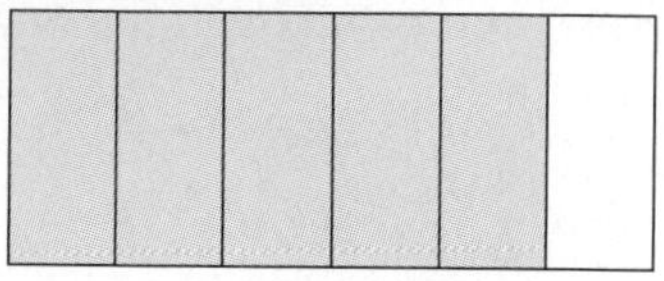

2. 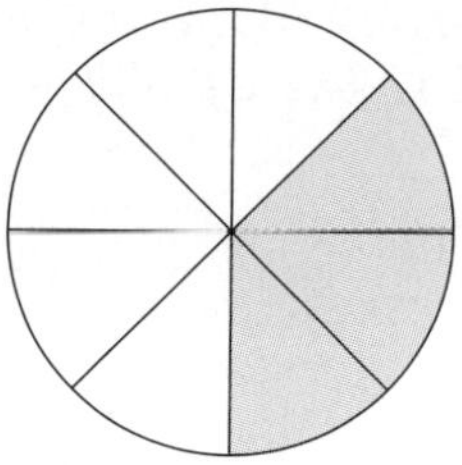

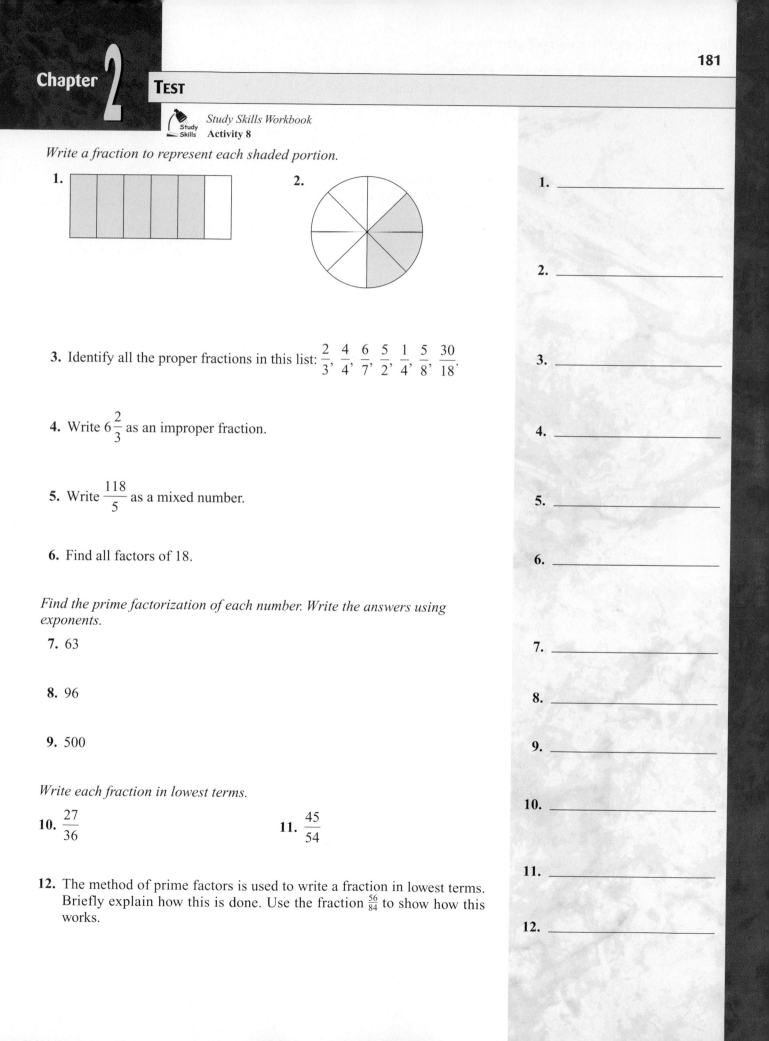

3. Identify all the proper fractions in this list: $\frac{2}{3}, \frac{4}{4}, \frac{6}{7}, \frac{5}{2}, \frac{1}{4}, \frac{5}{8}, \frac{30}{18}$.

4. Write $6\frac{2}{3}$ as an improper fraction.

5. Write $\frac{118}{5}$ as a mixed number.

6. Find all factors of 18.

Find the prime factorization of each number. Write the answers using exponents.

7. 63

8. 96

9. 500

Write each fraction in lowest terms.

10. $\frac{27}{36}$

11. $\frac{45}{54}$

12. The method of prime factors is used to write a fraction in lowest terms. Briefly explain how this is done. Use the fraction $\frac{56}{84}$ to show how this works.

1. ______________

2. ______________

3. ______________

4. ______________

5. ______________

6. ______________

7. ______________

8. ______________

9. ______________

10. ______________

11. ______________

12. ______________

13. ____________

14. ____________

15. ____________

16. ____________

17. ____________

18. ____________

19. ____________

20. ____________

21. *Estimate:* ____________

Exact: ____________

22. *Estimate:* ____________

Exact: ____________

23. *Estimate:* ____________

Exact: ____________

24. *Estimate:* ____________

Exact: ____________

25. *Estimate:* ____________

Exact: ____________

13. Explain how to multiply fractions. What additional step must be taken when dividing fractions?

Multiply or divide. Write answers in lowest terms, and as mixed numbers or whole numbers where possible.

14. $\frac{2}{5} \cdot \frac{3}{8}$

15. $36 \cdot \frac{4}{9}$

16. Find the area of a kitchen grill measuring $\frac{3}{4}$ yd by $\frac{1}{2}$ yd.

17. There are 4224 people in a survey of licensed drivers. If $\frac{7}{8}$ of these drivers have had no traffic citations within the last 2 years, find the number of drivers who have had no citations.

18. $\frac{3}{4} \div \frac{5}{6}$

19. $\dfrac{7}{\frac{4}{9}}$

20. The Lincoln Fun Run committee has acquired 120 quarts of Sports Drink to sell in $\frac{3}{5}$-quarts sports bottles at the race. Find the number of sports bottles that can be filled.

First estimate the answer. Then either multiply or divide to find the exact answer. Simplify all answers.

21. $4\frac{1}{8} \cdot 3\frac{1}{2}$

22. $1\frac{5}{6} \cdot 4\frac{1}{3}$

23. $9\frac{3}{5} \div 2\frac{1}{4}$

24. $\dfrac{8\frac{1}{2}}{1\frac{2}{3}}$

25. A new vaccine is synthesized at the rate of $2\frac{1}{2}$ grams per day. How many grams can be synthesized in $12\frac{1}{4}$ days?

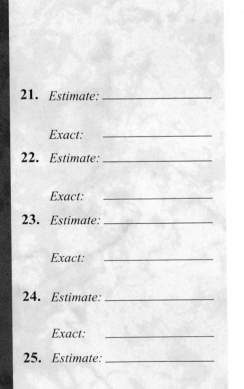

Cumulative Review Exercises CHAPTERS 1–2

Name the digit that has the given place value in each number.

1. 579
hundreds
tens

2. 8,621,785
millions
ten thousands

Add, subtract, multiply, or divide as indicated.

3. $\begin{array}{r} 43 \\ 11 \\ 85 \\ +\ 27 \\ \hline \end{array}$

4. $\begin{array}{r} 82{,}121 \\ 5\ 468 \\ 316 \\ +61{,}294 \\ \hline \end{array}$

5. $\begin{array}{r} 6537 \\ -\ 2085 \\ \hline \end{array}$

6. $\begin{array}{r} 4{,}819{,}604 \\ -\ 1{,}597{,}783 \\ \hline \end{array}$

7. $\begin{array}{r} 68 \\ \times\ 7 \\ \hline \end{array}$

8. $8 \cdot 3 \cdot 4$

9. $\begin{array}{r} 3784 \\ \times\ \ 573 \\ \hline \end{array}$

10. $\begin{array}{r} 563 \\ \times\ \ 800 \\ \hline \end{array}$

11. $\dfrac{63}{7}$

12. $18\overline{)136{,}458}$

13. $33{,}886 \div 4$

14. $492\overline{)10{,}850}$

Round each number to the nearest ten, nearest hundred, and nearest thousand.

	Ten	**Hundred**	**Thousand**
15. 5737	________	________	________
16. 76,271	________	________	________

Simplify each expression by using the order of operations.

17. $3^3 - 8 \cdot 3$

18. $\sqrt{36} - 2 \cdot 3 + 5$

Solve each application problem using the six problem-solving steps.

19. The manager of Starbucks purchased 8 cases of small-size coffee cups for $46 per case and 12 cases of large-size coffee cups for $52 per case. Find the total cost of the 20 cases of coffee cups.

20. Home blood-pressure monitors sell for as little as $20 and as much as $150. If Scott bought the cheapest model and Jenn the most expensive model, how much more did Jenn pay than Scott?

21. A typical adult loses 100 hairs a day out of approximately 120,000 hairs. If the lost hairs were not replaced, find the number of hairs remaining after two years. (1 year = 365 days).

22. The Maxey family reunion will cost \$4275. If the cost is to be divided evenly among 15 family members, find the cost for each family member.

23. The glass face for a picture frame measures $\frac{11}{12}$ ft by $\frac{3}{4}$ ft. Find its area.

24. The Municipal Utility District says that the cost of operating a hair dryer is $\frac{1}{5}$¢ per minute. Find the cost of operating the hair dryer for a half hour.

Write proper *or* improper *for each fraction.*

25. $\frac{2}{3}$ **26.** $\frac{6}{6}$ **27.** $\frac{9}{18}$

Write each mixed number as an improper fraction. Write each improper fraction as a whole or mixed number.

28. $3\frac{3}{8}$ **29.** $6\frac{2}{5}$ **30.** $\frac{14}{7}$ **31.** $\frac{103}{8}$

Find the prime factorization of each number. Write answers by using exponents.

32. 72 **33.** 126 **34.** 350

Simplify each expression.

35. $4^2 \cdot 2^2$ **36.** $2^3 \cdot 6^2$ **37.** $2^3 \cdot 4^2 \cdot 5$

Write each fraction in lowest terms.

38. $\frac{42}{48}$ **39.** $\frac{24}{36}$ **40.** $\frac{30}{54}$

Multiply or divide as indicated. Simplify all answers.

41. $\frac{1}{2} \cdot \frac{3}{4}$ **42.** $30 \cdot \frac{2}{3} \cdot \frac{3}{5}$ **43.** $7\frac{1}{2} \cdot 3\frac{1}{3}$

44. $\frac{3}{5} \div \frac{5}{8}$ **45.** $\frac{7}{8} \div 1\frac{1}{2}$ **46.** $3 \div 1\frac{1}{4}$

Adding and Subtracting Fractions

3

Most types of home improvement, such as carpentry, wallpapering a room, or hanging pictures on a wall, require the use of a tape measure. The inches on a tape measure are typically divided into smaller and smaller segments that represent $\frac{1}{2}$, $\frac{1}{4}$, $\frac{1}{8}$, and $\frac{1}{16}$ of an inch. In order to use a tape measure effectively, you need to be able to add and subtract common fractions and mixed numbers. (See Exercises 41 and 51 in Section 3.3 and Exercise 53 in Section 3.4.)

3.1 Adding and Subtracting Like Fractions

Objectives

1. Define like and unlike fractions.
2. Add like fractions.
3. Subtract like fractions.

In Chapter 2 we looked at the basics of fractions and then practiced with multiplication and division of common fractions and mixed numbers. In this chapter we will work with addition and subtraction of common fractions and mixed numbers.

1 Define like and unlike fractions. Fractions with the same denominators are **like fractions.** Fractions with different denominators are **unlike fractions.**

Example 1 Identifying Like and Unlike Fractions

(a) $\frac{3}{4}, \frac{1}{4}, \frac{5}{4}, \frac{6}{4},$ and $\frac{4}{4}$ are **like** fractions.

All denominators are the same.

(b) $\frac{7}{12}$ and $\frac{12}{7}$ are **unlike** fractions.

All denominators are different.

Work Problem 1 at the Side.

1 Next to each pair of fractions write *like* or *unlike*.

(a) $\frac{2}{3}$ $\frac{1}{3}$ ______

(b) $\frac{3}{4}$ $\frac{3}{8}$ ______

(c) $\frac{6}{15}$ $\frac{11}{15}$ ______

(d) $\frac{5}{12}$ $\frac{5}{7}$ ______

NOTE

Like fractions have the ***same*** denominator.

2 Add like fractions. The following figures show you how to add the fractions $\frac{2}{7}$ and $\frac{4}{7}$.

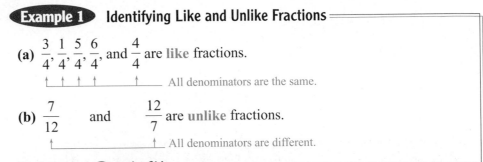

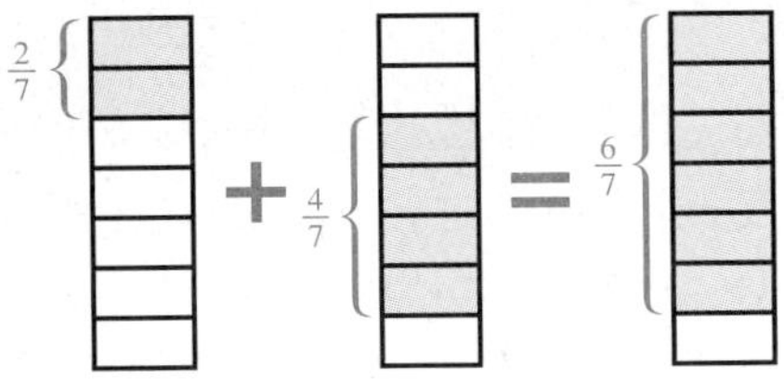

As the figures show,

$$\frac{2}{7} + \frac{4}{7} = \frac{6}{7}.$$

Add like fractions as follows.

Adding Like Fractions

Step 1 Add the numerators to find the numerator of the sum.

Step 2 Write the denominator of the like fractions as the denominator of the sum.

Step 3 Write the answer in lowest terms.

Answers

1. (a) like (b) unlike (c) like (d) unlike

Example 2 Adding Like Fractions

Add and write the answer in lowest terms.

(a) $\frac{1}{5}+\frac{2}{5}$

$$\frac{1}{5}+\frac{2}{5}=\overbrace{\frac{1+2}{5}}^{\text{Add numerators.}}=\frac{3}{5} \leftarrow \text{Same denominator}$$

(b) $\frac{1}{12}+\frac{7}{12}+\frac{1}{12}$

Step 1 $\quad \dfrac{\overbrace{1+7+1}^{\text{Add numerators.}}}{12}$

Step 2 $\quad =\dfrac{9}{12} \begin{matrix}\leftarrow \text{Sum} \\ \leftarrow \text{Same denominator}\end{matrix}$

Step 3 $\quad =\dfrac{9\div 3}{12\div 3}=\dfrac{3}{4}$ In lowest terms

CAUTION

Fractions may be added ***only*** if they have like denominators.

Work Problem 2 at the Side.

3 Subtract like fractions. The figures show $\frac{7}{8}$ broken into $\frac{4}{8}$ and $\frac{3}{8}$.

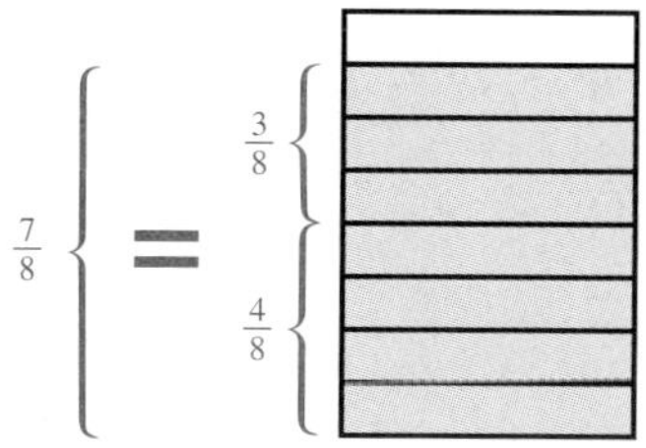

Subtracting $\frac{3}{8}$ from $\frac{7}{8}$ gives the answer $\frac{4}{8}$, or

$$\frac{7}{8}-\frac{3}{8}=\frac{4}{8}.$$

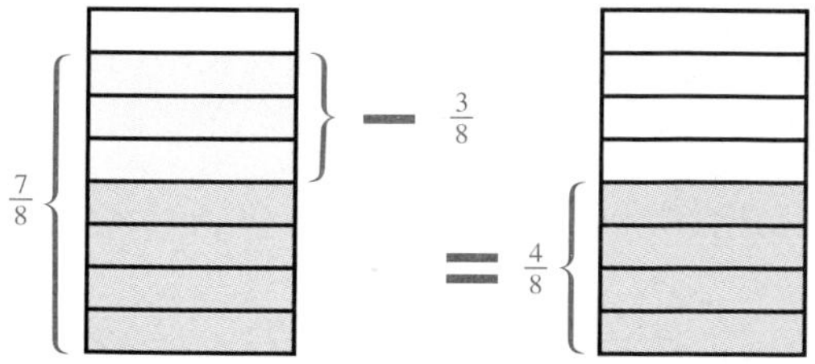

2 Add and write the answers in lowest terms.

(a) $\frac{1}{5}+\frac{3}{5}$

(b) $\begin{array}{r}\frac{1}{9}\\ +\frac{5}{9}\\ \hline\end{array}$

(c) $\frac{5}{8}+\frac{1}{8}$

(d) $\frac{3}{10}+\frac{1}{10}+\frac{4}{10}$

ANSWERS

2. (a) $\frac{4}{5}$ (b) $\frac{2}{3}$ (c) $\frac{3}{4}$ (d) $\frac{4}{5}$

3 Subtract and simplify.

(a) $\frac{7}{9} - \frac{5}{9}$

(b) $\frac{13}{16} - \frac{5}{16}$

(c) $\frac{15}{3} - \frac{5}{3}$

(d) $\frac{25}{32} - \frac{6}{32}$

Write $\frac{4}{8}$ in lowest terms.

$$\frac{7}{8} - \frac{3}{8} = \frac{4 \div 4}{8 \div 4} = \frac{1}{2}$$

The steps for subtracting like fractions are very similar to those for adding like fractions.

Subtracting Like Fractions

Step 1 Subtract the numerators to find the numerator of the difference.

Step 2 Write the denominator of the like fractions as the denominator of the difference.

Step 3 Write the answer in lowest terms.

Example 3 **Subtracting Like Fractions**

Subtract and simplify the answer.

(a) $\frac{15}{16} - \frac{3}{16}$

Step 1 $\frac{15}{16} - \frac{3}{16} = \frac{15 - 3}{16}$ (Subtract numerators.)

Step 2 $= \frac{12}{16}$ ← Difference; ← Same denominator

Step 3 $= \frac{12 \div 4}{16 \div 4} = \frac{3}{4}$ In lowest terms

(b) $\frac{13}{4} - \frac{6}{4}$

$$\frac{13}{4} - \frac{6}{4} = \frac{13 - 6}{4}$$ (Subtract numerators.) ← Same denominator

$$= \frac{7}{4}$$

Write as a mixed number.

$$\frac{7}{4} = 1\frac{3}{4}$$

CAUTION

Fractions may be subtracted ***only*** if they have like denominators.

Work Problem 3 at the Side.

ANSWERS

3. **(a)** $\frac{2}{9}$ **(b)** $\frac{1}{2}$ **(c)** $3\frac{1}{3}$ **(d)** $\frac{19}{32}$

3.1 EXERCISES

FOR EXTRA HELP: Student's Solutions Manual | MyMathLab.com | InterAct Math Tutorial Software | AW Math Tutor Center | www.mathxl.com | Digital Video Tutor CD 2 Videotape 6

Add and simplify the answer. See Example 2.

1. $\frac{3}{5} + \frac{1}{5}$

2. $\frac{5}{8} + \frac{2}{8}$

3. $\frac{2}{6} + \frac{3}{6}$

4. $\frac{9}{11} + \frac{1}{11}$

5. $\frac{1}{6} + \frac{1}{6}$

6. $\frac{1}{12} + \frac{1}{12}$

7. $\begin{array}{r} \frac{9}{10} \\ + \frac{3}{10} \\ \hline \end{array}$

8. $\begin{array}{r} \frac{13}{12} \\ + \frac{5}{12} \\ \hline \end{array}$

9. $\begin{array}{r} \frac{2}{9} \\ + \frac{1}{9} \\ \hline \end{array}$

10. $\frac{3}{14} + \frac{5}{14}$

11. $\frac{6}{20} + \frac{4}{20} + \frac{3}{20}$

12. $\frac{1}{7} + \frac{2}{7} + \frac{3}{7}$

13. $\frac{4}{15} + \frac{2}{15} + \frac{5}{15}$

14. $\frac{5}{11} + \frac{1}{11} + \frac{4}{11}$

15. $\frac{3}{8} + \frac{7}{8} + \frac{2}{8}$

16. $\frac{4}{9} + \frac{1}{9} + \frac{7}{9}$

17. $\frac{2}{54} + \frac{8}{54} + \frac{12}{54}$

18. $\frac{7}{64} + \frac{15}{64} + \frac{20}{64}$

Subtract and simplify the answer. See Example 3.

19. $\frac{7}{8} - \frac{4}{8}$

20. $\frac{2}{3} - \frac{1}{3}$

21. $\frac{10}{11} - \frac{4}{11}$

22. $\frac{4}{5} - \frac{3}{5}$

23. $\frac{9}{10} - \frac{3}{10}$

24. $\frac{7}{8} - \frac{3}{8}$

25. $\begin{array}{r} \frac{31}{21} \\ - \frac{7}{21} \\ \hline \end{array}$

26. $\begin{array}{r} \frac{43}{24} \\ - \frac{13}{24} \\ \hline \end{array}$

27. $\begin{array}{r} \frac{27}{40} \\ - \frac{19}{40} \\ \hline \end{array}$

28. $\begin{array}{r} \frac{38}{55} \\ - \frac{16}{55} \\ \hline \end{array}$

29. $\frac{47}{36} - \frac{5}{36}$

30. $\frac{76}{45} - \frac{21}{45}$

31. $\frac{73}{60} - \frac{7}{60}$

32. $\frac{181}{100} - \frac{31}{100}$

33. In your own words, write an explanation of either how to add or subtract like fractions. Consider using three steps in your explanation.

34. Describe in your own words the difference between *like* fractions and *unlike* fractions. Give three examples of each type.

Solve each application problem. Write answers in lowest terms.

35. In June, Robert and Maria Hernandez had saved $\frac{3}{8}$ of the amount needed for a down payment on their first home. By November, they had saved another $\frac{3}{8}$ of the amount needed. What fraction of the amount needed have they saved?

ALMOST THERE . . .
How much have the Hernandezes saved, expressed as a fraction?

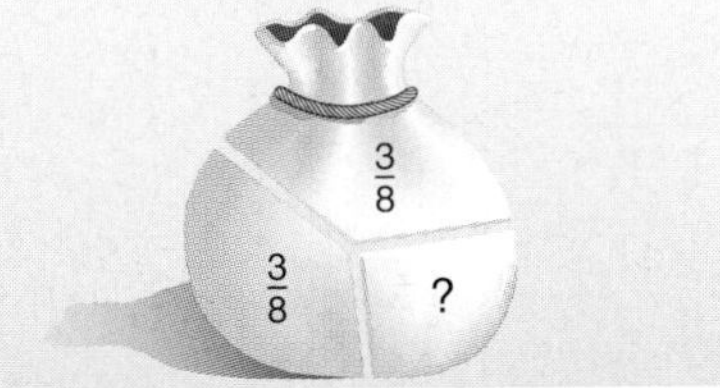

36. After an initial payment to a lotto winner, the state lottery commission still owed the lotto winner $\frac{17}{20}$ of his total winnings. If the state pays the lotto winner another $\frac{3}{20}$ of the winnings, what fraction is still owed?

37. The Gerards owe $\frac{5}{9}$ of a loan for last year's vacation. If they pay $\frac{2}{9}$ of it this month, what fraction of the loan will they still owe?

38. Captain Sisko has been ordered to investigate the wreckage of a plane crash. Sisko leads a search-and-rescue team $\frac{5}{12}$ mile down a ravine and then $\frac{1}{12}$ mile along a creek bed to the crash site. How far does the search-and-rescue team travel?

39. An organic farmer purchased $\frac{9}{10}$ acre of land one year and $\frac{3}{10}$ acre the next year. She then planted carrots on $\frac{7}{10}$ acre of the land and squash on the remainder. How much land is planted in squash?

40. A forester planted $\frac{5}{12}$ acre in seedlings in the morning and $\frac{11}{12}$ acre in the afternoon. If $\frac{7}{12}$ acre of seedlings were destroyed by frost, how many acres remained?

3.2 Least Common Multiples

Objectives

1. Find the least common multiple.
2. Find the least common multiple by using multiples of the largest number.
3. Find the least common multiple by using prime factorization.
4. Find the least common multiple by using an alternative method.
5. Write a fraction with an indicated denominator.

Only *like* fractions can be added or subtracted. Because of this, we must rewrite *unlike* fractions as *like* fractions before we can add or subtract.

1 Find the least common multiple. We can rewrite unlike fractions as like fractions by finding the *least common multiple* of the denominators.

Least Common Multiple

The **least common multiple (LCM)** of two whole numbers is the smallest whole number divisible by both those numbers.

Example 1 Finding the Least Common Multiple

Find the least common multiple of 6 and 9.

This list shows the multiples of 6.

$$\underbrace{6 \cdot 1}_{6,} \quad \underbrace{6 \cdot 2}_{12,} \quad \underbrace{6 \cdot 3}_{18,} \quad \underbrace{6 \cdot 4}_{24,} \quad \underbrace{6 \cdot 5}_{30,} \quad \underbrace{6 \cdot 6}_{36,} \quad \underbrace{6 \cdot 7}_{42,} \quad \underbrace{6 \cdot 8}_{48,} \ \ldots$$

(The three dots at the end of the list show that the list continues in the same pattern without stopping.) The next list shows multiples of 9.

$$\underbrace{9 \cdot 1}_{9,} \quad \underbrace{9 \cdot 2}_{18,} \quad \underbrace{9 \cdot 3}_{27,} \quad \underbrace{9 \cdot 4}_{36,} \quad \underbrace{9 \cdot 5}_{45,} \quad \underbrace{9 \cdot 6}_{54,} \quad \underbrace{9 \cdot 7}_{63,} \quad \underbrace{9 \cdot 8}_{72,} \ \ldots$$

The smallest number found in *both* lists is 18, so 18 is the **least common multiple** of 6 and 9; the number 18 is the smallest whole number divisible by both 6 and 9.

Multiples of 6: 6, 12, 18, 24, 30, 36, 42, 48, . . .
Multiples of 9: 9, 18, 27, 36, 45, 54, 63, 72, . . .

18 is the smallest number found in both lists. 18 is the least common multiple of 6 and 9.

Work Problem 1 at the Side.

1 (a) List the multiples of 5.

5, ___, ___, ___,

___, ___, ___,

___, . . .

(b) List the multiples of 8.

8, ___, ___, ___,

___, ___, ___, . . .

(c) Find the least common multiple of 5 and 8.

2 Find the least common multiple by using multiples of the largest number. There are several ways to find the least common multiple of a number. If the numbers are small, the least common multiple can often be found by inspection. Can you think of a number that can be divided evenly by both 3 and 4? What about 6 or 8, or perhaps 10 or 12? The number 12 will work; it is the least common multiple of the numbers 3 and 4. A method that works well to find the least common multiple is to write multiples of the larger number.

In this case, write the multiples of 4:

4, 8, 12, 16, 20, . . .

Now, check each multiple of 4 to see if it is divisible by 3.

4 is *not* divisible by 3.

8 is *not* divisible by 3.

12 *is* divisible by 3.

The first multiple of 4 that is divisible by 3 is 12, so 12 is the least common multiple of 3 and 4.

Answers

1. (a) 10, 15, 20, 25, 30, 35, 40, . . .
(b) 16, 24, 32, 40, 48, 56, . . .
(c) 40

2 Use multiples of the larger number to find the least common multiple in each set of numbers.

(a) 2 and 3

(b) 4 and 5

(c) 6 and 8

(d) 5 and 9

Example 2 **Finding the Least Common Multiple**

Use multiples of the larger number to find the least common multiple of 6 and 9.

We start by writing the first few multiples of 9.

Multiples of 9

9, 18, 27, 36, 45, 54, . . .

Now, we check each multiple of 9 to see if it is divisible by 6. The first multiple of 9 that is divisible by 6 is 18.

9, **18**, 27, 36, 45, 54, . . .

First multiple divisible by 6 ($18 \div 6 = 3$)

The least common multiple of the numbers 6 and 9 is 18.

Work Problem 2 at the Side.

3 **Find the least common multiple by using prime factorization.** Example 2 shows how to find the least common multiple of two numbers by making a list of the multiples of the *larger* number. Although this method works for smaller numbers, it is usually easier to find the least common multiple for larger numbers by using *prime factorization*, as shown in the next example.

3 Use prime factorization to find the LCM for each pair of numbers.

(a) 15 and 18

(b) 12 and 20

Example 3 **Applying Prime Factorization Knowledge**

Use prime factorization to find the least common multiple of 9 and 12.

We start by finding the prime factorization of each number.

$$9 = 3 \cdot 3$$
$$12 = 2 \cdot 2 \cdot 3$$

Factors of 9

$$\text{LCM} = 3 \cdot 3 \cdot 2 \cdot 2 = 36$$

Factors of 12

Check to see that 36 is divisible by 9 (yes) and by 12 (yes). The smallest whole number divisible by both 9 and 12 is 36.

CAUTION

Notice that we did *not* have to repeat the factors that 9 and 12 have in common. In this case, the **3** in $2 \cdot 2 \cdot \mathbf{3} = 12$ was *not* used because 3 is already included in $3 \cdot 3 = 9$.

Work Problem 3 at the Side.

Example 4 **Using Prime Factorization**

Find the least common multiple of 12, 18, and 20.

Continued on Next Page

ANSWERS

2. **(a)** 6 **(b)** 20 **(c)** 24 **(d)** 45

3. **(a)** $15 = 3 \cdot 5$ $\quad$ $18 = 2 \cdot 3 \cdot 3$ $\quad$ $\text{LCM} = 2 \cdot 3 \cdot 3 \cdot 5 = 90$ (15 above the factors $3 \cdot 5$; 18 below the factors $2 \cdot 3 \cdot 3$)

(b) $12 = 2 \cdot 2 \cdot 3$ $\quad$ $20 = 2 \cdot 2 \cdot 5$ $\quad$ $\text{LCM} = 2 \cdot 2 \cdot 3 \cdot 5 = 60$ (12 above the factors $2 \cdot 2 \cdot 3$; 20 below the factors $2 \cdot 2 \cdot 5$)

Find the prime factorization of each number. Then use the prime factors to build the LCM.

$12 = 2 \cdot 2 \cdot 3$

$18 = 2 \cdot 3 \cdot 3$

$20 = 2 \cdot 2 \cdot 5$

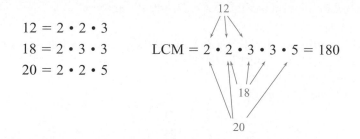

$\text{LCM} = 2 \cdot 2 \cdot 3 \cdot 3 \cdot 5 = 180$ (factors from 12, 18, and 20)

Check to see that 180 is divisible by 12 (yes), and by 18 (yes) and by 20 (yes). The smallest whole number divisible by 12, 18, and 20 is 180.
Note: We did *not* repeat the factors that 12, 18, and 20 have in common.

Work Problem 4 at the Side.

Example 5 Finding the Least Common Multiple

Find the least common multiple for each set of numbers.

(a) 5, 6, 35
Find the prime factorization for each number.

$5 = 5$

$6 = 2 \cdot 3$

$35 = 5 \cdot 7$

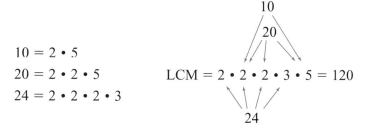

$\text{LCM} = 2 \cdot 3 \cdot 5 \cdot 7 = 210$ (factors from 5, 6, and 35)

The least common multiple of 5, 6, and 35 is 210.

(b) 10, 20, 24
Find the prime factorization for each number.

$10 = 2 \cdot 5$

$20 = 2 \cdot 2 \cdot 5$

$24 = 2 \cdot 2 \cdot 2 \cdot 3$

$\text{LCM} = 2 \cdot 2 \cdot 2 \cdot 3 \cdot 5 = 120$ (factors from 10, 20, and 24)

The least common multiple of 10, 20, and 24 is 120.

Work Problem 5 at the Side.

4 **Find the least common multiple by using an alternative method.** Some people like the following ***alternative method*** for finding the least common multiple for larger numbers. Try both methods, and ***use the one you prefer.*** As a review, a list of the first few prime numbers follows.

2, 3, 5, 7, 11, 13, 17

4 Find the least common multiple of the denominators in each set of fractions.

(a) $\frac{3}{8}$ and $\frac{6}{5}$

(b) $\frac{5}{6}$ and $\frac{1}{14}$

(c) $\frac{4}{9}$, $\frac{5}{18}$, and $\frac{7}{24}$

5 Find the least common multiple for each set of numbers.

(a) 12, 15

(b) 8, 9, 12

(c) 18, 20, 30

(d) 15, 20, 30, 40

ANSWERS

4. (a) $8 = 2 \cdot 2 \cdot 2$, $5 = 1 \cdot 5$; $\text{LCM} = 2 \cdot 2 \cdot 2 \cdot 5 = 40$

(b) $6 = 2 \cdot 3$, $14 = 2 \cdot 7$; $\text{LCM} = 2 \cdot 3 \cdot 7 = 42$

(c) $9 = 3 \cdot 3$, $18 = 2 \cdot 3 \cdot 3$, $24 = 2 \cdot 2 \cdot 2 \cdot 3$; $\text{LCM} = 2 \cdot 2 \cdot 2 \cdot 3 \cdot 3 = 72$

5. (a) 60 **(b)** 72 **(c)** 180 **(d)** 120

6 In the following problems, the divisions have already been worked out. Multiply the prime numbers on the left to find the least common multiple.

(a)
$$\begin{array}{r|rr} 2 & 6 & \cancel{15} \\ \hline 3 & 3 & 15 \\ \hline 5 & \cancel{1} & 5 \\ \hline & 1 & 1 \end{array}$$

(b)
$$\begin{array}{r|rr} 2 & 20 & 36 \\ \hline 2 & 10 & 18 \\ \hline 3 & \cancel{5} & 9 \\ \hline 3 & \cancel{5} & 3 \\ \hline 5 & 5 & \cancel{1} \\ \hline & 1 & 1 \end{array}$$

Example 6 Alternative Method for Finding the Least Common Multiple

Find the least common multiple of each set of numbers.

(a) 14 and 21

Start by trying to divide 14 and 21 by the preceding list of prime numbers, 2, 3, 5, 7, 11, 13, and 17.

Use the following shortcut.

Divide by 2, the first prime

$$\begin{array}{r|rr} 2 & 14 & 21 \\ \hline & 7 & 21 \end{array}$$

Because 21 cannot be divided evenly by 2, cross 21 out and bring it down.

Divide by 3, the second prime.

$$\begin{array}{r|rr} 2 & 14 & \cancel{21} \\ \hline 3 & \cancel{7} & 21 \\ \hline & 7 & 7 \end{array}$$

Since 7 cannot be divided evenly by the third prime, 5, skip 5 and divide by the next prime, 7.

Divide by 7, the fourth prime.

$$\begin{array}{r|rr} 2 & 14 & \cancel{21} \\ \hline 3 & \cancel{7} & 21 \\ \hline 7 & 7 & 7 \\ \hline & \mathbf{1} & \mathbf{1} \end{array}$$ All quotients are 1.

When all quotients are 1, multiply the prime numbers on the left side.

$$\text{least common multiple} = \mathbf{2 \cdot 3 \cdot 7 = 42}$$

The least common multiple of 14 and 21 is 42.

(b) 6, 15, 18

Divide by 2.

$$\begin{array}{r|rrr} 2 & 6 & \cancel{15} & 18 \\ \hline & 3 & 15 & 9 \end{array}$$ Cross out 15 and bring it down.

Divide by 3.

$$\begin{array}{r|rrr} 2 & 6 & \cancel{15} & 18 \\ \hline 3 & 3 & 15 & 9 \\ \hline & 1 & 5 & 3 \end{array}$$

Divide by 3 again, since the remaining 3 can be divided.

$$\begin{array}{r|rrr} 2 & 6 & \cancel{15} & 18 \\ \hline 3 & 3 & 15 & 9 \\ \hline 3 & \cancel{1} & \cancel{5} & 3 \\ \hline & 1 & 5 & 1 \end{array}$$

Finally, divide by 5.

$$\begin{array}{r|rrr} 2 & 6 & \cancel{15} & 18 \\ \hline 3 & 3 & 15 & 9 \\ \hline 3 & \cancel{1} & \cancel{5} & 3 \\ \hline 5 & \cancel{1} & 5 & \cancel{1} \\ \hline & 1 & 1 & 1 \end{array}$$ All quotients are 1.

Multiply the prime numbers on the left side.

$$\mathbf{2 \cdot 3 \cdot 3 \cdot 5} = 90 \leftarrow \text{Least common multiple}$$

Work Problem 6 at the Side.

ANSWERS

6. **(a)** 30 **(b)** 180

5 Write a fraction with an indicated denominator. Before we can add or subtract unlike fractions, we must find the least common multiple, which is then used as the denominator of the fractions.

Example 7 Writing a Fraction with an Indicated Denominator

Write the fraction $\frac{2}{3}$ with a denominator of 15.

Find a numerator, so that

$$\frac{2}{3} = \frac{?}{15}.$$

To find the new numerator, first divide 15 by 3.

$$\frac{2}{3} = \frac{?}{15} \qquad 15 \div 3 = 5$$

Multiply both numerator and denominator of the fraction $\frac{2}{3}$ by 5.

$$\frac{2}{3} = \frac{2 \cdot 5}{3 \cdot 5} = \frac{10}{15}$$

Work Problem 7 at the Side.

This process is just the opposite of writing a fraction in lowest terms. Check the answer by writing $\frac{10}{15}$ in lowest terms; you should get $\frac{2}{3}$.

Example 8 Writing Fractions with a New Denominator

Rewrite each fraction with the indicated denominator.

(a) $\frac{3}{8} = \frac{?}{48}$

Divide 48 by 8, getting 6. Now multiply both numerator and denominator of $\frac{3}{8}$ by 6.

$$\frac{3}{8} = \frac{3 \cdot 6}{8 \cdot 6} = \frac{18}{48}$$ Multiply numerator and denominator by 6.

That is, $\frac{3}{8} = \frac{18}{48}$. As a check, write $\frac{18}{48}$ in lowest terms.

(b) $\frac{5}{6} = \frac{?}{42}$

Divide 42 by 6, getting 7. Next, multiply both numerator and denominator of $\frac{5}{6}$ by 7.

$$\frac{5}{6} = \frac{5 \cdot 7}{6 \cdot 7} = \frac{35}{42}$$ Multiply numerator and denominator by 7.

This shows that $\frac{5}{6} = \frac{35}{42}$. As a check, write $\frac{35}{42}$ in lowest terms.

7 Find the least common multiple of each set of numbers.

(a) 4, 8, 12

(b) 25 and 30

(c) 9, 24

(d) 8, 21, 24

ANSWERS

7. (a) 24 **(b)** 150 **(c)** 72 **(d)** 168

NOTE

In Example 7, the fraction $\frac{2}{3}$ was multiplied by $\frac{5}{5}$. In Example 8, the fraction $\frac{3}{8}$ was multiplied by $\frac{6}{6}$ and the fraction $\frac{5}{6}$ was multiplied by $\frac{7}{7}$. The fractions $\frac{5}{5}$, $\frac{6}{6}$, and $\frac{7}{7}$ are all equal to 1.

$$\frac{5}{5} = 1 \qquad \frac{6}{6} = 1 \qquad \frac{7}{7} = 1$$

Recall that any number multiplied by 1 is the number itself.

Work Problem 8 at the Side.

8 Rewrite each fraction with the indicated denominator.

(a) $\frac{1}{3} = \frac{?}{12}$

(b) $\frac{3}{4} = \frac{?}{8}$

(c) $\frac{4}{5} = \frac{?}{25}$

(d) $\frac{6}{11} = \frac{?}{33}$

ANSWERS

8. (a) $\frac{4}{12}$ (b) $\frac{6}{8}$ (c) $\frac{20}{25}$ (d) $\frac{18}{33}$

3.2 Exercises

FOR EXTRA HELP: Student's Solutions Manual | MyMathLab.com | InterAct Math Tutorial Software | AW Math Tutor Center | www.mathxl.com | Digital Video Tutor CD 2 Videotape 6

Use multiples of the larger number to find the least common multiple in each set of numbers. See Examples 1 and 2.

1. 2 and 4

2. 6 and 12

3. 3 and 5

4. 3 and 7

5. 4 and 9

6. 4 and 10

7. 2 and 7

8. 6 and 8

9. 6 and 10

10. 12 and 16

11. 20 and 50

12. 25 and 75

Find the least common multiple of each set of numbers. Use any method. See Examples 2–6.

13. 4, 10

14. 8, 10

15. 18, 24

16. 9 and 15

17. 6, 9, 12

18. 20, 24, 30

19. 4, 6, 8, 10

20. 8, 9, 12, 18

21. 12, 15, 18, 20

22. 6, 9, 27, 36

23. 8, 12, 16, 36

24. 5, 6, 25, 30

Rewrite each fraction with a denominator of 24. See Examples 7 and 8.

25. $\frac{1}{3} =$

26. $\frac{3}{8} =$

27. $\frac{3}{4} =$

28. $\frac{5}{12} =$

29. $\frac{5}{6} =$

30. $\frac{7}{8} =$

Rewrite each fraction with the indicated denominators.

31. $\frac{1}{4} = \frac{}{8}$

32. $\frac{2}{3} = \frac{}{12}$

33. $\frac{5}{8} = \frac{}{24}$

34. $\frac{7}{10} = \frac{}{30}$

35. $\frac{7}{8} = \frac{}{32}$

36. $\frac{5}{12} = \frac{}{48}$

37. $\frac{5}{6} = \frac{}{66}$

38. $\frac{7}{8} = \frac{}{96}$

39. $\frac{8}{5} = \frac{}{20}$

40. $\frac{7}{8} = \frac{}{40}$

41. $\frac{9}{7} = \frac{}{56}$

42. $\frac{3}{2} = \frac{}{64}$

43. $\frac{8}{3} = \frac{}{51}$

44. $\frac{7}{6} = \frac{}{120}$

45. $\frac{8}{11} = \frac{}{132}$

46. $\frac{4}{15} = \frac{}{165}$

47. $\frac{3}{16} = \frac{}{144}$

48. $\frac{7}{16} = \frac{}{112}$

49. There are several methods for finding the least common multiple (LCM). Do you prefer the method using multiples of the largest number or the method using prime factorizations? Why? Would you ever use the other method?

50. Explain in your own words how to write a fraction with an indicated denominator. As part of your explanation, show how to change $\frac{3}{4}$ to a fraction having 12 as a denominator.

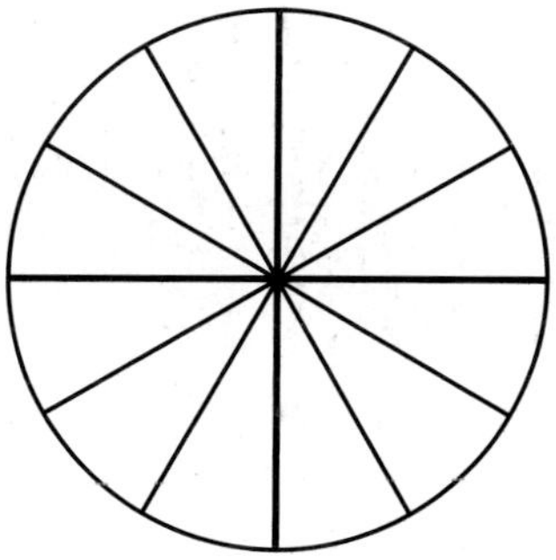

Find the least common multiple of the denominators of each pair of fractions.

51. $\frac{17}{800}, \frac{23}{3600}$

52. $\frac{53}{288}, \frac{115}{1568}$

53. $\frac{109}{1512}, \frac{23}{392}$

54. $\frac{61}{810}, \frac{37}{1170}$

RELATING CONCEPTS (Exercises 55–62) FOR INDIVIDUAL OR GROUP WORK

Most people think that addition and subtraction of fractions is more difficult than multiplication and division of fractions. This is probably because a common denominator must be used. ***Work Exercises 55–62 in order.***

55. Fractions with the same denominators are _______ fractions and fractions with different denominators are _______ fractions.

56. To subtract like fractions, first find the numerator of the difference by subtracting the ____________. Write the denominator of the like fractions as the ____________ of the difference. The answer is then written in ____________ terms.

57. The _______ common multiple (LCM) of two numbers is the ____________ whole number (smallest/largest) divisible by both those numbers.

58. The following shows the common multiples for both 8 and 10. What is the least common multiple for these two numbers?

Multiples of 8: 8, 16, 24, 32, 40, 48, 56, 64, 72, 80, 88, . . .

Multiples of 10: 10, 20, 30, 40, 50, 60, 70, 80, 90, . . .

Find the least common multiple for each set of numbers.

59. 9, 8, 18, 12

60. 25, 18, 30, 5

61. Explain why the least common multiple for 8, 3, 5, 4, and 10 is not 240. Find the least common multiple.

62. Demonstrate that the least common multiple of 55 and 1760 is 1760.

3.3 Adding and Subtracting Unlike Fractions

Objectives

1. Add unlike fractions.
2. Add fractions vertically.
3. Subtract unlike fractions.

Study Skills Workbook **Activity 9**

1 Add unlike fractions. In this section, we add and subtract unlike fractions. To add unlike fractions, we must first change them to like fractions (fractions with the same denominator). For example, the diagrams show $\frac{3}{8}$ and $\frac{1}{4}$.

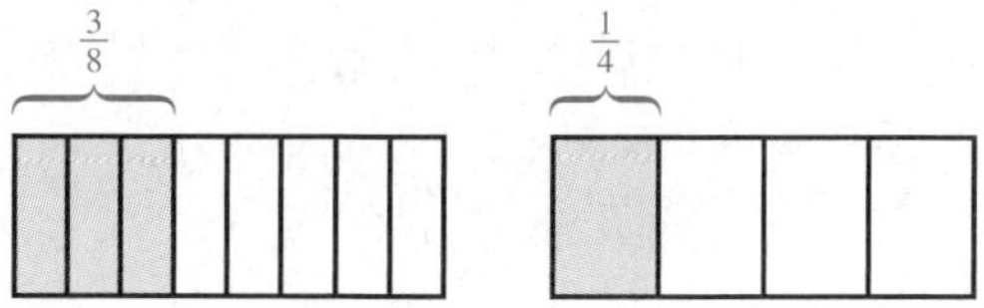

These fractions can be added by changing them to like fractions. Make like fractions by changing $\frac{1}{4}$ to the equivalent fraction $\frac{2}{8}$.

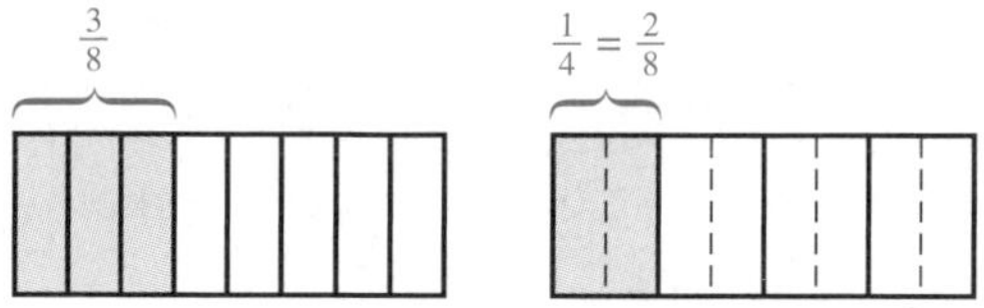

Next, add.

Becomes

$$\frac{3}{8} + \frac{1}{4} = \frac{3}{8} + \frac{2}{8} = \frac{5}{8}$$

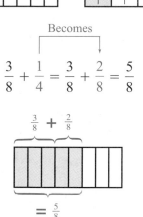

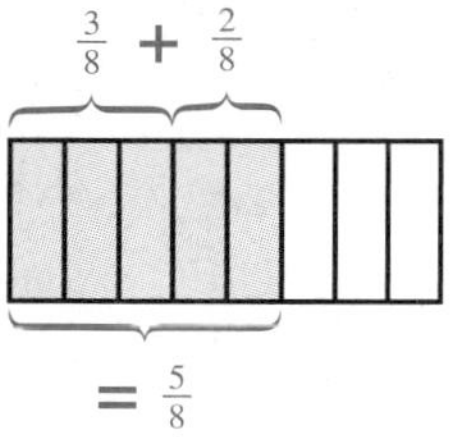

Use the following steps to add or subtract unlike fractions.

Adding or Subtracting Unlike Fractions

Step 1 Rewrite the *unlike fractions* as *like fractions* with the least common multiple as their new denominator. This new denominator is called the **least common denominator (LCD).**

Step 2 Add or subtract as with like fractions.

Step 3 Simplify the answer by writing in lowest terms and as a whole or mixed number where possible.

Example 1 **Adding Unlike Fractions**

Add $\frac{2}{3}$ and $\frac{1}{9}$.

The least common multiple of 3 and 9 is 9, so first write the fractions as like fractions with a denominator of 9. This is the ***least common denominator*** of 3 and 9.

Continued on Next Page

1 Add.

(a) $\frac{1}{2} + \frac{3}{8}$

(b) $\frac{3}{4} + \frac{1}{8}$

(c) $\frac{3}{5} + \frac{3}{10}$

(d) $\frac{1}{12} + \frac{5}{6}$

2 Add. Simplify all answers.

(a) $\frac{3}{10} + \frac{1}{5}$

(b) $\frac{5}{8} + \frac{1}{3}$

(c) $\frac{1}{10} + \frac{1}{3} + \frac{1}{6}$

ANSWERS

1. **(a)** $\frac{7}{8}$ **(b)** $\frac{7}{8}$ **(c)** $\frac{9}{10}$ **(d)** $\frac{11}{12}$

2. **(a)** $\frac{1}{2}$ **(b)** $\frac{23}{24}$ **(c)** $\frac{3}{5}$

Step 1 $\frac{2}{3} = \frac{?}{9}$

Divide 9 by 3, getting 3. Next, multiply numerator and denominator by 3.

$$\frac{2}{3} = \frac{2 \cdot 3}{3 \cdot 3} = \frac{6}{9}$$

Now, add the like fractions $\frac{6}{9}$ and $\frac{1}{9}$.

Becomes

Step 2 $\frac{2}{3} + \frac{1}{9} = \frac{6}{9} + \frac{1}{9} = \frac{6+1}{9} = \frac{7}{9}$

Step 3 Step 3 is not needed because $\frac{7}{9}$ is already in lowest terms.

Work Problem 1 at the Side.

Example 2 **Adding Fractions**

Add the following fractions using the three steps. Simplify all answers.

(a) $\frac{1}{3} + \frac{1}{6}$

The least common multiple of 3 and 6 is 6. Rewrite both fractions as fractions with a least common denominator of 6.

Rewritten as like fractions

Step 1 $\frac{1}{3} + \frac{1}{6} = \frac{2}{6} + \frac{1}{6}$

Add numerators.

Step 2 $\frac{2}{6} + \frac{1}{6} = \frac{2+1}{6} = \frac{3}{6}$

Step 3 $\frac{3}{6} = \frac{1}{2}$ ← In lowest terms

(b) $\frac{6}{15} + \frac{3}{10}$

The least common multiple of 15 and 10 is 30, so rewrite both fractions with a least common denominator of 30.

Rewritten as like fractions

Step 1 $\frac{6}{15} + \frac{3}{10} = \frac{12}{30} + \frac{9}{30}$

Add numerators.

Step 2 $\frac{12}{30} + \frac{9}{30} = \frac{12+9}{30} = \frac{21}{30}$

Step 3 $\frac{21}{30} = \frac{7}{10}$ ← In lowest terms

Work Problem 2 at the Side.

2 Add fractions vertically. Fractions can also be added vertically (up and down).

Example 3 Vertical Addition of Fractions

Add the following fractions vertically.

(a) Rewritten as like fractions

$$\begin{aligned} &\frac{3}{8} = \frac{3 \cdot 3}{8 \cdot 3} = \frac{9}{24} \\ +&\frac{7}{12} = \frac{7 \cdot 2}{12 \cdot 2} = \frac{14}{24} \\ &\qquad\qquad\qquad \frac{23}{24} \leftarrow \text{Add the numerators.} \end{aligned}$$

(b) Rewritten as like fractions

$$\begin{aligned} &\frac{2}{9} = \frac{2 \cdot 4}{9 \cdot 4} = \frac{8}{36} \\ +&\frac{1}{4} = \frac{1 \cdot 9}{4 \cdot 9} = \frac{9}{36} \\ &\qquad\qquad\qquad \frac{17}{36} \leftarrow \text{Add the numerators.} \end{aligned}$$

Work Problem 3 at the Side.

3 Add the following fractions vertically.

(a) $\begin{array}{r} \frac{3}{4} \\ +\frac{3}{16} \\ \hline \end{array}$

(b) $\begin{array}{r} \frac{5}{9} \\ +\frac{1}{3} \\ \hline \end{array}$

3 Subtract unlike fractions. The next example shows subtraction of unlike fractions.

Example 4 Subtracting Unlike Fractions

Subtract. Simplify all answers.

As with addition, rewrite unlike fractions with a least common denominator.

(a) $\frac{3}{4} - \frac{3}{8}$

Step 1 $\frac{3}{4} - \frac{3}{8} = \frac{6}{8} - \frac{3}{8}$ (Rewritten as like fractions)

Step 2 $\frac{6}{8} - \frac{3}{8} = \frac{6-3}{8} = \frac{3}{8}$ (Subtract numerators.)

Step 3 Not needed. ($\frac{3}{8}$ is in lowest terms.)

Continued on Next Page

Answers

3. (a) $\frac{15}{16}$ **(b)** $\frac{8}{9}$

4 Subtract. Simplify all answers.

(a) $\frac{5}{8} - \frac{1}{4}$

(b) $\frac{4}{5} - \frac{3}{4}$

(c) $\begin{array}{r} \frac{7}{8} \\ -\frac{2}{3} \\ \hline \end{array}$

ANSWERS

4. **(a)** $\frac{3}{8}$ **(b)** $\frac{1}{20}$ **(c)** $\frac{5}{24}$

(b) $\frac{3}{4} - \frac{5}{9}$

Rewritten as like fractions

Step 1 $\frac{\mathbf{3}}{\mathbf{4}} - \frac{\mathbf{5}}{\mathbf{9}} = \frac{\mathbf{27}}{\mathbf{36}} - \frac{\mathbf{20}}{\mathbf{36}}$

Subtract numerators.

Step 2 $\frac{27}{36} - \frac{20}{36} = \frac{27 - 20}{36} = \frac{7}{36}$

Step 3 Not needed. ($\frac{7}{36}$ is in lowest terms.)

Work Problem 4 at the Side.

3.3 EXERCISES

FOR EXTRA HELP: Student's Solutions Manual | MyMathLab.com | InterAct Math Tutorial Software | AW Math Tutor Center | www.mathxl.com | Digital Video Tutor CD 2 Videotape 6

Study Skills Workbook **Activity 9**

Add the following fractions. Simplify all answers. See Examples 1–3.

1. $\frac{1}{6}+\frac{2}{3}$

2. $\frac{1}{4}+\frac{1}{8}$

3. $\frac{2}{3}+\frac{2}{9}$

4. $\frac{3}{7}+\frac{1}{14}$

5. $\frac{9}{20}+\frac{3}{10}$

6. $\frac{5}{8}+\frac{1}{4}$

7. $\frac{3}{5}+\frac{3}{8}$

8. $\frac{5}{7}+\frac{3}{14}$

9. $\frac{2}{9}+\frac{5}{12}$

10. $\frac{5}{8}+\frac{1}{12}$

11. $\frac{1}{3}+\frac{3}{5}$

12. $\frac{2}{5}+\frac{3}{7}$

13. $\frac{1}{4}+\frac{2}{9}+\frac{1}{3}$

14. $\frac{3}{7}+\frac{2}{5}+\frac{1}{10}$

15. $\frac{3}{10}+\frac{2}{5}+\frac{3}{20}$

16. $\frac{1}{3}+\frac{3}{8}+\frac{1}{4}$

17. $\frac{4}{15}+\frac{1}{6}+\frac{1}{3}$

18. $\frac{5}{12}+\frac{2}{9}+\frac{1}{6}$

19. $\begin{array}{r}\frac{1}{3}\\ +\ \frac{1}{4}\\ \hline\end{array}$

20. $\begin{array}{r}\frac{7}{12}\\ +\ \frac{1}{8}\\ \hline\end{array}$

21. $\begin{array}{r}\frac{5}{12}\\ +\ \frac{1}{16}\\ \hline\end{array}$

22. $\begin{array}{r}\frac{3}{7}\\ +\ \frac{1}{3}\\ \hline\end{array}$

Subtract the following fractions. Simplify all answers. See Example 4.

23. $\frac{7}{8} - \frac{1}{2}$

24. $\frac{7}{8} - \frac{1}{4}$

25. $\frac{2}{3} - \frac{1}{6}$

26. $\frac{3}{4} - \frac{5}{8}$

27. $\frac{2}{3} - \frac{1}{5}$

28. $\frac{5}{6} - \frac{7}{9}$

29. $\frac{5}{12} - \frac{1}{4}$

30. $\frac{5}{7} - \frac{1}{3}$

31. $\frac{8}{9} - \frac{7}{15}$

32. $\begin{array}{r} \frac{4}{5} \\ -\frac{1}{3} \\ \hline \end{array}$

33. $\begin{array}{r} \frac{7}{8} \\ -\frac{4}{5} \\ \hline \end{array}$

34. $\begin{array}{r} \frac{5}{8} \\ -\frac{1}{3} \\ \hline \end{array}$

35. $\begin{array}{r} \frac{5}{12} \\ -\frac{1}{16} \\ \hline \end{array}$

36. $\begin{array}{r} \frac{7}{12} \\ -\frac{1}{3} \\ \hline \end{array}$

Solve each application problem.

37. To prepare her flower bed, Katie Lim ordered $\frac{1}{4}$ cubic yard of sand, $\frac{3}{8}$ cubic yard of mulch, and $\frac{1}{3}$ cubic yard of peat moss. How many cubic yards of material did she order?

38. Sergy Shibko has used his savings to repair his car. He spent $\frac{1}{4}$ of his savings for new tires, $\frac{1}{6}$ of it for brakes, $\frac{1}{10}$ of it for a tune-up, and $\frac{1}{12}$ of it for new belts and hoses. What fraction of his total savings has he spent?

39. The Weiner Works has $\frac{3}{4}$ acre of land. If $\frac{1}{6}$ acre must remain as a green belt and the remainder is buildable, find the amount of land that is buildable.

40. Della Daniel wants to open a day care center and has saved $\frac{2}{5}$ of the amount needed for start-up costs. If she saves another $\frac{1}{8}$ of the amount needed and then $\frac{1}{6}$ more, find the total portion of the start-up costs she has saved.

41. When installing cabinets, Cecil Feathers must be certain that the proper type and size of mounting screw is used. Find the total length of the screw shown.

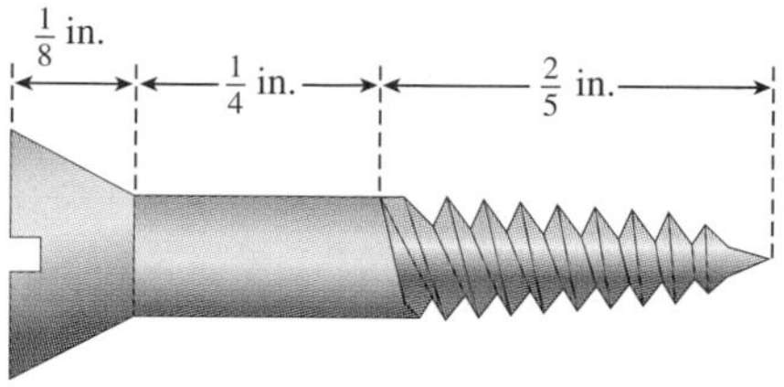

42. When installing a computer chassis, Carolyn Phelps must be certain that the proper type and size of bolt is used. Find the total length of the bolt shown.

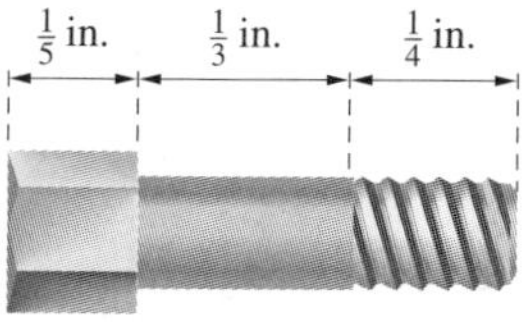

43. A hydraulic jack contains $\frac{7}{8}$ gallon of hydraulic fluid. A cracked seal resulted in a loss of $\frac{1}{6}$ gallon of fluid in the morning and another $\frac{1}{3}$ gallon in the afternoon. Find the amount of fluid remaining.

44. Adrian Ortega drives a tanker for the British Petroleum Company. He leaves the refinery with his tanker filled to $\frac{7}{8}$ of capacity. If he delivers $\frac{1}{4}$ of the tanker's capacity at the first stop and $\frac{1}{3}$ of the tanker's capacity at the second stop, find the fraction of the tanker's capacity remaining.

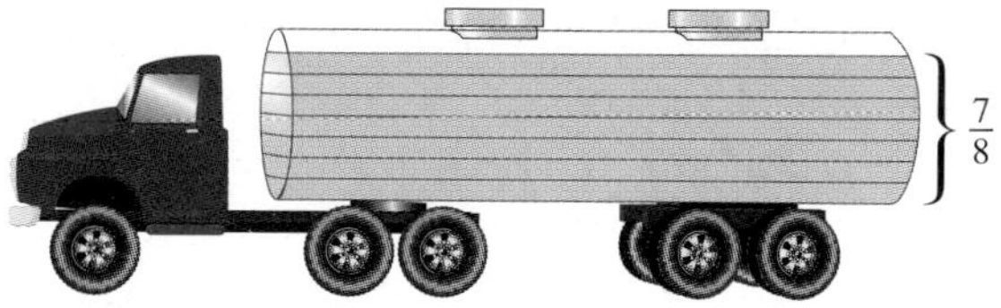

45. Step 1 in adding or subtracting unlike fractions is to rewrite the fractions so they have the least common multiple as a denominator. Explain in your own words why this is necessary.

46. Briefly list the three steps used for addition and subtraction of unlike fractions.

Refer to the circle graph to answer Exercises 47–52.

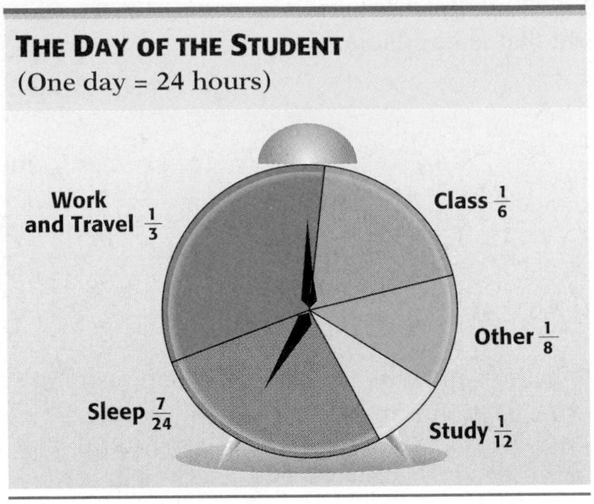

47. What fraction of the day was spent in class and study?

48. What fraction of the day was spent in work and travel and other?

49. In which activity was the greatest amount of time spent? How many hours did this activity take?

50. In which activity was the least amount of time spent? How many hours did this activity take?

51. Find the diameter of the hole in the mounting bracket shown. (The diameter is the distance across the center of the hole.)

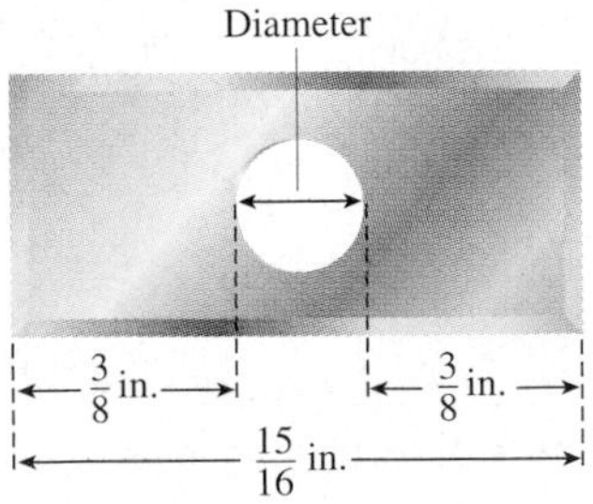

52. Chakotay is fitting a turquoise stone into a bear claw pendant. Find the diameter of the hole in the pendant. (The diameter is the distance across the center of the hole.)

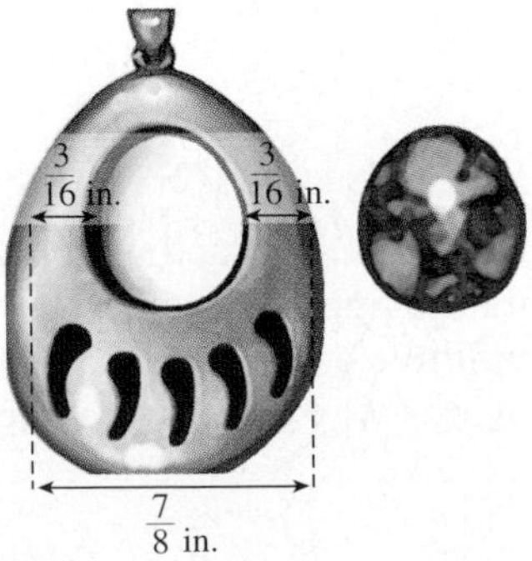

3.4 Adding and Subtracting Mixed Numbers

Objectives

1. Estimate an answer, then add or subtract mixed numbers.
2. Estimate an answer, then subtract mixed numbers by borrowing.
3. Add or subtract mixed numbers using an alternate method.

Recall that a mixed number is the sum of a whole number and a fraction. For example,

$$3\frac{2}{5} \quad \text{means} \quad 3 + \frac{2}{5}.$$

1 Estimate an answer, then add or subtract mixed numbers. Add or subtract mixed numbers by adding or subtracting the fraction parts and then the whole number parts. It is a good idea to estimate the answer first, as we did when multiplying and dividing mixed numbers in **Section 2.8.**

Work Problem 1 at the Side.

Example 1 **Adding and Subtracting Mixed Numbers**

First estimate the answer. Then add or subtract to find the exact answer.

(a) $16\frac{1}{8} + 5\frac{5}{8}$

Estimate:		*Exact:*
16	← Rounds to —	$16\frac{1}{8}$
+ 6	← Rounds to —	$+ 5\frac{5}{8}$
22		$21\frac{6}{8}$

Sum of whole numbers ↑ ↑ Sum of fractions

In lowest terms $\frac{6}{8}$ *is* $\frac{3}{4}$, so the final answer is $21\frac{3}{4}$. This is reasonable because it is close to the estimate of 22.

(b) $8\frac{5}{8} - 3\frac{1}{12}$

Estimate:		*Exact:*		
9	← Rounds to —	$8\frac{5}{8}$	=	$8\frac{15}{24}$
− 3	← Rounds to —	$-3\frac{1}{12}$	=	$-3\frac{2}{24}$
6				$5\frac{13}{24}$

Least common denominator (24)

Subtract whole numbers. ↑ ↑ Subtract fractions.

$5\frac{13}{24}$ is reasonable because it is close to the estimated answer of 6. Just as before, check by adding $5\frac{13}{24}$ and $3\frac{1}{12}$; the sum should be $8\frac{5}{8}$.

1 As a review of mixed numbers, convert each mixed number to an improper fraction and each improper fraction to a mixed number.

(a) $\frac{7}{4}$

(b) $\frac{8}{3}$

(c) $5\frac{3}{5}$

(d) $3\frac{7}{8}$

Answers

1. **(a)** $1\frac{3}{4}$ **(b)** $2\frac{2}{3}$ **(c)** $\frac{28}{5}$ **(d)** $\frac{31}{8}$

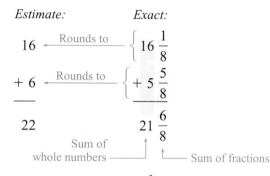

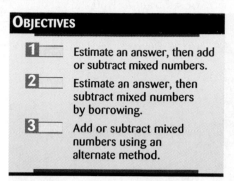

NOTE

When estimating, if the numerator is *half* of the denominator or *more*, round up the whole number part. If the numerator is *less* than *half* the denominator, leave the whole number part as it is.

Work Problem 2 at the Side.

When you add the fraction parts of mixed numbers, the sum may be greater than 1. If this happens, **carrying** from the fraction column to the whole number column is the best procedure. (You wouldn't want a whole number along with an improper fraction as the answer.)

Example 2 Carrying When Adding Mixed Numbers

First estimate, and then add $9\frac{5}{8} + 13\frac{7}{8}$.

Estimate:		*Exact:*
10	← Rounds to	$9\frac{5}{8}$
+ 14	← Rounds to	$+ 13\frac{7}{8}$
24		$22\frac{12}{8}$

Sum of whole numbers (22) — Sum of fractions ($\frac{12}{8}$)

The improper fraction $\frac{12}{8}$ can be written in lowest terms as $\frac{3}{2}$. Because $\frac{3}{2} = 1\frac{1}{2}$, the sum is

Becomes ($\frac{12}{8}$ → $1\frac{1}{2}$)

$$22\frac{12}{8} = 22 + \frac{12}{8} = 22 + 1\frac{1}{2} = 23\frac{1}{2}.$$

The estimate was 24, so the exact answer is reasonable.

NOTE

When adding mixed numbers, first add the fraction parts, then add the whole number parts. Then combine the two answers simplifying the fraction part when necessary.

Work Problem 3 at the Side.

2 Estimate an answer, then subtract mixed numbers by borrowing. Borrowing is sometimes necessary when subtracting mixed numbers.

Example 3 Borrowing When Subtracting Mixed Numbers

First estimate, and then subtract.

(a) $7 - 2\frac{5}{6}$

Continued on Next Page

2 First estimate, and then add or subtract to find the exact answer.

(a) *Estimate:* / *Exact:*

7	← Rounds to	$6\frac{7}{8}$
+ 2	← Rounds to	$+ 2\frac{1}{4}$
____		____

(b) $25\frac{3}{5} + 12\frac{3}{10}$

Estimate:

____ + ____ = ____

Exact:

____ + ____ = ____

(c) *Estimate:* / *Exact:*

	← Rounds to	$4\frac{7}{9}$
−	← Rounds to	$- 2\frac{2}{3}$
____		____

3 First estimate, and then add to find the exact answer.

(a) *Estimate:* / *Exact:*

	← Rounds to	$9\frac{3}{4}$
+	← Rounds to	$+ 7\frac{1}{2}$
____		____

(b) *Estimate:* / *Exact:*

	← Rounds to	$15\frac{4}{5}$
+	← Rounds to	$+ 12\frac{2}{3}$
____		____

ANSWERS

2. (a) $7 + 2 = 9; 9\frac{1}{8}$

(b) $26 + 12 = 38; 37\frac{9}{10}$

(c) $5 - 3 = 2; 2\frac{1}{9}$

3. (a) $10 + 8 = 18; 17\frac{1}{4}$

(b) $16 + 13 = 29; 28\frac{7}{15}$

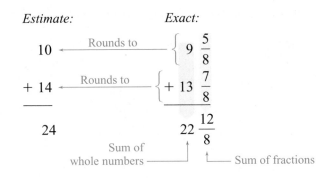

Estimate: *Exact:*

$$\begin{array}{rcl} 7 & \xleftarrow{\text{Rounds to}} & 7 \quad \leftarrow \text{There is no fraction here from which to subtract } \tfrac{5}{6}. \\ -\ 3 & \xleftarrow{\text{Rounds to}} & -2\frac{5}{6} \\ \hline 4 & & \end{array}$$

It is not possible to subtract $\frac{5}{6}$ without borrowing from the whole number 7 first.

$$\begin{aligned} 7 &= 6 + \mathbf{1} \quad \leftarrow \text{Borrow 1.} \\ &= 6 + \frac{\mathbf{6}}{\mathbf{6}} \quad \leftarrow 1 = \tfrac{6}{6} \\ &= 6\frac{6}{6} \end{aligned}$$

Now, subtract.

$$\begin{array}{rcr} 7 & = & 6\frac{6}{6} \\ -2\frac{5}{6} & = & -2\frac{5}{6} \\ \hline & & 4\frac{1}{6} \end{array}$$

The estimate was 4, so the exact answer is reasonable.

(b) $8\frac{1}{3} - 4\frac{3}{5}$

Estimate: *Exact:*

$$\begin{array}{rcrcr} 8 & \xleftarrow{\text{Rounds to}} & 8\frac{1}{3} & = & 8\frac{5}{15} \\ -\ 5 & \xleftarrow{\text{Rounds to}} & -4\frac{3}{5} & = & -4\frac{9}{15} \\ \hline 3 & & & & \end{array}$$

Least common denominator

It is not possible to subtract $\frac{9}{15}$ from $\frac{5}{15}$, so borrow from the whole number **8.**

$$\begin{aligned} 8\frac{5}{15} = 8 + \frac{5}{15} &= 7 + \mathbf{1} + \frac{5}{15} \quad \leftarrow \text{Borrow 1.} \\ &= 7 + \frac{\mathbf{15}}{\mathbf{15}} + \frac{5}{15} \quad \leftarrow 1 = \tfrac{15}{15}. \\ &= 7 + \frac{\mathbf{20}}{\mathbf{15}} \quad \leftarrow \tfrac{15}{15} + \tfrac{5}{15} \\ &= 7\frac{20}{15} \end{aligned}$$

Continued on Next Page

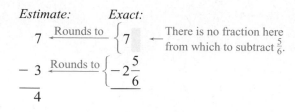

4 First estimate and then subtract to find the exact answer.

(a) *Estimate:* *Exact:*

Rounds to $7\frac{1}{3}$

$-$ Rounds to $-4\frac{5}{6}$

(b) *Estimate:* *Exact:*

Rounds to $4\frac{5}{8}$

$-$ Rounds to $-2\frac{15}{16}$

(c) *Estimate:* *Exact:*

Rounds to 15

$-$ Rounds to $-6\frac{4}{9}$

5 Add or subtract by changing mixed numbers to improper fractions. Write answers as mixed numbers in lowest terms.

(a) $3\frac{3}{8} + 2\frac{1}{2}$ **(b)** $5\frac{3}{5} + 3\frac{1}{2}$

(c) $6\frac{3}{4} - 4\frac{2}{3}$ **(d)** $9\frac{7}{8} - 4\frac{3}{4}$

Now, subtract.

$$\begin{array}{r} 8\frac{1}{3} = 8\frac{5}{15} = 7\frac{20}{15} \\ -4\frac{3}{5} = 4\frac{9}{15} = 4\frac{9}{15} \\ \hline 3\frac{11}{15} \end{array}$$

The exact answer is $3\frac{11}{15}$ (lowest terms), which is reasonable because it is close to the estimate of 3.

Work Problem 4 at the Side.

3 **Add or subtract mixed numbers using an alternate method.** An alternate method for adding or subtracting mixed numbers is to first change the mixed numbers to improper fractions. Then rewrite the unlike fractions as like fractions. Finally, add or subtract the numerators and write the answer in lowest terms.

Example 4 **Adding or Subtracting Mixed Numbers**

Add or subtract.

(a)
$$\begin{array}{rcccl} 2\frac{3}{8} & = & \frac{19}{8} & = & \frac{19}{8} \\ +3\frac{3}{4} & = & \frac{15}{4} & = & \frac{30}{8} \\ \hline & & & & \frac{49}{8} = 6\frac{1}{8} \end{array}$$

Least common denominator (8)
Change to improper fractions.
Answer as mixed number

(b)
$$\begin{array}{rcccl} 4\frac{2}{3} & = & \frac{14}{3} & = & \frac{70}{15} \\ -2\frac{1}{5} & = & \frac{11}{5} & = & \frac{33}{15} \\ \hline & & & & \frac{37}{15} = 2\frac{7}{15} \end{array}$$

Least common denominator (15)
Improper fractions
Answer as mixed number

Work Problem 5 at the Side.

NOTE

The advantage of this alternate method of adding or subtracting mixed numbers is that it eliminates the need to carry when adding or to borrow when subtracting. However, if the mixed numbers are large, then the numerators of the improper fractions become so large that they are difficult to work with. In such cases, you may want to keep the numbers as mixed numbers.

ANSWERS

4. (a) $7 - 5 = 2;\ 2\frac{1}{2}$ (b) $5 - 3 = 2;\ 1\frac{11}{16}$
(c) $15 - 6 = 9;\ 8\frac{5}{9}$

5. (a) $\frac{47}{8} = 5\frac{7}{8}$ (b) $\frac{91}{10} = 9\frac{1}{10}$
(c) $\frac{25}{12} = 2\frac{1}{12}$ (d) $\frac{41}{8} = 5\frac{1}{8}$

3.4 Exercises

FOR EXTRA HELP: Student's Solutions Manual · MyMathLab.com · InterAct Math Tutorial Software · AW Math Tutor Center · www.mathxl.com · Digital Video Tutor CD 2 Videotape 6

First estimate the answer. Then add to find the exact answer. Write answers as mixed numbers. See Examples 1 and 2.

1. *Estimate:* ___ + ___ (Rounds to) *Exact:* $6\frac{1}{3} + 2\frac{1}{2}$

2. *Estimate:* ___ + ___ *Exact:* $8\frac{3}{10} + 1\frac{3}{5}$

3. *Estimate:* ___ + ___ *Exact:* $7\frac{1}{3} + 4\frac{1}{6}$

4. *Estimate:* ___ + ___ *Exact:* $10\frac{1}{4} + 5\frac{5}{8}$

5. *Estimate:* ___ + ___ *Exact:* $\frac{5}{8} + 3\frac{7}{12}$

6. *Estimate:* ___ + ___ *Exact:* $12\frac{4}{5} + \frac{7}{10}$

7. *Estimate:* ___ + ___ *Exact:* $24\frac{5}{6} + 18\frac{5}{6}$

8. *Estimate:* ___ + ___ *Exact:* $14\frac{6}{7} + 15\frac{1}{2}$

9. *Estimate:* ___ + ___ *Exact:* $33\frac{3}{5} + 18\frac{1}{2}$

10. *Estimate:* ___ + ___ *Exact:* $18\frac{5}{8} + 6\frac{2}{3}$

11. *Estimate:* ___ + ___ *Exact:* $22\frac{3}{4} + 15\frac{3}{7}$

12. *Estimate:* ___ + ___ *Exact:* $7\frac{1}{4} + 25\frac{7}{8}$

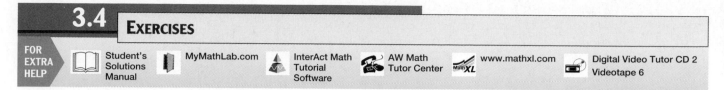

13. *Estimate:* *Exact:*

$$10\frac{3}{5} + 17\frac{7}{10} + 15\frac{8}{15}$$

14. *Estimate:* *Exact:*

$$14\frac{9}{10} + 8\frac{1}{4} + 13\frac{3}{5}$$

First estimate the answer. Then subtract to find the exact answer. Simplify all answers. See Examples 1 and 3.

15. *Estimate:* *Exact:* $12\frac{3}{8} - 10\frac{1}{4}$

16. *Estimate:* *Exact:* $14\frac{3}{4} - 11\frac{3}{8}$

17. *Estimate:* *Exact:* $12\frac{2}{3} - 1\frac{1}{5}$

18. *Estimate:* *Exact:* $11\frac{9}{20} - 4\frac{3}{5}$

19. *Estimate:* *Exact:* $28\frac{3}{10} - 6\frac{1}{15}$

20. *Estimate:* *Exact:* $15\frac{7}{20} - 6\frac{1}{8}$

21. *Estimate:* *Exact:* $17 - 6\frac{5}{8}$

22. *Estimate:* *Exact:* $22 - 4\frac{5}{8}$

23. *Estimate:* *Exact:* $18\frac{3}{4} - 5\frac{4}{5}$

24. *Estimate:* *Exact:*

$$\begin{array}{r} \\ - \underline{\quad} \end{array} \qquad \begin{array}{r} 17\frac{5}{8} \\ \underline{-\ 5\frac{2}{3}} \end{array}$$

25. *Estimate:* *Exact:*

$$\begin{array}{r} \\ - \underline{\quad} \end{array} \qquad \begin{array}{r} 16\frac{3}{8} \\ \underline{-\ 10\frac{1}{2}} \end{array}$$

26. *Estimate:* *Exact:*

$$\begin{array}{r} \\ - \underline{\quad} \end{array} \qquad \begin{array}{r} 20\frac{3}{5} \\ \underline{-\ 12\frac{7}{15}} \end{array}$$

Add or subtract by changing mixed numbers to improper fractions. Write answers as mixed numbers. See Example 4.

27. $\begin{array}{r} 3\frac{1}{2} \\ \underline{+\ 5\frac{7}{8}} \end{array}$

28. $\begin{array}{r} 8\frac{3}{4} \\ \underline{+\ 1\frac{5}{8}} \end{array}$

29. $\begin{array}{r} 4\frac{2}{3} \\ \underline{+\ 6\frac{5}{6}} \end{array}$

30. $\begin{array}{r} 7\frac{5}{12} \\ \underline{+\ 6\frac{2}{3}} \end{array}$

31. $\begin{array}{r} 2\frac{2}{3} \\ \underline{+\ 1\frac{1}{6}} \end{array}$

32. $\begin{array}{r} 4\frac{1}{2} \\ \underline{+\ 2\frac{3}{4}} \end{array}$

33. $\begin{array}{r} 3\frac{1}{4} \\ \underline{+\ 3\frac{2}{3}} \end{array}$

34. $\begin{array}{r} 2\frac{4}{5} \\ \underline{+\ 5\frac{1}{3}} \end{array}$

35. $\begin{array}{r} 1\frac{3}{8} \\ \underline{+\ 6\frac{3}{4}} \end{array}$

36. $\begin{array}{r} 1\frac{5}{12} \\ \underline{+\ 1\frac{7}{8}} \end{array}$

37. $\begin{array}{r} 3\frac{1}{2} \\ \underline{-\ 2\frac{2}{3}} \end{array}$

38. $\begin{array}{r} 4\frac{1}{4} \\ \underline{-\ 3\frac{7}{12}} \end{array}$

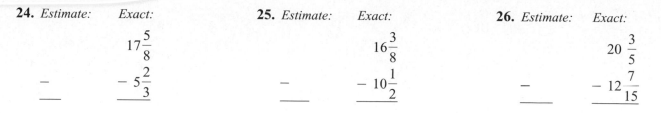

39. $\begin{array}{r} 5\frac{5}{8} \\ -\ 2\frac{3}{4} \\ \hline \end{array}$

40. $\begin{array}{r} 10\frac{1}{3} \\ -\ 6\frac{5}{6} \\ \hline \end{array}$

41. $\begin{array}{r} 7\frac{1}{4} \\ -\ 4\frac{2}{3} \\ \hline \end{array}$

42. $\begin{array}{r} 4\frac{1}{10} \\ -\ 3\frac{7}{8} \\ \hline \end{array}$

43. $\begin{array}{r} 9\frac{1}{5} \\ -\ 3\frac{3}{4} \\ \hline \end{array}$

44. $\begin{array}{r} 10\frac{2}{7} \\ -\ 5\frac{5}{14} \\ \hline \end{array}$

45. $\begin{array}{r} 6\frac{3}{7} \\ -\ 2\frac{2}{3} \\ \hline \end{array}$

46. $\begin{array}{r} 8\frac{2}{15} \\ -\ 6\frac{1}{2} \\ \hline \end{array}$

47. In your own words, explain the steps you would take to add two large mixed numbers.

48. When subtracting mixed numbers, explain when you need to borrow. Explain how to borrow using your own example.

First estimate the answer. Then solve each application problem.

49. The heaviest marine mammal in the world is the blue whale at $143\frac{3}{10}$ tons. The sixth heaviest is the humpback whale at $29\frac{1}{5}$ tons. Find the difference in their weight. (*Source: Top 10 of Everything,* 2000.)

Estimate:

Exact:

50. How much longer is the humpback whale, at $49\frac{1}{5}$ ft, than the southern elephant whale, at $21\frac{3}{10}$ ft? (*Source: Top 10 of Everything,* 2000.)

Estimate:

Exact:

51. The country with the lowest death rate is Qatar with $1\frac{3}{5}$ deaths per 1000 people each year. The death rate in the Marshall Islands is $3\frac{9}{10}$ per 1000 people. Find the difference in these death rates. (*Source:* United Nations.)

Estimate:

Exact:

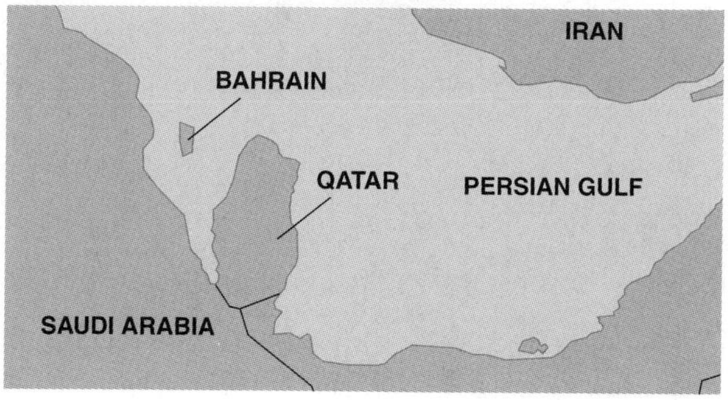

52. The death rate in the United States is $8\frac{4}{5}$ per 1000 people. The death rate in the United Arab Emirates is $2\frac{7}{10}$ per 1000 people. Find the difference in these death rates. (*Source:* United Nations.)

Estimate:

Exact:

53. A carpenter has two pieces of oak trim. One piece of trim is $12\frac{1}{2}$ ft long and the other is $8\frac{2}{3}$ ft long. How many feet of oak trim does he have in all?

Estimate:

Exact:

54. On Monday, $5\frac{3}{4}$ tons of cans were recycled, and $9\frac{3}{5}$ tons were recycled on Tuesday. How many tons were recycled on these two days?

Estimate:

Exact:

55. Andrea Abriani, a college student, works part-time at the Cyber Coffeehouse. She worked $3\frac{3}{8}$ hours on Monday, $5\frac{1}{2}$ hours on Tuesday, $4\frac{3}{4}$ hours on Wednesday, $3\frac{1}{4}$ hours on Thursday, and 6 hours on Friday. How many hours did she work altogether?

Estimate:

Exact:

56. On a recent vacation to Canada, Erin Gavin drove for $7\frac{3}{4}$ hours on the first day, $5\frac{1}{4}$ hours on the second day, $6\frac{1}{2}$ hours on the third day, and 9 hours on the fourth day. How many hours did she drive altogether?

Estimate:

Exact:

57. A craftsperson must attach a lead strip around all four sides of a stained glass window before it is installed. Find the length of lead stripping needed.

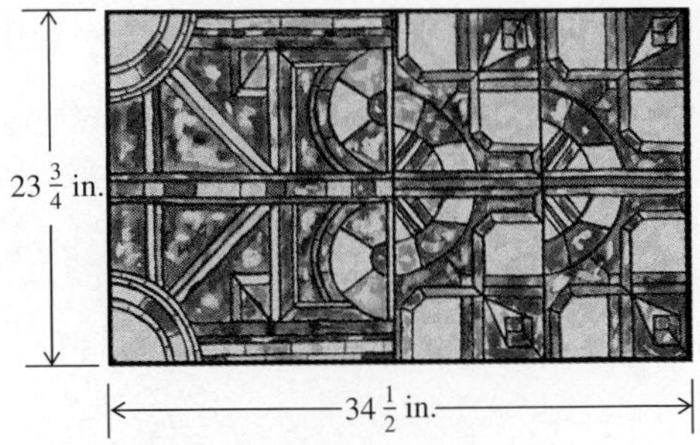

Estimate:

Exact:

58. To complete a custom order, Zak Morten of Home Depot must find the number of inches of brass trim needed to go around the four sides of the lamp base plate shown. Find the length of brass trim needed.

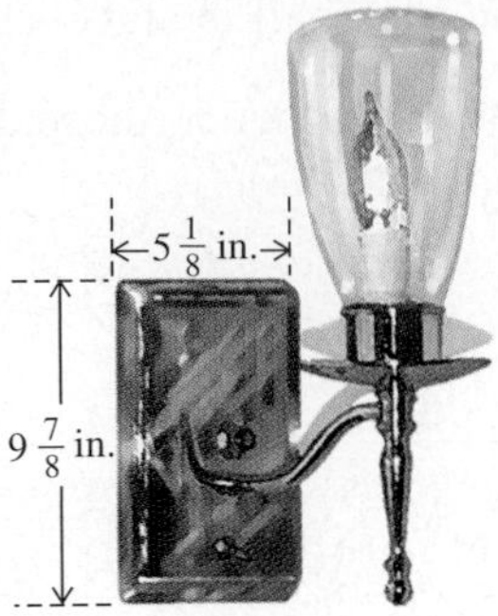

Estimate:

Exact:

59. A cement-truck driver has $8\frac{7}{8}$ cubic yards of concrete in a truck. If he unloads $2\frac{1}{2}$ cubic yards at the first stop, 3 cubic yards at the second stop, and $1\frac{3}{4}$ cubic yards at the third stop, how much concrete remains in the truck?

Estimate:

Exact:

60. Marv Levenson bought 15 yd of Italian silk fabric. He made two shirts with $3\frac{3}{4}$ yd of the material, a suit for his wife with $4\frac{1}{8}$ yd, and a jacket with $3\frac{7}{8}$ yd. Find the number of yards of material remaining.

Estimate:

Exact:

61. The exercise yard at the correction center has four sides and is surrounded by $527\frac{1}{24}$ ft of security fencing. If three sides of the yard measure $107\frac{2}{3}$ ft, $150\frac{3}{4}$ ft, and $138\frac{5}{8}$ ft, find the length of the fourth side.

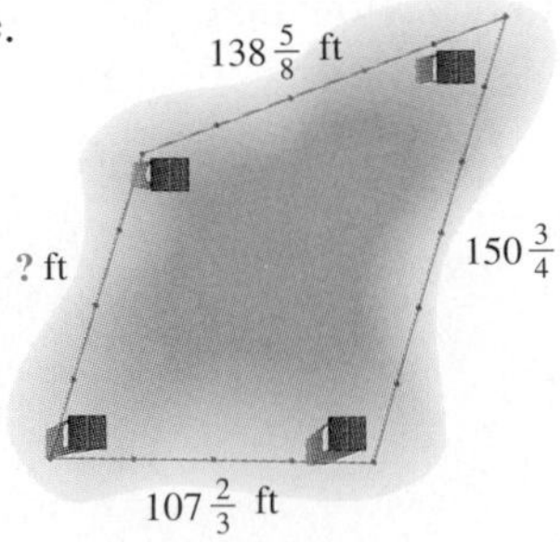

Estimate:

Exact:

62. Three sides of a parking lot are $108\frac{1}{4}$ ft, $162\frac{3}{8}$ ft, and $143\frac{1}{2}$ ft. If the distance around the lot is $518\frac{3}{4}$ ft, find the length of the fourth side.

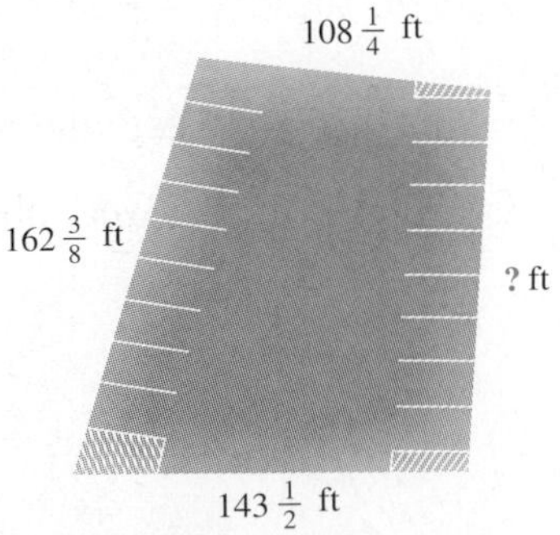

Estimate:

Exact:

63. A truck trailer is to be loaded with personal computers weighing $2\frac{5}{8}$ tons, computer monitors weighing $6\frac{1}{2}$ tons, computer printers weighing $1\frac{5}{6}$ tons, and assorted accessories weighing $3\frac{1}{4}$ tons. If the truck trailer weighs $7\frac{3}{8}$ tons empty, find the total weight after it has been loaded.

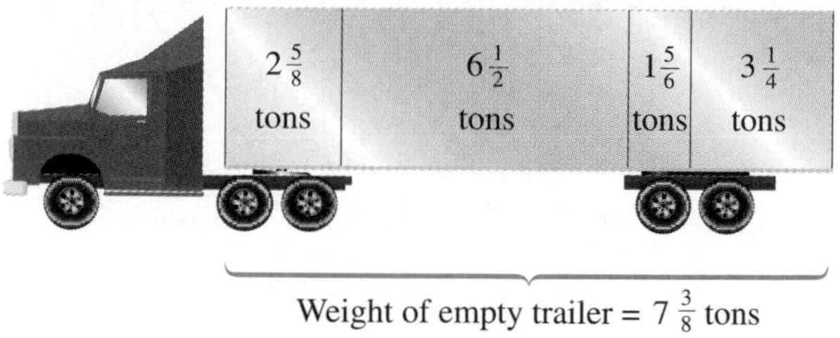

Estimate:

Exact:

64. Comet Auto Supply sold $16\frac{1}{2}$ cases of generic brand oil last week, $12\frac{1}{8}$ cases of Havoline oil, $8\frac{3}{4}$ cases of Valvoline oil, and $12\frac{5}{8}$ cases of Castrol oil. Find the total number of cases of oil sold during the week.

Estimate:

Exact:

Find the unknown length, labeled with a question mark, in each figure.

65.

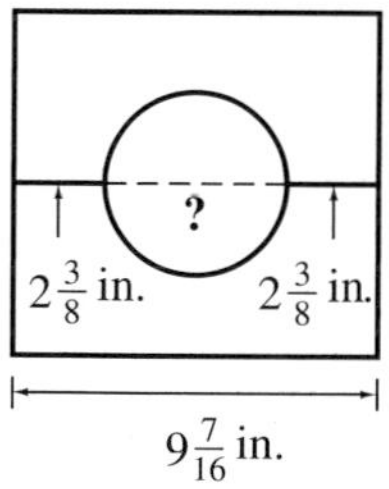

66.

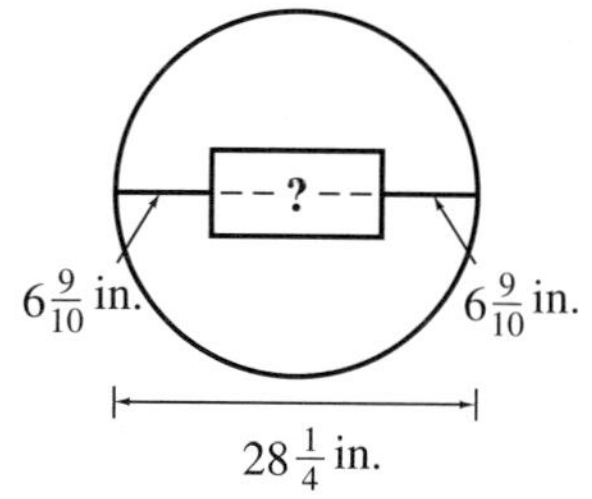

67.

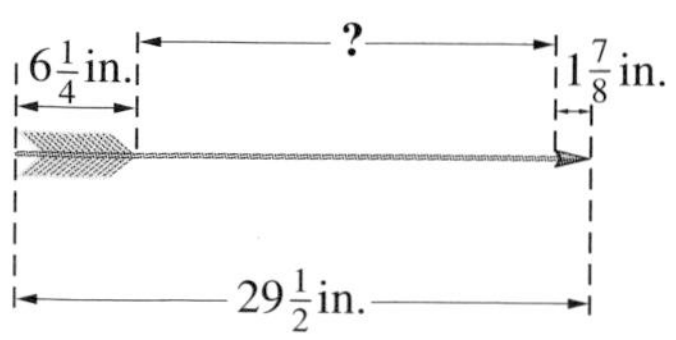

68.

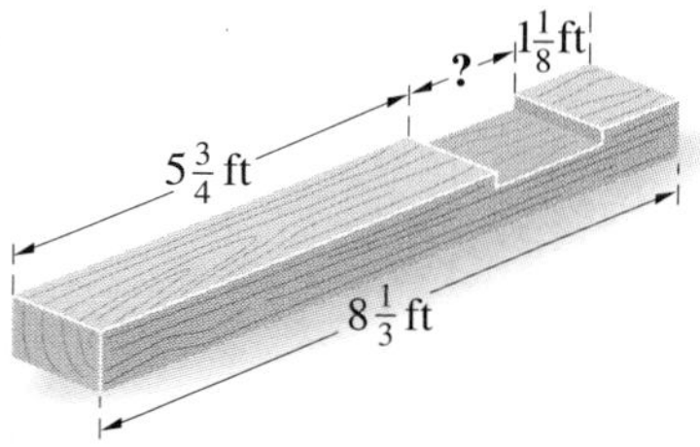

Relating Concepts (Exercises 69–74) For Individual or Group Work

Most fraction problems include fractions with different denominators.
Work Exercises 69–74 in order.

69. To add or subtract fractions, we must first rewrite them as like fractions. Rewrite each fraction with the indicated denominator.

(a) $\frac{5}{9} = \frac{}{54}$ **(b)** $\frac{7}{12} = \frac{}{48}$

(c) $\frac{5}{8} = \frac{}{40}$ **(d)** $\frac{11}{5} = \frac{}{120}$

70. When rewriting unlike fractions as like fractions with the least common multiple as a denominator, the new denominator is called the ____________ ____________ ____________, or LCD.

71. Add or subtract as indicated. Write answers in lowest terms.

(a) $\frac{5}{8} + \frac{1}{3}$ **(b)** $\frac{19}{20} - \frac{5}{12}$

(c) $\frac{7}{12} + \frac{3}{16} + \frac{3}{24}$ **(d)** $\frac{6}{7} - \frac{2}{3}$

72. A common method for adding or subtracting mixed numbers is to add or subtract the ____________ ____________ and then the whole number parts.

73. Another method for adding or subtracting mixed numbers is to first change the mixed numbers to ____________ fractions. After adding or subtracting, write the answer as a mixed number in lowest terms.

74. Add or subtract the following fractions as indicated. First use the method where you add or subtract fraction parts and then whole number parts. Then use the method where you change each mixed number to an improper fraction, add or subtract, and then write the answer as a mixed number in lowest terms. Do you get the same answer using both methods? Which method do you prefer?

(a) $4\frac{5}{8} + 3\frac{1}{4}$ **(b)** $12\frac{2}{5} - 8\frac{7}{8}$

3.5 ORDER RELATIONS AND THE ORDER OF OPERATIONS

There are times when we want to compare the size of two numbers. For example, we might want to know which is the greater amount, the larger size, or the longer distance.

Fractions, like whole numbers, can be located on a number line. For fractions, divide the space between whole numbers into equal parts.

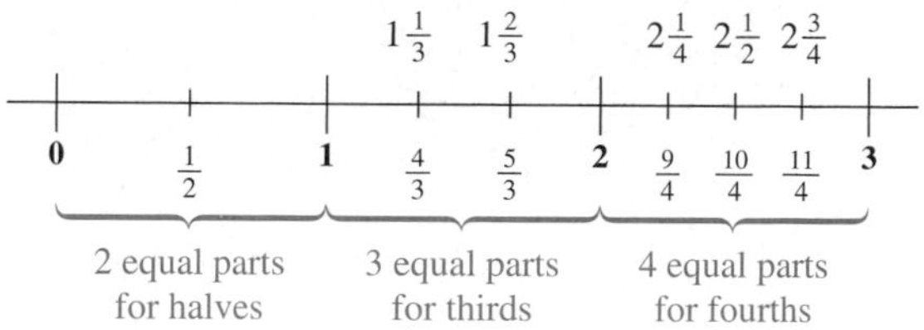

OBJECTIVES

1. Identify the greater of two fractions.
2. Use exponents with fractions.
3. Use the order of operations.

1 Identify the greater of two fractions. To compare the size of two numbers, place the two numbers on a number line and use the following rule.

Comparing the Size of Two Numbers

The number farther to the left on the number line is always less and the number farther to the right on the number line is always greater.

For example, on the preceding number line, $\frac{1}{2}$ is to the *left* of $\frac{4}{3}$ $(1\frac{1}{3})$, so $\frac{1}{2}$ is less than $\frac{4}{3}$ $(1\frac{1}{3})$.

Work Problem 1 at the Side.

Write *order relations* by using the following symbols.

Symbols Used to Show Order Relations

<	is less than
>	is greater than

Example 1 Using Less-Than and Greater-Than Symbols

Rewrite the following using < and > symbols.

(a) $\frac{1}{2}$ is less than $\frac{4}{3}$.

$\frac{1}{2}$ **is less than** $\frac{4}{3}$ is written as $\frac{1}{2} < \frac{4}{3}$.

(b) $\frac{9}{4}$ **is greater than** 1.

$\frac{9}{4}$ **is greater than** 1 is written as $\frac{9}{4} > 1$.

(c) $\frac{5}{3}$ is less than $\frac{11}{4}$.

$\frac{5}{3}$ **is less than** $\frac{11}{4}$ is written as $\frac{5}{3} < \frac{11}{4}$.

1 Locate each fraction on the number line.

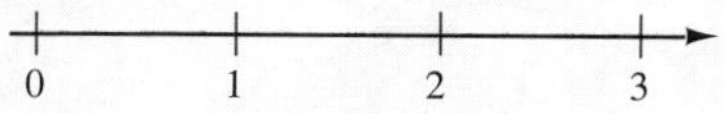

(a) $\frac{1}{4}$

(b) $\frac{2}{3}$

(c) $1\frac{1}{2}$

(d) $2\frac{3}{4}$

ANSWERS

1.

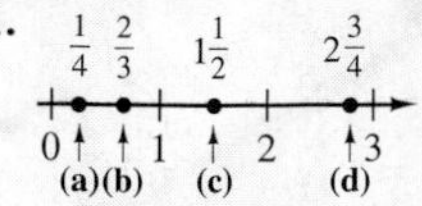

2 Use the number line in the text to help you write < or > in each blank to make a true statement.

(a) $1 \_\_\_\_ \frac{5}{4}$

(b) $\frac{8}{3} \_\_\_\_ \frac{3}{2}$

(c) $0 \_\_\_\_ 1$

(d) $\frac{17}{8} \_\_\_\_ \frac{8}{4}$

3 Write < or > in each blank to make a true statement.

(a) $\frac{2}{3} \_\_\_\_ \frac{5}{8}$

(b) $\frac{5}{9} \_\_\_\_ \frac{13}{8}$

(c) $\frac{3}{4} \_\_\_\_ \frac{5}{6}$

(d) $\frac{9}{10} \_\_\_\_ \frac{14}{15}$

ANSWERS
2. (a) < **(b)** > **(c)** < **(d)** >
3. (a) > **(b)** < **(c)** < **(d)** <

NOTE

A number line is a very useful tool when working with order relations.

Work Problem 2 at the Side.

The fraction $\frac{7}{8}$ represents 7 of 8 equivalent parts, while $\frac{3}{8}$ means 3 of 8 equivalent parts. Because $\frac{7}{8}$ represents more of the equivalent parts, $\frac{7}{8}$ is greater than $\frac{3}{8}$, or

$$\frac{7}{8} > \frac{3}{8}.$$

To identify the greater fraction, use the following steps.

Identifying the Greater Fraction

Step 1 Write the fractions as like fractions (same denominators).

Step 2 Compare the numerators. The fraction with the greater numerator is the greater fraction.

Example 2 Identifying the Greater Fraction

Determine which fraction in each pair is greater.

(a) $\frac{7}{8}, \frac{9}{10}$

First, write the fractions as like fractions. The least common multiple for 8 and 10 is 40, so

$$\frac{7}{8} = \frac{7 \cdot 5}{8 \cdot 5} = \frac{35}{40} \quad \text{and} \quad \frac{9}{10} = \frac{9 \cdot 4}{10 \cdot 4} = \frac{36}{40}.$$

Look at the numerators. Because 36 is greater than 35, $\frac{36}{40}$ is greater than $\frac{35}{40}$. Because $\frac{36}{40}$ is equivalent to $\frac{9}{10}$,

$$\frac{9}{10} > \frac{7}{8} \quad \text{or} \quad \frac{7}{8} < \frac{9}{10}.$$

The greater fraction is $\frac{9}{10}$.

(b) $\frac{8}{5}, \frac{23}{15}$

The least common multiple of 5 and 15 is 15.

$$\frac{8}{5} = \frac{8 \cdot 3}{5 \cdot 3} = \frac{24}{15} \quad \text{and} \quad \frac{23}{15} = \frac{23}{15}$$

This shows that $\frac{8}{5}$ is greater than $\frac{23}{15}$, or

$$\frac{8}{5} > \frac{23}{15}.$$

Work Problem 3 at the Side.

2 Use exponents with fractions. Exponents were used in Chapter 1 to write repeated multiplication. For example,

$$\underbrace{3^2}_{\text{Exponent}} = \underbrace{3 \cdot 3}_{\text{Two factors of 3}} = 9 \quad \text{and} \quad \underbrace{5^3}_{\text{Exponent}} = \underbrace{5 \cdot 5 \cdot 5}_{\text{Three factors of 5}} = 125.$$

The next example shows exponents used with fractions.

Example 3 Using Exponents with Fractions

Simplify.

(a) $\left(\frac{1}{2}\right)^3$

$$\left(\frac{1}{2}\right)^3 = \overbrace{\frac{1}{2} \cdot \frac{1}{2} \cdot \frac{1}{2}}^{\text{Three factors of } \frac{1}{2}} = \frac{1}{8}$$

(b) $\left(\frac{5}{8}\right)^2$

$$\left(\frac{5}{8}\right)^2 = \overbrace{\frac{5}{8} \cdot \frac{5}{8}}^{\text{Two factors of } \frac{5}{8}} = \frac{25}{64}$$

(c) $\left(\frac{3}{4}\right)^2 \cdot \left(\frac{2}{3}\right)^3$

$$\left(\frac{3}{4}\right)^2 \cdot \left(\frac{2}{3}\right)^3 = \left(\frac{3}{4} \cdot \frac{3}{4}\right) \cdot \left(\frac{2}{3} \cdot \frac{2}{3} \cdot \frac{2}{3}\right)$$

$$= \frac{\overset{1}{\cancel{3}} \cdot \overset{1}{\cancel{3}} \cdot \overset{1}{\cancel{2}} \cdot \overset{1}{\cancel{2}} \cdot \overset{1}{\cancel{2}}}{\underset{2}{\cancel{4}} \cdot \underset{\underset{1}{\cancel{2}}}{\cancel{4}} \cdot \underset{1}{\cancel{3}} \cdot \underset{1}{\cancel{3}} \cdot 3} \quad \text{Divide out all the common factors.}$$

$$= \frac{1}{6}$$

Work Problem 4 at the Side.

3 Use the order of operations. Recall the ***order of operations*** from Chapter 1.

Order of Operations

1. Do all operations inside *parentheses or other grouping symbols.*
2. Simplify any expressions with *exponents* and find any *square roots.*
3. *Multiply* or *divide* proceeding from left to right.
4. *Add* or *subtract* proceeding from left to right.

4 Simplify.

(a) $\left(\frac{1}{4}\right)^2$

(b) $\left(\frac{3}{8}\right)^2$

(c) $\left(\frac{1}{2}\right)^3 \cdot \left(\frac{2}{3}\right)^2$

(d) $\left(\frac{1}{5}\right)^2 \cdot \left(\frac{5}{3}\right)^2$

ANSWERS

4. **(a)** $\frac{1}{16}$ **(b)** $\frac{9}{64}$ **(c)** $\frac{1}{18}$ **(d)** $\frac{1}{9}$

5 Simplify by using the order of operations.

(a) $\frac{5}{9} - \frac{3}{4} \cdot \frac{2}{3}$

(b) $\frac{3}{5} \cdot \left(\frac{3}{4} - \frac{1}{3}\right)$

(c) $\frac{7}{8} \cdot \frac{2}{3} - \left(\frac{1}{2}\right)^2$

(d) $\dfrac{\left(\frac{5}{6}\right)^2}{\frac{4}{3}}$

ANSWERS

5. (a) $\frac{1}{18}$ **(b)** $\frac{1}{4}$ **(c)** $\frac{1}{3}$ **(d)** $\frac{25}{48}$

The next example shows how to apply the order of operations with fractions.

Example 4 Using Order of Operations with Fractions

Simplify by using the order of operations.

(a) $\frac{1}{3} + \frac{1}{2} \cdot \frac{4}{5}$

Multiply first.

$$\frac{1}{3} + \frac{1}{\cancel{2}_1} \cdot \frac{\cancel{4}^2}{5} = \frac{1}{3} + \frac{2}{5}$$

Next, add. The least common denominator of 3 and 5 is 15.

$$\frac{1}{3} + \frac{2}{5} = \frac{5}{15} + \frac{6}{15} = \frac{11}{15}$$

(b) $\frac{3}{8} \cdot \left(\frac{1}{2} + \frac{1}{3}\right)$

$$\frac{3}{8} \cdot \left(\frac{1}{2} + \frac{1}{3}\right) = \frac{3}{8} \cdot \underbrace{\left(\frac{3}{6} + \frac{2}{6}\right)}_{\text{Work inside parentheses first.}}$$

$$= \frac{3}{8} \cdot \frac{5}{6}$$

$$= \frac{\cancel{3}^1}{8} \cdot \frac{5}{\cancel{6}_2} \quad \text{Divide numerator and denominator by 3.}$$

$$= \frac{5}{16} \quad \text{Multiply.}$$

(c) $\left(\frac{2}{3}\right)^2 - \frac{4}{5} \cdot \frac{1}{2}$

$$\left(\frac{2}{3}\right)^2 - \frac{4}{5} \cdot \frac{1}{2} = \frac{4}{9} - \frac{4}{5} \cdot \frac{1}{2}$$

First simplify the expression with the exponent. $\frac{2}{3} \cdot \frac{2}{3}$ is $\frac{4}{9}$.

$$= \frac{4}{9} - \frac{\cancel{4}^2}{5} \cdot \frac{1}{\cancel{2}_1} \quad \text{Next, multiply.}$$

$$= \frac{4}{9} - \frac{2}{5}$$

$$= \frac{20}{45} - \frac{18}{45} \quad \text{Subtract last. (Least common denominator is 45.)}$$

$$= \frac{2}{45}$$

Work Problem 5 at the Side.

3.5 EXERCISES

FOR EXTRA HELP: Student's Solutions Manual | MyMathLab.com | InterAct Math Tutorial Software | AW Math Tutor Center | www.mathxl.com | Digital Video Tutor CD 2 Videotape 6

Locate each fraction in Exercises 1–12 on the following number line. See Margin Problem 1.

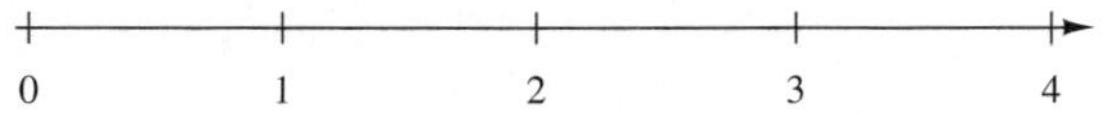

1. $\frac{1}{2}$ **2.** $\frac{1}{4}$ **3.** $\frac{3}{2}$ **4.** $\frac{5}{4}$ **5.** $\frac{7}{3}$ **6.** $\frac{11}{4}$

7. $2\frac{1}{6}$ **8.** $3\frac{4}{5}$ **9.** $\frac{7}{2}$ **10.** $\frac{7}{8}$ **11.** $3\frac{1}{4}$ **12.** $1\frac{7}{8}$

Write < or > to make a true statement. See Examples 1 and 2.

13. $\frac{1}{2}$ ____ $\frac{3}{8}$ **14.** $\frac{5}{8}$ ____ $\frac{3}{4}$ **15.** $\frac{5}{6}$ ____ $\frac{11}{12}$ **16.** $\frac{13}{18}$ ____ $\frac{5}{6}$

17. $\frac{5}{12}$ ____ $\frac{3}{8}$ **18.** $\frac{7}{15}$ ____ $\frac{9}{20}$ **19.** $\frac{7}{12}$ ____ $\frac{11}{18}$ **20.** $\frac{17}{24}$ ____ $\frac{5}{6}$

21. $\frac{11}{18}$ ____ $\frac{5}{9}$ **22.** $\frac{13}{15}$ ____ $\frac{8}{9}$ **23.** $\frac{37}{50}$ ____ $\frac{13}{20}$ **24.** $\frac{7}{12}$ ____ $\frac{11}{20}$

Simplify. See Example 3.

25. $\left(\frac{1}{2}\right)^2$ **26.** $\left(\frac{2}{3}\right)^2$ **27.** $\left(\frac{5}{7}\right)^2$

28. $\left(\frac{7}{8}\right)^2$ **29.** $\left(\frac{3}{4}\right)^2$ **30.** $\left(\frac{3}{5}\right)^3$

31. $\left(\frac{4}{5}\right)^3$ **32.** $\left(\frac{4}{7}\right)^3$ **33.** $\left(\frac{3}{2}\right)^4$

34. $\left(\frac{4}{3}\right)^4$ **35.** $\left(\frac{3}{4}\right)^4$ **36.** $\left(\frac{2}{3}\right)^5$

37. Describe in your own words what a number line is, and draw a picture of one. Be sure to include how it works and how it can be used.

38. You have used the order of operations with whole numbers and again with fractions. List from memory the steps in the order of operations.

Use the order of operations to simplify each expression. See Example 4.

39. $3 + 4 - 2^2$

40. $2^2 + 5 \cdot 1$

41. $2 \cdot 3^2 - \frac{4}{2}$

42. $5 \cdot 2^3 - \frac{6}{2}$

43. $\left(\frac{1}{2}\right)^2 \cdot 4$

44. $\left(\frac{1}{4}\right)^2 \cdot 4$

45. $\left(\frac{3}{4}\right)^2 \cdot \left(\frac{1}{3}\right)$

46. $\left(\frac{2}{3}\right)^3 \cdot \left(\frac{1}{2}\right)$

47. $\left(\frac{4}{5}\right)^2 \cdot \left(\frac{5}{6}\right)^2$

48. $\left(\frac{5}{8}\right)^2 \cdot \left(\frac{4}{25}\right)^2$

49. $6 \cdot \left(\frac{2}{3}\right)^2 \cdot \left(\frac{1}{2}\right)^3$

50. $9 \cdot \left(\frac{1}{3}\right)^3 \cdot \left(\frac{4}{3}\right)^2$

51. $\frac{3}{5} \cdot \frac{1}{3} + \frac{2}{5} \cdot \frac{3}{4}$

52. $\frac{1}{4} \cdot \frac{3}{4} + \frac{3}{8} \cdot \frac{4}{3}$

53. $\frac{1}{2} + \left(\frac{1}{2}\right)^2 - \frac{3}{8}$

54. $\frac{2}{3} + \left(\frac{1}{3}\right)^2 - \frac{5}{9}$

55. $\left(\frac{1}{3}+\frac{1}{6}\right)\cdot\frac{1}{2}$

56. $\left(\frac{3}{5}-\frac{3}{20}\right)\cdot\frac{4}{3}$

57. $\frac{9}{8}\div\left(\frac{2}{3}+\frac{1}{12}\right)$

58. $\frac{6}{5}\div\left(\frac{3}{5}-\frac{3}{10}\right)$

59. $\left(\frac{7}{8}-\frac{3}{4}\right)\div\frac{3}{2}$

60. $\left(\frac{4}{5}-\frac{3}{10}\right)\div\frac{4}{5}$

61. $\frac{3}{8}\cdot\left(\frac{1}{4}+\frac{1}{2}\right)\cdot\frac{32}{3}$

62. $\frac{1}{3}\cdot\left(\frac{4}{5}-\frac{3}{10}\right)\cdot\frac{4}{2}$

63. $\left(\frac{3}{4}\right)^2-\left(\frac{1}{2}-\frac{1}{6}\right)\div\frac{4}{3}$

64. $\left(\frac{2}{3}\right)^2-\left(\frac{5}{8}-\frac{1}{2}\right)\div\frac{3}{2}$

65. $\left(\frac{7}{8}-\frac{1}{4}\right)-\left(\frac{3}{4}\right)^2\cdot\frac{2}{3}$

66. $\left(\frac{5}{6}-\frac{7}{12}\right)-\left(\frac{1}{3}\right)^2\cdot\frac{3}{4}$

67. $\left(\frac{3}{4}\right)^2\cdot\left(\frac{2}{3}-\frac{5}{9}\right)-\frac{1}{4}\cdot\frac{1}{8}$

68. $\left(\frac{2}{3}\right)^2\cdot\left(\frac{1}{2}-\frac{1}{8}\right)-\frac{2}{3}\cdot\frac{1}{8}$

Solve each application problem.

69. In a study of workplace security, $\frac{9}{25}$ of the employees thought that more lighting on the grounds and parking lot was most important, while $\frac{7}{20}$ of the employees thought that limited public access was most important. Which group represents the greater number of employees? (*Source:* Society for Human Resource Management.)

$\frac{9}{25}$

$\frac{7}{20}$

70. When adults were surveyed about conserving and recycling, $\frac{3}{4}$ of them said they conserve water while $\frac{19}{25}$ of them said they use reusable containers at home. Which group represents the greater number of adults? (*Source:* The NPD Group of the Rechargeable Battery Recycling Corporation.)

RELATING CONCEPTS (Exercises 71–80) FOR INDIVIDUAL OR GROUP WORK

Problems involving order relations and order of operations commonly occur in problem solving. ***Work Exercises 71–80 in order.***

71. When comparing the size of two numbers, we use the symbol ______ for **is less than** and the symbol ______ for **is greater than.**

72. (a) To identify the greater of two or more fractions, we must first write the fractions as ____________ fractions and then compare the ____________. The fraction with the greater ____________ is the greater fraction.

(b) Write four pairs of fractions, all with different denominators, using the symbols for **less than** and **greater than.**

73. Fill in the blanks to complete the order of operations.

1. Do all operations inside ____________ and other grouping symbols.
2. Simplify any expressions with ____________ and find any ____________ roots.
3. ____________ or ____________ proceeding from left to right.
4. ____________ or ____________ proceeding from left to right.

74. Use the order of operations to simplify the following.

$$\left(\frac{2}{3}\right)^2 - \left(\frac{4}{5} - \frac{3}{10}\right) \div \frac{5}{4}$$

Simplify, then place the results on the number line.

75. $\left(\frac{2}{3}\right)^2$

76. $\left(\frac{3}{8}\right)^2$

77. $\left(\frac{8}{7}\right)^3$

78. $\left(\frac{5}{4}\right)^4$

79. $4 + 2 - 2^2$

80. $\left(\frac{2}{3}\right)^2 - \left(\frac{4}{5} - \frac{3}{10}\right) \div \frac{5}{4}$

Summary Exercises on FRACTIONS

Write proper *or* improper *for each fraction.*

1. $\frac{7}{8}$ **2.** $\frac{6}{5}$ **3.** $\frac{8}{8}$ **4.** $\frac{9}{10}$

Write each fraction in lowest terms.

5. $\frac{24}{30}$ **6.** $\frac{175}{200}$ **7.** $\frac{15}{35}$ **8.** $\frac{115}{235}$

Add, subtract, multiply, or divide as indicated. Simplify all answers.

9. $\frac{3}{4} \cdot \frac{2}{3}$ **10.** $\frac{7}{12} \cdot \frac{9}{14}$ **11.** $56 \cdot \frac{5}{8}$

12. $\frac{5}{8} \div \frac{3}{4}$ **13.** $\frac{35}{45} \div \frac{10}{15}$ **14.** $21 \div \frac{3}{8}$

15. $\frac{7}{8} + \frac{2}{3}$ **16.** $\frac{5}{8} + \frac{3}{4} + \frac{7}{16}$ **17.** $\frac{7}{12} + \frac{5}{6} + \frac{2}{3}$

18. $\frac{5}{6} - \frac{3}{4}$ **19.** $\frac{7}{8} - \frac{5}{12}$ **20.** $\frac{4}{5} - \frac{2}{3}$

First estimate the answer. Then add or subtract to find the exact answer.

21. *Exact:*

$3\frac{1}{2} \cdot 2\frac{1}{4}$

Estimate:

___ • ___ = ___

22. *Exact:*

$5\frac{3}{8} \cdot 3\frac{1}{4}$

Estimate:

___ • ___ = ___

23. *Exact:*

$8 \cdot 5\frac{2}{3} \cdot 2\frac{3}{8}$

Estimate:

___ • ___ • ___ = ___

24. *Exact:*

$4\frac{3}{8} \div 3\frac{3}{4}$

Estimate:

___ ÷ ___ = ___

25. *Exact:*

$6\frac{7}{8} \div 2$

Estimate:

___ ÷ ___ = ___

26. *Exact:*

$4\frac{5}{8} \div \frac{3}{4}$

Estimate:

___ ÷ ___ = ___

27. *Estimate:* *Exact:*

$$\begin{array}{r} \underline{\quad} \xleftarrow{\text{Rounds to}} 4\frac{3}{8} \\ + \underline{\quad} \xleftarrow{\text{Rounds to}} +\,3\frac{1}{2} \\ \hline \end{array}$$

28. *Estimate:* *Exact:*

$$\begin{array}{r} \\ + \\ \hline \end{array} \qquad \begin{array}{r} 21\frac{3}{4} \\ +\,8\frac{5}{12} \\ \hline \end{array}$$

29. *Estimate:* *Exact:*

$$\begin{array}{r} \\ + \\ \hline \end{array} \qquad \begin{array}{r} 14\frac{3}{5} \\ +\,10\frac{2}{3} \\ \hline \end{array}$$

30. *Estimate:* *Exact:*

$$\begin{array}{r} \\ - \\ \hline \end{array} \qquad \begin{array}{r} 7\frac{1}{2} \\ -\,2\frac{3}{5} \\ \hline \end{array}$$

31. *Estimate:* *Exact:*

$$\begin{array}{r} \\ - \\ \hline \end{array} \qquad \begin{array}{r} 14 \\ -\,7\frac{3}{8} \\ \hline \end{array}$$

32. *Estimate:* *Exact:*

$$\begin{array}{r} \\ - \\ \hline \end{array} \qquad \begin{array}{r} 31\frac{5}{6} \\ -\,22\frac{7}{12} \\ \hline \end{array}$$

Simplify by using the order of operations.

33. $\left(\frac{2}{3} - \frac{1}{4}\right) \cdot \frac{1}{5}$

34. $\frac{3}{4} \div \left(\frac{1}{2} + \frac{1}{3}\right)$

35. $\frac{2}{3} + \left(\frac{2}{3}\right)^2 - \frac{5}{6}$

Find the least common multiple of each set of numbers.

36. 8, 10

37. 9, 18, 24

38. 4, 12, 21

Write each fraction by using the indicated denominator.

39. $\frac{5}{6} = \frac{\quad}{42}$

40. $\frac{3}{7} = \frac{\quad}{28}$

41. $\frac{11}{12} = \frac{\quad}{60}$

Write < or > to make a true statement.

42. $\frac{7}{8}$ ____ $\frac{15}{16}$

43. $\frac{16}{20}$ ____ $\frac{23}{30}$

44. $\frac{11}{15}$ ____ $\frac{7}{10}$

Chapter 3

Summary

Key Terms

3.1	**like fractions**	Fractions with the same denominator are called *like fractions*.
	unlike fractions	Fractions with different denominators are called *unlike fractions*.
3.2	**least common multiple**	Given two or more whole numbers, the least common multiple is the smallest whole number that is divisible by all the numbers.
	LCM	The abbreviation for *least common multiple* is LCM.
3.3	**least common denominator**	When unlike fractions are rewritten as like fractions having the least common multiple as the denominator, the new denominator is the least common denominator.
	LCD	The abbreviation for *least common denominator* is LCD.
3.4	**carrying**	Carrying is the method used when the sum of the fractions of mixed numbers is greater than 1. Carry from the fraction to the whole number.
	borrowing	Borrowing is the method used in subtracting mixed numbers when the fraction part of the minuend is less than the fraction part of the subtrahend.

Test Your Word Power

See how well you have learned the vocabulary in this chapter. Answers follow the Quick Review.

1. **Like fractions** are
   (a) fractions that are equivalent
   (b) fractions that are not equivalent
   (c) fractions that have the same numerator
   (d) fractions that have the same denominator.

2. Two or more fractions are **unlike fractions** if
   (a) they are not equivalent
   (b) they have different numerators
   (c) they have different denominators
   (d) they are improper fractions.

3. The **least common multiple** is
   (a) the smallest whole number that is divisible by each of two or more numbers
   (b) the smallest numerator
   (c) the smallest denominator
   (d) the smallest whole number that is not divisible by a group of numbers.

4. The abbreviation **LCM** stands for
   (a) the largest common multiple
   (b) the longest common multiple
   (c) the most likely common multiplier
   (d) the least common multiple.

5. The **least common denominator** is
   (a) needed when multiplying fractions
   (b) needed when dividing fractions
   (c) the least common multiple of the denominators in a fraction problem
   (d) any denominator that is common to a group of fractions.

6. The abbreviation **LCD** stands for
   (a) the largest common denominator
   (b) the least common denominator
   (c) the least common divisor
   (d) the most likely common denominator.

Quick Review

Concepts	Examples
3.1 Adding Like Fractions Add numerators and write in lowest terms.	$\frac{3}{4}+\frac{1}{4}+\frac{5}{4}=\frac{3+1+5}{4}=\frac{9}{4}=2\frac{1}{4}$
3.1 Subtracting Like Fractions Subtract numerators and write in lowest terms.	$\frac{7}{8}-\frac{5}{8}=\frac{7-5}{8}=\frac{2\div 2}{8\div 2}=\frac{1}{4}$
3.2 Finding the Least Common Multiple Method of using multiples of the larger number: List the first few multiples of the larger number. Check each one until you find the multiple that is divisible by the smaller number.	$\frac{1}{3}+\frac{1}{4}$ 4, 8, 12, 16, . . . ← Multiples of 4 12: First multiple divisible by 3 ($12 \div 3 = 4$) The least common multiple of the numbers 3 and 4 is 12.
3.2 Finding the Least Common Multiple Method of prime numbers: First find the prime factorization of each number. Then use the prime factors to build the least common multiple.	$9 = 3 \cdot 3$ (Factors of 9) $15 = 3 \cdot 5$ (Factors of 15) $\text{LCM} = 3 \cdot 3 \cdot 5 = 45$ The least common multiple (LCM) is 45.
3.3 Adding Unlike Fractions *Step 1* Find the least common multiple (LCM). *Step 2* Rewrite fractions with the least common multiple as the denominator. *Step 3* Add numerators, placing the sum over the common denominator and simplify the answer.	$\frac{1}{3}+\frac{1}{4}+\frac{1}{10}$ LCM = 60 $\frac{1}{3}=\frac{20}{60}$, $\frac{1}{4}=\frac{15}{60}$, $\frac{1}{10}=\frac{6}{60}$ $\frac{20}{60}+\frac{15}{60}+\frac{6}{60}=\frac{41}{60}$
3.3 Subtracting Unlike Fractions *Step 1* Find the least common multiple (LCM). *Step 2* Rewrite fractions with the least common multiple as the denominator. *Step 3* Subtract numerators, placing the difference over the common denominator and simplifying the answer.	$\frac{5}{8}-\frac{1}{3}$ LCM = 24 $\frac{5}{8}=\frac{15}{24}$, $\frac{1}{3}=\frac{8}{24}$ $\frac{15}{24}-\frac{8}{24}=\frac{7}{24}$

Concepts	*Examples*
3.4 Adding Mixed Numbers Round the numbers and estimate the answer. Then find the exact answer. **1.** Add fractions using a common denominator. **2.** Add whole numbers. **3.** Combine the sums of whole numbers and fractions, simplifying the fraction part when necessary. Compare the exact answer to the estimate to see if it is reasonable.	*Estimate:* 10 (Rounds to $9\frac{2}{3}$), + 7 (Rounds to $+6\frac{3}{4}$), total 17 *Exact:* $9\frac{2}{3} = 9\frac{8}{12}$; $+6\frac{3}{4} = 6\frac{9}{12}$; $15\frac{17}{12} = 16\frac{5}{12}$ The exact answer is reasonable because it is close to the estimate of 17.
3.4 Subtracting Mixed Numbers Round the numbers and estimate the answer. Then find the exact answer. **1.** Subtract fractions, using borrowing if necessary. **2.** Subtract whole numbers. **3.** Combine the differences of whole numbers and fractions, simplifying the fraction parts when necessary. Compare the exact answer to the estimate to see if it is reasonable.	*Estimate:* 9 (Rounds to $8\frac{5}{8}$), − 4 (Rounds to $-3\frac{11}{12}$), total 5 *Exact:* $8\frac{5}{8} = 8\frac{15}{24} = 7\frac{39}{24}$; $-3\frac{11}{12} = 3\frac{22}{24} = 3\frac{22}{24}$; $4\frac{17}{24}$ The exact answer is reasonable because it is close to the estimate of 5.
3.4 Adding or Subtracting Mixed Numbers Using an Alternate Method **1.** Change the mixed numbers to improper fractions. **2.** Rewrite the unlike fractions as like fractions. **3.** Add or subtract the numerators and simplify the anwer.	Add. $2\frac{2}{3} = \frac{8}{3} = \frac{64}{24}$ $+1\frac{3}{8} = \frac{11}{8} = +\frac{33}{24}$ (Least common denominator) $\frac{97}{24} = 4\frac{1}{24}$ Answer as mixed number Improper fractions Subtract. $8\frac{2}{3} = \frac{26}{3} = \frac{104}{12}$ $-5\frac{3}{4} = \frac{23}{4} = -\frac{69}{12}$ (Least common denominator) $\frac{35}{12} = 2\frac{11}{12}$ Answer as mixed number Improper fractions

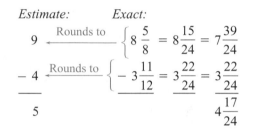

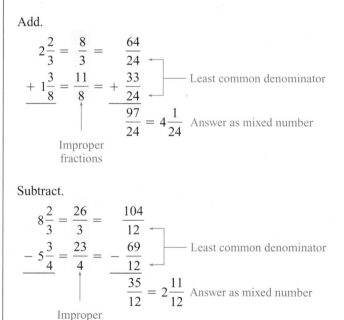

Concepts	*Examples*
3.5 Identifying the Larger of Two Fractions With unlike fractions, change to like fractions first. The fraction with the greater numerator is the greater fraction. Use these symbols: $<$ is less than $>$ is greater than	Identify the greater fraction. $\frac{7}{8}, \frac{9}{10}$ $\frac{7}{8} = \frac{7 \cdot 5}{8 \cdot 5} = \frac{35}{40}$ $\frac{9}{10} = \frac{9 \cdot 4}{10 \cdot 4} = \frac{36}{40}$ $\frac{35}{40}$ is smaller than $\frac{36}{40}$, so $\frac{7}{8} < \frac{9}{10}$ or $\frac{9}{10} > \frac{7}{8}$. $\frac{9}{10}$ is greater.
3.5 Using the Order of Operations with Fractions Follow the order of operations. **1.** Do all operations inside parentheses or other grouping symbols. **2.** Simplify any expressions with exponents and find any square roots. **3.** Multiply or divide from left to right. **4.** Add or subtract from left to right.	Simplify by using the order of operations. $\frac{1}{2} \cdot \frac{2}{3} - \left(\frac{1}{4}\right)^2$ Simplify fraction with exponent. $= \frac{1}{\cancel{2}_1} \cdot \frac{\cancel{2}^1}{3} - \frac{1}{16}$ Next, multiply. $= \frac{1}{3} - \frac{1}{16}$ $= \frac{16}{48} - \frac{3}{48}$ Change to common denominator and subtract. $= \frac{13}{48}$

Answers to Test Your Word Power

1. (d) *Example:* Because the fractions $\frac{3}{8}$ and $\frac{10}{8}$ both have 8 as a denominator, they are like fractions.

2. (c) *Example:* The fractions $\frac{2}{3}$ and $\frac{3}{4}$ are unlike fractions because they have different denominators.

3. (a) *Example:* The least common multiple of 4 and 5 is 20 because 20 is the smallest number into which 4 and 5 will divide evenly. **4. (d)** *Example:* LCM is the abbreviation for least common multiple.

5. (c) *Example:* The least common denominator of the fractions $\frac{2}{3}$ and $\frac{1}{2}$ is 6 because 6 is the least common multiple of 3 and 2. When written using the least common denominator, $\frac{2}{3}$ and $\frac{1}{2}$ become $\frac{4}{6}$ and $\frac{3}{6}$, respectively.

6. (b) *Example:* LCD is the abbreviation for least common denominator.

Chapter 3 Review Exercises

[3.1] *Add or subtract. Write answers in lowest terms.*

1. $\frac{5}{9} + \frac{2}{9}$

2. $\frac{2}{7} + \frac{3}{7}$

3. $\frac{1}{8} + \frac{3}{8} + \frac{2}{8}$

4. $\frac{5}{16} - \frac{3}{16}$

5. $\frac{3}{10} - \frac{1}{10}$

6. $\frac{5}{12} - \frac{3}{12}$

7. $\frac{36}{62} - \frac{10}{62}$

8. $\frac{68}{75} - \frac{43}{75}$

Solve each application problem. Write answers in lowest terms.

9. Nurse Suzie Brasher screened $\frac{7}{16}$ of her patients in her first hour on duty and $\frac{5}{16}$ of her patients in the second hour. What fraction of her patients did she screen in the two hours?

10. The Koats for Kids committee members completed $\frac{5}{8}$ of their Web-page design in the morning and $\frac{3}{8}$ in the afternoon. How much less did they complete in the afternoon than in the morning?

[3.2] *Find the least common multiple of each set of numbers.*

11. 4, 3

12. 5, 4

13. 10, 12, 20

14. 3, 8, 4

15. 6, 8, 5, 15

16. 15, 9, 20

Rewrite each fraction using the indicated denominator.

17. $\frac{2}{3} = \frac{}{12}$

18. $\frac{3}{8} = \frac{}{56}$

19. $\frac{2}{5} = \frac{}{25}$

20. $\frac{5}{9} = \frac{}{81}$

21. $\frac{4}{5} = \frac{}{40}$

22. $\frac{5}{16} = \frac{}{64}$

[3.1–3.3] *Add or subtract. Write answers in lowest terms.*

23. $\frac{1}{2} + \frac{1}{3}$

24. $\frac{1}{5} + \frac{3}{10} + \frac{3}{8}$

25. $\frac{5}{12} + \frac{5}{24}$

26. $\frac{2}{3} - \frac{1}{4}$

27. $\frac{7}{8} - \frac{1}{3}$

28. $\frac{11}{12} - \frac{4}{9}$

Solve each application problem.

29. The master gardener for a public garden used $\frac{3}{8}$ sack of fertilizer on the lawn, $\frac{1}{4}$ sack of fertilizer on the trees, and $\frac{1}{3}$ sack of fertilizer on the flower beds. How much of the fertilizer did he use?

30. Rachel is planning a birthday bash for Ross. Monica has raised $\frac{2}{5}$ of the amount needed through a bake sale; Joey has earned $\frac{1}{3}$ of the amount needed from an acting job, and Phoebe has raised another $\frac{1}{4}$ singing at the local coffee-house. Find the portion of the total that has been raised.

[3.4] *First estimate the answer. Then add or subtract to find the exact answer. Write answers as mixed numbers.*

31. *Estimate:* $+$ ____ *Exact:*

Rounds to ← $18\frac{5}{8}$

Rounds to ← $+\ 13\frac{3}{4}$

32. *Estimate:* $+$ ____ *Exact:* $22\frac{2}{3}$, $+\ 15\frac{4}{9}$

33. *Estimate:* $+$ ____ *Exact:* $12\frac{3}{5}$, $8\frac{5}{8}$, $+\ 10\frac{5}{16}$

34. *Estimate:* $-$ ____ *Exact:* $31\frac{3}{4}$, $-\ 14\frac{2}{3}$

35. *Estimate:* $-$ ____ *Exact:* 34, $-\ 15\frac{2}{3}$

36. *Estimate:* $-$ ____ *Exact:* $215\frac{7}{16}$, $-\ 136$

Add or subtract by changing mixed numbers to improper fractions. Simplify all answers.

37. $5\frac{2}{5} + 3\frac{7}{10}$

38. $4\frac{3}{4} + 5\frac{2}{3}$

39. $5 - 1\frac{3}{4}$

40. $6\frac{1}{2} - 4\frac{5}{6}$

41. $8\frac{1}{3} - 2\frac{5}{6}$

42. $5\frac{5}{12} - 2\frac{5}{8}$

First estimate the answer. Then solve each application problem.

43. A lab had $14\frac{2}{3}$ gallons of distilled water. If $5\frac{1}{2}$ gallons were used in the morning and $6\frac{3}{4}$ gallons were used in the afternoon, find the number of gallons remaining.

Estimate:

Exact:

44. The Boys and Girls Clubs of America collected $28\frac{2}{3}$ tons of newspapers on Saturday and $24\frac{3}{4}$ tons on Sunday. Find the total weight of the newspapers collected.

Estimate:

Exact:

45. In a recent bass-fishing derby, Matt Urban caught three largemouth bass weighing $8\frac{7}{8}$ lb, $9\frac{1}{3}$ lb, and $6\frac{3}{4}$ lb. Find their total weight.

Estimate:

Exact:

46. A developer wants to build a shopping center. She bought two parcels of land, one, $1\frac{11}{16}$ acres, and the other, $2\frac{3}{4}$ acres. If she needs a total of $8\frac{1}{2}$ acres for the center, how much additional land does she need to buy?

Estimate:

Exact:

[3.5] *Locate each fraction in Exercises 47–50 on the following number line.*

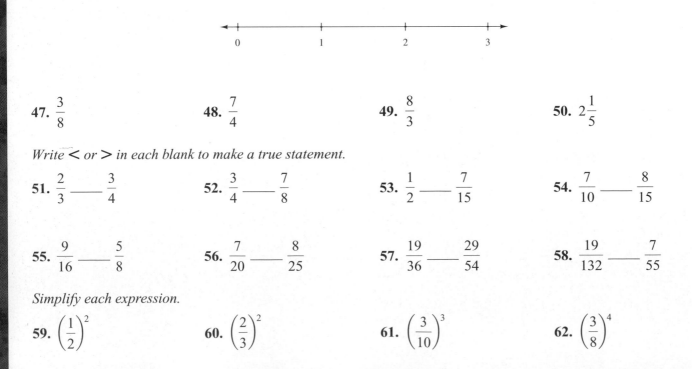

47. $\frac{3}{8}$ **48.** $\frac{7}{4}$ **49.** $\frac{8}{3}$ **50.** $2\frac{1}{5}$

Write < or > in each blank to make a true statement.

51. $\frac{2}{3}$ ____ $\frac{3}{4}$ **52.** $\frac{3}{4}$ ____ $\frac{7}{8}$ **53.** $\frac{1}{2}$ ____ $\frac{7}{15}$ **54.** $\frac{7}{10}$ ____ $\frac{8}{15}$

55. $\frac{9}{16}$ ____ $\frac{5}{8}$ **56.** $\frac{7}{20}$ ____ $\frac{8}{25}$ **57.** $\frac{19}{36}$ ____ $\frac{29}{54}$ **58.** $\frac{19}{132}$ ____ $\frac{7}{55}$

Simplify each expression.

59. $\left(\frac{1}{2}\right)^2$ **60.** $\left(\frac{2}{3}\right)^2$ **61.** $\left(\frac{3}{10}\right)^3$ **62.** $\left(\frac{3}{8}\right)^4$

Simplify by using the order of operations.

63. $6 \cdot \left(\frac{1}{3}\right)^2$

64. $\left(\frac{2}{3}\right)^2 \cdot 15$

65. $\left(\frac{3}{4}\right)^2 \cdot \left(\frac{8}{9}\right)^2$

66. $\frac{7}{8} \div \left(\frac{1}{8} + \frac{3}{4}\right)$

67. $\left(\frac{1}{2}\right)^2 \cdot \left(\frac{1}{4} + \frac{1}{2}\right)$

68. $\left(\frac{1}{4}\right)^3 + \left(\frac{5}{8} + \frac{3}{4}\right)$

Mixed Review Exercises

Simplify by using the order of operations as necessary. Write answers in lowest terms and as whole or as mixed numbers when possible.

69. $\frac{5}{6} - \frac{1}{6}$

70. $\frac{7}{8} - \frac{3}{4}$

71. $\frac{29}{32} - \frac{5}{16}$

72. $\frac{1}{4} + \frac{1}{8} + \frac{5}{16}$

73. $6\frac{2}{3} - 4\frac{1}{2}$

74. $9\frac{1}{2} + 16\frac{3}{4}$

75. $7 - 1\frac{5}{8}$

76. $2\frac{3}{5} + 8\frac{5}{8} + \frac{5}{16}$

77. $32\frac{5}{12} - 17$

78. $\frac{7}{22} + \frac{3}{22} + \frac{3}{11}$

79. $\left(\frac{1}{4}\right)^2 \cdot \left(\frac{2}{5}\right)^3$

80. $\frac{3}{8} \div \left(\frac{1}{2} + \frac{1}{4}\right)$

81. $\left(\frac{2}{3}\right)^2 \cdot \left(\frac{1}{3} + \frac{1}{6}\right)$

82. $\left(\frac{2}{3}\right)^3 + \left(\frac{2}{3} - \frac{5}{9}\right)$

Write $<$ or $>$ in each blank to make a true statement.

83. $\frac{2}{3}$ ____ $\frac{7}{12}$

84. $\frac{8}{9}$ ____ $\frac{15}{8}$

85. $\frac{17}{30}$ ____ $\frac{36}{60}$

86. $\frac{5}{8}$ ____ $\frac{17}{30}$

Find the least common multiple of each set of numbers.

87. 12, 18

88. 6, 8, 10, 12

89. 9, 14, 21

Rewrite each fraction using the indicated denominator.

90. $\frac{2}{3} = \frac{}{27}$

91. $\frac{9}{12} = \frac{}{144}$

92. $\frac{4}{5} = \frac{}{75}$

First estimate the answer. Then solve each application problem.

93. A cement contractor needs $13\frac{1}{2}$ ft of wire mesh for a concrete walkway and $22\frac{3}{8}$ ft of wire mesh for a driveway. If the contractor has a roll of wire from which he is cutting that is $92\frac{3}{4}$ ft long, find the number of feet of wire remaining after the two jobs have been completed.

Estimate:

Exact:

94. The Horticulture Club used $3\frac{5}{8}$ gallons of insecticide on one crop of plants and $8\frac{1}{2}$ gallons on another crop. If their original supply of insecticide was 15 gallons, find the number of gallons remaining.

Estimate:

Exact:

Chapter 3 Test

Study Skills Workbook
Activity 10

Add or subtract. Write answers in lowest terms.

1. $\frac{3}{8} + \frac{3}{8}$

2. $\frac{3}{16} + \frac{5}{16}$

3. $\frac{7}{10} - \frac{3}{10}$

4. $\frac{7}{12} - \frac{5}{12}$

Find the least common multiple of each set of numbers.

5. 2, 3, 4

6. 6, 3, 5, 15

7. 6, 9, 27, 36

Add or subtract. Write answers in lowest terms.

8. $\frac{3}{8} + \frac{1}{4}$

9. $\frac{2}{9} + \frac{5}{12}$

10. $\frac{7}{8} - \frac{2}{3}$

11. $\frac{2}{5} - \frac{3}{8}$

First estimate the answer. Then add or subtract to find the exact answer; simplify exact answers.

12. $7\frac{2}{3} + 4\frac{5}{6}$

13. $16\frac{2}{5} - 11\frac{2}{3}$

14. $18\frac{3}{4} + 9\frac{2}{5} + 12\frac{1}{3}$

15. $24 - 18\frac{3}{8}$

1. ______

2. ______

3. ______

4. ______

5. ______

6. ______

7. ______

8. ______

9. ______

10. ______

11. ______

12. *Estimate:* ______

Exact: ______

13. *Estimate:* ______

Exact: ______

14. *Estimate:* ______

Exact: ______

15. *Estimate:* ______

Exact: ______

16. ______

16. Most students say that "addition and subtraction of fractions is more difficult than multiplication and division of fractions." Why do you think they say this? Do you agree with these students?

17. ______

17. Devise and explain a method of estimating an answer to addition and subtraction problems involving mixed numbers. Might your estimated answer vary from the exact answer? If it did, what would the estimation accomplish?

First estimate the answer. Then solve each application problem.

18. *Estimate:* ______

Exact: ______

18. A professional football player trains with body building, jogging, and wind sprints. He trains $5\frac{2}{3}$ hours on Monday, $4\frac{3}{4}$ hours on Tuesday, $3\frac{5}{6}$ hours on Wednesday, $7\frac{1}{3}$ hours on Thursday, and $5\frac{1}{6}$ hours on Friday. Find the total number of hours that he trained.

19. *Estimate:* ______

Exact: ______

19. A painting contractor arrived at a 6-unit apartment complex with $147\frac{1}{2}$ gallons of exterior paint. If his crew sprayed $68\frac{1}{2}$ gallons on the wood siding, rolled $37\frac{3}{8}$ gallons on the masonry exterior, and brushed $5\frac{3}{4}$ gallons on the trim, find the number of gallons of paint remaining.

Write $<$ or $>$ to make a true statement.

20. $\frac{3}{4}$ ___ $\frac{17}{24}$

21. $\frac{19}{24}$ ___ $\frac{17}{36}$

20. ______

21. ______

Simplify. Use the order of operations as needed.

22. $\left(\frac{1}{3}\right)^3 \cdot 54$

23. $\left(\frac{3}{4}\right)^2 - \left(\frac{7}{8} \cdot \frac{1}{3}\right)$

22. ______

23. ______

24. $\left(\frac{7}{8} - \frac{7}{16}\right) \cdot 4$

25. $\frac{5}{6} + \frac{4}{3} \cdot \frac{3}{8}$

24. ______

25. ______

Cumulative Review Exercises CHAPTERS 1–3

For each number, name the digit that has the given place value.

1. 871
hundreds
ones

2. 5,629,428
millions
thousands

Round each number to the nearest ten, nearest hundred, and nearest thousand.

	Ten	Hundred	Thousand
3. 1438	______	______	______
4. 59,803	______	______	______

Use front end rounding to estimate each answer. Then add, subtract, multiply, or divide to find the exact answer.

5. *Estimate:* *Exact:*

Estimate		Exact
	Rounds to	2361
	Rounds to	386
	Rounds to	47,304
+	Rounds to	+ 29,728

6. *Estimate:* − ______ *Exact:* 24,276 − 9 887

7. *Estimate:* × ______ *Exact:* 4468 × 280

8. *Estimate:*) ______ *Exact:* 35)112,385

Add, subtract, multiply, or divide as indicated.

9. 4 + 8 + 5 + 9

10. 375,899 + 521,742 + 357,968

11. 1687 − 1096

12. 3,896,502 − 1,094,807

13. 3 × 8 × 5

14. 6 × 3 × 7

15. 5 × 8 × 4

16. 57 × 8

17. 962 × 384

18. 340 × 50

19. 8)1080

20. 13,467 ÷ 5

21. 506)16,358

Use front end rounding to estimate the answer to each application problem. Then find the exact answer.

22. The Americans with Disabilities Act provides the single parking space design shown. Find the perimeter of (distance around) this parking space, including the accessible aisle.

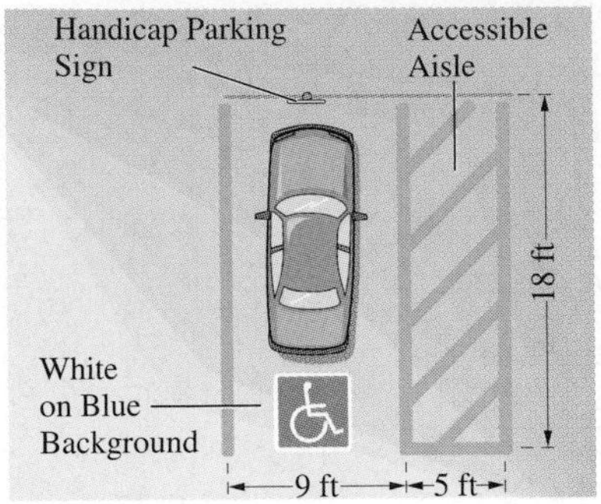

Estimate:

Exact:

23. The single parking space design in Exercise 22 measures 18 ft by 14 ft. Find its area.

Estimate:

Exact:

24. Find the number of 32 ounce (1 quart) cartons of orange juice that can be filled with 40,320 ounces of orange juice.

Estimate:

Exact:

25. A dentist's drill makes 3600 revolutions each minute. How many revolutions would it make in 60 minutes of operation?

Estimate:

Exact:

Round the mixed number in each problem to the nearest whole number and estimate the answer. Then find the exact answer.

26. The top of a rectangular pool table is $1\frac{3}{4}$ yd by $2\frac{2}{3}$ yd. Find its area.

Estimate:

Exact:

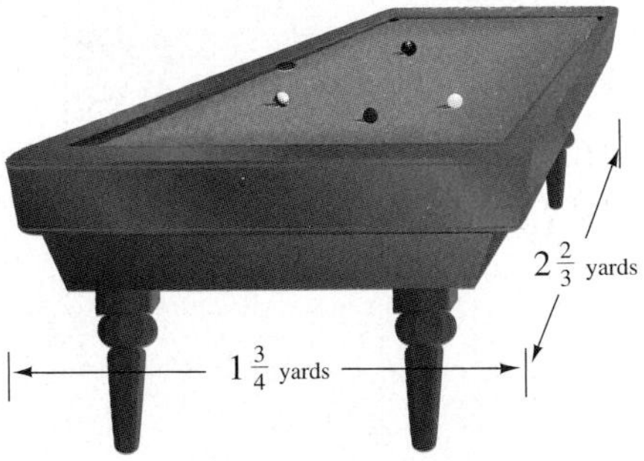

27. A wilderness park is $4\frac{5}{8}$ miles wide by $6\frac{2}{3}$ miles long. How many square miles are in the wilderness park?

Estimate:

Exact:

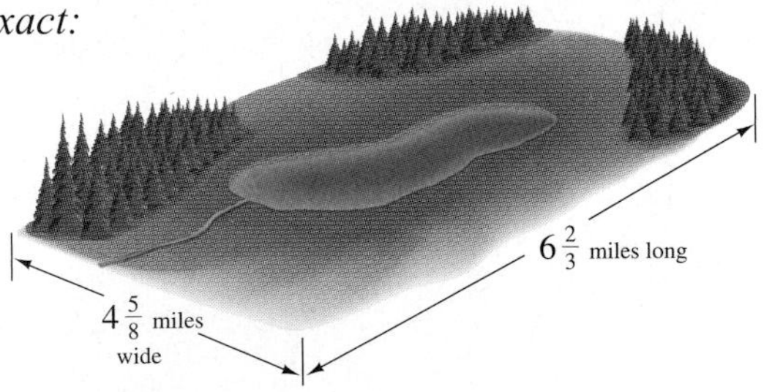

28. Larry Foxworthy cuts, splits, and delivers firewood. If his truck, when fully loaded, holds $5\frac{1}{4}$ cords of firewood, find the number of cords he could deliver in $3\frac{1}{2}$ loads.

Estimate:

Exact:

29. The Sears Tower in Chicago is 110 stories tall. The total height of the building is $1536\frac{7}{8}$ ft, including a flagpole at the top of the building that is $82\frac{1}{2}$ ft tall. Find the height of the building itself. (*Source: Top 10 of Everything,* 1999.)

Estimate:

Exact:

Find the prime factorization of each number. Write answers by using exponents.

30. 18

31. 200

32. 1225

Simplify.

33. $2^4 \cdot 3^2$

34. $3^3 \cdot 2^2$

35. $6^2 \cdot 3^3$

Find each square root.

36. $\sqrt{36}$

37. $\sqrt{81}$

38. $\sqrt{144}$

Simplify by using the order of operations.

39. $5^2 - 2 \cdot 8$

40. $\sqrt{25} + 5 \cdot 9 - 6$

41. $\left(\frac{4}{5} - \frac{2}{3}\right) \cdot \frac{2}{3}$

42. $\frac{3}{4} \div \left(\frac{1}{3} + \frac{1}{2}\right)$

43. $\frac{7}{8} + \left(\frac{3}{4}\right)^2 - \frac{3}{8}$

Write proper *or* improper *for each fraction.*

44. $\frac{5}{6}$

45. $\frac{6}{6}$

46. $\frac{11}{10}$

Write each fraction in lowest terms.

47. $\frac{25}{60}$

48. $\frac{84}{96}$

49. $\frac{63}{70}$

Add, subtract, multiply, or divide as indicated. Simplify all answers.

50. $\frac{3}{4} \cdot \frac{2}{3}$

51. $\frac{3}{8} \cdot \frac{5}{6}$

52. $42 \cdot \frac{7}{8}$

53. $\frac{5}{8} \div \frac{3}{8}$

54. $\frac{25}{40} \div \frac{10}{35}$

55. $9 \div \frac{2}{3}$

56. $\frac{3}{4} + \frac{1}{7}$

57. $\frac{3}{8} + \frac{3}{16} + \frac{1}{4}$

58. $\frac{11}{18} - \frac{5}{12}$

First estimate the answer. Then add or subtract to find the exact answer. Write exact answers as mixed numbers.

59. *Estimate:* *Exact:*

$$\begin{array}{lcr} \underline{\quad} & \xleftarrow{\text{Rounds to}} & 3\frac{3}{8} \\ + \underline{\quad} & \xleftarrow{\text{Rounds to}} & +\ 4\frac{1}{2} \\ \hline \end{array}$$

60. *Estimate:* *Exact:*

$$\begin{array}{lr} \ & 21\frac{7}{8} \\ + & +\ 4\frac{5}{12} \\ \hline \end{array}$$

61. *Estimate:* *Exact:*

$$\begin{array}{lr} \ & 5 \\ - & -2\frac{3}{8} \\ \hline \end{array}$$

Find the least common multiple of each set of numbers.

62. 8, 12

63. 3, 8, 15

64. 12, 16, 18

Rewrite each fraction using the indicated denominator.

65. $\frac{4}{5} = \frac{}{45}$

66. $\frac{7}{9} = \frac{}{72}$

67. $\frac{9}{15} = \frac{}{135}$

68. $\frac{5}{7} = \frac{}{84}$

Locate each fraction in Exercises 69–72 on the following number line.

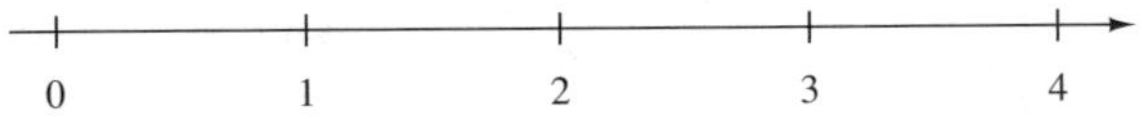

69. $\frac{3}{4}$

70. $\frac{1}{9}$

71. $\frac{5}{3}$

72. $\frac{10}{3}$

Write < or > in each blank to make a true statement.

73. $\frac{3}{5}$ ____ $\frac{5}{8}$

74. $\frac{17}{20}$ ____ $\frac{3}{4}$

75. $\frac{7}{12}$ ____ $\frac{11}{18}$

Decimals

4

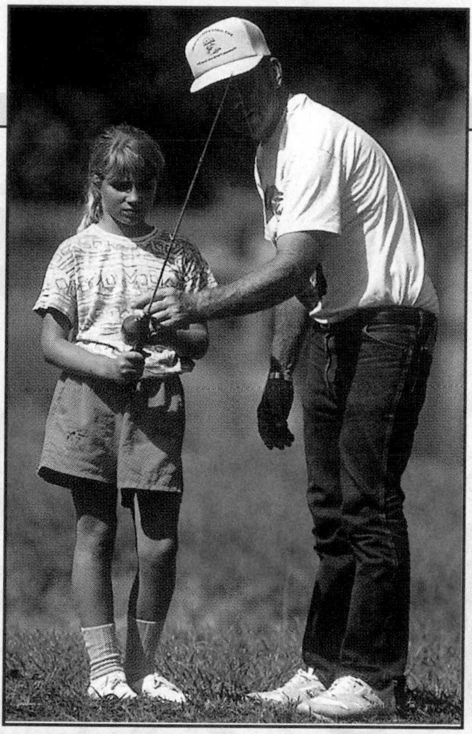

Freshwater fishing is America's third most popular recreational activity, and saltwater fishing is ninth. (*Source:* Sporting Goods Manufacturers Association.) In Section 4.3, Exercises 47–50, this father will use decimal numbers when he pays for his daughter's equipment. But will decimals help him and his daughter to catch their limit? (See also Section 4.1, Exercises 59–62, and Section 4.6, Exercises 65–68.)

4.1 Reading and Writing Decimals

Objectives

1. Write parts of a whole using decimals.
2. Identify the place value of a digit.
3. Read and write decimals in words.
4. Write decimals as fractions or mixed numbers.

1 There are 10 dimes in one dollar. Each dime is $\frac{1}{10}$ of a dollar. Write the yellow shaded portion of each dollar as a fraction, as a decimal, and in words.

(a)

(b)

(c)

Answers

1. (a) $\frac{1}{10}$; 0.1; one tenth

(b) $\frac{3}{10}$; 0.3; three tenths

(c) $\frac{9}{10}$; 0.9; nine tenths

Fractions are used to represent parts of a whole. In this chapter, **decimals** are used as another way to show parts of a whole. For example, our money system is based on decimals. One dollar is divided into 100 equivalent parts. One cent ($0.01) is one of the parts, and a dime ($0.10) is 10 of the parts. Metric measurement (see Chapter 7) is also based on decimals.

1 Write parts of a whole using decimals. Decimals are used when a whole is divided into 10 equivalent parts, or into 100 or 1000 or 10,000 equivalent parts. In other words, decimals are fractions with denominators that are a power of 10. For example, the square below is cut into 10 equivalent parts. Written as a fraction, each part is $\frac{1}{10}$ of the whole. Written as a decimal, each part is 0.1. Both $\frac{1}{10}$ and 0.1 are read as "*one tenth.*"

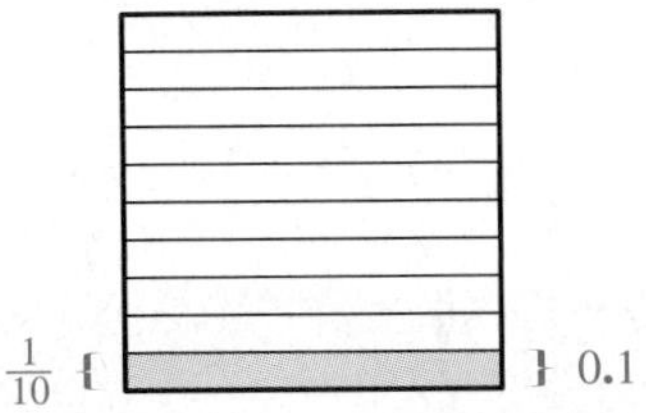

One-tenth of the square is shaded.

The dot in 0.1 is called the **decimal point.**

0.1
↑
Decimal point

The square at the right has 7 of its 10 parts shaded.

Written as a *fraction,* $\frac{7}{10}$ of the square is shaded.

Written as a *decimal,* 0.7 of the square is shaded.

Both $\frac{7}{10}$ and 0.7 are read as "*seven tenths.*"

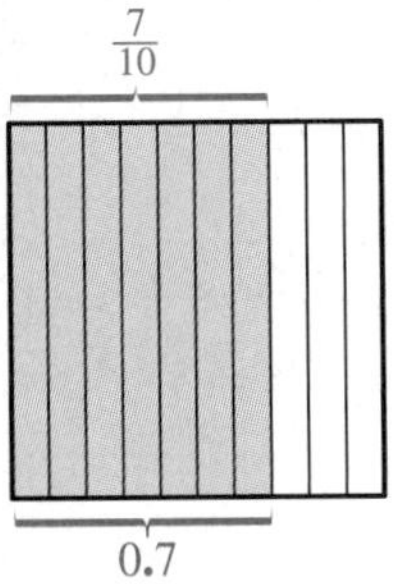

Seven-tenths of the square is shaded.

Work Problem 1 at the Side.

The square below is cut into 100 equivalent parts. Written as a *fraction,* each part is $\mathbf{\frac{1}{100}}$ of the whole.

Written as a decimal, each part is **0.01** of the whole.

Both $\frac{1}{100}$ and 0.01 are read as "one hundredth."

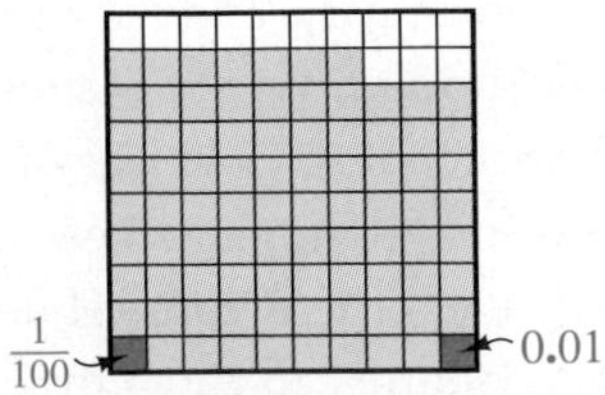

The square above has 87 parts shaded.

Written as a fraction, $\mathbf{\frac{87}{100}}$ of the total area is shaded.

Written as a decimal, **0.87** of the total area is shaded.

Both $\frac{87}{100}$ and 0.87 are read as "*eighty-seven hundredths.*"

Work Problem 2 at the Side.

Example 1 shows several numbers written as both fractions and decimals.

Example 1 Using the Decimal Forms of Fractions

	Fraction	Decimal	Read As
(a)	$\frac{4}{10}$	0.4	four tenths
(b)	$\frac{9}{100}$	0.09	nine hundredths
(c)	$\frac{71}{100}$	0.71	seventy-one hundredths
(d)	$\frac{8}{1000}$	0.008	eight thousandths
(e)	$\frac{45}{1000}$	0.045	forty-five thousandths
(f)	$\frac{832}{1000}$	0.832	eight hundred thirty-two thousandths

Work Problem 3 at the Side.

2 Identify the place value of a digit. The decimal point separates the *whole number part* from the *fractional part* in a decimal number. In the chart below, you see that the **place value names** for fractional parts are similar to those on the whole number side, but end in "*ths*."

Decimal Place Value Chart

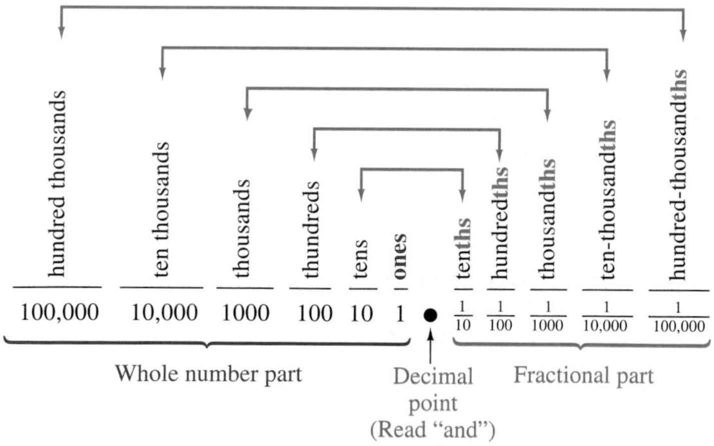

Notice that the **ones** place is at the center. (There is no "oneths" place.) Also notice that each place is 10 times the value of the place to its right.

2 Write the portion of each square that is shaded as a fraction, as a decimal, and in words.

(a)

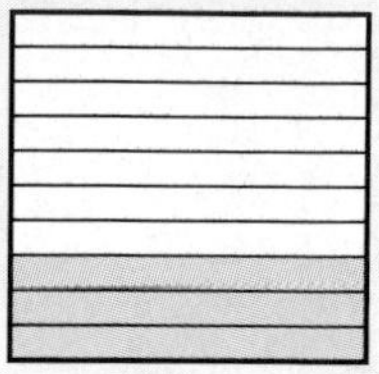

(b)

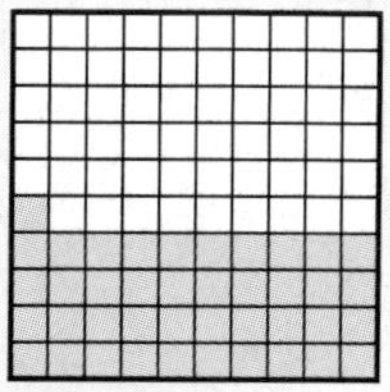

3 Write each decimal as a fraction.

(a) 0.7

(b) 0.2

(c) 0.03

(d) 0.69

(e) 0.047

(f) 0.351

ANSWERS

2. (a) $\frac{3}{10}$; 0.3; three tenths

(b) $\frac{41}{100}$; 0.41; forty-one hundredths

3. (a) $\frac{7}{10}$ **(b)** $\frac{2}{10}$ **(c)** $\frac{3}{100}$

(d) $\frac{69}{100}$ **(e)** $\frac{47}{1000}$ **(f)** $\frac{351}{1000}$

4 Identify the place value of each digit.

(a) 971.54

(b) 0.4

(c) 5.60

(d) 0.0835

5 Tell how to read each decimal in words.

(a) 0.6

(b) 0.46

(c) 0.05

(d) 0.409

(e) 0.0003

(f) 0.0703

(g) 0.088

CAUTION

In this chapter, if a number does *not* have a decimal point, it is a *whole number.* A whole number has no fractional part. If you want to show the decimal point in a whole number, it is just to the *right* of the digit in the ones place. For example:

8 = 8. (↑ Decimal point) 306 = 306. (↑ Decimal point)

Example 2 Identifying the Place Value of a Digit

Identify the place value of each digit.

(a) 178.36

hundreds	tens	ones		tenths	hundredths
1	7	8	.	3	6

(b) 0.00935

ones		tenths	hundredths	thousandths	ten-thousandths	hundred-thousandths
0	.	0	0	9	3	5

Notice in Example 2(b) that we do *not* use commas on the right side of the decimal point.

Work Problem 4 at the Side.

3 Read and write decimals in words. A decimal is read according to its form as a fraction. We read 0.9 as "nine tenths" because 0.9 is the same as $\frac{9}{10}$. Notice that 0.9 ends in the tenths place.

ones		tenths
0	.	9

We read 0.02 as "two hundredths" because 0.02 is the same as $\frac{2}{100}$. Notice that 0.02 ends in the hundredths place.

ones		tenths	hundredths
0	.	0	2

Example 3 Reading Decimal Numbers

Tell how to read each decimal in words.

(a) 0.3

Because $0.3 = \frac{3}{10}$, read the decimal as three tenths.

(b) 0.49 Read it as: forty-nine hundredths.

(c) 0.08 Read it as: eight hundredths.

(d) 0.918 Read it as: nine hundred eighteen thousandths.

(e) 0.0106 Read it as: one hundred six ten-thousandths.

Work Problem 5 at the Side.

ANSWERS

4. (a)

hundreds	tens	ones		tenths	hundredths
9	7	1	.	5	4

(b)

ones		tenths
0	.	4

(c)

ones		tenths	hundredths
5	.	6	0

(d)

ones		tenths	hundredths	thousandths	ten-thousandths
0	.	0	8	3	5

5. (a) six tenths
(b) forty-six hundredths
(c) five hundredths
(d) four hundred nine thousandths
(e) three ten-thousandths
(f) seven hundred three ten-thousandths
(g) eighty-eight thousandths

Reading Decimal Numbers

Step 1 Read any whole number part to the *left* of the decimal point as you normally would.

Step 2 Read the decimal point as "*and*."

Step 3 Read the part of the number to the *right* of the decimal point as if it were an ordinary whole number.

Step 4 Finish with the place value name of the rightmost digit; these names all end in "*ths*."

NOTE

If there is *no whole number part,* you will use only Steps 3 and 4.

Example 4 Reading Decimals

Read each decimal.

(a) 16.9 — 9 is in tenths place.

sixteen **and** nine **tenths**

16.9 is read "sixteen and nine tenths."

(b) 482.35 — 5 is in hundredths place.

four hundred eighty-two **and** thirty-five **hundredths**

482.35 is read "four hundred eighty-two and thirty-five hundredths."

3 is in thousandths place.

(c) 0.063 is "sixty-three **thousandths**." (No whole number part.)

(d) 11.1085 is "eleven **and** one thousand eighty-five **ten-thousandths**."

CAUTION

Use "and" *only* when reading a decimal point. A common mistake is to read the whole number 405 as "four hundred *and* five." But there is *no decimal point* shown in 405, so it is read "four hundred five."

Work Problem 6 at the Side.

4 Write decimals as fractions or mixed numbers. Knowing how to read decimals will help you when writing decimals as fractions.

Writing Decimals as Fractions or Mixed Numbers

Step 1 The digits to the right of the decimal point are the numerator of the fraction.

Step 2 The denominator is 10 for tenths, 100 for hundredths, 1000 for thousandths, 10,000 for ten-thousandths, and so on.

Step 3 If the decimal has a whole number part, the fraction will be a mixed number with the same whole number part.

6 Tell how to read each decimal in words.

(a) 3.8

(b) 15.001

(c) 0.0073

(d) 64.309

ANSWERS

6. (a) three and eight tenths
(b) fifteen and one thousandth
(c) seventy-three ten-thousandths
(d) sixty-four and three hundred nine thousandths

7 Write each decimal as a fraction or mixed number.

(a) 0.7

(b) 12.21

(c) 0.101

(d) 0.007

(e) 1.3717

8 Write each decimal as a fraction or mixed number in lowest terms.

(a) 0.5

(b) 12.6

(c) 0.85

(d) 3.05

(e) 0.225

(f) 420.0802

ANSWERS

7. (a) $\frac{7}{10}$ (b) $12\frac{21}{100}$ (c) $\frac{101}{1000}$ (d) $\frac{7}{1000}$ (e) $1\frac{3717}{10{,}000}$

8. (a) $\frac{1}{2}$ (b) $12\frac{3}{5}$ (c) $\frac{17}{20}$ (d) $3\frac{1}{20}$ (e) $\frac{9}{40}$ (f) $420\frac{401}{5000}$

Example 5 Writing Decimals as Fractions or Mixed Numbers

Write each decimal as a fraction or mixed number.

(a) 0.19

The digits to the right of the decimal point, 19, are the numerator of the fraction. The denominator is 100 for hundredths because the right-most digit is in the hundredths place.

$$0.19 = \frac{19}{100} \leftarrow \text{100 for hundredths}$$

(Hundredths place)

(b) 0.863

$$0.863 = \frac{863}{1000} \leftarrow \text{1000 for thousandths}$$

(Thousandths place)

(c) 4.0099

The whole number part stays the same.

$$4.0099 = 4\frac{99}{10{,}000} \leftarrow \text{10,000 for ten-thousandths}$$

(Ten-thousandths place)

Work Problem 7 at the Side.

CAUTION

After you write a decimal as a fraction or a mixed number, make sure the fraction is in lowest terms.

Example 6 Writing Decimals as Fractions or Mixed Numbers in Lowest Terms

Write each decimal as a fraction or mixed number in lowest terms.

(a) $0.4 = \frac{4}{10} \leftarrow$ 10 for tenths

Write $\frac{4}{10}$ in lowest terms. $\frac{4}{10} = \frac{4 \div 2}{10 \div 2} = \frac{2}{5} \leftarrow$ Lowest terms

(b) $0.75 = \frac{75}{100} = \frac{75 \div 25}{100 \div 25} = \frac{3}{4} \leftarrow$ Lowest terms

(c) $18.105 = 18\frac{105}{1000} = 18\frac{105 \div 5}{1000 \div 5} = 18\frac{21}{200} \leftarrow$ Lowest terms

(d) $42.8085 = 42\frac{8085}{10{,}000} = 42\frac{8085 \div 5}{10{,}000 \div 5} = 42\frac{1617}{2000} \leftarrow$ Lowest terms

Work Problem 8 at the Side.

Calculator Tip In this book you'll notice that we use a 0 in the ones place for decimal fractions. We write 0.45 instead of just .45, to emphasize that there is no whole number. Your scientific (not graphing) calculator shows these 0s also. Enter (.) (4) (5). Notice that the display automatically shows 0.45 even though you did not press 0. For comparison, enter the whole number 45 by pressing (4) (5) (+) and notice where the decimal point is shown in the display. (It automatically appears to the *right* of the 5.)

4.1 Exercises

FOR EXTRA HELP: Student's Solutions Manual · MyMathLab.com · InterAct Math Tutorial Software · AW Math Tutor Center · www.mathxl.com · Digital Video Tutor CD 2 Videotape 7

Identify the digit that has the given place value. See Example 2.

1. 70.489
tens
ones
tenths

2. 135.296
ones
tenths
tens

3. 0.2518
hundredths
thousandths
ten-thousandths

4. 0.9347
hundredths
thousandths
ten-thousandths

5. 93.01472
thousandths
ten-thousandths
tenths

6. 0.51968
tenths
ten-thousandths
hundredths

7. 314.658
tens
tenths
hundreds

8. 51.325
tens
tenths
hundredths

9. 149.0832
hundreds
hundredths
ones

10. 3458.712
hundreds
hundredths
tenths

11. 6285.7125
thousands
thousandths
hundredths

12. 5417.6832
thousands
thousandths
ones

Write the decimal number that has the specified place values. See Example 2.

13. 0 ones, 5 hundredths, 1 ten, 4 hundreds, 2 tenths

14. 7 tens, 9 tenths, 3 ones, 6 hundredths, 8 hundreds

15. 3 thousandths, 4 hundredths, 6 ones, 2 ten-thousandths, 5 tenths

16. 8 ten-thousandths, 4 hundredths, 0 ones, 2 tenths, 6 thousandths

17. 4 hundredths, 4 hundreds, 0 tens, 0 tenths, 5 thousandths, 5 thousands, 6 ones

18. 7 tens, 7 tenths, 6 thousands, 6 thousandths, 3 hundreds, 3 hundredths, 2 ones

Write each decimal as a fraction or mixed number in lowest terms. See Examples 1, 5, and 6.

19. 0.7 **20.** 0.1 **21.** 13.4 **22.** 9.8 **23.** 0.25

24. 0.55 **25.** 0.66 **26.** 0.33 **27.** 10.17 **28.** 31.99

29. 0.06 **30.** 0.08 **31.** 0.205 **32.** 0.805

33. 5.002 **34.** 4.008 **35.** 0.686 **36.** 0.492

Tell how to read each decimal in words. See Examples 3 and 4.

37. 0.5 **38.** 0.2

39. 0.78 **40.** 0.55

41. 0.105 **42.** 0.609

43. 12.04 **44.** 86.09

45. 1.075 **46.** 4.025

Write each decimal in numbers. See Examples 3 and 4.

47. six and seven tenths

48. eight and twelve hundredths

49. thirty-two hundredths

50. one hundred eleven thousandths

51. four hundred twenty and eight thousandths

52. two hundred and twenty-four thousandths

53. seven hundred three ten-thousandths

54. eight hundred and six hundredths

55. seventy-five and thirty thousandths

56. sixty and fifty hundredths

57. Anne read the number 4302 as "four thousand three hundred and two." Explain what is wrong with the way Anne read the number.

58. Jerry read the number 9.0106 as "nine and one hundred and six ten-thousandths." Explain the error he made.

The dad on the first page of this chapter needs to select the correct fishing line for his daughter's reel. Fishing line is sold according to how many pounds of "pull" the line can withstand before breaking. Use the table to answer Exercises 59–62. Write all fractions in lowest terms. (Note: The diameter of the fishing line is its thickness.)

RELATING FISHING LINE DIAMETER TO TEST STRENGTH

Test Strength (pounds)	Average Diameter (inches)
4	0.008
8	0.010
12	0.013
14	0.014
17	0.015
20	0.016

Source: Berkley Outdoor Technologies Group.

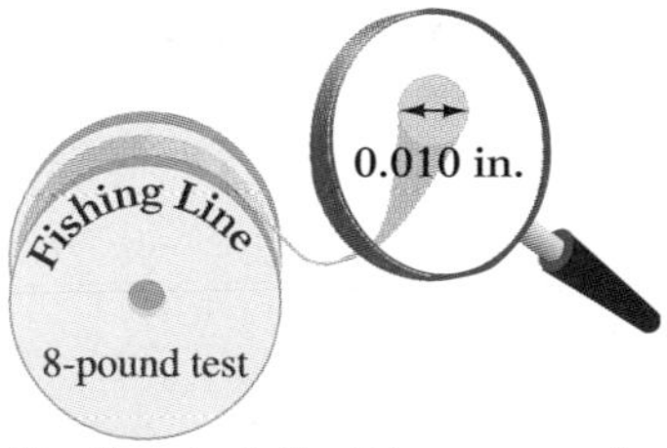

The diameter is the distance across the end of the line, or its thickness.

59. Write the diameter of 8-pound test line in words and as a fraction.

60. Write the diameter of 17-pound test line in words and as a fraction.

61. What is the test strength of the line with a diameter of $\frac{13}{1000}$ inch?

62. What is the test strength of the line with a diameter of sixteen thousandths inch?

Suppose your job is to take phone orders for precision parts. Use the table, and in Exercises 63–68, write the correct part number that matches what you hear the customer say over the phone. In Exercises 67–68, write the words you would say to the customer.

Part Number	Size in Centimeters
3-A	0.06
3-B	0.26
3-C	0.6
3-D	0.86
4-A	1.006
4-B	1.026
4-C	1.06
4-D	1.6
4-E	1.602

63. "Please send the six-tenths centimeter bolt."

Part number _______.

64. "The part missing from our order was the one and six hundredths size."

Part number _______.

65. "The size we need is one and six thousandths centimeters."

Part number _______.

66. "Do you still stock the twenty-six hundredths centimeter bolt?"

Part number _______.

67. "What size is part number 4-E?" Write your answer in words.

68. "What size is part number 4-B?" Write your answer in words.

RELATING CONCEPTS (Exercises 69–76) FOR INDIVIDUAL OR GROUP WORK

Use your knowledge of place value to ***work Exercises 69–76 in order.***

69. Look back at the decimal place value chart on page 249. What do you think would be the names of the next four places to the *right* of hundred-thousandths? What information did you use to come up with these names?

70. A common mistake is to think that the first place to the right of the decimal point is "oneths" and the second place is "tenths." Why might someone make that mistake? How would you explain why there is no "oneths" place?

71. Use your answer to Exercise 69 to write 0.72436955 in words.

72. Use your answer to Exercise 69 to write 0.000678554 in words.

73. Write 8006.500001 in words.

74. Write 20,060.000505 in words.

75. Write this decimal in numbers:

Three hundred two thousand forty ten-millionths.

76. Write this decimal in numbers:

nine billion, eight hundred seventy-six million, five hundred forty-three thousand, two hundred ten and one hundred million two hundred thousand three hundred billionths.

4.2 ROUNDING DECIMALS

OBJECTIVES

1. Learn the rules for rounding decimals.
2. Round decimals to any given place.
3. Round money amounts to the nearest cent or nearest dollar.

Section 1.7 showed how to round whole numbers. For example, 89 rounded to the nearest ten is 90, and 8512 rounded to the nearest hundred is 8500.

1 Learn the rules for rounding decimals. It is also important to be able to **round** decimals. For example, a store is selling 2 candy mints for \$0.75 but you want only one mint. The price of each mint is \$0.75 ÷ 2, which is \$0.375, but you cannot pay part of a cent. Is \$0.375 closer to \$0.37 or to \$0.38? Actually, it's exactly halfway between. When this happens in everyday situations, the rule is to round *up*. The store will charge you \$0.38 for the mint.

Rounding Decimals

Step 1 Find the place to which the rounding is being done. Draw a "cut-off" line ***after*** that place to show that you are cutting off and dropping the rest of the digits.

Step 2 Look ***only*** at the ***first*** digit you are cutting off.

Step 3(a) If this digit is ***4 or less,*** the part of the number you are keeping ***stays the same.***

Step 3(b) If this digit is ***5 or more,*** you must ***round up*** the part of the number you are keeping.

Step 4 You can use the "≈" sign to indicate that the rounded number is now an approximation (close, but not exact). "≈" means "is approximately equal to."

CAUTION

Do ***not*** move the decimal point when rounding.

2 Round decimals to any given place. The following examples show you how to round decimals.

Example 1 Rounding a Decimal Number

Round 14.39652 to the nearest thousandth. Is it closer to 14.396 or to 14.397?

Step 1 Draw a "cut-off" line after the thousandths place.

14.396 | 52 — You are cutting off the 5 and 2. They will be dropped.

Thousandths ↑ (the 6)

Step 2 Look *only* at the *first* digit you are cutting off. Ignore the other digits you are cutting off.

14.396 | 52 — Look only at the 5. Ignore the 2.

Continued on Next Page

1 Round to the nearest thousandth.

(a) 0.33492

(b) 8.00851

(c) 265.42038

(d) 10.70180

Step 3 If the first digit you are cutting off is 5 or more, round up the part of the number you are keeping.

```
  1 4 . 3 9 6 | 5 2     First digit cut is 5 or more,
+   0 . 0 0 1           so round up by adding
  -----------           1 thousandth to the part
  1 4 . 3 9 7           you are keeping.
```

So, 14.39652 rounded to the nearest thousandth is 14.397. We can write 14.39652 ≈ 14.397.

CAUTION

When rounding whole numbers in **Section 1.7,** you kept all the digits but changed some to 0s. With decimals, you cut off and *drop the extra digits.* In Example 1 above, 14.39652 rounds to 14.397 ***not*** 14.39700.

Work Problem 1 at the Side.

In Example 1, the rounded number 14.397 had *three decimal places.* **Decimal places** are the number of digits to the *right* of the decimal point. The first decimal place is tenths, the second is hundredths, the third is thousandths, and so on.

Example 2 Rounding Decimals to Different Places

Round to the place indicated.

(a) 5.3496 to the nearest tenth (Is it closer to 5.3 or to 5.4?)

Step 1 Draw a cut-off line after the tenths place.

```
5 . 3 | 4 9 6     You are cutting off the 4, 9, and 6.
    ↑             They will be dropped.
 Tenths
```

Step 2

```
5 . 3 | 4 9 6     Look only at the 4.
          ‿
  Ignore these digits.
```

Step 3

```
5 . 3 | 4 9 6     First digit cut is 4 or less,
                  so the part you are keeping
                  stays the same.
5 . 3  ← Stays the same
```

5.3496 rounded to the nearest tenth is 5.3 (*one* decimal place for *tenths*). We can write 5.3496 ≈ 5.3. Notice that 5.3496 does ***not*** round to 5.3000, which would be ten-thousandths.

(b) 0.69738 to the nearest hundredth (Is it closer to 0.69 or to 0.70?)

Step 1

```
0 . 6 9 | 7 3 8     Draw a cut-off line after the
      ↑             hundredths place.
 Hundredths
```

Step 2

```
0 . 6 9 | 7 3 8     Look only at the 7.
```

Continued on Next Page

ANSWERS
1. (a) 0.335 (b) 8.009 (c) 265.420 (d) 10.702

Step 3

$$0.69|738$$ ← First digit cut is 5 or more, so round up by adding 1 hundredth to the part you are keeping.

$$\begin{array}{r} 0.69 \\ +\ 0.01 \\ \hline 0.70 \end{array}$$

0.69 ← Keep this part.
+ 0.01 ← To round up, add 1 hundredth.
0.70 ← 9 + 1 is 10; write 0 and carry 1 to the 6 in the tenths place.

0.69738 rounded to the nearest hundredth is 0.70. Hundredths is *two* decimal places so you *must* write the 0 in the hundredths place. We can write 0.69738 ≈ 0.70.

(c) 0.01806 to the nearest thousandth (Is it closer to 0.018 or to 0.019?)

$$0.018|06$$ ← First digit cut is 4 or less, so the part you are keeping stays the same.

0.018

0.01806 rounded to the nearest thousandth is 0.018 (*three* decimal places for *thousandths*). We can write 0.01806 ≈ 0.018.

(d) 57.976 to the nearest tenth (Is it closer to 57.9 or to 58.0?)

57.9|76 ← First digit cut is 5 or more, so round up by adding 1 tenth to the part you are keeping.

$$\begin{array}{r} 57.9 \\ +\ 0.1 \\ \hline 58.0 \end{array}$$

← 9 + 1 is 10; write the 0 and carry 1 to the 7 in the ones place.

57.976 rounded to the nearest tenth is 58.0. We can write 57.976 ≈ 58.0. You *must* write the 0 in the tenths place to show that the number was rounded to the nearest tenth.

CAUTION

Check that your rounded answer shows ***exactly*** the number of decimal places called for, even if a 0 is in that place. See Examples 2(b) and 2(d). Be sure your answer shows *one* decimal place if you rounded to *tenths, two* decimal places for *hundredths,* or *three* decimal places for *thousandths.*

Work Problem 2 at the Side.

3 Round money amounts to the nearest cent or nearest dollar. In many everyday situations, such as shopping in a store, money amounts are rounded to the nearest cent. There are 100 cents in a dollar.

$$\text{Each cent is } \frac{1}{100} \text{ of a dollar.}$$

Another way to write $\frac{1}{100}$ is 0.01. So rounding to the *nearest cent* is the same as rounding to the *nearest hundredth of a dollar.*

2 Round to the place indicated.

(a) 0.8988 to the nearest hundredth

(b) 5.8903 to the nearest hundredth

(c) 11.0299 to the nearest thousandth

(d) 0.545 to the nearest tenth

ANSWERS
2. (a) 0.90 **(b)** 5.89 **(c)** 11.030 **(d)** 0.5

3 Round each money amount to the nearest cent.

(a) \$14.595

(b) \$578.0663

(c) \$0.849

(d) \$0.0548

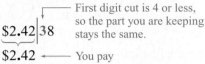

ANSWERS
3. **(a)** \$14.60 **(b)** \$578.07 **(c)** \$0.85 **(d)** \$0.05

Example 3 Rounding to the Nearest Cent

Round each money amount to the nearest cent.

(a) \$2.4238 (Is it closer to \$2.42 or to \$2.43?)

\$2.42|38 ← First digit cut is 4 or less, so the part you are keeping stays the same.

\$2.42 ← You pay

(b) \$0.695 (Is it closer to \$0.69 or to \$0.70?)

\$0.69|5 ← 5 or more; round up

\$0.69
+ \$0.01 ← To round up, add 1 hundredth (1 cent).
\$0.70 ← You pay

Work Problem 3 at the Side.

It is also common to round money amounts to the nearest dollar. For example, you can do that on your federal and state income tax returns to make the calculations easier.

Example 4 Rounding to the Nearest Dollar

Round to the nearest dollar.

(a) \$48.69 (Is it closer to \$48 or to \$49?)

\$48.|69 ← First digit cut is 5 or more, so round up by adding \$1.

\$48
+ 1
\$49

CAUTION

\$48.69 rounded to the nearest dollar is \$49. Write the answer as \$49 to show that the rounding is to the *nearest dollar.* Writing \$49.00 would show rounding to the nearest *cent.*

(b) \$594.36 (Is it closer to \$594 or to \$595?)

\$594.|36 ← First digit cut is 4 or less, so the part you keep stays the same.

\$594

\$594.36 rounded to the nearest dollar is \$594.

(c) \$349.88 (Is it closer to \$349 or to \$350?)

\$349.|88 ← 5 or more, so round up by adding \$1.

\$349
+ 1
\$350

\$349.88 rounded to the nearest dollar is \$350.

Continued on Next Page

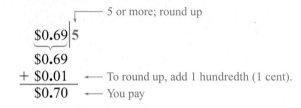

(d) \$2689.50 rounded to the nearest dollar is \$2690.

(e) \$0.61 rounded to the nearest dollar is \$1.

Calculator Tip Accountants and other people who work with money amounts often set their calculators to automatically round to two decimal places (nearest cent) or to round to zero decimal places (nearest dollar). Your calculator may have this feature.

Work Problem 4 at the Side.

4 Round to the nearest dollar.

(a) \$29.10

(b) \$136.49

(c) \$990.91

(d) \$5949.88

(e) \$49.60

(f) \$0.55

(g) \$1.08

ANSWERS

4. **(a)** \$29 **(b)** \$136 **(c)** \$991 **(d)** \$5950 **(e)** \$50 **(f)** \$1 **(g)** \$1

Focus on **Real-Data Applications**

Lawn Fertilizer

Gotta Be Green

A lot's being said about personal responsibility these days, and the idea seems to be ending up on the front lawn—literally! Each spring, homeowners across the country gear up to green up their lawns, and the increased use of fertilizer has a lot of environmentalists concerned about the potential effects of chemical runoff into nearby rivers and streams.

Each year, according to a study conducted by the University of Minnesota's Department of Agriculture, each household in the Minneapolis/St. Paul metro area uses an average of 36 pounds of lawn fertilizer. That adds up to 25,529,295 pounds, or 12,765 tons. Add to that another 193,000 pounds of weed killer and you're looking at the total picture for keeping it green in the Twin Cities.

Source: Minneapolis Star Tribune.

1. According to the article,

(a) How many pounds of lawn fertilizer are used each year in the *entire metro area?*

(b) Do a division on your calculator to find the number of *households* in the metro area.

(c) Would it make sense to round your answer to part (b)? If so, how would you round it?

(d) How many pounds of *weed killer* are used each year in the entire metro area?

(e) Do a division on your calculator to find the number of pounds of *weed killer* used by each household in the metro area. Round your answer to the nearest hundredth.

2. There are 2000 pounds in one ton.

(a) Find the number of tons equivalent to 25,529,295 pounds of fertilizer.

(b) Does your answer match the figure given in the article? If not, what did the author of the article do to get 12,765 tons?

(c) Is the author's figure accurate? Why or why not?

(d) Find the number of tons equivalent to 193,000 pounds of weed killer.

(e) How can you write the division problem $193{,}000 \div 2{,}000$ in a simpler way? Explain what you did.

3. According to the article,

(a) "each household in the Minneapolis/St. Paul metro area uses an average of 36 pounds of lawn fertilizer" each year. What mathematical operation do you do to find an average?

(b) When the calculations were done to find the average, the answer was probably not *exactly* 36 pounds. List six different values that are *less than* 36 that would round to 36. List two values with one decimal place; two values with two decimal places, and two values with three decimal places.

(c) List six different values that are *greater than* 36 that would round to 36. List two values each with one, two, and three decimal places.

(d) What is the *smallest* number that can be rounded to 36? What is the *largest* number?

4.2 Exercises

FOR EXTRA HELP Student's Solutions Manual | MyMathLab.com | InterAct Math Tutorial Software | AW Math Tutor Center | www.mathxl.com | Digital Video Tutor CD 2 Videotape 7

Round each number to the place indicated. See Examples 1 and 2.

1. 16.8974 to the nearest tenth

2. 193.845 to the nearest hundredth

3. 0.95647 to the nearest thousandth

4. 96.81584 to the nearest ten-thousandth

5. 0.799 to the nearest hundredth

6. 0.952 to the nearest tenth

7. 3.66062 to the nearest thousandth

8. 1.5074 to the nearest hundredth

9. 793.988 to the nearest tenth

10. 476.1196 to the nearest thousandth

11. 0.09804 to the nearest ten-thousandth

12. 176.004 to the nearest tenth

13. 48.512 to the nearest one

14. 3.385 to the nearest one

15. 9.0906 to the nearest hundredth

16. 30.1290 to the nearest thousandth

17. 82.000151 to the nearest ten-thousandth

18. 0.400594 to the nearest ten-thousandth

Nardos is grocery shopping. The store will round the amount she pays for each item to the nearest cent. Write the rounded amounts. See Example 3.

19. Soup is three cans for $2.45, so one can is $0.81666. Nardos pays _______.

20. Orange juice is two cartons for $2.69, so one carton is $1.345. Nardos pays _______.

21. Facial tissue is four boxes for $4.89, so one box is $1.2225. Nardos pays _______.

22. Muffin mix is three packages for $1.75, so one package is $0.58333. Nardos pays _______.

23. Candy bars are six for $2.99, so one bar is $0.4983. Nardos pays _______.

24. Spaghetti is four boxes for $3.59, so one box is $0.8975. Nardos pays _______.

As she gets ready to do her income tax return, Ms. Chen rounds each amount to the nearest dollar. Write the rounded amounts. See Example 4.

25. Income from job, $48,649.60

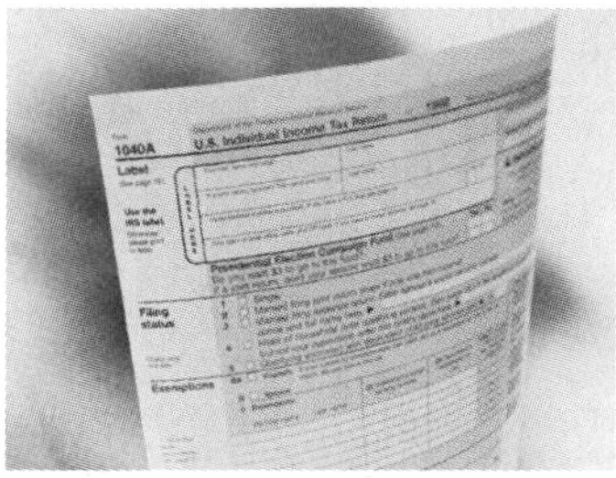

26. Income from interest on bank account, $69.58

27. Union dues, $310.08

28. Federal withholding, $6064.49

29. Donations to charity, $848.91

30. Medical expenses, $609.38

Round each money amount as indicated.

31. \$499.98 to the nearest dollar.

32. \$9899.59 to the nearest dollar.

33. \$0.996 to the nearest cent.

34. \$0.09929 to the nearest cent.

35. \$999.73 to the nearest dollar.

36. \$9999.80 to the nearest dollar.

The table lists speed records for various types of transportation. Use the table to answer Exercises 37–40.

Record	Speed (miles per hour)
Land speed record (specially built car)	763.04
Motorcycle speed record (specially adapted motorcycle)	322.15
Fastest train (regular passenger service)	162.7
Fastest X-15 (military jet)	4520
Boeing 737-300 airplane (regular passenger service)	495
Indianapolis 500 auto race (fastest average winning speed)	185.984
Daytona 500 auto race (fastest average winning speed)	177.602

Source: The Top 10 of Everything 2000.

37. Round these speed records to the nearest tenth.

(a) Indianapolis 500 average winning speed

(b) Land speed record

38. Round these speed records to the nearest hundredth.

(a) Daytona 500 average winning speed

(b) Indianapolis 500 average winning speed

39. Round these speed records to the nearest whole number.

(a) Motorcycle **(b)** Train

40. Round these speed records to the nearest hundred.

(a) X-15 military jet **(b)** Boeing 737-300 airplane

RELATING CONCEPTS (Exercises 41–44) FOR INDIVIDUAL OR GROUP WORK

Use your knowledge about rounding money amounts to ***work Exercises 41–44 in order.***

41. Explain what happens when you round \$0.499 to the nearest dollar. Why does this happen?

42. Look again at Exercise 41. How else could you round \$0.499 that would be more helpful? What kind of guideline does this suggest about rounding to the nearest dollar?

43. Explain what happens when you round \$0.0015 to the nearest cent. Why does this happen?

44. Suppose you want to know which of these amounts is less, so you round them both to the nearest cent.

\$0.5968 \$0.6014

Explain what happens. Describe what you could do instead of rounding to the nearest cent.

4.3 Adding and Subtracting Decimals

Objectives

1. Add decimals.
2. Subtract decimals.
3. Estimate the answer when adding or subtracting decimals.

1 Add decimals. When adding or subtracting *whole* numbers **(Sections 1.2 and 1.3),** you lined up the numbers in columns so that you were adding ones to ones, tens to tens, and so on. A similar idea applies to adding or subtracting *decimal* numbers. With decimals, you line up the decimal points to make sure you are adding tenths to tenths, hundredths to hundredths, and so on.

Adding and Subtracting Decimals

Step 1 Write the numbers in columns with the decimal points lined up.

Step 2 If necessary, write in 0s so both numbers have the same number of decimal places. Then add or subtract as if they were whole numbers.

Step 3 Line up the decimal point in the answer directly below the decimal points in the problem.

Example 1 Adding Decimal Numbers

Add.

(a) 16.92 and 48.34

Step 1 Write the numbers in columns with the decimal points lined up.

```
  tens ones . tenths hundredths
     1    6 .   9     2
 +   4    8 .   3     4
```

Decimal points are lined up.

Step 2 Add as if these were whole numbers.

```
   1 1
   16.92
 + 48.34
 -------
   65.26
```

Step 3 Decimal point in answer is lined up under decimal points in problem.

(b) 5.897 + 4.632 + 12.174

Write the numbers vertically with decimal points lined up. Then add.

```
   11  21
    5.897
    4.632
 + 12.174
 --------
   22.703
```

Decimal points are lined up.

Work Problem 1 at the Side.

1 Find each sum.

(a) 2.86 + 7.09

(b) 13.761 + 8.325

(c) 0.319 + 56.007 + 8.252

(d) 39.4 + 0.4 + 177.2

In Example 1(a), both numbers had *two* decimal places (two digits to the right of the decimal point). In Example 1(b), all the numbers had *three decimal places* (three digits to the right of the decimal point). That made it easy to add tenths to tenths, hundredths to hundredths, and so on.

Answers

1. **(a)** 9.95 **(b)** 22.086 **(c)** 64.578 **(d)** 217.0

2 Find each sum.

(a) 6.54 + 9.8

(b) 0.831 + 222.2 + 10

(c) 8.64 + 39.115 + 3.0076

(d) 5 + 429.823 + 0.76

ANSWERS
2. **(a)** 16.34 **(b)** 233.031 **(c)** 50.7626 **(d)** 435.583

If the number of decimal places does *not* match, you can write in 0s as placeholders to make them match. This is shown in Example 2.

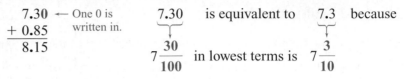

Example 2 **Writing 0s as Placeholders before Adding**

Add.

(a) 7.3 + 0.85

There are two decimal places in 0.85 (tenths and hundredths), so write a 0 in the hundredths place in 7.3 so that it has two decimal places also.

$$\begin{array}{r} 7.3\mathbf{0} \\ +\ 0.85 \\ \hline 8.15 \end{array} \leftarrow \text{One 0 is written in.}$$

7.30 is equivalent to 7.3 because $7\frac{30}{100}$ in lowest terms is $7\frac{3}{10}$

(b) 6.42 + 9 + 2.576

Write in 0s so that all the addends have three decimal places. Notice how the whole number 9 is written with the decimal point at the *far right* side. (If you put the decimal point on the *left* side of the 9, you would turn it into the decimal fraction 0.9.)

$$\begin{array}{r} 6.42\mathbf{0} \\ 9.\mathbf{000} \\ +\ 2.576 \\ \hline 17.996 \end{array}$$

6.420 ← One 0 is written in.
9.000 ← 9 is a whole number; decimal point and three 0s are written in.
2.576 ← No 0s are needed.
Decimal points are lined up.

NOTE

Writing 0s to the right of a *decimal* number does *not* change the value of the number.

Work Problem 2 at the Side.

2 Subtract decimals. Subtraction of decimals is done in much the same way as addition of decimals. You can check the answers to subtraction problems using addition, as you did with whole numbers (see **Section 1.3**).

Example 3 **Subtracting Decimal Numbers**

Subtract. Check your answers using addition.

(a) 15.82 from 28.93

Step 1

$$\begin{array}{r} 28.93 \\ -\ 15.82 \end{array}$$

Line up decimal points. Then you will be subtracting hundredths from hundredths and tenths from tenths.

Step 2

$$\begin{array}{r} 28.93 \\ -\ 15.82 \\ \hline 13\ 11 \end{array}$$

Both numbers have two decimal places; no need to write in 0s.

Subtract as if they were whole numbers.

Continued on Next Page

Step 3

$$\begin{array}{r} 28.93 \\ -\ 15.82 \\ \hline 13.11 \end{array}$$

Decimal point in answer is lined up.

Check the answer by adding 13.11 and 15.82. If the subtraction is done correctly, the sum will be 28.93.

(b) 146.35 minus 58.98
Borrowing is needed here.

$$\begin{array}{r} \scriptstyle 0\ 13\ 15\ \ 12\ 15 \\ \not{1}\not{4}\not{6}.\not{3}\not{5} \\ -\ 58.98 \\ \hline 87.37 \end{array}$$

Line up decimal points.

Check the answer by adding 87.37 and 58.98. If you did the subtraction correctly, the sum will be 146.35. (If it *isn't,* you need to rework the problem.)

Work Problem 3 at the Side.

Example 4 Writing 0s as Placeholders before Subtracting

Subtract.

(a) 16.5 from 28.362
Use the same steps as in Example 3 above, remembering to write in 0s so both numbers have three decimal places.

$$\begin{array}{r} 28.362 \\ -\ 16.500 \\ \hline 11.862 \end{array}$$

Line up decimal points.
16.500 ← Write two 0s.
11.862 ← Subtract as usual.

Check the answer by adding.

$$\begin{array}{r} 16.500 \\ +\ 11.862 \\ \hline 28.362 \end{array}$$

← Matches minuend in original problem.

(b) 59.7 − 38.914

$$\begin{array}{r} 59.700 \\ -\ 38.914 \\ \hline 20.786 \end{array}$$

59.700 ← Write two 0s
20.786 ← Subtract as usual.

(c) 12 less 5.83

$$\begin{array}{r} 12.00 \\ -\ 5.83 \\ \hline 6.17 \end{array}$$

12.00 ← Write a decimal point and two 0s.
6.17 ← Subtract as usual.

Work Problem 4 at the Side.

3 Estimate the answer when adding or subtracting decimals. A common error in working decimal problems by hand is to misplace the decimal point in the answer. Or, when using a calculator, you may accidentally press the wrong key. **Estimating** the answer will help you avoid these mistakes. Start by using *front end rounding* on each number (as you did in **Section 1.7**). Here are several examples. Notice that in the rounded numbers only the left-most digit is something other than 0.

3.25	rounds to	3	6.812	rounds to	7
532.6	rounds to	500	26.397	rounds to	30
7094.2	rounds to	7000	351.24	rounds to	400

3 Subtract. Check your answers using addition.

(a) 22.7 from 72.9

(b) 6.425 from 11.813

(c) 20.15 − 19.67

4 Subtract. Check your answers using addition.

(a) 18.651 from 25.3

(b) 5.816 − 4.98

(c) 40 less 3.66

(d) 1 − 0.325

ANSWERS
3. (a) 50.2; 50.2 + 22.7 = 72.9
(b) 5.388; 5.388 + 6.425 = 11.813
(c) 0.48; 0.48 + 19.67 = 20.15
4. (a) 6.649; 6.649 + 18.651 = 25.3
(b) 0.836; 0.836 + 4.98 = 5.816
(c) 36.34; 36.34 + 3.66 = 40
(d) 0.675; 0.675 + 0.325 = 1

5 First, use front end rounding and estimate each answer. Then add or subtract to find the exact answer.

(a) 2.83 + 5.009 + 76.1

(b) 11.365 from 58

(c) 398.81 + 47.658 + 4158.7

(d) Find the difference between 12.837 meters and 46.091 meters.

(e) \$19.28 plus \$1.53

Example 5 Estimating Decimal Answers

Use front end rounding to round each number. Then add or subtract the rounded numbers to get an estimated answer. Finally, find the exact answer.

(a) Add 194.2 and 6.825.

Estimate:		*Exact:*
200	← Rounds to	**194.200**
+ **7**	← Rounds to	+ **6.825**
207		201.025

The estimate goes out to the hundreds place (three places to the *left* of the decimal point), and so does the exact answer. Therefore, the decimal point is probably in the correct place in the exact answer.

(b) \$69.42 + \$13.78

Estimate:		*Exact:*	
\$70	← Rounds to	\$69.42	
+ 10	← Rounds to	+ 13.78	
\$80		\$83.20	← Answer is close to estimate, so the problem is probably set up correctly.

(c) Find the difference between 0.92 ft and 8 ft.

Use subtraction to find the difference between two numbers. The larger number, 8, is written on top.

Estimate:		*Exact:*	
8	← Rounds to	8.00	← Write a decimal point and two 0s.
− 1	← Rounds to	− 0.92	
7		7.08 ft	← Answer is close to estimate.

(d) Subtract 1.8614 from 7.3

Estimate:		*Exact:*	
7	← Rounds to	7.3000	← Write three 0s.
− 2	← Rounds to	− 1.8614	
5		5.4386	← Answer is close to estimate.

Work Problem 5 at the Side.

Calculator Tip If you are *adding* numbers, you can enter them in any order on your calculator. Try these; jot down the answers.

9.82 (+) 1.86 (=) ______ 1.86 (+) 9.82 (=) ______

The answers are the same because addition is *commutative.* (See **Section 1.2.**) But subtraction is *not* commutative. It *does* matter which number you enter first. Try these:

9.82 (−) 1.86 (=) ______ 1.86 (−) 9.82 (=) ______

The second answer has a negative sign (−) next to it. A negative number is less than 0. If it's in your checkbook, you'd be "in the hole" by \$7.96. (See **Section 9.1** for more about negative numbers.)

ANSWERS

5. (a) 3 + 5 + 80 = 88; 83.939
(b) 60 − 10 = 50; 46.635
(c) 400 + 50 + 4000 = 4450; 4605.168
(d) 50 − 10 = 40; 33.254 meters
(e) \$20 + \$2 = \$22; \$20.81

4.3 EXERCISES

FOR EXTRA HELP: Student's Solutions Manual; MyMathLab.com; InterAct Math Tutorial Software; AW Math Tutor Center; www.mathxl.com; Digital Video Tutor CD 3 Videotapes 7 and 8

Find each sum or difference. See Examples 1–4.

1. 5.69 + 11.79

2. 372.1 − 33.7

3. 24.008 − 0.995

4. 0.7759 + 9.8883

5. 8.263 − 0.5

6. 47.658 − 20.9

7. 76.5 + 0.506

8. 1.87 + 9.749

9. 21 − 0.896

10. 9 − 1.183

11. 0.4 − 0.291

12. 0.35 − 0.088

13. 39.76005 + 182 + 4.799 + 98.31 + 5.9999

14. 489.76 + 0.9993 + 38 + 8.55087 + 80.697

This drawing of a human skeleton shows the average length of the longest bones, in inches. Use the drawing to answer Exercises 15–18.

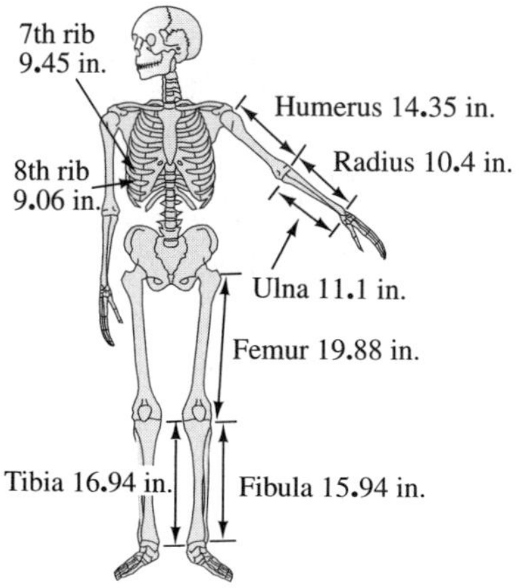

Source: *The Top 10 of Everything 2000.*

15. (a) What is the combined length of the humerus and radius bones?

(b) What is the difference in the lengths of these two bones?

16. (a) What is the total length of the femur and tibia bones?

(b) How much longer is the femur than the tibia?

17. (a) Find the sum of the lengths of the humerus, ulna, femur, and tibia.

(b) How much shorter is the 8th rib than the 7th rib?

18. (a) What is the difference in the lengths of the two bones in the lower arm?

(b) What is the difference in the lengths of the two bones in the lower leg?

19. Explain and correct the error that a student made when he added 0.72 + 6 + 39.5 this way:

$$\begin{array}{r} 0.72 \\ 6 \\ +\ 39.50 \\ \hline 40.28 \end{array}$$

20. Explain the difference between saying "subtract 2.9 from 8" and saying "2.9 minus 8."

Use front end rounding to round each number. Then add or subtract the rounded numbers to get an estimated answer. Finally, find the exact answer. See Example 5.

21. *Estimate:* *Exact:*

$$\begin{array}{r} \$\ \_\_\_ \\ -\ \_\_\_ \\ \hline \$\ \_\_\_ \end{array} \qquad \begin{array}{r} \$19.74 \\ -\ 6.58 \\ \hline \$\ \_\_\_ \end{array}$$

22. *Estimate:* *Exact:*

$$\begin{array}{r} \$\ \_\_\_ \\ -\ \_\_\_ \\ \hline \$\ \_\_\_ \end{array} \qquad \begin{array}{r} \$27.96 \\ -\ 8.39 \\ \hline \$\ \_\_\_ \end{array}$$

23. *Estimate:* *Exact:*

$$\begin{array}{r} \_\_\_ \\ \_\_\_ \\ +\ \_\_\_ \\ \hline \end{array} \qquad \begin{array}{r} 392.7 \\ 0.865 \\ +\ 21.08 \\ \hline \end{array}$$

24. *Estimate:* *Exact:*

$$\begin{array}{r} \_\_\_ \\ \_\_\_ \\ +\ \_\_\_ \\ \hline \end{array} \qquad \begin{array}{r} 38.55 \\ 7.716 \\ +\ 0.6 \\ \hline \end{array}$$

25. What is 8.6 less 3.751?

Estimate: *Exact:*

26. What is 31.7 less 4.271?

Estimate: *Exact:*

27. *Estimate:* *Exact:*

$$\begin{array}{r} \_\_\_ \\ \_\_\_ \\ +\ \_\_\_ \\ \hline \end{array} \qquad \begin{array}{r} 62.8173 \\ 539.99 \\ +\ 5.629 \\ \hline \end{array}$$

28. *Estimate:* *Exact:*

$$\begin{array}{r} \_\_\_ \\ \_\_\_ \\ +\ \_\_\_ \\ \hline \end{array} \qquad \begin{array}{r} 332.607 \\ 12.5 \\ +\ 823.3949 \\ \hline \end{array}$$

Use your estimation skills to pick the most reasonable answer for each example. Do ***not*** *solve the problems. Circle your choice.*

29. 12 − 11.725

2.75 0.275 27.5

30. 20 − 1.37

0.1863 1.863 18.63

31. 6.5 + 0.007

6.507 0.6507 65.07

32. 9.67 + 0.09

0.976 9.76 0.00976

33. 456.71 − 454.9

18.1 181 1.81

34. 803.25 − 0.6

802.65 0.80265 8.0265

35. 6004.003 + 52.7172

60.567202 605.67202 6056.7202

36. 128.35 + 97.0093

2253.593 225.3593 0.2253593

First use front end rounding to round each number and estimate the answer. Then find the exact answer.

37. The cost of Julie's tennis racket, with tax, is \$41.09. She gave the clerk two \$20 bills and a \$10 bill. What amount of change did Julie receive?

Estimate:

Exact:

38. The tallest known land mammal is a prehistoric ancestor of the rhino measuring 6.4 meters. Find the combined heights of these NBA basketball stars: Charles Barkley at 1.98 meters, Karl Malone at 2.06 meters, and David Robinson at 2.16 meters. Is their combined height greater or less than the prehistoric rhino? (*Source:* Harper's Index and NBA.)

Estimate:

Exact:

39. Find the difference between 1.981 in. and 2 in.

Estimate:

Exact:

40. Find the difference between 13.582 meters and 28 meters.

Estimate:

Exact:

41. The U.S. population in 2000 was approximately 281.42 million. The U.S. Bureau of the Census estimates that it will be 393.9 million in the year 2050. The increase in population during that 50-year period is how many millions of people? (*Source:* U.S. Bureau of the Census.)

Estimate:

Exact:

42. At a bakery, Sue Chee bought \$7.42 worth of muffins and \$10.09 worth of croissants for a staff party and a \$0.69 cookie for herself. How much money did she spend altogether?

Estimate:

Exact:

43. Namiko is comparing two boxes of chicken nuggets. One box weighs 9.85 ounces and the other weighs 10.5 ounces. What is the difference in the weight of the two boxes?

Estimate:

Exact:

44. Sammy works in a veterinarian's office. He weighed two newborn kittens. One was 3.9 ounces and the other was 4.05 ounces. What was the difference in the weight of the two kittens?

Estimate:

Exact:

Find the perimeter of (distance around) each figure by adding the lengths of the sides.

45.

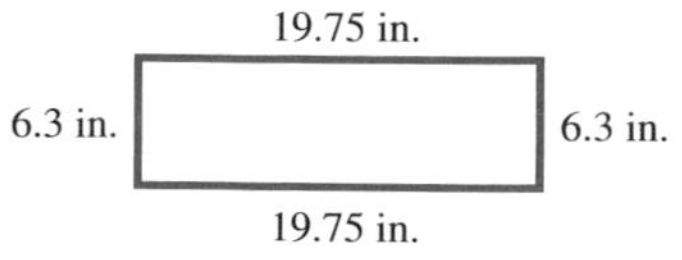

Estimate:

Exact:

46.

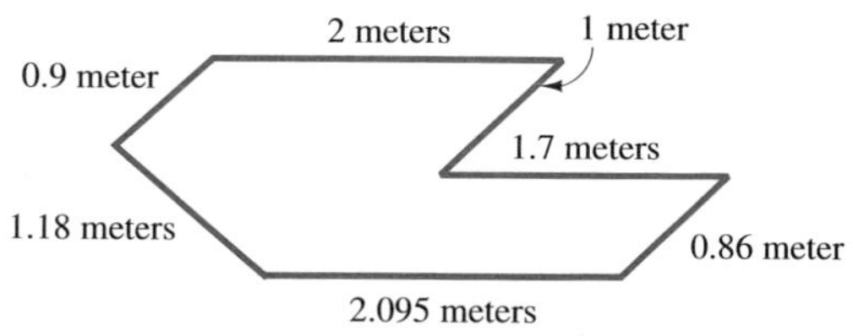

Estimate:

Exact:

The dad and daughter buying fishing equipment on the first page of this chapter brought along the store's sale insert from the Sunday paper. Use the information on sale prices to answer Exercises 47–50.

Fishing Opener Sale
Catch your limit of savings!
Bobbers 3 for 87¢
4-, 6-, or 8-pound test line
330 yd for $5.32
110 yd for $2.72
No-See Line
Environmentally safe tin split shot $1.57
Small tackle boxes
One tray $6.38
Two trays $7.96
Leaded split shot 94¢
Spinning reels: $8.96, $13.47, $17.96, $29.46
Lightweight rods: $6.96, $12.97, $19.94

Source: Wal-Mart.

47. What is the difference in price between the most expensive and least expensive spinning reel?

Estimate:

Exact:

48. How much more would 330 yd of line cost than three bobbers?

Estimate:

Exact:

49. What is the total cost of the middle-priced rod, the second most expensive reel, a one-tray tackle box, 110 yd of line, and a package of environmentally safe split shot?

Estimate:

Exact:

50. Dad also bought his daughter a cap for $8.49, SPF45 sunscreen for $6.97, and a child-size flotation vest for $19.99. How much did he spend on these items?

Estimate:

Exact:

Olivia Sanchez kept track of her expenses for one month. Use her list to answer Exercises 51–56.

Monthly Expenses	
Rent	$994
Car payment	$190.78
Car repairs, gas	$205
Cable TV	$39.95
Internet access	$19.95
Electricity	$40.80
Telephone	$57.32
Groceries	$186.81
Entertainment	$97.75
Clothing, laundry	$107

51. What were Olivia's total expenses for the month?

52. How much did Olivia pay for telephone, cable TV, and Internet access?

53. What was the difference in the amounts spent for groceries and for the car payment?

54. Compare the amount Olivia spent on entertainment to the amount spent on car repairs and gas. What is the difference?

55. How much more did Olivia spend on rent than on all her car expenses?

56. How much less did Olivia spend on clothing and laundry than on all her car expenses?

Find the length of the dashed line in each rectangle or circle.

57.

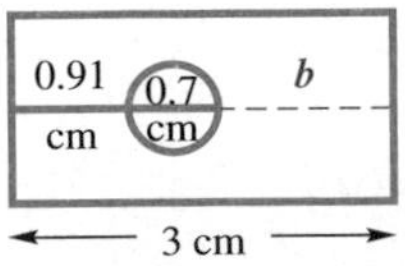
0.91 cm
0.7 cm
b
3 cm

58.

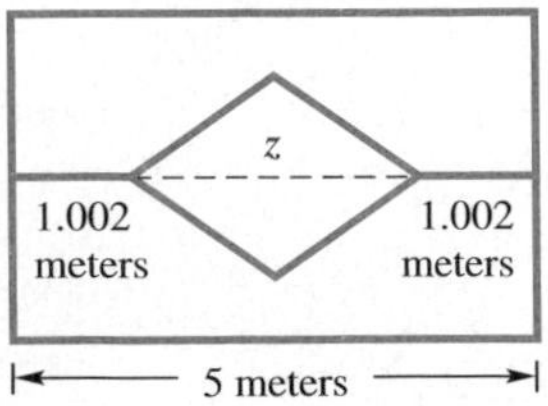
z
1.002 meters
1.002 meters
5 meters

59.

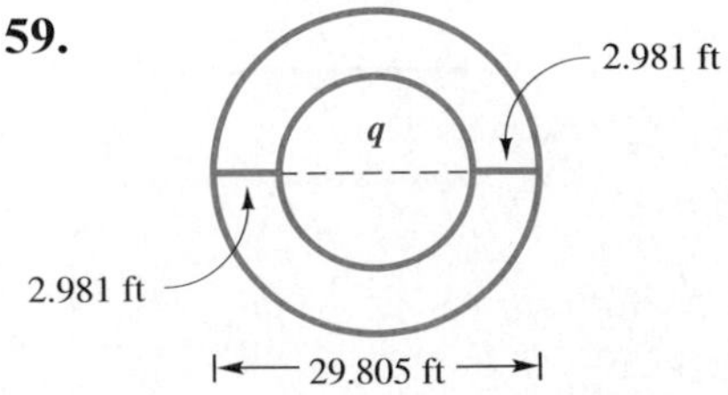
2.981 ft
q
2.981 ft
29.805 ft

4.4 Multiplying Decimals

1 Multiply decimals. The decimals 0.3 and 0.07 can be multiplied by writing them as fractions.

$$\underset{\text{1 decimal place}}{0.3} \times \underset{\text{2 decimal places}}{0.07} = \frac{3}{10} \times \frac{7}{100} = \frac{3 \times 7}{10 \times 100} = \frac{21}{1000} = \underset{\text{3 decimal places}}{0.021}$$

1 decimal place + 2 decimal places → 3 decimal places

Can you see a way to multiply decimals without writing them as fractions? Try these steps. Remember that each number in a multiplication problem is called a *factor,* and the answer is called the *product.*

Multiplying Decimals

Step 1 Multiply the numbers (the factors) as if they were whole numbers.

Step 2 Find the *total* number of decimal places in *both* factors.

Step 3 Write the decimal point in the product (the answer) so it has the same number of decimal places as the total from Step 2. You may need to write in extra 0s on the left side of the product to get the correct number of decimal places.

NOTE

When multiplying decimals, you do ***not*** need to line up decimal points. (You ***do*** need to line up decimal points when adding or subtracting decimals.)

Example 1 Multiplying Decimal Numbers

Multiply: 8.34 times 4.2.

Step 1 Multiply the numbers as if they were whole numbers.

```
   8.3 4
 ×   4.2
 -------
  1 6 6 8
 3 3 3 6
 -------
 3 5 0 2 8
```

Step 2 Count the total number of decimal places in both factors.

```
   8.3 4   ← 2 decimal places
 ×   4.2   ← 1 decimal place
 -------     3 total decimal places
  1 6 6 8
 3 3 3 6
 -------
 3 5 0 2 8
```

Step 3 Count over 3 places in the product and write the decimal point. Count from *right to left.*

```
   8.3 4   ← 2 decimal places
 ×   4.2   ← 1 decimal place
 -------     3 total decimal places
  1 6 6 8
 3 3 3 6
 -------
 3 5.0 2 8 ← 3 decimal places in product
```

Count over 3 places from right to left to position the decimal point.

Work Problem 1 at the Side.

Objectives

1. Multiply decimals.
2. Estimate the answer when multiplying decimals.

1 Multiply.

(a) 2.6 × 0.4

(b) 45.2 × 0.25

(c) 0.104 ← 3 decimal places
× 7 ← 0 decimal places
← 3 decimal places in the product

(d) 3.18 × 2.23

(e) 611 × 3.7

Answers

1. (a) 1.04 (b) 11.300 (c) 0.728 (d) 7.0914 (e) 2260.7

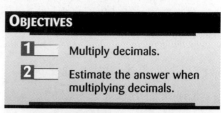

2 Multiply.

(a) 0.04×0.09

(b) $0.2 \cdot 0.008$

(c) $(0.063)(0.04)$

(d) $0.0081 \cdot 0.003$

(e) $(0.11)(0.0005)$

3 First use front end rounding and estimate the answer. Then find the exact answer.

(a) $(11.62)(4.01)$

(b) $(5.986)(33)$

(c) $8.31 \cdot 4.2$

(d) 58.6×17.4

ANSWERS

2. **(a)** 0.0036 **(b)** 0.0016 **(c)** 0.00252 **(d)** 0.0000243 **(e)** 0.000055

3. **(a)** $(10)(4) = 40$; 46.5962 **(b)** $(6)(30) = 180$; 197.538 **(c)** $8 \cdot 4 = 32$; 34.902 **(d)** $60 \times 20 = 1200$; 1019.64

Example 2 Writing 0s as Placeholders in the Product

Multiply 0.042 by 0.03.

Start by multiplying, then count decimal places.

```
    0.0 4 2 ← 3 decimal places
×   0.0 3   ← 2 decimal places
      1 2 6 ← 5 decimal places needed in product
```

After multiplying, the answer has only three decimal places, but five are needed. So write two 0s on the *left* side of the answer.

```
   0.0 4 2              0.0 4 2 ← 3 decimal places
×  0.0 3            ×   0.0 3   ← 2 decimal places
   0 0 1 2 6          .0 0 1 2 6 ← 5 decimal places
   ↑ ↑
```

Write two 0s on *left* side of answer. — Now count over 5 places and write in the decimal point.

The final product is 0.00126, which has five decimal places.

Work Problem 2 at the Side.

2 **Estimate the answer when multiplying decimals.** If you are doing multiplication problems by hand, estimating the answer helps you check that the decimal point is in the right place. When you are using a calculator, estimating helps you catch an error like pressing the ÷ key instead of the × key.

Example 3 Estimating before Multiplying

First estimate (76.34)(12.5) using front end rounding to round each number. Then find the exact answer.

```
Estimate:                    Exact:
  80   ← Rounds to ——        7 6.3 4 ← 2 decimal places
× 10   ← Rounds to ——    ×     1 2.5 ← 1 decimal place
 800                         3 8 1 7 0   3 decimal places needed
                           1 5 2 6 8      in product.
                           7 6 3 4
                           9 5 4.2 5 0
```

Both the estimate and the exact answer go out to the hundreds place, so the decimal point in 954.250 is probably in the correct place.

Work Problem 3 at the Side.

Calculator Tip When working with money amounts, you may need to write a 0 in your answer. For example, try multiplying $3.54 × 5 on your calculator. Write down the result.

3.54 × 5 = ______

Notice that the result is 17.7, which is *not* the way to write a money amount. You have to write the 0 in the hundredths place: $17.70 is correct. The calculator does not show the "extra" 0 because:

$$17.70 \text{ or } 17\frac{70}{100} \quad \text{reduces to} \quad 17\frac{7}{10} \text{ or } 17.7.$$

So keep an eye on your calculator—it doesn't know when you're working with money amounts.

4.4 EXERCISES

FOR EXTRA HELP: Student's Solutions Manual | MyMathLab.com | InterAct Math Tutorial Software | AW Math Tutor Center | www.mathxl.com | Digital Video Tutor CD 3 Videotape 8

Multiply. See Example 1.

1. 0.042×3.2

2. 0.571×2.9

3. 21.5×7.4

4. 85.4×3.5

5. 23.4×0.666

6. 0.896×0.799

7. $\$51.88 \times 665$

8. $\$736.75 \times 118$

Use the fact that $72 \times 6 = 432$ *to help you answer Exercises 9–16 by simply counting decimal places. See Examples 1 and 2.*

9. $72 \times 0.6 =$ 4 3 2

10. $7.2 \times 6 =$ 4 3 2

11. $(7.2)(0.06) =$ 4 3 2

12. $(0.72)(0.6) =$ 4 3 2

13. $0.72 \times 0.06 =$ 4 3 2

14. $72 \times 0.0006 =$ 4 3 2

15. $0.0072 \times 0.6 =$ 4 3 2

16. $0.072 \times 0.006 =$ 4 3 2

Multiply. See Example 2.

17. $(0.006)(0.0052)$

18. $(0.0052)(0.009)$

19. $0.003 \cdot 0.002$

20. $0.0079 \cdot 0.006$

RELATING CONCEPTS (Exercises 21–22) FOR INDIVIDUAL OR GROUP WORK

Look for patterns in the multiplications as you ***work Exercises 21 and 22 in order.***

21. Do these multiplications:

(5.96)(10)	(3.2)(10)
(0.476)(10)	(80.35)(10)
(722.6)(10)	(0.9)(10)

What pattern do you see? Write a "rule" for multiplying by 10. What do you think the rule is for multiplying by 100? by 1000? Write the rules and try them out on the numbers above.

22. Do these multiplications:

(59.6)(0.1)	(3.2)(0.1)
(0.476)(0.1)	(80.35)(0.1)
(65)(0.1)	(523)(0.1)

What pattern do you see? Write a "rule" for multiplying by 0.1. What do you think the rule is for multiplying by 0.01? by 0.001? Write the rules and try them out on the numbers above.

First use front end rounding to round each number and estimate the answer. Then find the exact answer. See Example 3.

23. *Estimate:*

$\times$ ____

Rounds to
Rounds to

Exact:

$$\begin{array}{r} 39.6 \\ \times\ 4.8 \\ \hline \end{array}$$

24. *Estimate:*

$\times$ ____

Exact:

$$\begin{array}{r} 18.7 \\ \times\ 2.3 \\ \hline \end{array}$$

25. *Estimate:*

$\times$ ____

Exact:

$$\begin{array}{r} 37.1 \\ \times\ 42 \\ \hline \end{array}$$

26. *Estimate:*

$\times$ ____

Exact:

$$\begin{array}{r} 5.08 \\ \times\ 71 \\ \hline \end{array}$$

27. *Estimate:*

$\times$ ____

Exact:

$$\begin{array}{r} 6.53 \\ \times\ 4.6 \\ \hline \end{array}$$

28. *Estimate:*

$\times$ ____

Exact:

$$\begin{array}{r} 7.51 \\ \times\ 8.2 \\ \hline \end{array}$$

29. *Estimate:*

$\times$ ____

Exact:

$$\begin{array}{r} 2.809 \\ \times\ 6.85 \\ \hline \end{array}$$

30. *Estimate:*

$\times$ ____

Exact:

$$\begin{array}{r} 73.52 \\ \times\ 22.34 \\ \hline \end{array}$$

Even with most of the problem missing, you can tell whether or not these answers are reasonable. Circle reasonable *or* unreasonable. *If the answer is unreasonable, move the decimal point, or insert a decimal point, to make the answer reasonable.*

31. How much was his car payment? \$18.90

reasonable

unreasonable, should be ____________

32. How many hours did she work today? 25 hours

reasonable

unreasonable, should be ____________

33. How tall is her son? 60.5 in.

reasonable

unreasonable, should be ____________

34. How much does he pay for rent now? \$6.92

reasonable

unreasonable, should be ____________

35. What is the price of one gallon of milk? \$319

reasonable

unreasonable, should be ____________

36. How long is the living room? 16.8 feet

reasonable

unreasonable, should be ____________

37. How much did the baby weigh? 0.095 pounds

reasonable

unreasonable, should be ____________

38. What was the sale price of the jacket? \$1.49

reasonable

unreasonable, should be ____________

Solve each application problem. If the problem involves money, round to the nearest cent, when necessary.

39. LaTasha worked 50.5 hours over the last two weeks. She earns \$18.73 per hour. How much did she make?

40. Michael's time card shows 42.2 hours at \$10.03 per hour. What are his gross earnings?

41. Sid needs 0.6 meter of canvas material to make a carry-all bag that fits on his wheelchair. If canvas is \$4.09 per meter, how much will Sid spend? (*Note:* \$4.09 *per* meter means \$4.09 for *one* meter.)

42. How much will Mrs. Nguyen pay for 3.5 yards of lace trim that costs \$0.87 per yard?

43. Michelle filled the tank of her pickup truck with regular unleaded gas. Use the information shown on the pump to find how much she paid for gas.

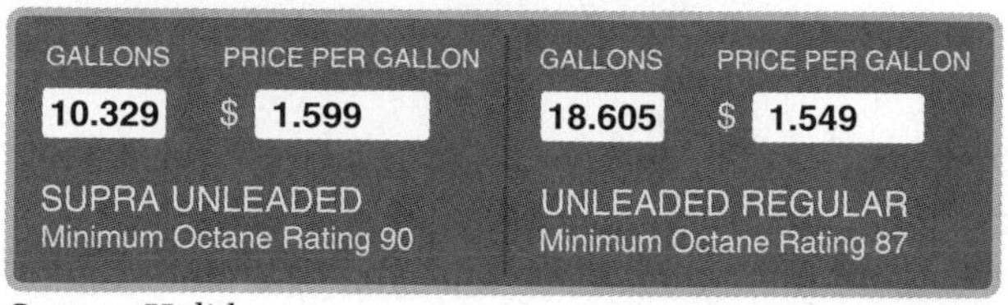

Source: Holiday.

44. Ground beef and spicy chicken wings are on sale. Juma bought 1.7 pounds of wings. Use the information in the ad to find the amount she paid.

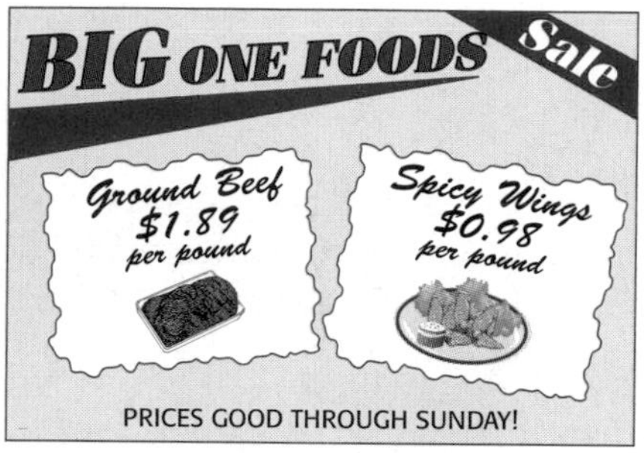

45. Ms. Rolack is a real estate broker who helps people sell their homes. Her fee is 0.07 times the price of the home. What was her fee for selling a \$175,300 home?

46. Alex Rodriguez, shortstop for the Seattle Mariners, had a batting average of 0.316 in the 2000 season. If he went to bat 554 times, how many hits did he make? (*Hint:* Multiply his batting average by the number of times at bat.) Round to the nearest whole number. (*Source: World Almanac*, 2001.)

47. Judy Lewis pays \$28.96 per month for basic cable TV. How much will she pay for cable over one year? How much would she pay in a year for the deluxe cable package that costs \$59.95 per month?

48. Chuck's car payment is \$220.27 per month for three years. How much will he pay altogether?

49. Paper for the copy machine at the library costs \$0.015 per sheet. How much will the library pay for 5100 sheets?

50. A student group collected 2200 pounds of plastic as a fund-raiser. How much will they make if the recycling center pays \$0.142 per pound?

51. The National Aquarium in Baltimore charges \$11.95 for adults, \$10.50 for seniors, and \$7.50 for children. How much will a mother with four children spend for her family and three senior relatives? (*Source:* Lyon Group.)

52. (Complete Exercise 51 first.) How much *less* would the same family spend at the Texas State Aquarium, which charges \$8 for adults, \$5.75 for seniors, and \$4.50 for children? (*Source:* Lyon Group.)

53. Ms. Sanchez paid \$29.95 a day to rent a car, plus \$0.29 per mile. Find the cost of her rental for a four-day trip of 926 miles.

54. The Bell family rented a motor home for \$375 per week plus \$0.35 per mile. What was the rental cost for their three-week vacation trip of 2650 miles?

55. Barry bought 16.5 meters of rope at \$0.47 per meter and three meters of wire at \$1.05 per meter. How much change did he get from three \$5 bills?

56. Susan bought a VCR that cost \$229.88. She paid \$45 down and \$37.98 per month for six months. How much could she have saved by paying cash?

Use the information from the Look Smart mail order catalog to answer Exercises 57–60.

Knit Shirt Ordering Information		
43–2A	Short sleeve, solid colors	\$14.75 each
43–2B	Short sleeve, stripes	\$16.75 each
43–3A	Long sleeve, solid colors	\$18.95 each
43–3B	Long sleeve, stripes	\$21.95 each
Extra-large size, add \$2 per shirt.		
Monogram, \$4.95 each. Gift box, \$5 each.		

Total Price of All Items (excluding monograms and gift boxes)	Shipping, Packing, and Handling
\$0–25.00	\$3.50
\$25.01–75.00	\$5.95
\$75.01–125.00	\$7.95
\$125.01+	\$9.95
Shipping to each additional address add \$4.25.	

57. Find the total cost of ordering four long-sleeve, solid-color shirts and two short-sleeve, striped shirts, all in the extra-large size, and all shipped to your home.

58. What is the total cost of eight long-sleeve shirts, five in solid colors and three striped? Include the cost of shipping the solid shirts to your home and the striped shirts to your brother's home.

59. (a) What is the total cost, including shipping, of sending three short-sleeve, solid-color shirts, with monograms, in a gift box to your aunt for her birthday?

(b) How much did the monograms, gift box, and shipping add to the cost of your gift?

60. (a) Suppose you order one of each type of shirt for yourself, adding a monogram on each of the solid-color shirts. At the same time, you order three long-sleeved striped shirts, in the extra-large size, shipped to your dad in a gift box. Find the total cost of your order.

(b) What is the difference in total cost (excluding shipping) between the shirts for yourself and the gift for your dad?

4.5 DIVIDING DECIMALS

There are two kinds of decimal division problems; those in which a decimal is divided by a whole number, and those in which a decimal is divided by a decimal. First recall the parts of a division problem from **Section 1.5.**

$$\begin{array}{rl} 8 & \leftarrow \text{Quotient} \\ \text{Divisor} \rightarrow 4\overline{)33} & \leftarrow \text{Dividend} \\ \underline{32} & \\ 1 & \leftarrow \text{Remainder} \end{array}$$

OBJECTIVES

1. Divide a decimal by a whole number.
2. Divide a decimal by a decimal.
3. Estimate the answer when dividing decimals.
4. Use the order of operations with decimals.

1 Divide a decimal by a whole number. When the divisor is a whole number, use these steps.

Dividing Decimals by Whole Numbers

Step 1 Write the decimal point in the quotient (answer) directly above the decimal point in the dividend.

Step 2 Divide as if both numbers were whole numbers.

Example 1 Dividing Decimals by Whole Numbers

Divide.

(a) 21.93 by 3

(21.93: Dividend; 3: Divisor)

Rewrite the division problem. $3\overline{)21.93}$

Step 1 Write the decimal point in the quotient directly above the decimal point in the dividend. $3\overline{)21.93}$ (Decimal points lined up)

Step 2 Divide as if the numbers were whole numbers.

$$\begin{array}{r} 7.31 \\ 3\overline{)21.93} \end{array}$$

Check by multiplying the quotient times the divisor.

$$\begin{array}{r} 7.31 \\ \times \quad 3 \\ \hline 21.93 \end{array}$$

Matches, so 7.31 is correct.

The quotient (answer) is 7.31.

(b) $9\overline{)470.7}$

(9: Divisor; 470.7: Dividend)

Write the decimal point in the quotient above the decimal point in the dividend. Then divide as if the numbers were whole numbers.

Decimal points lined up

$$\begin{array}{r} 52.3 \\ 9\overline{)470.7} \\ \underline{45} \\ 20 \\ \underline{18} \\ 2\,7 \\ \underline{2\,7} \\ 0 \end{array} \qquad \textit{Check:} \quad \begin{array}{r} 52.3 \\ \times \quad 9 \\ \hline 470.7 \end{array}$$

Matches

The quotient is 52.3.

Work Problem 1 at the Side.

1 Divide. Check your answers by multiplying.

(a) $4\overline{)93.6}$

(b) $6\overline{)6.804}$

(c) $11\overline{)278.3}$

(d) $0.51835 \div 5$

(e) $213.45 \div 15$

ANSWERS

1. **(a)** 23.4; (23.4)(4) = 93.6
(b) 1.134; (1.134)(6) = 6.804
(c) 25.3; (25.3)(11) = 278.3
(d) 0.10367; (0.10367)(5) = 0.51835
(e) 14.23; (14.23)(15) = 213.45

2 Divide. Check your answers by multiplying.

(a) $5\overline{)6.4}$

(b) $30.87 \div 14$

(c) $\dfrac{259.5}{30}$

(d) $0.3 \div 8$

Example 2 Writing Extra 0s to Complete a Division

Divide 1.5 by 8.

Keep dividing until the remainder is 0, or until the digits in the quotient begin to repeat in a pattern. In Example 1(b), you ended up with a remainder of 0. But sometimes you run out of digits in the dividend before that happens. If so, write extra 0s on the right side of the dividend so you can continue dividing.

$$\begin{array}{r} 0.1 \\ 8\overline{)1.5} \\ \underline{8} \\ 7 \end{array}$$

1.5 ← All digits have been used.

7 ← Remainder is not yet 0.

Write a 0 after the 5 in the dividend so you can continue dividing. Keep writing more 0s in the dividend if needed. Recall that writing 0s to the *right* of a decimal number does *not* change its value.

$$\begin{array}{r} 0.1875 \\ 8\overline{)1.5000} \\ \underline{8} \\ 70 \\ \underline{64} \\ 60 \\ \underline{56} \\ 40 \\ \underline{40} \\ 0 \end{array}$$

1.5**000** ← Three 0s needed to complete the division.

0 ← Stop dividing when the remainder is 0.

Check:

$$\begin{array}{r} 0.1875 \\ \times \quad 8 \\ \hline 1.5000 \end{array}$$

Matches dividend, so 0.1875 is correct.

Calculator Tip When *multiplying* numbers, you can enter them in any order because multiplication is commutative (see **Section 1.4**). But division is *not* commutative. It *does* matter which number you enter first. Try Example 2 both ways; jot down your answers.

1.5 (÷) 8 (=) ______ 8 (÷) 1.5 (=) ______

Notice that the first answer, 0.1875, matches the result from Example 2. But the second answer is much different: 5.333333333. Be careful to enter the dividend first.

CAUTION

When dividing decimals, notice that the dividend may *not* be the larger number, as it was in whole numbers. In Example 2 the dividend is 1.5, which is *smaller* than 8.

Work Problem 2 at the Side.

The next example shows a quotient (answer) that must be rounded because you will never get a remainder of 0.

ANSWERS

2. **(a)** 1.28; (1.28)(5) = 6.40 or 6.4
(b) 2.205; (2.205)(14) = 30.870 or 30.87
(c) 8.65; (8.65)(30) = 259.50 or 259.5
(d) 0.0375; (0.0375)(8) = 0.3000 or 0.3

Example 3 Rounding a Decimal Quotient

Divide 4.7 by 3. Round the quotient to the nearest thousandth.

Write extra 0s in the dividend so you can continue dividing.

```
      1.5 6 6 6
  3 ) 4.7 0 0 0  ← Three 0s added so far
      3
      1 7
      1 5
        2 0
        1 8
          2 0
          1 8
            2 0
            1 8
              2  ← Remainder is still not 0.
```

Notice that the digit 6 in the answer is repeating. It will continue to do so. The remainder will *never be 0*. There are two ways to show that the answer is a **repeating decimal** that goes on forever. You can write three dots after the answer, or you can write a bar above the digits that repeat (in this case, the 6).

$1.5666\ldots$ (Three dots) or $1.5\overline{6}$ ← Bar above repeating digit

When repeating decimals occur, round the answer according to the directions in the problem. In this example, to round to thousandths, divide out one *more* place, to ten-thousandths.

$$4.7 \div 3 = 1.5666\ldots \quad \text{rounds to} \quad 1.567$$

Check the answer by multiplying 1.567 by 3. Because 1.567 is a rounded answer, the check will not give exactly 4.7, but it should be very close.

$$(1.567)(3) = 4.701$$ ← Does not equal exactly 4.7 because 1.567 was rounded.

CAUTION

When checking answers that you've rounded, the check will *not* match the dividend exactly, but it should be very close.

Work Problem 3 at the Side.

2 Divide a decimal by a decimal. To divide by a *decimal* divisor, first change the divisor to a whole number. Then divide as before. To see how this is done, write the problem in fraction form. Here is an example.

$$1.2\overline{)6.36} \quad \text{can be written} \quad \frac{6.36}{1.2}$$

In **Section 3.2** you learned that multiplying the numerator and denominator by the same number gives an equivalent fraction. We want the divisor (1.2) to be a whole number. Multiplying by 10 will accomplish that.

$$\frac{6.36}{1.2} = \frac{6.36 \cdot 10}{1.2 \cdot 10} = \frac{63.6}{12}$$

Decimal divisor → 1.2; 12 ← Whole number divisor

3 Divide. Round answers to the nearest thousandth. If it is a repeating decimal, also write the answer using a bar. Check your answers by multiplying.

(a) $13\overline{)267.01}$

(b) $6\overline{)20.5}$

(c) $\frac{10.22}{9}$

(d) $16.15 \div 3$

(e) $116.3 \div 7$

Answers

3. **(a)** 20.539 (rounded); no repeating digits visible on calculator; $(20.539)(13) = 267.007$
(b) 3.417 (rounded); $3.41\overline{6}$; $(3.417)(6) = 20.502$
(c) 1.136 (rounded); $1.13\overline{5}$; $(1.136)(9) = 10.224$
(d) 5.383 (rounded); $5.38\overline{3}$; $(5.383)(3) = 16.149$
(e) 16.614 (rounded); starts repeating in eighth decimal place as $16.6\overline{142857}$; $(16.614)(7) = 116.298$

4 Divide. If the quotient does not come out even, round to the nearest hundredth.

(a) $0.2\overline{)1.04}$

(b) $0.06\overline{)1.8072}$

(c) $0.005\overline{)32}$

(d) 8.1 ÷ 0.025

(e) $\dfrac{7}{1.3}$

(f) 5.3091 ÷ 6.2

ANSWERS

4. (a) 5.2 **(b)** 30.12 **(c)** 6400 **(d)** 324 **(e)** 5.38 (rounded) **(f)** 0.86 (rounded)

The short way to multiply by 10 is to move the decimal point *one place* to the *right* in both the divisor and the dividend.

$$1.2\overline{)6.36} \quad \text{is equivalent to} \quad 12\overline{)63.6}$$

NOTE

Moving the decimal points the ***same*** number of places in ***both*** the divisor and dividend will ***not*** change the answer.

Dividing by Decimals

Step 1 Count the number of decimal places in the divisor and move the decimal point that many places to the *right.* (This changes the divisor to a whole number.)

Step 2 Move the decimal point in the dividend the *same* number of places to the *right.* (Write in extra 0s if needed.)

Step 3 Write the decimal point in the quotient directly above the decimal point in the dividend. Then divide as usual.

Example 4 **Dividing by Decimals**

(a) $0.003\overline{)27.69}$

Move the decimal point in the divisor *three* places to the *right* so 0.003 becomes the whole number 3. To move the decimal point in the dividend the same number of places, write in an extra 0.

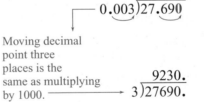

$$0.003\overline{)27.690}$$

Move decimal points in divisor and dividend. Then line up decimal point in answer.

Moving decimal point three places is the same as multiplying by 1000. → $\begin{array}{r} 9230. \\ 3\overline{)27690.} \end{array}$ Divide as usual.

(b) Divide 5 by 4.2. Round to the nearest hundredth.

Move the decimal point in the divisor one place to the right so 4.2 becomes the whole number 42. The decimal point in the dividend starts on the right side of 5 and is also moved one place to the right.

$$\begin{array}{r} 1.190 \\ 4.2\overline{)5.0000} \\ 42 \\ \hline 80 \\ 42 \\ \hline 380 \\ 378 \\ \hline 20 \end{array}$$

← To round to hundredths, divide out one *more* place, to thousandths.

Round the quotient. It is 1.19 (rounded to the nearest hundredth).

Work Problem 4 at the Side.

3 Estimate the answer when dividing decimals. Estimating the answer to a division problem helps you catch errors. Compare the estimate to your exact answer. If they are very different, do the division again.

Example 5 Estimating before Dividing

First use front end rounding to round each number and estimate the answer. Then divide to find the exact answer.

$$580.44 \div 2.8$$

Here is how one student solved this problem. She rounded 580.44 to 600 and rounded 2.8 to 3 to estimate the answer.

Estimate:

$$\begin{array}{r} 200 \\ 3\overline{)600} \end{array}$$

Exact:

$$\begin{array}{r} 27.3 \\ 2.8\overline{)580.44} \\ \underline{56} \\ 204 \\ \underline{196} \\ 84 \\ \underline{84} \\ 0 \end{array}$$

Very different; need to rework the problem.

Notice that the estimate, which is in the hundreds, is very different from the exact answer, which is only in the tens. This tells the student that she needs to rework the problem. Can you find the error? (The exact answer should be 207.3, which fits with the estimate of 200.)

Work Problem 5 at the Side.

4 Use the order of operations with decimals. Use the order of operations when a decimal problem involves more than one operation, as you did with whole numbers in **Section 1.8.**

Order of Operations

1. Do all operations inside *parentheses* or *other grouping symbols.*
2. Simplify any expressions with *exponents* and find any *square roots.*
3. *Multiply* or *divide,* proceeding from left to right.
4. *Add* or *subtract,* proceeding from left to right.

Example 6 Using the Order of Operations

Use the order of operations to simplify each expression.

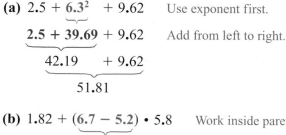

(a) $2.5 + 6.3^2 + 9.62$ Use exponent first.

$2.5 + 39.69 + 9.62$ Add from left to right.

$42.19 + 9.62$

51.81

(b) $1.82 + (6.7 - 5.2) \cdot 5.8$ Work inside parentheses.

$1.82 + 1.5 \cdot 5.8$ Multiply next.

$1.82 + 8.7$ Add last.

10.52

Continued on Next Page

5 Decide whether each answer is reasonable by rounding the numbers and estimating the answer. If the exact answer is *not* reasonable, find and correct the error.

(a) $42.75 \div 3.8 = 1.125$

Estimate:

(b) $807.1 \div 1.76 = 458.580$ to nearest thousandth

Estimate:

(c) $48.63 \div 52 = 93.519$ to nearest thousandth

Estimate:

(d) $9.0584 \div 2.68 = 0.338$

Estimate:

ANSWERS

5. (a) Estimate is $40 \div 4 = 10$; answer not reasonable, should be 11.25

(b) Estimate is $800 \div 2 = 400$; answer is reasonable.

(c) Estimate is $50 \div 50 = 1$; answer is not reasonable, should be 0.935.

(d) Estimate is $9 \div 3 = 3$; answer is not reasonable; should be 3.38.

6 Use the order of operations to simplify each expression.

(a) $4.6 - 0.79 + 1.5^2$

(b) $3.64 \div 1.3 \cdot 3.6$

(c) $0.08 + 0.6 \cdot (3 - 2.99)$

(d) $10.85 - 2.3 \times 5.2 \div 3.2$

ANSWERS
6. (a) 6.06 (b) 10.08 (c) 0.086
(d) 7.1125

(c) $\underbrace{3.7^2} - 1.8 \times 5.1 \div 1.5$ Use exponent first.

$13.69 - \underbrace{\mathbf{1.8 \times 5.1}} \div 1.5$ Multiply and divide from left to right.

$13.69 - \underbrace{\mathbf{9.18 \div 1.5}}$

$\underbrace{13.69 - 6.12}$ Subtract last.

7.57

Work Problem 6 at the Side.

Calculator Tip Most scientific calculators that have parentheses keys ( ) can handle calculations like those in Example 6 just by entering the numbers in the order given. For example, the keystrokes for Example 6(b) are:

Parentheses

1.82 (+) (() 6.7 (−) 5.2 ()) (×) 5.8 (=) Answer is 10.52.

Standard, four-function calculators generally do not have parentheses keys and will *not* give the correct answer if you simply enter the numbers in the order given.

Check the instruction manual that came with your calculator for information on "order of calculations" to see if your machine has the rules for order of operations built into it. For a quick check, try entering this problem:

2 (+) 2 (×) 2 (=)

If the result is 6, the calculator follows the order of operations. If the result is 8, it does *not* have the rules built into it. To see why this test works, do the calculations by hand.

Use order of operations.

$2 + \underbrace{2 \times 2}$

$\underbrace{2 + 4}$

6 ← Correct

Work from left to right.

$\underbrace{2 + 2} \times 2$

$\underbrace{4 \times 2}$

8 ← Incorrect

4.5 Exercises

FOR EXTRA HELP — Student's Solutions Manual | MyMathLab.com | InterAct Math Tutorial Software | AW Math Tutor Center | www.mathxl.com | Digital Video Tutor CD 3 Videotape 8

Divide. See Examples 1 and 4.

1. $7\overline{)27.3}$

2. $8\overline{)50.4}$

3. $\dfrac{4.23}{9}$

4. $\dfrac{1.62}{6}$

5. $0.05\overline{)20.01}$

6. $0.08\overline{)16.04}$

7. $1.5\overline{)54}$

8. $2.4\overline{)132}$

Use the fact that $108 \div 18 = 6$ *to work Exercises 9–12 simply by moving decimal points. See Examples 2 and 4.*

9. $0.108 \div 1.8$

10. $10.8 \div 18$

11. $0.018\overline{)108}$

12. $0.18\overline{)1.08}$

Divide. Round quotients to the nearest hundredth if necessary. See Examples 3 and 4.

13. $4.6\overline{)116.38}$

14. $2.6\overline{)4.992}$

15. $\dfrac{3.1}{0.006}$

16. $\dfrac{1.7}{0.09}$

Divide. Round quotients to the nearest thousandth. See Example 4.

17. $240 \div 9.88$

18. $7643 \div 5.36$

19. $0.034\overline{)342.81}$

20. $0.043\overline{)1748.4}$

Relating Concepts (Exercises 21–22) For Individual or Group Work

Look back at your work in Exercises 21 and 22 in Section 4.4. Then ***work Exercises 21 and 22 in order.***

21. Do these division problems:

$3.77 \div 10$	$9.1 \div 10$
$0.886 \div 10$	$30.19 \div 10$
$406.5 \div 10$	$6625.7 \div 10$

What pattern do you see? Write a "rule" for dividing by 10. What do you think the rule is for dividing by 100? by 1000? Write the rules and try them out on the numbers above.

22. Do these division problems:

$40.2 \div 0.1$	$7.1 \div 0.1$
$0.339 \div 0.1$	$15.77 \div 0.1$
$46 \div 0.1$	$873 \div 0.1$

What pattern do you see? Write a "rule" for dividing by 0.1. What do you think the rule is for dividing by 0.01? by 0.001? Write the rules and try them out on the numbers above.

Decide whether each answer is reasonable or unreasonable by rounding the numbers and estimating the answer. If the exact answer is not reasonable, find the correct answer. See Example 5.

23. $37.8 \div 8 = 47.25$

Estimate:

24. $345.6 \div 3 = 11.52$

Estimate:

25. $54.6 \div 48.1 = 1.135$

Estimate:

26. $2428.8 \div 4.8 = 50.6$

Estimate:

27. $307.02 \div 5.1 = 6.2$

Estimate:

28. $395.415 \div 5.05 = 78.3$

Estimate:

29. $9.3 \div 1.25 = 0.744$

Estimate:

30. $78 \div 14.2 = 0.182$

Estimate:

Solve each application problem. Round money answers to the nearest cent, if necessary.

31. Alfred has discovered that Batman's favorite brand of superhero tights are on sale. He's been told to buy only one pair for Robin. How much will he pay for one pair?

32. The bookstore has a special price on notepads. How much did Randall pay for one notepad?

33. It will take 21 equal monthly payments for Aimee to pay off her charge account balance of $408.66. How much is she paying each month?

34. Marcella Anderson bought 2.6 meters of suede fabric for $18.19. How much did she pay per meter?

35. Adrian Webb bought 619 bricks to build a barbecue pit, paying $185.70. Find the cost per brick. (*Hint:* Cost *per* brick means the cost for *one* brick.)

36. Lupe Wilson is a newspaper distributor. Last week she paid the newspaper $130.51 for 842 copies. Find the cost per copy.

37. Darren Jackson earned $356.80 for 40 hours of work. Find his earnings per hour.

38. At a record manufacturing company, 400 records cost $289. Find the cost per record.

39. It took 16.35 gallons of gas to fill Kim's car gas tank. She had driven 346.2 miles since her last fill-up. How many miles per gallon did her car get? Round to the nearest tenth.

40. Mr. Rodriquez pays $53.19 each month to Household Finance. How many months will it take him to pay off $1436.13?

Use the table of longest long jumps (through the year 2000) to answer Exercises 41–46. To find an average, add up the values you are interested in and then divide the sum by the number of values. Round your answer to the nearest hundredth. Some of the other exercises may require subtraction or multiplication.

Athlete	Country	Year	Length (meters)
M. Powell	U.S.	1991	8.95
B. Beamon	U.S.	1968	8.90
C. Lewis	U.S.	1991	8.87
R. Emmiyan	USSR	1987	8.86
L. Myricks	U.S.	1988	8.74
E. Walder	U.S.	1994	8.74
I. Pedroso	Cuba	1995	8.71
K. Streete-Thompson	U.S.	1994	8.63
J. Beckford	Jamaica	1997	8.62

Source: www.Olympics.com

41. Find the average length of the long jumps made by U.S. athletes.

42. Find the average length of all the long jumps listed in the table.

43. How much longer was the second-place jump than the third-place jump?

44. If the first-place athlete made six jumps of the same length, what was the total distance jumped?

45. What was the total length jumped by the top three athletes?

46. How much less was the last-place jump than the next-to-last-place jump?

Use the order of operations to simplify each expression. See Example 6.

47. $7.2 - 5.2 + 3.5^2$

48. $6.2 + 4.3^2 - 9.72$

49. $38.6 + 11.6 \cdot (13.4 - 10.4)$

50. $2.25 - 1.06 \cdot (4.85 - 3.95)$

51. $8.68 - 4.6 \cdot 10.4 \div 6.4$

52. $25.1 + 11.4 \div 7.5 \cdot 3.75$

53. $33 - 3.2 \cdot (0.68 + 9) - 1.3^2$

54. $0.6 + (1.89 + 0.11) \div 0.004 \cdot 0.5$

Solve each application problem.

55. Soup is on sale at six cans for \$3.25, or you can purchase individual cans for \$0.57. How much will you save per can if you buy six cans? Round to the nearest cent.

56. Nadia's diet says she can eat 3.5 ounces of chicken nuggets. The package weighs 10.5 ounces and contains 15 nuggets. How many nuggets can Nadia eat?

57. In 1998, the U.S. Treasury spent about \$386,500,000 to print 9,200,000,000 pieces of paper money. How much did it cost to print each piece, to the nearest cent? (*Source:* U.S. Department of the Treasury.)

9,200,000,000 pieces of paper money printed in 1998

58. Mach 1 is the speed of sound. Dividing a vehicle's speed by the speed of sound gives its speed on the Mach scale. In 1997, a specially built car with two 110,000-horsepower engines broke the world land speed record by traveling 763.035 miles per hour. The speed of sound that day was 748.11 miles per hour. What was the car's Mach speed, to the nearest hundredth? (*Source:* Associated Press.)

General Mills will give a school 10¢ for each box top logo from its cereals and other products. A school can earn up to \$10,000 per year. Use this information to answer Exercises 59–62. Round your answers to the nearest whole number when necessary. (*Source:* General Mills.)

59. How many box tops would a school need to collect in one year to earn the maximum amount?

60. (Complete Exercise 59 first.) If a school has 550 children, how many box tops would each child need to collect in one year to reach the maximum?

61. How many box tops would need to be collected during each of the 38 weeks in the school year to reach the maximum amount?

62. How many box tops would each of the 550 children need to collect during each of the 38 weeks of school to reach the maximum amount?

4.6 Writing Fractions as Decimals

Objectives

1. Write a fraction as a decimal.
2. Compare the size of fractions and decimals.

Writing fractions as equivalent decimals can help you do calculations more easily or compare the size of two numbers.

1 Write a fraction as a decimal. Recall that a fraction is one way to show division (see **Section 1.5**). For example, $\frac{3}{4}$ means $3 \div 4$. If you are doing the division by hand, write it as $4\overline{)3}$. When you do the division, you will get the decimal equivalent of $\frac{3}{4}$, which is 0.75.

Writing Fractions as Decimals

Step 1 Divide the numerator of the fraction by the denominator.

Step 2 If necessary, round the answer to the place indicated.

Work Problem 1 at the Side.

1 Rewrite each fraction so you could do the division by hand. Do *not* complete the division.

(a) $\frac{1}{9}$ is written $9\overline{)}$

(b) $\frac{2}{3}$ is written $\overline{)}$

(c) $\frac{5}{4}$ is written $\overline{)}$

(d) $\frac{3}{10}$ is written $\overline{)}$

(e) $\frac{21}{16}$ is written $\overline{)}$

(f) $\frac{1}{50}$ is written $\overline{)}$

Example 1 Writing Fractions or Mixed Numbers as Decimals

(a) Write $\frac{1}{8}$ as a decimal.

$\frac{1}{8}$ means $1 \div 8$. Write it as $8\overline{)1}$. The decimal point in the dividend is on the right side of the 1. Write extra 0s in the dividend so you can continue dividing until the remainder is 0.

$$\frac{1}{8} = 1 \div 8 = 8\overline{)1} \quad \Rightarrow \quad \begin{array}{r} 0.125 \\ 8\overline{)1.000} \\ \underline{8} \\ 20 \\ \underline{16} \\ 40 \\ \underline{40} \\ 0 \end{array}$$

Decimal points lined up; Three extra 0s needed; Remainder is 0.

Therefore, $\frac{1}{8} = 0.125$. To check this, write 0.125 as a fraction, then change it to lowest terms.

$$0.125 = \frac{125}{1000} \qquad \text{In lowest terms} \qquad \frac{125 \div \mathbf{125}}{1000 \div \mathbf{125}} = \frac{1}{8} \leftarrow \text{Original fraction}$$

Calculator Tip When changing fractions to decimals on your calculator, enter the numbers from the top down. Remember that the order in which you enter the numbers *does* matter in division. Example 1(a) works like this:

$\frac{1}{8}\downarrow$ Top down Enter 1 (÷) 8 (=) Answer is 0.125.

What happens if you enter 8 (÷) 1 (=)? Do you see why that cannot possibly be correct? (Answer: $8 \div 1 = 8$, and a proper fraction like $\frac{1}{8}$ *cannot* be equivalent to a whole number.)

Continued on Next Page

Answers

1. (a) $9\overline{)1}$ (b) $3\overline{)2}$ (c) $4\overline{)5}$ (d) $10\overline{)3}$ (e) $16\overline{)21}$ (f) $50\overline{)1}$

2 Write each fraction or mixed number as a decimal.

(a) $\frac{1}{4}$

(b) $2\frac{1}{2}$

(c) $\frac{5}{8}$

(d) $4\frac{3}{5}$

(e) $\frac{7}{8}$

(b) Write $2\frac{3}{4}$ as a decimal.

One method is to divide 3 by 4 to get 0.75 for the fraction part. Then add the whole number part to 0.75.

$$\frac{3}{4} \Rightarrow \begin{array}{r} 0.75 \\ 4\overline{)3.00} \\ \underline{2\,8} \\ 20 \\ \underline{20} \\ 0 \end{array} \qquad \begin{array}{rl} 2.00 & \leftarrow \text{Whole number part} \\ \text{Fraction part} \rightarrow \; +\,0.75 & \\ \hline 2.75 & \end{array}$$

So, $2\frac{3}{4} = 2.75$ (Whole number parts match.)

Check: $2.75 = 2\frac{75}{100} = 2\frac{3}{4}$ ← Lowest terms

A second method is to first write $2\frac{3}{4}$ as an improper fraction and then divide numerator by denominator.

$$2\frac{3}{4} = \frac{11}{4}$$

$$\frac{11}{4} = 11 \div 4 = 4\overline{)11} \Rightarrow \begin{array}{r} 2.7\,5 \\ 4\overline{)1\,1.0\,0} \\ \underline{8} \\ 3\,0 \\ \underline{2\,8} \\ 2\,0 \\ \underline{2\,0} \\ 0 \end{array} \leftarrow \text{Two extra 0s needed}$$

Whole number parts match.

So, $2\frac{3}{4} = 2.75$

$\frac{3}{4}$ is equivalent to $\frac{75}{100}$ or 0.75.

Work Problem 2 at the Side.

Example 2 Writing a Fraction as a Decimal with Rounding

Write $\frac{2}{3}$ as a decimal and round to the nearest thousandth.

$\frac{2}{3}$ means $2 \div 3$. To round to thousandths, divide out one *more* place, to ten-thousandths.

$$\frac{2}{3} = 2 \div 3 = 3\overline{)2} \Rightarrow \begin{array}{r} 0.6666 \\ 3\overline{)2.0000} \\ \underline{1\,8} \\ 20 \\ \underline{18} \\ 20 \\ \underline{18} \\ 20 \\ \underline{18} \\ 2 \end{array} \leftarrow \text{Four 0s needed for ten-thousandths}$$

Written as a repeating decimal, $\frac{2}{3} = 0.\overline{6}$. ← Bar above repeating digit.

Rounded to the nearest thousandth, $\frac{2}{3} \approx 0.667$.

Answers

2. **(a)** 0.25 **(b)** 2.5 **(c)** 0.625 **(d)** 4.6 **(e)** 0.875

Calculator Tip Try Example 2 on your calculator. Enter 2 (÷) 3. Which answer do you get?

`0.666666667` or `0.6666666`

Many scientific calculators will show a 7 as the last digit. Because the 6s keep on repeating forever, the calculator automatically rounds in the last decimal place it has room to show. If you have a 10-digit display space, the calculator is rounding like this:

0.6666666666 (11 digits) rounds to 0.666666667.

Other calculators, especially standard, four-function ones, may *not* round. They just cut off, or *truncate,* the extra digits. Such a calculator would show 0.6666666 in the display.

Would this difference in calculators show up when changing $\frac{1}{3}$ to a decimal? Why not? (Answer: The repeating digit is 3, which is 4 or less, so it stays as 3 whether it's rounded or not.)

Work Problem 3 at the Side.

3 Write as decimals. Round to the nearest thousandth.

(a) $\frac{1}{3}$

(b) $2\frac{7}{9}$

(c) $\frac{10}{11}$

(d) $\frac{3}{7}$

(e) $3\frac{5}{6}$

2 Compare the size of fractions and decimals.

You can use a number line to compare fractions and decimals. For example, the number line below shows the space between 0 and 1. The locations of some commonly used fractions are marked, along with their decimal equivalents.

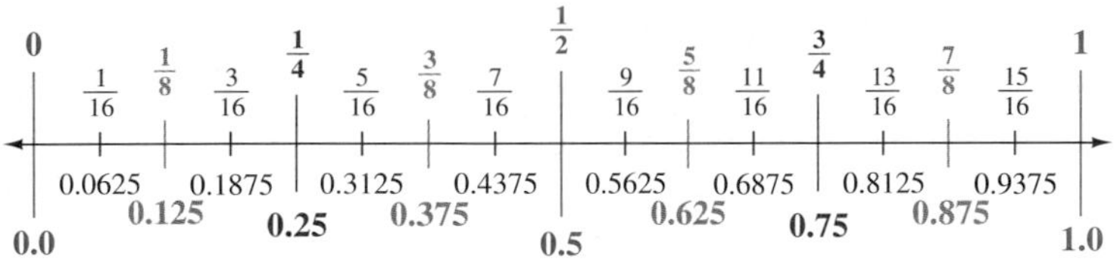

The next number line shows the locations of some commonly used fractions between 0 and 1 that are equivalent to *repeating* decimals. The decimal equivalents use a bar above repeating digits.

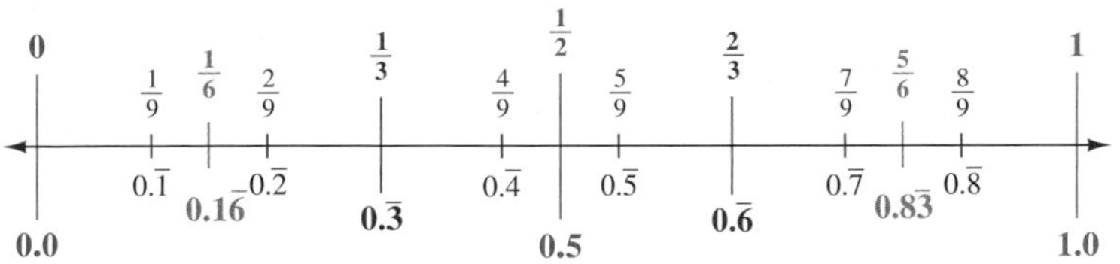

Example 3 Using a Number Line to Compare Numbers

Use the number lines above to decide whether to write >, <, or = in the blank between each pair of numbers.

(a) 0.6875 ________ 0.625

You learned in **Section 3.5** that the number farther to the right on the number line is the greater number. On the first number line, 0.6875 is to the *right* of 0.625, so use the > symbol.

0.6875 **is greater than** 0.625 0.6875 > 0.625

Continued on Next Page

ANSWERS

3. All answers are rounded.
(a) 0.333 **(b)** 2.778 **(c)** 0.909 **(d)** 0.429 **(e)** 3.833

4 Use the number lines on page 291 to help you decide whether to write <, >, or = in each blank.

(a) 0.4375 ________ 0.5

(b) 0.75 ________ 0.6875

(c) 0.625 ________ 0.0625

(d) $\frac{2}{8}$ ________ 0.375

(e) $0.8\overline{3}$ ________ $\frac{5}{6}$

(f) $\frac{1}{2}$ ________ $0.\overline{5}$

(g) $0.\overline{1}$ ________ $0.1\overline{6}$

(h) $\frac{8}{9}$ ________ $0.\overline{8}$

(i) $0.\overline{7}$ ________ $\frac{4}{6}$

(j) $\frac{1}{4}$ ________ 0.25

5 Arrange in order from smallest to largest.

(a) 0.7, 0.703, 0.7029

(b) 6.39, 6.309, 6.4, 6.401

(c) 1.085, $1\frac{3}{4}$, 0.9

(d) $\frac{1}{4}$, $\frac{2}{5}$, $\frac{3}{7}$, 0.428

ANSWERS

4. (a) < **(b)** > **(c)** > **(d)** < **(e)** = **(f)** < **(g)** < **(h)** = **(i)** > **(j)** =

5. (a) 0.7, 0.7029, 0.703
(b) 6.309, 6.39, 6.4, 6.401
(c) 0.9, 1.085, $1\frac{3}{4}$
(d) $\frac{1}{4}$, $\frac{2}{5}$, 0.428, $\frac{3}{7}$

(b) $\frac{3}{4}$ ________ 0.75

On the first number line, $\frac{3}{4}$ and 0.75 are at the same point on the number line. They are equivalent.

$$\frac{3}{4} = 0.75$$

(c) 0.5 ________ $0.\overline{5}$

On the second number line, 0.5 is to the *left* of $0.\overline{5}$ (which is actually 0.555 . . .) so use the < symbol.

0.5 is less than $0.\overline{5}$ $\qquad 0.5 < 0.\overline{5}$

(d) $\frac{2}{6}$ ________ $0.\overline{3}$

Write $\frac{2}{6}$ in lowest terms as $\frac{1}{3}$.

On the second number line you can see that $\frac{1}{3} = 0.\overline{3}$.

Work Problem 4 at the Side.

Fractions can also be compared by first writing each one as a decimal. The decimals can then be compared by writing each one with the same number of decimal places.

Example 4 Arranging Numbers in Order

Write each group of numbers in order, from smallest to largest.

(a) 0.49 0.487 0.4903

It is easier to compare decimals if they are all tenths, or all hundredths, and so on. Because 0.4903 has four decimal places (ten-thousandths), write 0s to the right of 0.49 and 0.487 so they also have four decimal places. Writing 0s to the right of a decimal number does *not* change its value (see **Section 4.3**). Then find the smallest and largest number of ten-thousandths.

0.49 = 0.4900 = **4900** ten-thousandths ← 4900 is in the middle.
0.487 = 0.4870 = **4870** ten-thousandths ← 4870 is the smallest.
0.4903 = **4903** ten-thousandths ← 4903 is the largest.

From smallest to largest, the correct order is shown below.

0.487 0.49 0.4903

(b) $2\frac{5}{8}$ 2.63 2.6

Write $2\frac{5}{8}$ as $\frac{21}{8}$ and divide $8\overline{)21}$ to get the decimal form, 2.625. Then, because 2.625 has three decimal places, write 0s so all the numbers have three decimal places.

$2\frac{5}{8}$ = 2.625 = 2 and **625** thousandths ← 625 is in the middle.
2.63 = 2.630 = 2 and **630** thousandths ← 630 is the largest.
2.6 = 2.600 = 2 and **600** thousandths ← 600 is the smallest.

From smallest to largest, the correct order is shown below.

2.6 $2\frac{5}{8}$ 2.63

Work Problem 5 at the Side.

4.6 EXERCISES

FOR EXTRA HELP: Student's Solutions Manual; MyMathLab.com; InterAct Math Tutorial Software; AW Math Tutor Center; www.mathxl.com; Digital Video Tutor CD 3 Videotape 8

Write each fraction or mixed number as a decimal. Round to the nearest thousandth if necessary. See Examples 1 and 2.

1. $\frac{1}{2}$ **2.** $\frac{1}{4}$ **3.** $\frac{3}{4}$ **4.** $\frac{1}{10}$ **5.** $\frac{3}{10}$

6. $\frac{7}{10}$ **7.** $\frac{9}{10}$ **8.** $\frac{4}{5}$ **9.** $\frac{3}{5}$ **10.** $\frac{2}{5}$

11. $\frac{7}{8}$ **12.** $\frac{3}{8}$ **13.** $2\frac{1}{4}$ **14.** $1\frac{1}{2}$ **15.** $14\frac{7}{10}$

16. $23\frac{3}{5}$ **17.** $3\frac{5}{8}$ **18.** $2\frac{7}{8}$ **19.** $\frac{1}{3}$ **20.** $\frac{2}{3}$

21. $\frac{5}{6}$ **22.** $\frac{1}{6}$ **23.** $1\frac{8}{9}$ **24.** $5\frac{4}{7}$

RELATING CONCEPTS (Exercises 25–28) FOR INDIVIDUAL OR GROUP WORK

Use your knowledge of fractions and decimals to ***work Exercises 25–28 in order.***

25. (a) Explain how you can tell that Keith made an error *just by looking at his final answer.* Here is his work.

$$\frac{5}{9} = 5\overline{)9.0}\text{ (quotient } 1.8) \quad \text{so} \quad \frac{5}{9} = 1.8$$

(b) Show the correct way to change $\frac{5}{9}$ to a decimal. Explain why your answer makes sense.

26. (a) How can you prove to Sandra that $2\frac{7}{20}$ is *not* equivalent to 2.035? Here is her work.

$$2\frac{7}{20} = 20\overline{)7.00}\text{ (quotient } 0.35) \quad \text{so} \quad 2\frac{7}{20} = 2.035$$

(b) What is the correct answer? Show how to prove that it is correct.

27. Ving knows that $\frac{3}{8} = 0.375$. How can he write $1\frac{3}{8}$ as a decimal *without* having to do a division? How can he write $3\frac{3}{8}$ as a decimal? $295\frac{3}{8}$? Explain your answer.

28. Iris has found a shortcut for writing mixed numbers as decimals:

$$2\frac{7}{10} = 2.7 \qquad 1\frac{13}{100} = 1.13.$$

Does her shortcut work for all mixed numbers? Explain when it works and why it works.

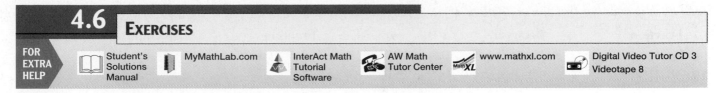

Find each decimal or fraction equivalent. Write fractions in lowest terms.

	Fraction	Decimal		Fraction	Decimal
29.	________	0.4	**30.**	________	0.75
31.	________	0.625	**32.**	________	0.111
33.	________	0.35	**34.**	________	0.9
35.	$\frac{7}{20}$	________	**36.**	$\frac{1}{40}$	________
37.	________	0.04	**38.**	________	0.52
39.	________	0.15	**40.**	________	0.85
41.	$\frac{1}{5}$	________	**42.**	$\frac{1}{8}$	________
43.	________	0.09	**44.**	________	0.02

Solve each application problem.

45. The average length of a newborn baby is 20.8 in. Charlene's baby is 20.08 in. long. Is her baby longer or shorter than the average? By how much?

46. The patient in room 830 is supposed to get 8.3 milligrams of medicine. She was actually given 8.03 milligrams. Did she get too much or too little medicine? What was the difference?

47. The label on the bottle of vitamins says that each capsule contains 0.5 gram of calcium. When checked, each capsule had 0.505 gram of calcium. Was there too much or too little calcium? What was the difference?

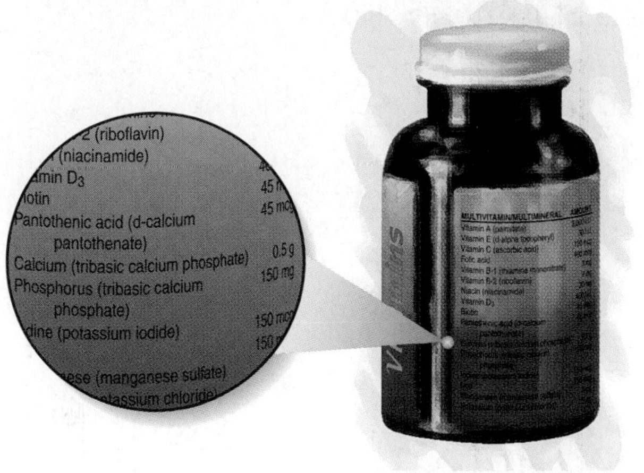

48. The glass mirror of the Hubble telescope had to be repaired in space in 1993 because it would not focus properly. The problem was that the mirror's outer edge had a thickness of 0.6248 centimeter when it was supposed to be 0.625 centimeter. Was the edge too thick or too thin? By how much? (*Source:* NASA.)

49. Precision Medical Parts makes an artificial heart valve that must measure between 0.998 centimeter and 1.002 centimeters. Circle the lengths that are acceptable:

1.01 cm, 0.9991 cm, 1.0007 cm, 0.99 cm.

50. The mice in a medical experiment must start out weighing between 2.95 ounces and 3.05 ounces. Circle the weights that can be used:

3.0 ounces, 2.995 ounces, 3.055 ounces, 3.005 ounces.

51. Ginny Brown hoped her crops would get $3\frac{3}{4}$ in. of rain this month. The newspaper said the area received 3.8 in. of rain. Was that more or less than Ginny had hoped for? By how much?

52. The mice in the experiment in Exercise 50 gained $\frac{3}{8}$ ounce. They were expected to gain 0.3 ounce. Was their actual gain more or less than expected? By how much?

Arrange each group of numbers in order, from smallest to largest. See Example 4.

53. 0.54, 0.5455, 0.5399

54. 0.76, 0.7, 0.7006

55. 5.8, 5.79, 5.0079, 5.804

56. 12.99, 12.5, 13.0001, 12.77

57. 0.628, 0.62812, 0.609, 0.6009

58. 0.27, 0.281, 0.296, 0.3

59. 5.8751, 4.876, 2.8902, 3.88

60. 0.98, 0.89, 0.904, 0.9

61. 0.043, 0.051, 0.006, $\frac{1}{20}$

62. 0.629, $\frac{5}{8}$, 0.65, $\frac{7}{10}$

63. $\frac{3}{8}$, $\frac{2}{5}$, 0.37, 0.4001

64. 0.1501, 0.25, $\frac{1}{10}$, $\frac{1}{5}$

Four boxes of fishing line are in the sale bin. The thicker the line, the stronger it is. The diameter of the fishing line is its thickness. Use the information on the boxes to answer Exercises 65–68.

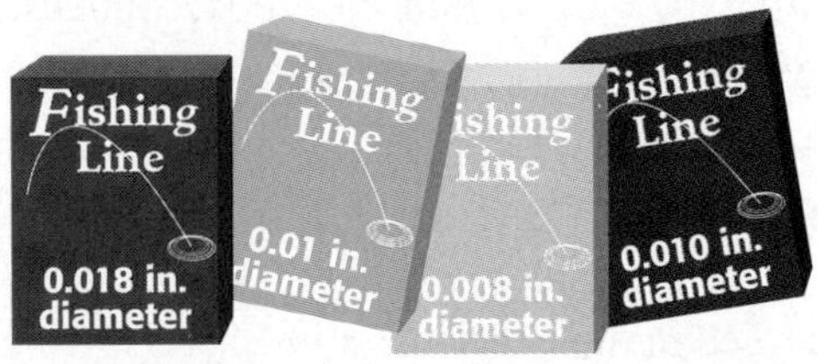

65. Which color box has the strongest line?

66. Which color box has the line with the least strength?

67. What is the difference in line diameter between the weakest and strongest line?

68. What is the difference in line diameter between the blue and purple boxes?

Some rulers for technical occupations show each inch divided into tenths. Use this scale drawing for Exercises 69–74. Change the measurements on the drawing to decimals and round them to the nearest tenth of an inch.

69. Length (a) is ______________.

70. Length (b) is ______________.

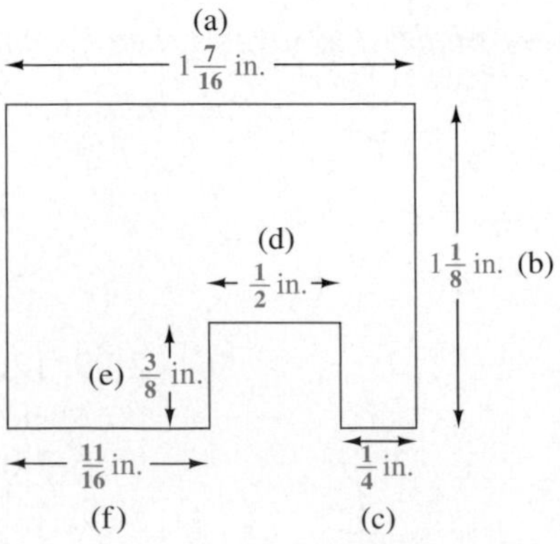

71. Length (c) is ______________.

72. Length (d) is ______________.

73. Length (e) is ______________.

74. Length (f) is ______________.

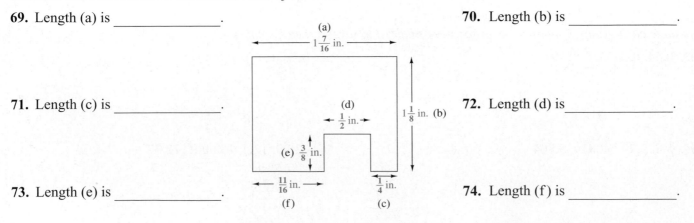

Chapter 4

SUMMARY

KEY TERMS

4.1	**decimals**	Decimals, like fractions, are used to show parts of a whole.
	decimal point	The dot that is used to separate the whole number part from the fractional part of a decimal number is the decimal point.
	place value	A place value is assigned to each place to the right or left of the decimal point. Whole numbers, such as ones and tens, are to the *left* of the decimal point. Fractional parts, such as tenths and hundredths, are to the *right* of the decimal point.
4.2	**rounding**	Rounding is "cutting off" a number after a certain place, such as rounding to the nearest hundredth. The rounded number is less accurate than the original number. You can use the symbol "$\approx$" to mean "approximately equal to."
	decimal places	The number of digits to the *right* of the decimal point are the decimal places; for example, 6.37 has two decimal places, 4.706 has three decimal places.
4.3	**estimating**	Estimating is the process of rounding the numbers in a problem and getting an approximate answer. This helps you check that the decimal point is in the correct place in the exact answer.
4.5	**repeating decimal**	A repeating decimal like the 6 in 0.1666 . . . has one or more digits that repeat forever. Use three dots to indicate that it is a repeating decimal; it never terminates (ends). Or, you can also write the number with a bar above the repeating digits, as in $0.1\overline{6}$.

TEST YOUR WORD POWER

See how well you have learned the vocabulary in this chapter. Answers follow the Quick Review.

1. **Decimal numbers** are like fractions in that they both
   (a) must be written in lowest terms
   (b) need common denominators
   (c) have decimal points
   (d) represent parts of a whole.

2. **Decimal places** refer to
   (a) the digits from 0 to 9
   (b) digits to the left of the decimal point
   (c) digits to the right of the decimal point
   (d) the number of 0s in a decimal number.

3. When a decimal number is **rounded,** it
   (a) always ends in 0
   (b) has the same number of decimal places as the original number
   (c) is less accurate than the original number
   (d) is less than one whole.

4. The **decimal point**
   (a) separates the whole number part from the fractional part
   (b) is always moved when finding a quotient
   (c) separates tenths from hundredths
   (d) is at the far left side of a whole number.

5. The number $0.\overline{3}$ is an example of
   (a) an estimate
   (b) a repeating decimal
   (c) a rounded number
   (d) a truncated number.

6. The **place value names** on the right side of the decimal point are
   (a) ones, tens, hundreds, and so on
   (b) ones, tenths, hundredths, and so on
   (c) zero, one, two, three, four, and so on
   (d) tenths, hundredths, thousandths, and so on.

Quick Review

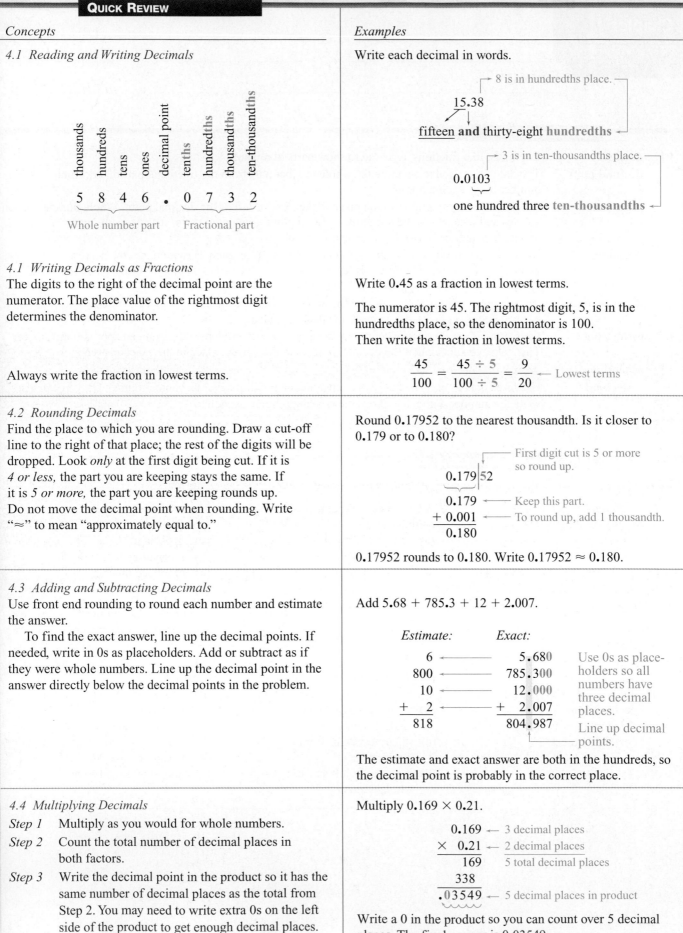

Concepts	*Examples*
4.1 Reading and Writing Decimals thousands, hundreds, tens, ones, decimal point, tenths, hundredths, thousandths, ten-thousandths 5 8 4 6 **.** 0 7 3 2 5846 = Whole number part; 0732 = Fractional part	Write each decimal in words. 15.38 (8 is in hundredths place.) fifteen **and** thirty-eight **hundredths** 0.0103 (3 is in ten-thousandths place.) one hundred three **ten-thousandths**
4.1 Writing Decimals as Fractions The digits to the right of the decimal point are the numerator. The place value of the rightmost digit determines the denominator. Always write the fraction in lowest terms.	Write 0.45 as a fraction in lowest terms. The numerator is 45. The rightmost digit, 5, is in the hundredths place, so the denominator is 100. Then write the fraction in lowest terms. $\frac{45}{100} = \frac{45 \div 5}{100 \div 5} = \frac{9}{20}$ ← Lowest terms
4.2 Rounding Decimals Find the place to which you are rounding. Draw a cut-off line to the right of that place; the rest of the digits will be dropped. Look *only* at the first digit being cut. If it is *4 or less,* the part you are keeping stays the same. If it is *5 or more,* the part you are keeping rounds up. Do not move the decimal point when rounding. Write "≈" to mean "approximately equal to."	Round 0.17952 to the nearest thousandth. Is it closer to 0.179 or to 0.180? 0.179|52 (First digit cut is 5 or more so round up.) 0.179 ← Keep this part. + 0.001 ← To round up, add 1 thousandth. 0.180 0.17952 rounds to 0.180. Write 0.17952 ≈ 0.180.
4.3 Adding and Subtracting Decimals Use front end rounding to round each number and estimate the answer. To find the exact answer, line up the decimal points. If needed, write in 0s as placeholders. Add or subtract as if they were whole numbers. Line up the decimal point in the answer directly below the decimal points in the problem.	Add 5.68 + 785.3 + 12 + 2.007. *Estimate:* 6 + 800 + 10 + 2 = 818 *Exact:* 5.680 + 785.300 + 12.000 + 2.007 = 804.987 Use 0s as placeholders so all numbers have three decimal places. Line up decimal points. The estimate and exact answer are both in the hundreds, so the decimal point is probably in the correct place.
4.4 Multiplying Decimals *Step 1* Multiply as you would for whole numbers. *Step 2* Count the total number of decimal places in both factors. *Step 3* Write the decimal point in the product so it has the same number of decimal places as the total from Step 2. You may need to write extra 0s on the left side of the product to get enough decimal places.	Multiply 0.169 × 0.21. 0.169 ← 3 decimal places × 0.21 ← 2 decimal places 169 (5 total decimal places) 338 .03549 ← 5 decimal places in product Write a 0 in the product so you can count over 5 decimal places. The final answer is 0.03549.

Concepts	*Examples*
4.5 Dividing by Decimals *Step 1* Change the divisor to a whole number by moving the decimal point to the right. *Step 2* Move the decimal point in the dividend the same number of places to the right. *Step 3* Write the decimal point in the quotient directly above the decimal point in the dividend. *Step 4* Divide as with whole numbers.	Divide 52.8 by 0.75. $$\begin{array}{r} 70.4 \\ 0.75\overline{)52.800} \\ \underline{525} \\ 300 \\ \underline{300} \\ 0 \end{array}$$ Move decimal point two places to the right in divisor and dividend. Write 0s in the dividend so you can move the decimal point and continue dividing until the remainder is 0. To check your answer, multiply 70.4 times 0.75. If the result matches the dividend (52.8), you solved the problem correctly.
4.6 Writing Fractions as Decimals Divide the numerator by the denominator. If necessary, round to the place indicated.	Write $\frac{1}{8}$ as a decimal. $\frac{1}{8}$ means $1 \div 8$. Write it as $8\overline{)1}$. The decimal point is on the right side of 1. $$\begin{array}{r} 0.125 \\ 8\overline{)1.000} \\ \underline{8} \\ 20 \\ \underline{16} \\ 40 \\ \underline{40} \\ 0 \end{array}$$ ← Decimal point and three 0s written in so you can continue dividing. Therefore, $\frac{1}{8}$ is equivalent to 0.125.
4.6 Comparing the Size of Fractions and Decimals *Step 1* Write any fractions as decimals. *Step 2* Write 0s so that all the numbers being compared have the same number of decimal places. *Step 3* Use < to mean "is less than," > to mean "is greater than," or list the numbers from smallest to largest.	Arrange in order from smallest to largest. $0.505 \quad \frac{1}{2} \quad 0.55$ $0.505 = 505$ thousandths $\frac{1}{2} = 0.5 = 0.500 = 500$ thousandths ← 500 is smallest. $0.55 = 0.550 = 550$ thousandths ← 550 is largest. (smallest) $\frac{1}{2}$ 0.505 0.55 (largest)

ANSWERS TO TEST YOUR WORD POWER

1. (d) *Example:* For 0.7, the whole is cut into ten parts, and you are interested in 7 of the parts. **2. (c)** *Examples:* The number 6.87 has two decimal places; 0.309 has three decimal places. **3. (c)** *Example:* When 0.815 is rounded to 0.8, it is accurate only to the nearest tenth, while the original number was accurate to the nearest thousandth. **4. (a)** *Example:* In 5.42, the whole number part is 5 ones, and the decimal part is 42 hundredths. **5. (b)** *Example:* The bar above the 3 in $0.\overline{3}$ indicates that the 3 repeats forever. **6. (d)** *Example:* In 6.219, the 2 is in the tenths place, the 1 is in the hundredths place, and the 9 is in the thousandths place.

Focus on

Real-Data Applications

Dollar-Cost Averaging

To gain the most benefit from investing in the stock market, you should purchase shares when stock prices are low and sell stocks when prices are high. Unfortunately, it is very difficult to predict whether prices will rise or fall from one day to the next. Financial advisors recommend using *dollar-cost averaging* instead of trying to guess how the market will change. By investing the same amount of money at regular intervals, such as the beginning of each month, you will be buying more shares when the stock prices are low and fewer shares when stock prices are high. Of course, *dollar-cost averaging* does not guarantee that you will make a profit. However, over time, your shares should cost less than the market average.

Many financial Web sites report current and historical data on stock performance. Suppose you invested $100 at the first of each month in 2000 and purchased shares of Microsoft at its closing price. The table shows data on Microsoft closing prices during 2000.

	Amount Invested	Price per Share	Number of Shares Bought
January	$100	97.875	($100 ÷ 97.8750) = 1.0217 shares
February	$100	89.375	
March	$100	106.25	
April	$100	69.75	
May	$100	62.5625	
June	$100	80	
July	$100	69.8125	
August	$100	69.8125	
September	$100	60.3125	
October	$100	68.875	
November	$100	57.375	
December	$100	43.375	
Total Investment			

Source: www.biz.yahoo

1. Calculate the total amount invested and enter the value in the table.
2. Calculate the number of shares bought each month. Round the number of shares to the nearest ten-thousandth. The calculation for January is shown as an example.
3. Calculate the average market price per share. (*Hint:* Add the monthly prices per share and divide by 12.)
4. Calculate the average price based on *dollar-cost averaging.* (*Hint:* Divide the total amount invested by the total number of shares.)
5. Rank the following scenarios in order of which was the best investment (most profit or least loss) at the end of December 2000. Show the value of each investment as a basis for your answer.
   (a) $1200 invested in Microsoft in January 2000.
   (b) $1200 invested in Microsoft using *dollar-cost averaging.*
   (c) $1200 invested in Microsoft in December 2000.

Chapter 4 Review Exercises

[4.1] *Name the digit that has the given place value.*

1. 243.059
   tenths
   hundredths
2. 0.6817
   ones
   tenths
3. $5824.39
   hundreds
   hundredths
4. 896.503
   tenths
   tens
5. 20.73861
   tenths
   ten-thousandths

Write each decimal as a fraction or mixed number in lowest terms.

6. 0.5
7. 0.75
8. 4.05
9. 0.875
10. 0.027
11. 27.8

Write each decimal in words.

12. 0.8
13. 400.29
14. 12.007
15. 0.0306

Write each decimal in numbers.

16. eight and three tenths
17. two hundred five thousandths
18. seventy and sixty-six ten-thousandths
19. thirty hundredths

[4.2] *Round each decimal to the place indicated.*

20. 275.635 to the nearest tenth
21. 72.789 to the nearest hundredth
22. 0.1604 to the nearest thousandth
23. 0.0905 to the nearest thousandth
24. 0.98 to the nearest tenth

Round each money amount to the nearest cent.

25. $15.8333
26. $0.698
27. $17,625.7906

Round each income or expense item to the nearest dollar.

28. Income from pancake breakfast was $350.48.
29. Members paid $129.50 in dues.
30. Refreshments cost $99.61.
31. Bank charges were $29.37.

[4.3] *First use front end rounding to round each number and estimate the answer. Then find the exact answer.*

32. *Estimate:* + ____ *Exact:*

$$\begin{array}{r} 5.81 \\ 423.96 \\ +\ 15.09 \\ \hline \end{array}$$

33. *Estimate:* + ____ *Exact:*

$$\begin{array}{r} 75.6 \\ 1.29 \\ 122.045 \\ 0.88 \\ +\ 33.7 \\ \hline \end{array}$$

34. *Estimate:* − ____ *Exact:*

$$\begin{array}{r} 308.5 \\ -\ 17.8 \\ \hline \end{array}$$

35. *Estimate:* − ____ *Exact:*

$$\begin{array}{r} 9.2 \\ -\ 7.9316 \\ \hline \end{array}$$

36. American's favorite recreational activity in 1999 was swimming. About 95.1 million people went swimming at least once during the year. Bicycling was second with 56.3 million people. How many more people went swimming than bicycling? (*Source:* Sporting Goods Manufacturers Association.)

Estimate:

Exact:

37. Today, Jasmin wrote a check to the day care center for $215.53 and a check for $44.47 at the grocery store. What was the total of the two checks?

Estimate:

Exact:

38. Joey spent $1.59 for toothpaste, $5.33 for a gift, and $18.94 for a toaster. He gave the clerk three $10 bills. How much change did he get?

Estimate:

Exact:

39. Roseanne is training for a wheelchair race. She raced 2.3 kilometers on Monday, 4 kilometers on Wednesday, and 5.25 kilometers on Friday. How far did she race altogether?

Estimate:

Exact:

[4.4] *First use front end rounding to round each number and estimate the answer. Then find the exact answer.*

40. *Estimate:* × ____ *Exact:*

$$\begin{array}{r} 6.138 \\ \times\ \ 3.7 \\ \hline \end{array}$$

41. *Estimate:* × ____ *Exact:*

$$\begin{array}{r} 42.9 \\ \times\ 3.3 \\ \hline \end{array}$$

Multiply.

42. (5.6)(0.002)

43. (0.071)(0.005)

[4.5] *Decide whether each answer is reasonable by rounding the numbers and estimating the answer. If the exact answer is not reasonable, find the correct answer.*

44. $706.2 \div 12 = 58.85$

Estimate:

45. $26.6 \div 2.8 = 0.95$

Estimate:

Divide. Round to the nearest thousandth if necessary.

46. $3\overline{)43.4}$

47. $\dfrac{72}{0.06}$

48. $0.00048 \div 0.0012$

[4.4–4.5] *Solve each application problem.*

49. Adrienne worked 46.5 hours this week. Her hourly wage is \$14.24 for the first 40 hours and 1.5 times that rate over 40 hours. Find her total earnings to the nearest dollar.

50. A book of 12 tickets costs \$35.89 at the State Fair midway. What is the cost per ticket, to the nearest cent?

51. Stock in MathTronic sells for \$3.75 per share. Kenneth is thinking of investing \$500. How many whole shares could he buy?

52. Ground beef is on sale at \$0.89 per pound. How much will Ms. Lee pay for a 3.5 pound package, to the nearest cent?

Use the order of operations to simplify each expression.

53. $3.5^2 + 8.7 \cdot 1.95$

54. $11 - 3.06 \div (3.95 - 0.35)$

[4.6] *Write each fraction or mixed number as a decimal. Round to the nearest thousandth if necessary.*

55. $3\frac{4}{5}$

56. $\frac{16}{25}$

57. $1\frac{7}{8}$

58. $\frac{1}{9}$

Arrange each group of numbers in order, from smallest to largest.

59. 3.68, 3.806, 3.6008

60. 0.215, 0.22, 0.209, 0.2102

61. $0.17, \frac{3}{20}, \frac{1}{8}, 0.159$

Mixed Review Exercises

Add, subtract, multiply, or divide as indicated.

62. $89.19 + 0.075 + 310.6 + 5$

63. 72.8×3.5

64. $1648.3 \div 0.46$
Round to thousandths.

65. $30 - 0.9102$

66. $(4.38)(0.007)$

67. $0.005\overline{)0.047}$

68. $72.105 + 8.2 + 95.37$

69. $81.36 \div 9$

70. $(5.6 - 1.22) + 4.8 \cdot 3.15$

71. 0.455×18

72. $(1.6)(0.58)$

73. $0.218\overline{)7.63}$

74. $21.059 - 20.8$

75. $18.3 - 3^2 \div 0.5$

Use the information in the ad to answer Exercises 76–80. Round money answers to the nearest cent. (Disregard any sales tax.)

Grand Opening Sale!
Save on Clothing for the Entire Family
Jeans for Teens
only 19.95 each
women's sizes $24.99
Athletic Shoes
regularly priced
$89.99 to $149.50
NOW just $71 to $119.60!
Men's socks NOW 3 pairs for $8.99
Children's socks 6 pairs for $5
Hurry in — TWO DAYS ONLY!!

76. How much would one pair of men's socks cost?

77. How much more would one pair of men's socks cost than one pair of children's socks?

78. How much would Fernando pay for a dozen pair of men's socks?

79. How much would Akiko pay for five pairs of teen jeans and four pairs of women's jeans?

80. What is the difference between the cheapest sale price for athletic shoes and the highest regular price?

To decrease your risk of clogged arteries, it is recommended that you get at least 2 milligrams of vitamin B-6 each day. Use the information in the table to answer Exercises 81–82.

Sources of Vitamin B-6 (amounts in milligrams)	
1 cup orange juice	0.11
1 banana	0.66
1 baked potato with skin	0.7
$\frac{1}{2}$ cup strawberries	0.45
3 ounces skinless chicken	0.5
3 ounces water-packed tuna	0.3
$\frac{1}{2}$ cup chickpeas	0.57

Source: Harvard Health Letter.

81. (a) Which food item has the highest amount of vitamin B-6?

(b) Which food item has the lowest amount?

(c) What is the difference in the amount of vitamin B-6 between the food items with the highest and lowest amounts?

82. (a) Suppose you ate a banana, 3 ounces of skinless chicken, and 1 cup of strawberries. How many milligrams of vitamin B-6 would you get?

(b) Did you get more or less than the recommended amount? By how much?

Chapter 4 Test

Study Skills Workbook
Activity 11

Write each decimal as a fraction or mixed number in lowest terms.

1. 18.4 **2.** 0.075

Write each decimal in words.

3. 60.007 **4.** 0.0208

Round each decimal to the place indicated.

5. 725.6089 to the nearest tenth

6. 0.62951 to the nearest thousandth

7. $1.4945 to the nearest cent

8. $7859.51 to the nearest dollar

First use front end rounding to round each number and estimate the answer. Then find the exact answer.

9. $7.6 + 82.0128 + 39.59$

10. $79.1 - 3.602$

11. $5.79 \cdot 1.2$

12. $20.04 \div 4.8$

Add, subtract, multiply, or divide as indicated.

13. $53.1 + 4.631 + 782 + 0.031$

14. $670 - 0.996$

15. $(0.0069)(0.007)$

16. $0.15\overline{)72}$

1. ______________

2. ______________

3. ______________

4. ______________

5. ______________

6. ______________

7. ______________

8. ______________

9. *Estimate:* ______________
Exact: ______________

10. *Estimate:* ______________
Exact: ______________

11. *Estimate:* ______________
Exact: ______________

12. *Estimate:* ______________
Exact: ______________

13. ______________

14. ______________

15. ______________

16. ______________

17. ______________

17. Write $2\frac{5}{8}$ as a decimal. Round to the nearest thousandth, if necessary.

Arrange in order from smallest to largest.

18. ______________

18. $0.44, 0.451, \frac{9}{20}, 0.4506$

Use the order of operations to simplify this expression.

19. ______________

19. $6.3^2 - 5.9 + 3.4 \cdot 0.5$

Solve each application problem.

20. ______________

20. Jennifer had \$71.15 in her checking account. Yesterday her account earned \$0.95 interest for the month, and she deposited a paycheck for \$390.77. The bank charged her \$16 for new checks. What is the new balance in her account?

21. ______________

21. During her career, Jackie Joyner-Kersee won three Olympic medals in the long jump event. In 1988 she jumped 7.4 meters in Seoul. In 1992 she jumped 7.07 meters in Barcelona, and in 1996 she jumped 7 meters in Atlanta. In which year did she have the longest jump? How much longer was it than the second best jump? (*Source:* Associated Press.)

22. ______________

22. Mr. Yamamoto bought 1.85 pounds of cheese at \$2.89 per pound. How much did he pay for the cheese, to the nearest cent?

23. ______________

23. Loren's baby had a temperature of 102.7 degrees. Later in the day it was 99.9 degrees. How much had the baby's temperature dropped?

24. ______________

24. Pat bought 3.4 meters of fabric. She paid \$15.47. What was the cost per meter?

25. ______________

25. Write your own application problem using decimals. Make it different from problems 20–24. Then show how to solve your problem.

Cumulative Review Exercises CHAPTERS 1–4

Name the digit that has the given place value.

1. 19,076,542
hundreds
millions
ones

2. 83.0754
tenths
thousandths
tens

Round each number as indicated.

3. 499,501 to the nearest thousand

4. 602.4937 to the nearest hundredth

5. $709.60 to the nearest dollar

6. $0.0528 to the nearest cent

First use front end rounding to round each number and estimate the answer. Then find the exact answer.

7. *Estimate:* $+$ ____ *Exact:* $3672 + 589 + 9078$

8. *Estimate:* $+$ ____ *Exact:* $4.06 + 15.7 + 0.923$

9. *Estimate:* $-$ ____ *Exact:* $5018 - 1809$

10. *Estimate:* $-$ ____ *Exact:* $51.6 - 7.094$

11. *Estimate:* $\times$ ____ *Exact:* 3317×166

12. *Estimate:* $\times$ ____ *Exact:* 6.82×7.3

13. *Estimate:*) ____ *Exact:* $46\overline{)123{,}740}$

14. *Estimate:*) ____ *Exact:* $8.4\overline{)37.8}$

15. *Estimate:* ____ • ____ = ____ *Exact:* $1\frac{9}{10} \cdot 3\frac{3}{4}$

16. *Estimate:* ____ ÷ ____ = ____ *Exact:* $2\frac{1}{3} \div \frac{5}{6}$

17. *Estimate:* ____ + ____ = ____ *Exact:* $1\frac{4}{5} + 1\frac{2}{3}$

18. *Estimate:* ____ − ____ = ____ *Exact:* $4\frac{1}{2} - 1\frac{7}{8}$

Add, subtract, multiply, or divide as indicated.

19. $10 - 0.329$

20. $2\frac{3}{5} \cdot \frac{5}{9}$

21. $9 + 72{,}417 + 799$

22. $11\frac{1}{5} \div 8$

23. $5006 - 92$

24. $0.7 + 85 + 7.903$

25. Write your answer using R for the remainder.

$7\overline{)2831}$

26. $\frac{5}{6} + \frac{7}{8}$

27. 332×704

28. $(0.006)(5.44)$

29. 3.2×2.5

30. $25.2 \div 0.56$

31. $\frac{2}{3} \div 5\frac{1}{6}$

32. $5\frac{1}{4} - 4\frac{7}{12}$

33. $4.7 \div 9.3$
Round to nearest hundredth.

Use the order of operations to simplify each expression.

34. $10 - 4 \div 2 \cdot 3$

35. $\sqrt{36} + 3 \cdot 8 - 4^2$

36. $\frac{2}{3} \cdot \left(\frac{7}{8} - \frac{1}{2}\right)$

37. $0.9^2 + 10.6 \div 0.53$

38. $4^3 \cdot 3^2$

39. $\sqrt{196}$

40. Find the prime factorization of 200. Write your answer using exponents.

41. Write 40.035 in words.

42. Write three hundred six ten-thousandths in numbers.

Write each decimal as a fraction or mixed number in lowest terms.

43. 0.125

44. 3.08

Write each fraction or mixed number as a decimal. Round to the nearest thousandth if necessary.

45. $2\frac{3}{5}$

46. $\frac{7}{11}$

47. Write $<$ or $>$ in the blank to make a true statement: $\frac{5}{8}$ _____ $\frac{4}{9}$.

Arrange each group of numbers in order, from smallest to largest.

48. 7.005, 7.5005, 7.5, 7.505

49. $\frac{7}{8}$, 0.8, $\frac{21}{25}$, 0.8015

Use the information in the table to answer Exercises 50–53. All measurements are in inches.

Children's Hats	**Head Size**	**Order Size:**
Measure around head above eyebrow ridges.	$16\frac{1}{2}''$–$18''$	XXS
	$18\frac{1}{4}''$–$19''$	XS
	$19\frac{1}{4}''$–$20''$	S
	$20\frac{1}{4}''$–$21\frac{1}{8}''$	M
	$21\frac{1}{2}''$–$22\frac{1}{4}''$	L

Source: Lands' End.

50. If the distance around your child's head is $21\frac{1}{16}$ in., which hat size should you order?

51. Find the difference between the smaller and larger measurements for each of the hat sizes.

52. Change the measurements for the medium (M) size to decimals. Then find the difference in the measurements. Show why this answer is equivalent to your answer from Exercise 51.

53. Did you prefer doing the subtraction using fractions (as in Exercise 51) or using decimals (as in Exercise 52)? Explain your reasoning.

First round the numbers and estimate the answer to each application problem. Then find the exact answer.

54. Lameck had two \$10 bills. He spent \$7.96 on gasoline and \$0.87 for a candy bar at the convenience store. How much money does he have left?

Estimate:

Exact:

55. Manuela's daughter is 50 in. tall. Last year she was $46\frac{5}{8}$ in. tall. How much has she grown? (When estimating, round to the nearest whole number.)

Estimate:

Exact:

56. Sharon records textbooks on tape for students who are blind. Her hourly wage is \$11.63. How much did she earn working 16.5 hours last week, to the nearest cent?

Estimate:

Exact:

57. The Farnsworth Elementary School has eight classrooms with 22 students in each one and 12 classrooms with 26 students in each one. How many students attend the school?

Estimate:

Exact:

58. Toshihiro bought $2\frac{1}{3}$ yd of cotton fabric and $3\frac{7}{8}$ yd of wool fabric. How many yards did he buy in all?

Estimate:

Exact:

59. Kimberly had \$29.44 in her checking account. She wrote a check for \$40 and deposited a \$220.06 paycheck into her account, but not in time to prevent an \$18 overdraft charge. What is the new balance in her account?

Estimate:

Exact:

60. Paulette bought 2.7 pounds of grapes for \$2.56. What was the cost per pound, to the nearest cent?

Estimate:

Exact:

61. Carter Community College received a \$78,000 grant from a local computer company to help students pay tuition for computer classes. How much money could be given to each of 107 students? Round to the nearest dollar.

Estimate:

Exact:

Use the information in the table to answer Exercises 62–64.

Animal	Average Weight of Animal (ounces)	Average Weight of Food Eaten Each Day (ounces)
Hamster	3.5	0.4
Queen bee	0.004	?
Hummingbird	?	0.07

Source: NCTM News Bulletin.

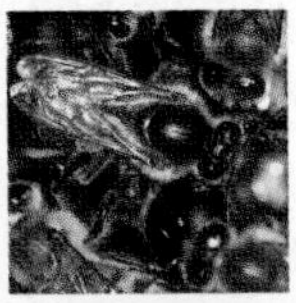

62. (a) In how many days will a hamster eat enough food to equal its body weight? Round to the nearest whole number of days.

(b) If a 140-pound woman ate her body weight of food in the same number of days as the hamster, how much would she eat each day? Round to the nearest tenth.

63. While laying eggs, a queen bee eats eighty times her weight each day. Use this information to fill in one of the missing values in the table.

64. A hummingbird's body weight is about $1\frac{3}{5}$ times the weight of its daily food intake. Find its body weight using decimal numbers and using fractions. Then prove that the two answers are equivalent.

5 Ratio and Proportion

Nearly $\frac{1}{3}$ of the people in the United States own a cell phone. (*Source: Scientific American.*) Everyone likes to talk, but no one likes to pay the bills! Now you can get the best possible deal on cellular phone service by finding unit rates. (See Section 5.2, Exercises 47–50.)

5.1 Ratios

Objectives

1. Write ratios as fractions.
2. Solve ratio problems involving decimals or mixed numbers.
3. Solve ratio problems after converting units.

A **ratio** compares two quantities. You can compare two numbers, such as 8 and 4, or two measurements that have the same type of units, such as 3 days and 12 days.

Ratios can help you see important relationships. For example, if the ratio of your monthly expenses to your monthly income is 10 to 9, then you are spending $10 for every $9 you earn, going deeper into debt.

1 **Write ratios as fractions.** A ratio can be written in three ways.

Writing a Ratio

The ratio of $7 **to** $3 can be written:

7 **to** 3 or 7:3 or $\frac{7}{3}$ ← Fraction bar indicates "**to**."

":" indicates "**to**."

Writing a ratio as a fraction is the most common method, and the one we will use here. All three ways are read, "the ratio of 7 **to** 3." The word **to** separates the quantities being compared.

Writing a Ratio as a Fraction

Order is important when writing a ratio. The quantity mentioned **first** is the **numerator**. The quantity mentioned **second** is the **denominator**. For example:

The ratio of **5** to **12** is written $\frac{5}{12}$.

Example 1 **Writing Ratios**

The Anasazi, ancestors of the Pueblo Indians, built multistory apartment towns in New Mexico about 1100 years ago. A room might measure 14 ft long, 11 ft wide, and 15 ft high.

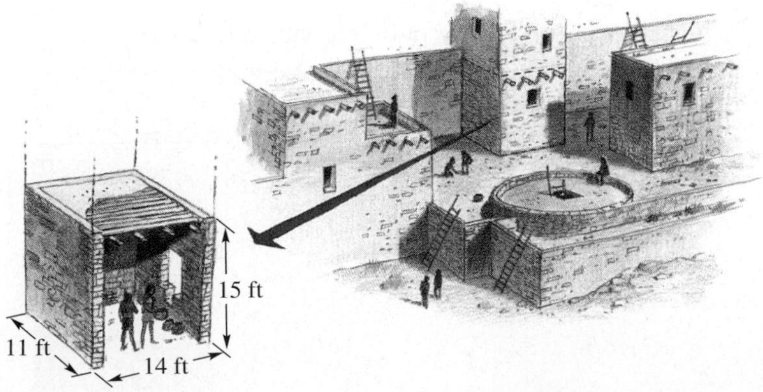

Continued on Next Page

Write each ratio as a fraction, using these room measurements.

(a) Ratio of length to width

The ratio of **length to width** is $\frac{14 \text{ ft}}{11 \text{ ft}} = \frac{14}{11}$.

Numerator (mentioned first) — Denominator (mentioned second)

You can divide out common *units* just like you divided out common *factors* when writing fractions in lowest terms. (See **Section 2.4.**)

(b) Ratio of width to height

The ratio of width to height is $\frac{11 \text{ ft}}{15 \text{ ft}} = \frac{11}{15}$.

CAUTION

Remember, the *order* of the numbers is important in a ratio. Look for the words "ratio of a to b." Write the ratio as $\frac{a}{b}$, ***not*** $\frac{b}{a}$. The quantity mentioned first is the numerator.

Work Problem 1 at the Side.

Any ratio can be written as a fraction. Therefore, you can write a ratio in *lowest terms,* just as you do with any fraction.

Example 2 Writing Ratios in Lowest Terms

Write each ratio in lowest terms.

(a) 60 days to 20 days

The ratio is $\frac{60}{20}$. Write this ratio in lowest terms by dividing the numerator and denominator by 20.

$$\frac{60}{20} = \frac{60 \div \mathbf{20}}{20 \div \mathbf{20}} = \frac{3}{1} \leftarrow \text{Ratio in lowest terms}$$

CAUTION

In the fractions chapters you would have rewritten $\frac{3}{1}$ as 3. But a *ratio* compares *two* quantities, so you need to keep both parts of the ratio and write it as $\frac{3}{1}$.

(b) 50 ounces of medicine to 120 ounces of medicine

The ratio is $\frac{50}{120}$. Divide the numerator and denominator by 10.

$$\frac{50}{120} = \frac{50 \div \mathbf{10}}{120 \div \mathbf{10}} = \frac{5}{12} \leftarrow \text{Ratio in lowest terms}$$

Continued on Next Page

1 Shane spent \$14 on meat, \$5 on milk, and \$7 on fresh fruit. Write each ratio as a fraction.

(a) The ratio of amount spent on fruit to amount spent on milk.

(b) The ratio of amount spent on milk to amount spent on meat.

(c) The ratio of amount spent on meat to amount spent on milk.

ANSWERS

1. (a) $\frac{7}{5}$ **(b)** $\frac{5}{14}$ **(c)** $\frac{14}{5}$

(c) 15 people in a large van to 6 people in a small van

$$\text{The ratio is } \frac{15}{6} = \frac{15 \div 3}{6 \div 3} = \frac{5}{2} \leftarrow \left\{\begin{matrix}\text{Ratio in} \\ \text{lowest terms}\end{matrix}\right.$$

NOTE

Although $\frac{5}{2} = 2\frac{1}{2}$, ratios are *not* written as mixed numbers. Nevertheless, in Example 2(c), the ratio $\frac{5}{2}$ does mean the large van holds $2\frac{1}{2}$ times as many people as the small van.

Work Problem 2 at the Side.

2 Solve ratio problems involving decimals or mixed numbers. Sometimes a ratio compares two decimal numbers or two fractions. It is easier to understand if we rewrite the ratio as a ratio of two whole numbers.

Example 3 Using Decimal Numbers in a Ratio

The price of a Sunday newspaper increased from \$1.50 to \$1.75. Find the ratio of the increase in price to the original price.

The words increase in price are mentioned first, so the increase will be the numerator. How much did the price go up? Use subtraction.

$$\begin{array}{ccccc} \text{new price} & - & \text{original price} & = & \text{increase} \\ \$1.75 & - & \$1.50 & = & \$0.25 \end{array}$$

The words the original price are mentioned second, so the original price of \$1.50 is the denominator.

The ratio of increase in price **to** original price is

$$\frac{0.25}{1.50} \begin{matrix}\leftarrow \text{increase} \\ \leftarrow \text{original price}\end{matrix}$$

Now rewrite the ratio as a ratio of whole numbers. Recall that if you multiply both the numerator and denominator of a fraction by the same number, you get an equivalent fraction. The decimals in this example are hundredths, so multiply by 100 to get whole numbers. (If the decimals are tenths, multiply by 10. If thousandths, multiply by 1000.) Then write the ratio in lowest terms.

$$\frac{0.25}{1.50} = \frac{0.25 \cdot 100}{1.50 \cdot 100} = \underbrace{\frac{25}{150}}_{\text{Ratio as two whole numbers}} = \frac{25 \div 25}{150 \div 25} = \frac{1}{6} \leftarrow \left\{\begin{matrix}\text{Ratio in} \\ \text{lowest terms}\end{matrix}\right.$$

Work Problem 3 at the Side.

Example 4 Using Mixed Numbers in Ratios

Write each ratio as a comparison of whole numbers in lowest terms.

(a) 2 days to $2\frac{1}{4}$ days

Write the ratio as follows. Divide out the common units.

$$\frac{2 \text{ \sout{days}}}{2\frac{1}{4} \text{ \sout{days}}} = \frac{2}{2\frac{1}{4}}$$

Continued on Next Page

2 Write each ratio as a fraction in lowest terms.

(a) 9 hours to 12 hours

(b) 100 meters to 50 meters

(c) Write the ratio of width to length for this rectangle.

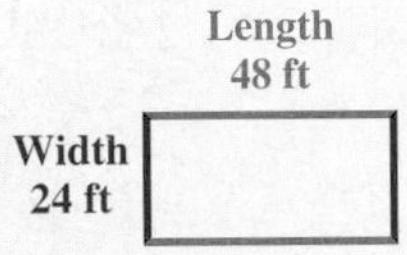

3 Write each ratio as a ratio of whole numbers in lowest terms.

(a) The price of Tamar's favorite brand of lipstick increased from \$5.50 to \$7.00. Find the ratio of the increase in price to the original price.

(b) Last week, Lance worked 4.5 hours each day. This week he cut back to 3 hours each day. Find the ratio of the decrease in hours to the original number of hours.

ANSWERS

2. (a) $\frac{3}{4}$ **(b)** $\frac{2}{1}$ **(c)** $\frac{1}{2}$

3. (a) $\frac{1.50 \times 100}{5.50 \times 100} = \frac{150 \div 50}{550 \div 50} = \frac{3}{11}$

(b) $\frac{1.5 \times 10}{4.5 \times 10} = \frac{15 \div 15}{45 \div 15} = \frac{1}{3}$

Next, write 2 as $\frac{2}{1}$ and $2\frac{1}{4}$ as the improper fraction $\frac{9}{4}$.

$$\frac{2}{2\frac{1}{4}} = \frac{\frac{2}{1}}{\frac{9}{4}}$$

Now rewrite the problem in horizontal format, using the "÷" symbol for division. Finally, multiply by the reciprocal of the divisor, as you did in **Section 2.7.**

Reciprocal

$$\frac{\frac{2}{1}}{\frac{9}{4}} = \frac{2}{1} \div \frac{9}{4} = \frac{2}{1} \cdot \frac{4}{9} = \frac{8}{9}$$

The ratio, in lowest terms, is $\frac{8}{9}$.

(b) $3\frac{1}{4}$ to $1\frac{1}{2}$

Write the ratio as $\frac{3\frac{1}{4}}{1\frac{1}{2}}$. Then write $3\frac{1}{4}$ and $1\frac{1}{2}$ as improper fractions.

$$3\frac{1}{4} = \frac{13}{4} \quad \text{and} \quad 1\frac{1}{2} = \frac{3}{2}$$

The ratio is shown below.

$$\frac{3\frac{1}{4}}{1\frac{1}{2}} = \frac{\frac{13}{4}}{\frac{3}{2}}$$

Rewrite as a division problem in horizontal format, using the "÷" symbol. Then multiply by the reciprocal of the divisor.

$$\frac{13}{4} \div \frac{3}{2} = \frac{13}{\cancel{4}_{2}} \cdot \frac{\cancel{2}^{1}}{3} = \frac{13}{6} \leftarrow \begin{cases}\text{Ratio in}\\ \text{lowest terms}\end{cases}$$

Work Problem 4 at the Side.

3 Solve ratio problems after converting units. When a ratio compares measurements, both measurements must be in the *same* units. For example, *feet* must be compared to *feet, hours* to *hours, pints* to *pints,* and *inches* to *inches.*

Example 5 Ratio Applications Using Measurement

(a) Write the ratio of the length of the board on the left to the length of the board on the right. Compare in inches.

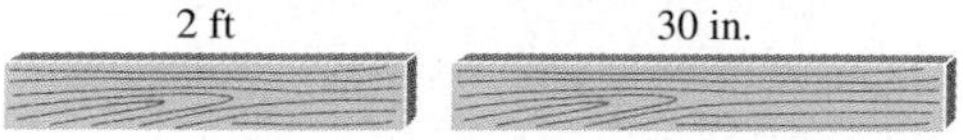

First, express 2 ft in inches. Because 1 ft has 12 in., 2 ft is

$$2 \cdot 12 \text{ in.} = 24 \text{ in.}$$

Continued on Next Page

4 Write each ratio as a ratio of whole numbers in lowest terms.

(a) $3\frac{1}{2}$ to 4

(b) $5\frac{5}{8}$ pounds to $3\frac{3}{4}$ pounds

(c) $3\frac{1}{2}$ in. to $\frac{7}{8}$ in.

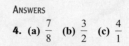

ANSWERS

4. (a) $\frac{7}{8}$ **(b)** $\frac{3}{2}$ **(c)** $\frac{4}{1}$

5 Write each ratio as a fraction in lowest terms. (*Hint:* Recall that it is usually easier to write the ratio using the smaller measurement unit.)

(a) 9 in. to 6 ft

(b) 2 days to 8 hours

(c) 7 yd to 14 ft

(d) 3 quarts to 3 gallons

(e) 25 minutes to 2 hours

(f) 4 pounds to 12 ounces

ANSWERS

5. (a) $\frac{1}{8}$ (b) $\frac{6}{1}$ (c) $\frac{3}{2}$ (d) $\frac{1}{4}$ (e) $\frac{5}{24}$ (f) $\frac{16}{3}$

The length of the board on the left is 24 in., so the ratio of the lengths is

$$\frac{24\text{ in.}}{30\text{ in.}} = \frac{24}{30}$$

Write the ratio in lowest terms

$$\frac{24}{30} = \frac{24 \div 6}{30 \div 6} = \frac{4}{5} \leftarrow \begin{cases}\text{Ratio in}\\ \text{lowest terms}\end{cases}$$

The shorter board on the left is $\frac{4}{5}$ the length of the longer board on the right.

NOTE

Notice that we wrote the ratio using the smaller unit (inches are smaller than feet). Using the smaller unit will help you avoid working with fractions. If we wrote the ratio using feet, then

$$30\text{ in.} = 2\frac{1}{2}\text{ ft.}$$

So the ratio in feet is shown below.

$$\frac{2\text{ ft}}{2\frac{1}{2}\text{ ft}} = \frac{2}{1} \div \frac{5}{2} = \frac{2}{1} \bullet \frac{2}{5} = \frac{4}{5} \leftarrow \text{Same result.}$$

The ratio is the same, but it takes more steps to get the answer. Using the smaller unit is usually easier.

(b) Write the ratio of 28 days to 3 weeks.

Since it is easier to write the ratio using the smaller measurement unit, compare in *days* because days are shorter than weeks.

First express 3 weeks in days. Because 1 week has 7 days, 3 weeks is

$$3 \bullet 7\text{ days} = 21\text{ days.}$$

So the ratio in days is shown below.

$$\frac{28\text{ days}}{21\text{ days}} = \frac{28}{21} = \frac{28 \div 7}{21 \div 7} = \frac{4}{3} \leftarrow \text{Lowest terms}$$

The following table will help you set up ratios that compare measurements. You will work with these measurements again in Chapter 7.

Measurement Comparisons

Length	Capacity (Volume)
1 foot = 12 inches	1 pint = 2 cups
1 yard = 3 feet	1 quart = 2 pints
1 mile = 5280 feet	1 gallon = 4 quarts
Weight	**Time**
1 pound = 16 ounces	1 minute = 60 seconds
1 ton = 2000 pounds	1 hour = 60 minutes
	1 day = 24 hours
	1 week = 7 days

Work Problem 5 at the Side.

5.1 Exercises

FOR EXTRA HELP: Student's Solutions Manual | MyMathLab.com | InterAct Math Tutorial Software | AW Math Tutor Center | www.mathxl.com | Digital Video Tutor CD 3 Videotape 9

Write each ratio as a fraction in lowest terms. See Examples 1 and 2.

1. 8 to 9

2. 11 to 15

3. \$100 to \$50

4. 35¢ to 7¢

5. 30 minutes to 90 minutes

6. 9 pounds to 36 pounds

7. 80 miles to 50 miles

8. 300 people to 450 people

9. 6 hours to 16 hours

10. 45 books to 35 books

Write each ratio as a ratio of whole numbers in lowest terms. See Examples 3 and 4.

11. \$4.50 to \$3.50

12. \$0.08 to \$0.06

13. 15 to $2\frac{1}{2}$

14. 5 to $1\frac{1}{4}$

15. $1\frac{1}{4}$ to $1\frac{1}{2}$

16. $2\frac{1}{3}$ to $2\frac{2}{3}$

Write each ratio as a fraction in lowest terms. For help, use the table of measurement relationships on page 316. See Example 5.

17. 4 ft to 30 in.

18. 8 ft to 4 yd

19. 5 minutes to 1 hour

20. 8 quarts to 5 pints

21. 15 hours to 2 days

22. 3 pounds to 6 ounces

23. 5 gallons to 5 quarts

24. 3 cups to 3 pints

The table shows the number of greeting cards that Americans buy for various occasions. Use the information to answer Exercises 25–30. Write each ratio as a fraction in lowest terms.

Holiday/Event	Cards Sold
Valentine's Day	900 million
Mother's Day	150 million
Father's Day	95 million
Graduation	60 million
Thanksgiving	30 million
Halloween	25 million

Source: Hallmark Cards.

25. Find the ratio of Thanksgiving cards to graduation cards.

26. Find the ratio of Halloween cards to Mother's Day cards.

27. Find the ratio of Valentine's Day cards to Halloween cards.

28. Find the ratio of Mother's Day cards to Father's Day cards.

29. Explain how you might use the information in the table if you owned a shop selling gifts and greeting cards.

30. Why is the ratio of Valentine's Day cards to graduation cards $\frac{15}{1}$? Give two possible reasons.

The bar graph shows the number of Americans who play various instruments. Use the graph to complete Exercises 31–34. Write each ratio as a fraction in lowest terms.

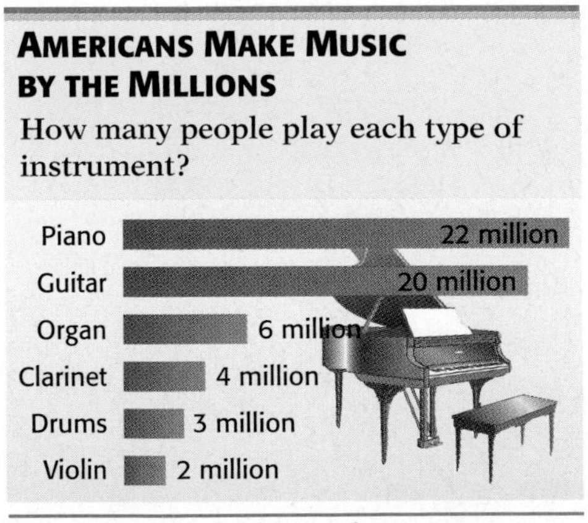

Source: *America by the Numbers.*

31. Write six ratios that compare the least popular instrument to each of the other instruments.

32. Which two instruments give each of these ratios: **(a)** $\frac{5}{1}$; **(b)** $\frac{2}{1}$. There may be more than one correct answer.

33. Why might the ratio of guitar players to drum players be $\frac{20}{3}$? Give two possible explanations.

34. If you ran a music school, explain how you might use the information in the graph to decide how many teachers to hire.

Use the circle graph of one family's monthly budget to complete Exercises 35–36. Write each ratio as a fraction in lowest terms.

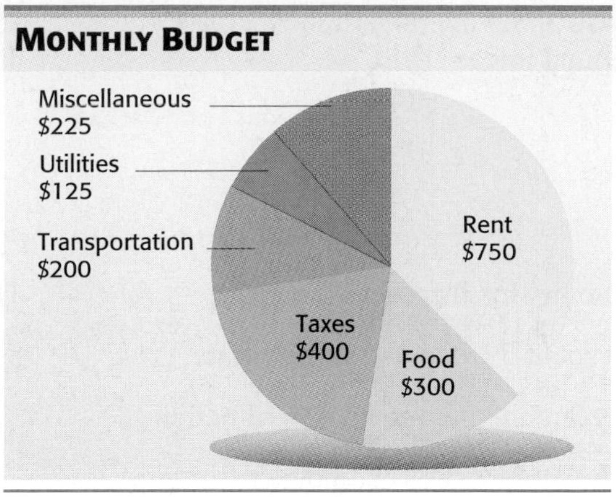

35. Find the ratio of
 (a) rent to utilities
 (b) rent to food
 (c) rent and utilities to total expenses.

36. Find the ratio of
 (a) taxes to rent
 (b) food to transportation
 (c) taxes and food to rent and utilities.

For each figure, find the ratio of the length of the longest side to the length of the shortest side. Write each ratio as a fraction in lowest terms.

37.

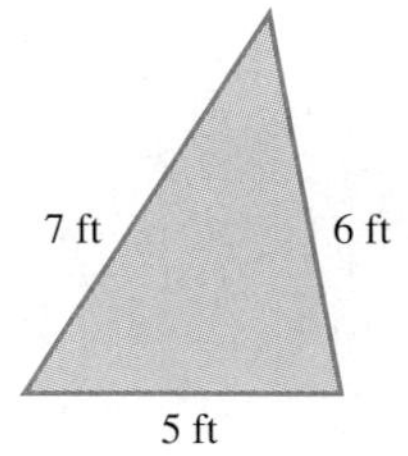

38.

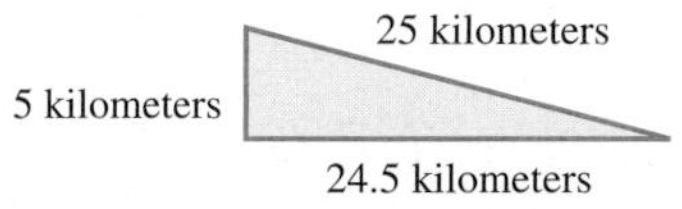

39.

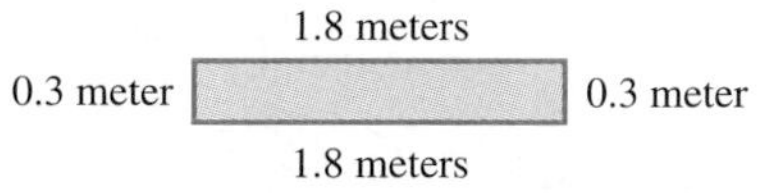

40.

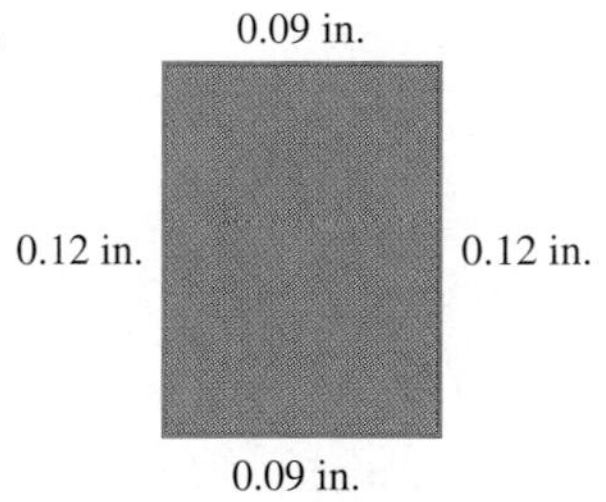

41.

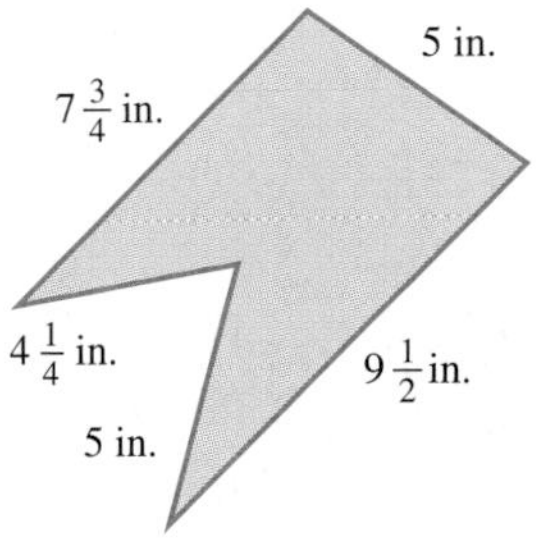

42.

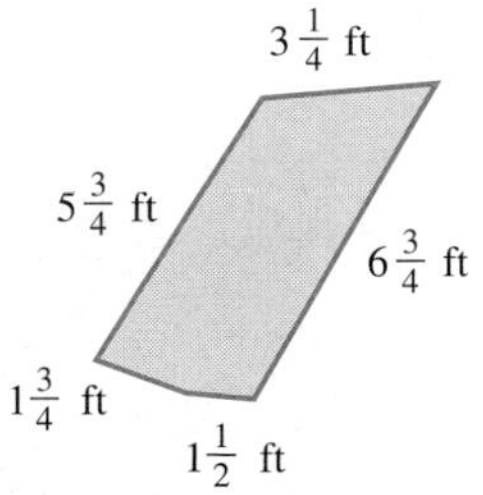

Write each ratio as a fraction in lowest terms.

43. The price of oil recently went from \$10.80 to \$13.50 per case of 12 quarts. Find the ratio of the increase in price to the original price.

44. The price of an antibiotic decreased from \$8.80 to \$5.60 for a bottle of 100 tablets. Find the ratio of the decrease in price to the original price.

45. The first time a movie was made in Minnesota, the cast and crew spent $59\frac{1}{2}$ days filming winter scenes. The next year, another movie was filmed in $8\frac{3}{4}$ weeks. Find the ratio of the first movie's filming time to the second movie's time. Compare in weeks.

46. The percheron, a large draft horse, measures about $5\frac{3}{4}$ ft at the shoulder. The prehistoric ancestor of the horse measured only $15\frac{3}{4}$ in. at the shoulder. Find the ratio of the percheron's height to its prehistoric ancestor's height. Compare in inches. (*Source: Eyewitness Books: Horse.*)

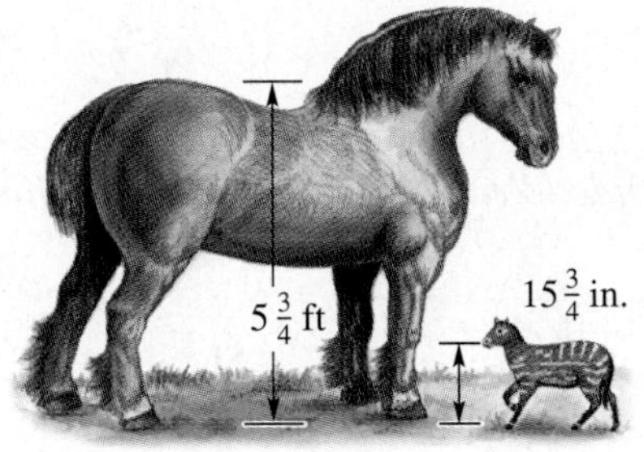

RELATING CONCEPTS (Exercises 47–50) FOR INDIVIDUAL OR GROUP WORK

Use your knowledge of ratios to ***work Exercises 47–50 in order.***

47. In this painting, what is the ratio of the length of the longest side to the length of the shortest side? What other measurements could the painting have and still maintain the same ratio?

48. The ratio of my son's age to my daughter's age is 4 to 5. One possibility is that my son is 4 years old and my daughter is 5 years old. Find six other possibilities that fit the 4 to 5 ratio.

49. Amelia said that the ratio of her age to her mother's age is 5 to 3. Is this possible? Explain your answer.

50. Would you prefer that the ratio of your income to your friend's income be 1 to 3 or 3 to 1? Explain your answer.

5.2 Rates

A *ratio* compares two measurements with the same type of units, such as 9 feet to 12 feet (both length measurements). But many of the comparisons we make use measurements with different types of units, such as the following.

160 dollars **for** 8 hours **(money to time)**

450 miles **on** 15 gallons **(distance to capacity)**

This type of comparison is called a **rate.**

Objectives

1. Write rates as fractions.
2. Find unit rates.
3. Find the best buy based on cost per unit.

1 Write rates as fractions. Suppose you hiked 18 miles **in** 4 hours. The *rate* at which you hiked can be written as a fraction in lowest terms.

$$\frac{18 \text{ miles}}{4 \text{ hours}} = \frac{18 \text{ miles} \div 2}{4 \text{ hours} \div 2} = \frac{9 \text{ miles}}{2 \text{ hours}} \leftarrow \text{Lowest terms}$$

In a rate, you often find these words separating the quantities you are comparing.

in **for** **on** **per** **from**

CAUTION

When writing a rate, always include the units, such as miles, hours, dollars, and so on. Because the units in a rate are different, the units do *not* divide out.

Example 1 Writing Rates in Lowest Terms

Write each rate as a fraction in lowest terms.

(a) 5 gallons of chemical **for** $60.

$$\frac{5 \text{ gallons} \div 5}{60 \text{ dollars} \div 5} = \frac{1 \text{ gallon}}{12 \text{ dollars}}$$

Write the units: gallons and dollars

(b) $1500 wages **in** 10 weeks

$$\frac{1500 \text{ dollars} \div 10}{10 \text{ weeks} \div 10} = \frac{150 \text{ dollars}}{1 \text{ week}}$$

(c) 2225 miles **on** 75 gallons of gas

$$\frac{2225 \text{ miles} \div 25}{75 \text{ gallons} \div 25} = \frac{89 \text{ miles}}{3 \text{ gallons}}$$

Work Problem 1 at the Side.

1 Write each rate as a fraction in lowest terms.

(a) $6 for 30 packages

(b) 500 miles in 10 hours

(c) 4 teachers for 90 students

(d) 1270 bushels on 30 acres

2 Find unit rates. When the *denominator* of a rate is 1, it is called a **unit rate.** We use unit rates frequently. For example, you earn $12.75 for *1 hour* of work. This unit rate is written:

$12.75 **per** hour or $12.75/hour.

Or, you drive 28 miles on *1 gallon* of gas. This unit rate is written

28 miles **per** gallon or 28 miles/gallon.

Use **per** or a slash mark (/) when writing unit rates.

Answers

1. (a) $\frac{\$1}{5 \text{ packages}}$ **(b)** $\frac{50 \text{ miles}}{1 \text{ hour}}$ **(c)** $\frac{2 \text{ teachers}}{45 \text{ students}}$ **(d)** $\frac{127 \text{ bushels}}{3 \text{ acres}}$

2 Find each unit rate.

(a) \$4.35 for 3 pounds of cheese

(b) 304 miles on 9.5 gallons of gas

(c) \$850 in 5 days

(d) 24-pound turkey for 15 people

Example 2 Finding Unit Rates

Find each unit rate.

(a) 337.5 miles on 13.5 gallons of gas
Write the rate as a fraction.

$$\frac{337.5 \text{ miles}}{13.5 \text{ gallons}} \leftarrow \text{The fraction bar indicates division.}$$

Divide 337.5 by 13.5 to find the unit rate.

$$13.5\overline{)337.5} = 25.$$

$$\frac{337.5 \text{ miles} \div 13.5}{13.5 \text{ gallons} \div 13.5} = \frac{25 \text{ miles}}{1 \text{ gallon}}$$

The unit rate is 25 miles **per** gallon, or 25 miles/gallon.

(b) 549 miles in 18 hours

$$\frac{549 \text{ miles}}{18 \text{ hours}} \qquad \text{Divide: } 18\overline{)549.0} = 30.5$$

The unit rate is 30.5 miles/hour.

(c) \$810 in 6 days

$$\frac{810 \text{ dollars}}{6 \text{ days}} \qquad \text{Divide: } 6\overline{)810} = 135$$

The unit rate is \$135/day.

Work Problem 2 at the Side.

3 Find the best buy based on cost per unit. When shopping for groceries, household supplies, and health and beauty items, you will find many different brands and package sizes. You can save money by finding the lowest *cost per unit.*

Cost per Unit

Cost per unit is a rate that tells how much you pay for *one* item or *one* unit. Examples are \$1.65 per gallon, \$47 per shirt, and \$2.98 per pound.

Example 3 Determining the Best Buy

The local store charges the following prices for pancake syrup. Find the best buy.

Continued on Next Page

ANSWERS
2. (a) \$1.45/pound (b) 32 miles/gallon (c) \$170/day (d) 1.6 pounds/person

The best buy is the container with the *lowest* cost per unit. All the containers are measured in *ounces* (oz), so you first need to find the *cost per ounce* for each one. Divide the price of the container by the number of ounces in it. Round to the nearest thousandth, if necessary.

Size	Cost per Unit (Rounded)
12 ounces	$\frac{\$1.28}{12 \text{ ounces}} \approx \$0.107$ per ounce (highest)
24 ounces	$\frac{\$1.81}{24 \text{ ounces}} \approx \$0.075$ per ounce (lowest)
36 ounces	$\frac{\$2.73}{36 \text{ ounces}} \approx \$0.076$ per ounce

The lowest cost per ounce is \$0.075, so the 24-ounce container is the best buy.

NOTE

Earlier we rounded money amounts to the nearest hundredth (nearest cent). But when comparing unit costs, rounding to the nearest thousandth will help you see the difference between very similar unit costs. Notice that the 24-ounce and 36-ounce syrup containers would both have rounded to \$0.08 per ounce if we had rounded to hundredths.

Work Problem 3 at the Side.

Calculator Tip When using a calculator to find unit prices, remember that division is *not* commutative. In Example 3 you wanted to find cost per ounce. Let the *order* of the *words* help you enter the numbers in the correct order.

cost	per	ounce		
↓	↓	↓		
Enter the cost	*Per* means divide	Enter number of ounces		
↓	↓	↓		
\$2.73	÷	36	=	**0.076 (rounded)**

If you entered 36 ÷ 2.73 =, you'd get the number of *ounces* per *dollar.* How could you use that information to find the best buy? (*Answer:* The best buy would be to get the greatest number of ounces per dollar.)

Finding the best buy is sometimes a complicated process. Things that affect the cost per unit can include "cents off" coupons and differences in how much use you'll get out of each unit.

3 Find the best buy (lowest cost per unit) for each purchase.

(a) 2 quarts for \$3.25
3 quarts for \$4.95
4 quarts for \$6.48

(b) 6 cans of cola for \$1.99
12 cans of cola for \$3.49
24 cans of cola for \$7

ANSWERS

3. (a) 4 quarts, at \$1.62 per quart
(b) 12 cans, at \$0.291 per can (rounded)

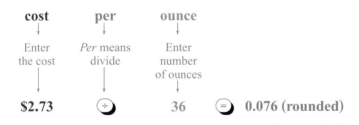

4 Solve each problem.

(a) Some batteries claim to last longer than others. If you believe these claims, which brand is the best buy?

Four-pack of AA-size batteries for \$2.79

One AA-size battery for \$1.19; lasts twice as long

(b) Which tube of toothpaste is the better buy? You have a coupon for 85¢ off Brand C and a coupon for 20¢ off Brand D.

Brand C is \$3.89 for 6 ounces.

Brand D is \$1.59 for 2.5 ounces.

Example 4 Solving Best Buy Applications

Solve each application problem.

(a) There are many brands of liquid laundry detergent. If you feel they all do a good job of cleaning your clothes, you can base your purchase on cost per unit. But some brands are "concentrated" so you can use less detergent for each load of clothes. Which of the choices shown below is the best buy?

try
SUDZY
to clean your clothes!
50 fluid ounces for \$3.99
Does same number of
washloads as the old
64-ounce bottle!

WHITE-O
gets out ALL the stains!
One gallon
(128 ounces)
for \$9.89
Does twice
the washloads
of the old
gallon bottle!

To find Sudzy's unit cost, divide \$3.99 by 64 ounces, not 50 ounces. You're getting as many clothes washed as if you bought 64 ounces. Similarly, to find White-O's unit cost, divide \$9.89 by 256 ounces (twice 128 ounces, or 2 • 128 ounces = 256 ounces).

$$\text{Sudzy} \quad \frac{\$3.99}{64 \text{ ounces}} \approx \$0.062 \text{ per ounce}$$

$$\text{White-O} \quad \frac{\$9.89}{256 \text{ ounces}} \approx \$0.039 \text{ per ounce}$$

White-O has the lower cost per ounce and is the better buy. (However, if you try it and it really doesn't get out all the stains, Sudzy may be worth the extra cost.)

(b) "Cents-off" coupons also affect the best buy. Suppose you are looking at these choices for "extra strength" aspirin.

Brand X is \$2.29 for 50 tablets.

Brand Y is \$10.75 for 200 tablets.

You have a 40¢ coupon for Brand X and a 75¢ coupon for Brand Y. Which choice is the best buy?

To find the better buy, first subtract the coupon amounts, then divide to find the lower cost per ounce.

$$\text{Brand X costs } \$2.29 - \$0.40 = \$1.89$$

$$\frac{\$1.89}{50 \text{ tablets}} \approx \$0.038 \text{ per tablet}$$

$$\text{Brand Y costs } \$10.75 - \$0.75 = \$10.00$$

$$\frac{\$10.00}{200 \text{ tablets}} = \$0.05 \text{ per tablet}$$

Brand X has the lower cost per tablet and is the better buy.

Work Problem 4 at the Side.

Answers

4. (a) One battery that lasts twice as long (like getting two) is the better buy. The cost per unit is \$0.595 per battery. The four-pack is \$0.698 per battery (rounded).

(b) Brand C with the 85¢ coupon is the better buy at \$0.507 per ounce (rounded). Brand D with the 20¢ coupon is \$0.556 per ounce.

5.2 Exercises

FOR EXTRA HELP: Student's Solutions Manual | MyMathLab.com | InterAct Math Tutorial Software | AW Math Tutor Center | www.mathxl.com | Digital Video Tutor CD 3 Videotape 9

Write each rate as a fraction in lowest terms. See Example 1.

1. 10 cups for 6 people

2. \$12 for 30 pens

3. 15 feet in 35 seconds

4. 100 miles in 30 hours

5. 14 people for 28 dresses

6. 12 wagons for 48 horses

7. 25 letters in 5 minutes

8. 68 pills for 17 people

9. \$63 for 6 visits

10. 25 doctors for 310 patients

11. 72 miles on 4 gallons

12. 132 miles on 8 gallons

Find each unit rate. See Example 2.

13. \$60 in 5 hours

14. \$2500 in 20 days

15. 50 eggs from 10 chickens

16. 36 children from 12 families

17. 7.5 pounds for 6 people

18. 44 bushels from 8 trees

19. \$413.20 for 4 days

20. \$74.25 for 9 hours

Earl kept the following record of the gas he bought for his car. For each entry, find the number of miles he traveled and the unit rate. Round your answers to the nearest tenth.

	Date	Odometer at Start	Odometer at End	Miles Traveled	Gallons Purchased	Miles per Gallon
21.	2/4	27,432.3	27,758.2		15.5	
22.	2/9	27,758.2	28,058.1		13.4	
23.	2/16	28,058.1	28,396.7		16.2	
24.	2/20	28,396.7	28,704.5		13.3	

Source: Author's car records.

Find the best buy (based on the cost per unit) for each item. See Example 3. (Source: Piggly Wiggly.)

25. Black pepper

26. Shampoo

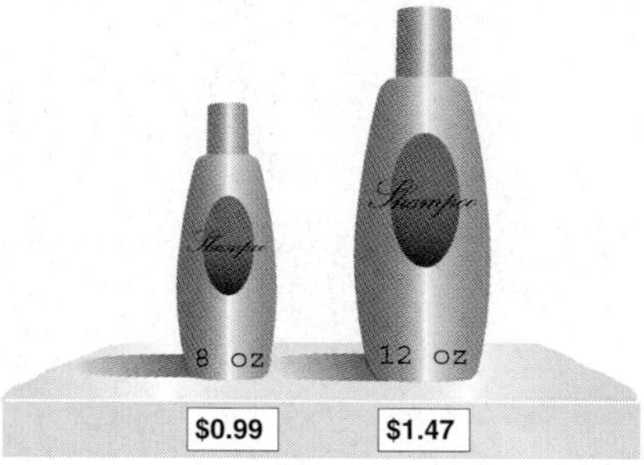

27. Cereal
13 ounces for $2.80
15 ounces for $3.15
18 ounces for $3.98

28. Soup
2 cans for $0.95
3 cans for $1.45
5 cans for $2.29

29. Chunky peanut butter
12 ounces for $1.29
18 ounces for $1.79
28 ounces for $3.39
40 ounces for $4.39

30. Pork and beans
8 ounces for $0.37
16 ounces for $0.77
21 ounces for $0.99
31 ounces for $1.50

31. Suppose you are choosing between two brands of chicken noodle soup. Brand A is $0.48 per can and Brand B is $0.58 per can. But Brand B has more chunks of chicken in it. Which soup is the better buy? Explain your choice.

32. A small bag of potatoes costs $0.19 per pound. A large bag costs $0.15 per pound. But there are only two people in your family, so half the large bag would probably rot before you use it up. Which bag is the better buy? Explain.

Solve each application problem. See Examples 2–4.

33. Makesha lost 10.5 pounds in six weeks. What was her rate of loss in pounds per week?

34. Enrique's taco recipe uses four pounds of meat to feed 10 people. Give the rate in pounds per person.

35. Russ works 7 hours to earn $85.82. What is his pay rate per hour?

36. Find the cost of 1 gallon of gas if 18 gallons cost $26.28.

37. Ms. Keskinen bought 150 shares of stock for $1725. Find the cost of one share.

38. A company pays $6450 in dividends for the 2500 shares of its stock. Find the dividend per share.

39. In the 2000 Olympics, Michael Johnson ran the 400-meter event in a record time of approximately 44 seconds (actually 43.84 seconds). Give his rate in seconds per meter and in meters per second. Use 44 seconds as the time. (*Source:* www.Olympics.com)

40. Sofia can clean and adjust five hearing aids in four hours. Give her rate in hearing aids per hour and in hours per hearing aid.

41. A long-distance phone service advertised that all calls were 5¢ a minute, with a 50¢ minimum per completed call. (*Source:* VarTech Telecom, Inc.) Find the actual cost per minute for: **(a)** a three-minute call, **(b)** a four-minute call, and **(c)** a six-minute call. Round your answers to the nearest tenth of a cent.

42. Another long-distance phone service advertised a rate of 7¢ a minute, with no minimum per call, plus a $5.95 monthly fee. (*Source:* AT&T.) Find the actual cost per minute during one month if you make long-distance calls totaling **(a)** 10 minutes, **(b)** 30 minutes, or **(c)** 60 minutes. Round your answers to the nearest cent.

43. If you believe the claims that some batteries last longer, which is the better buy?

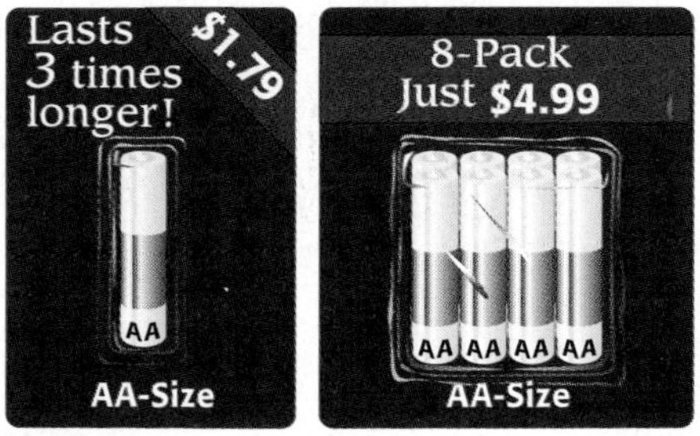

44. Which is the better buy, assuming these laundry detergents both clean equally well?

45. Three brands of cornflakes are available. Brand G is priced at \$2.39 for 10 ounces. Brand K is \$3.99 for 20.3 ounces and Brand P is \$3.39 for 16.5 ounces. You have a coupon for 50¢ off Brand P and a coupon for 60¢ off Brand G. Which cereal is the best buy based on cost per unit?

46. Two brands of facial tissue are available. Brand K is on special at three boxes of 175 tissues each for \$5. Brand S is priced at \$1.29 per box of 125 tissues. You have a coupon for 20¢ off one box of Brand S and a coupon for 45¢ off one box of Brand K. How can you get the best buy on one box of tissue?

RELATING CONCEPTS (Exercises 47–50) FOR INDIVIDUAL OR GROUP WORK

On the first page of this chapter, we said that unit rates can help you get the best deal on cell phone service. Use the information in the table to ***work Exercises 47–50 in order.***

Company	Anytime Minutes	Weekend Minutes	Monthly Charge	Other Costs
Verizon	500	1000	\$35	\$6.95 monthly access fee
Qwest	400	1000	\$39.99	\$25 one-time activation fee
VoiceStream	75	250	\$19.99	
Sprint PCS	250	250	\$29.95	

Source: Advertisements appearing in *Minneapolis Star Tribune*.

Notes:

1. All companies require that you sign up for 12 months of service and charge a \$150 fee if you quit early.
2. Weekend minutes can only be used from midnight Friday to midnight Sunday.
3. Unused minutes cannot be carried over to the next month.

47. If you sign up for one year of service with Qwest, how much will the activation fee increase your average monthly charge?

48. Find the actual average cost per minute for each company, including "other costs." Assume you use all the minutes and no more. Decide how to round your answers so you can find the best buy.

49. How many hours is 1000 minutes? Over the course of a year, what is the average number of "weekend days" per month? How many hours would you have to talk on each "weekend day" to use up 1000 minutes per month? (Round answers to nearest tenth.)

50. Suppose that after two months you canceled your service because you found that you only used 100 minutes per month. Under those conditions, find the actual cost per minute for each company, to the nearest cent.

5.3 PROPORTIONS

1 Write proportions. A **proportion** states that two ratios (or rates) are equivalent. For example,

$$\frac{\$20}{4 \text{ hours}} = \frac{\$40}{8 \text{ hours}}$$

is a proportion that says the rate $\frac{\$20}{4 \text{ hours}}$ is equivalent to the rate $\frac{\$40}{8 \text{ hours}}$. As the amount of money doubles, the number of hours also doubles. This proportion is read:

20 dollars **is to** 4 hours **as** 40 dollars **is to** 8 hours.

Example 1 Writing Proportions

Write each proportion.

(a) 6 ft is to 11 ft **as** 18 ft is to 33 ft.

$$\frac{6 \not{ft}}{11 \not{ft}} = \frac{18 \not{ft}}{33 \not{ft}} \quad \text{so} \quad \frac{6}{11} = \frac{18}{33}$$ The common units (ft) divide out and are not written.

(b) \$9 is to 6 liters **as** \$3 is to 2 liters.

$$\frac{\$9}{6 \text{ liters}} = \frac{\$3}{2 \text{ liters}}$$ Units must be written.

Work Problem 1 at the Side.

2 Determine whether proportions are true or false. There are two ways to see whether a proportion is true. One way is to *write both of the ratios in lowest terms.*

Example 2 Writing Both Ratios in Lowest Terms

Are the following proportions true?

(a) $\frac{5}{9} = \frac{18}{27}$

Write each ratio in lowest terms.

$\frac{5}{9}$ ← Already in lowest terms $\qquad \frac{18 \div 9}{27 \div 9} = \frac{2}{3}$ ← Lowest terms

Because $\frac{5}{9}$ is *not* equivalent to $\frac{2}{3}$, the proportion is *false.*

(b) $\frac{16}{12} = \frac{28}{21}$

Write each ratio in lowest terms.

$$\frac{16 \div 4}{12 \div 4} = \frac{4}{3} \quad \text{and} \quad \frac{28 \div 7}{21 \div 7} = \frac{4}{3}$$

Both ratios are equivalent to $\frac{4}{3}$, so the proportion is *true.*

Work Problem 2 at the Side.

OBJECTIVES

1. Write proportions.
2. Determine whether proportions are true or false.
3. Find cross products.

1 Write each proportion.

(a) \$7 is to 3 cans as \$28 is to 12 cans

(b) 9 meters is to 16 meters as 18 meters is to 32 meters

(c) 5 is to 7 as 35 is to 49

(d) 10 is to 30 as 60 is to 180

2 Determine whether each proportion is true or false by writing both ratios in lowest terms.

(a) $\frac{6}{12} = \frac{15}{30}$

(b) $\frac{20}{24} = \frac{3}{4}$

(c) $\frac{25}{40} = \frac{30}{48}$

(d) $\frac{35}{45} = \frac{12}{18}$

(e) $\frac{21}{45} = \frac{56}{120}$

ANSWERS

1. (a) $\frac{\$7}{3 \text{ cans}} = \frac{\$28}{12 \text{ cans}}$ **(b)** $\frac{9}{16} = \frac{18}{32}$
(c) $\frac{5}{7} = \frac{35}{49}$ **(d)** $\frac{10}{30} = \frac{60}{180}$

2. (a) true **(b)** false **(c)** true **(d)** false **(e)** true

3 **Find cross products.** Another way to test whether the ratios in a proportion are equivalent is to compare *cross products.*

Using Cross Products to Determine Whether a Proportion Is True

To see whether a proportion is true, first multiply along one diagonal, then multiply along the other diagonal, as shown here.

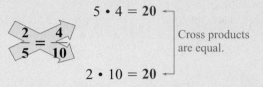

$$5 \cdot 4 = \mathbf{20}$$

$$\frac{2}{5} = \frac{4}{10}$$

$$2 \cdot 10 = \mathbf{20}$$

Cross products are equal.

In this case the **cross products** are both 20. When cross products are *equal,* the proportion is *true.* If the cross products are *unequal,* the proportion is *false.*

The cross products test is based on rewriting both fractions with the common denominator of $5 \cdot 10$, or 50.

$$\frac{2 \cdot \mathbf{10}}{5 \cdot \mathbf{10}} = \frac{20}{50} \quad \text{and} \quad \frac{4 \cdot \mathbf{5}}{10 \cdot \mathbf{5}} = \frac{20}{50}$$

We see that $\frac{2}{5}$ and $\frac{4}{10}$ are equivalent because both can be rewritten as $\frac{20}{50}$. The cross product test takes a shortcut by comparing only the two numerators ($20 = 20$).

Example 3 **Using Cross Products**

Use cross products to see whether each proportion is true or false.

(a) $\frac{3}{5} = \frac{12}{20}$

Multiply along one diagonal and then along the other diagonal.

$$5 \cdot 12 = \mathbf{60}$$

$$\frac{3}{5} = \frac{12}{20}$$

$$3 \cdot 20 = \mathbf{60}$$

Equal

The cross products are *equal,* so the proportion is *true.*

Continued on Next Page

(b) $\dfrac{2\frac{1}{3}}{3\frac{1}{3}} = \dfrac{9}{16}$

Cross multiply.

Changed to improper fractions

$$3\frac{1}{3} \cdot 9 = \frac{10}{\cancel{3}_1} \cdot \frac{\cancel{9}^3}{1} = \frac{30}{1} = \mathbf{30}$$

$$\frac{2\frac{1}{3}}{3\frac{1}{3}} = \frac{9}{16} \qquad \text{Unequal}$$

$$2\frac{1}{3} \cdot 16 = \frac{7}{3} \cdot \frac{16}{1} = \frac{112}{3} = \mathbf{37\frac{1}{3}}$$

The cross products are *unequal,* so the proportion is *false.*

NOTE

The numbers in a proportion do *not* have to be whole numbers. They can be fractions, mixed numbers, decimal numbers, and so on.

Work Problem 3 at the Side.

3 Cross multiply to see whether each proportion is true or false.

(a) $\dfrac{5}{9} = \dfrac{10}{18}$

(b) $\dfrac{32}{15} = \dfrac{16}{8}$

(c) $\dfrac{10}{17} = \dfrac{20}{34}$

(d) $\dfrac{\mathbf{2.4}}{\mathbf{6}} = \dfrac{\mathbf{5}}{\mathbf{12}}$ $\quad 6 \cdot 5 =$ $\quad 2.4 \cdot 12 =$

(e) $\dfrac{3}{4.25} = \dfrac{24}{34}$

(f) $\dfrac{1\frac{1}{6}}{2\frac{1}{3}} = \dfrac{4}{8}$

ANSWERS

3. **(a)** true **(b)** false **(c)** true
(d) $6 \cdot 5 = 30$; $2.4 \cdot 12 = 28.8$; false
(e) true **(f)** true

Solve a proportion for an unknown number by using the following steps.

Finding an Unknown Number in a Proportion

Step 1 Find the cross products.

Step 2 Show that the cross products are equivalent.

Step 3 Divide both products by the number multiplied by x (the number next to x).

Step 4 Check by writing the solution in the proportion and finding the cross products.

Example 1 **Solving for Unknown Numbers**

Find the unknown number in each proportion. Round to hundredths, if necessary.

(a) $\frac{16}{x} = \frac{32}{20}$

Recall that ratios can be rewritten in lowest terms. If desired, you can do that before finding the cross products. In this example, write $\frac{32}{20}$ in lowest terms $\left(\frac{8}{5}\right)$ to get $\frac{16}{x} = \frac{8}{5}$.

Step 1 $\frac{\mathbf{16}}{\boldsymbol{x}} = \frac{\mathbf{8}}{\mathbf{5}}$ — $x \cdot 8$ and $16 \cdot 5$ — Find the cross products.

Step 2 $x \cdot 8 = 16 \cdot 5$ ← Show that cross products are equivalent.

$x \cdot 8 = 80$

Step 3 $\frac{x \cdot \overset{1}{\cancel{8}}}{\underset{1}{\cancel{8}}} = \frac{80}{8}$ ← Divide each side by 8.

$x = 10$ ← Find x. (No rounding necessary.)

Step 4 Write the solution in the proportion and check by finding cross products.

x is 10. → $\frac{\mathbf{16}}{\mathbf{10}} = \frac{\mathbf{8}}{\mathbf{5}}$ — $10 \cdot 8 = \mathbf{80}$ and $16 \cdot 5 = \mathbf{80}$ — Equal; proportion is true.

The cross products are equal, so 10 is the correct solution.

NOTE

It is not necessary to write the ratios in lowest terms before solving. However, if you do, you will have smaller numbers to work with.

Continued on Next Page

(b) $\frac{7}{12} = \frac{15}{x}$

$$\frac{7}{12} = \frac{15}{x} \quad \begin{matrix} 12 \cdot 15 = 180 \\ 7 \cdot x \end{matrix}$$ Find the cross products.

Show that cross products are equivalent.

$$7 \cdot x = 180$$

Divide each side by 7.

$$\frac{\overset{1}{\cancel{7}} \cdot x}{\underset{1}{\cancel{7}}} = \frac{180}{7}$$

$$x \approx 25.71 \leftarrow \text{Rounded to nearest hundredth}$$

When the division does not come out even, check for directions on how to round your answer. Divide out one more place, then round.

$$7\overline{)180.000} = 25.714$$ ← Divide out to thousandths. Round to hundredths.

Write the solution in the proportion and check by finding the cross products.

$$\frac{7}{12} = \frac{15}{25.71} \quad \begin{matrix} 12 \cdot 15 = \mathbf{180} \\ 7 \cdot 25.71 = \mathbf{179.97} \end{matrix}$$ Very close, but not equal

The cross products are slightly different because you rounded the value of x. However, they are close enough to see that the problem was done correctly and 25.71 is the solution.

Work Problem 1 at the Side.

2 Find the unknown number in a proportion with mixed numbers or decimals. The following examples show how to work with mixed numbers or decimals in a proportion.

Example 2 Solving Proportions with Mixed Numbers and Decimals

Find the unknown number in each proportion.

(a) $\frac{2\frac{1}{5}}{6} = \frac{x}{10}$

$$\frac{2\frac{1}{5}}{6} = \frac{x}{10} \quad \begin{matrix} 6 \cdot x \\ 2\frac{1}{5} \cdot 10 \end{matrix}$$ Find the cross products.

Find $2\frac{1}{5} \cdot 10$.

$$2\frac{1}{5} \cdot 10 = \frac{11}{5} \cdot \frac{10}{1} = \frac{11}{\underset{1}{\cancel{5}}} \cdot \frac{\overset{2}{\cancel{10}}}{1} = \frac{22}{1} = 22$$

Changed to improper fraction

Continued on Next Page

1 Find the unknown numbers. Round to hundredths, if necessary. Check your answers by finding the cross products.

(a) $\frac{1}{2} = \frac{x}{12}$

(b) $\frac{6}{10} = \frac{15}{x}$

(c) $\frac{28}{x} = \frac{21}{9}$

(d) $\frac{x}{8} = \frac{3}{5}$

(e) $\frac{14}{11} = \frac{x}{3}$

Answers

1. (a) $x = 6$ **(b)** $x = 25$
(c) $x = 12$ **(d)** $x = 4.8$
(e) $x \approx 3.82$ (rounded to nearest hundredth)

2 Find the unknown numbers. Round to hundredths on the decimal problems, if necessary. Check your answers by finding the cross products.

(a) $\dfrac{3\frac{1}{4}}{2} = \dfrac{x}{8}$

(b) $\dfrac{x}{3} = \dfrac{1\frac{2}{3}}{5}$

(c) $\dfrac{0.06}{x} = \dfrac{0.3}{0.4}$

(d) $\dfrac{2.2}{5} = \dfrac{13}{x}$

(e) $\dfrac{x}{6} = \dfrac{0.5}{1.2}$

(f) $\dfrac{0}{2} = \dfrac{x}{7.092}$

Show that the cross products are equivalent.

$$6 \cdot x = 22$$

Divide each side by 6.

$$\frac{\overset{1}{\not{6}} \cdot x}{\underset{1}{\not{6}}} = \frac{22}{6}$$

Write the answer as a mixed number in lowest terms.

$$x = \frac{22 \div 2}{6 \div 2} = \frac{11}{3} = 3\frac{2}{3}$$

Write the solution in the proportion and check by finding the cross products.

$$\frac{2\frac{1}{5}}{6} = \frac{3\frac{2}{3}}{10}$$

$$6 \cdot 3\frac{2}{3} = \frac{\overset{2}{\not{6}}}{1} \cdot \frac{11}{\underset{1}{\not{3}}} = \frac{22}{1} = \mathbf{22}$$

$$2\frac{1}{5} \cdot 10 = \frac{11}{\underset{1}{\not{5}}} \cdot \frac{\overset{2}{\not{10}}}{1} = \frac{22}{1} = \mathbf{22}$$

Equal

The cross products are equal, so $3\frac{2}{3}$ is the correct solution.

(b) $\dfrac{1.5}{0.6} = \dfrac{2}{x}$

Show that cross products are equivalent.

$$1.5 \cdot x = \underbrace{0.6 \cdot 2}$$

$$1.5 \cdot x = 1.2$$

Divide each side by 1.5.

$$\frac{\overset{1}{\not{1.5}} \cdot x}{\underset{1}{\not{1.5}}} = \frac{1.2}{1.5}$$

$$x = \frac{1.2}{1.5}$$

Complete the division.

$$1.5\overline{)1.20} = .8$$

So the unknown number is 0.8. Check by finding the cross products.

$$\frac{1.5}{0.6} = \frac{2}{0.8}$$

$$0.6 \cdot 2 = \mathbf{1.2}$$

$$1.5 \cdot 0.8 = \mathbf{1.2}$$

Equal

The cross products are equal, so 0.8 is the correct solution.

Work Problem 2 at the Side.

ANSWERS

2. (a) $x = 13$ **(b)** $x = 1$ **(c)** $x = 0.08$
(d) $x \approx 29.55$
(rounded to nearest hundredth)
(e) $x = 2.5$ **(f)** $x = 0$

5.4 Exercises

FOR EXTRA HELP

Student's Solutions Manual | MyMathLab.com | InterAct Math Tutorial Software | AW Math Tutor Center | www.mathxl.com | Digital Video Tutor CD 3 Videotape 9

Find the unknown number in each proportion. Round your answers to hundredths, if necessary. Check your answers by finding the cross products. See Examples 1 and 2.

1. $\frac{1}{3} = \frac{x}{12}$

2. $\frac{x}{6} = \frac{15}{18}$

3. $\frac{15}{10} = \frac{3}{x}$

4. $\frac{5}{x} = \frac{20}{8}$

5. $\frac{x}{11} = \frac{32}{4}$

6. $\frac{12}{9} = \frac{8}{x}$

7. $\frac{42}{x} = \frac{18}{39}$

8. $\frac{49}{x} = \frac{14}{18}$

9. $\frac{x}{25} = \frac{4}{20}$

10. $\frac{6}{x} = \frac{4}{8}$

11. $\frac{8}{x} = \frac{24}{30}$

12. $\frac{32}{5} = \frac{x}{10}$

13. $\frac{99}{55} = \frac{44}{x}$

14. $\frac{x}{12} = \frac{101}{147}$

15. $\frac{0.7}{9.8} = \frac{3.6}{x}$

16. $\frac{x}{3.6} = \frac{4.5}{6}$

17. $\frac{250}{24.8} = \frac{x}{1.75}$

18. $\frac{4.75}{17} = \frac{43}{x}$

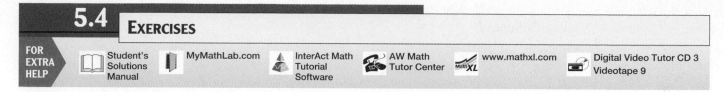

Find the unknown number in each proportion. Write your answers as whole or mixed numbers when possible. See Example 2.

19. $\dfrac{15}{1\frac{2}{3}} = \dfrac{9}{x}$

20. $\dfrac{x}{\frac{3}{10}} = \dfrac{2\frac{2}{9}}{1}$

21. $\dfrac{2\frac{1}{3}}{1\frac{1}{2}} = \dfrac{x}{2\frac{1}{4}}$

22. $\dfrac{1\frac{5}{6}}{x} = \dfrac{\frac{3}{14}}{\frac{6}{7}}$

Solve each proportion two different ways. First change all the numbers to decimal form and solve. Then change all the numbers to fraction form and solve; write your answers in lowest terms.

23. $\dfrac{\frac{1}{2}}{x} = \dfrac{2}{0.8}$

24. $\dfrac{\frac{3}{20}}{0.1} = \dfrac{0.03}{x}$

25. $\dfrac{x}{\frac{3}{50}} = \dfrac{0.15}{1\frac{4}{5}}$

26. $\dfrac{8\frac{4}{5}}{1\frac{1}{10}} = \dfrac{x}{0.4}$

RELATING CONCEPTS (Exercises 27–28) FOR INDIVIDUAL OR GROUP WORK

Work Exercises 27–28 in order. *First prove that the proportions are* **not** *true. Then create four true proportions for each exercise by changing one number at a time.*

27. $\dfrac{10}{4} = \dfrac{5}{3}$

28. $\dfrac{6}{8} = \dfrac{24}{30}$

5.5 Solving Application Problems with Proportions

Objective

1 Use proportions to solve application problems.

1 Use proportions to solve application problems. Proportions can be used to solve a wide variety of problems. Watch for problems in which you are given a ratio or rate and then asked to find part of a corresponding ratio or rate. Remember that a ratio or rate compares two quantities and often includes one of the following indicator words.

in **for** **on** **per** **from** **to**

Use the six problem-solving steps you learned in **Section 1.10.**

Step 1 **Read** the problem.

Step 2 **Work out a plan.**

Step 3 **Estimate** a reasonable answer.

Step 4 **Solve** the problem.

Step 5 **State the answer.**

Step 6 **Check** your work.

Example 1 Solving a Proportion Application

Mike's car can travel 163 **miles** **on** 6.4 **gallons** of gas. How far can it travel on a full tank of 14 **gallons** of gas? Round to the nearest whole mile.

Step 1 **Read** the problem. The problem asks for the number of miles the car can travel on 14 gallons of gas.

Step 2 **Work out a plan.** Decide what is being compared. This example compares **miles** to **gallons**. Write a proportion using the two rates. Be sure that *both* rates compare miles to gallons in the same order. In other words, miles is in both numerators and gallons is in both denominators. Use a letter to represent the unknown number.

Step 3 **Estimate** a reasonable answer. To estimate the answer, notice that 14 gallons is a little more than *twice as much* as 6.4 gallons, so the car should travel a little more than *twice as far*; 2 • 163 miles = 326 miles.

Step 4 **Solve** the problem. With the proportion set up correctly, solve for the unknown number.

Matching units

This rate compares **miles** to **gallons**. $\rightarrow$ $$\frac{163 \text{ miles}}{6.4 \text{ gallons}} = \frac{x \text{ miles}}{14 \text{ gallons}}$$ $\leftarrow$ This rate compares **miles** to **gallons**.

Matching units

Ignore the units while finding the cross products and dividing each side by 6.4.

$$6.4 \cdot x = \underbrace{163 \cdot 14}$$ Show that cross products are equivalent.

$$6.4 \cdot x = 2282$$

$$\frac{\overset{1}{\cancel{6.4}} \cdot x}{\underset{1}{\cancel{6.4}}} = \frac{2282}{6.4}$$ Divide each side by 6.4.

$$x = 356.5625$$

Continued on Next Page

1 Set up and solve a proportion for each problem.

(a) If 2 pounds of fertilizer will cover 50 square feet of garden, how many pounds are needed for 225 square feet?

(b) A U.S. map has a scale of 1 inch to 75 miles. Lake Superior is 4.75 inches long on the map. What is the lake's actual length in miles?

(c) Cough syrup is to be given at the rate of 30 milliliters for each 100 pounds of body weight. How much should be given to a 34-pound child? Round to the nearest whole milliliter.

ANSWERS

1. (a) $\frac{2 \text{ pounds}}{50 \text{ square feet}} = \frac{x \text{ pounds}}{225 \text{ square feet}}$
$x = 9$ pounds

(b) $\frac{1 \text{ inch}}{75 \text{ miles}} = \frac{4.75 \text{ inches}}{x \text{ miles}}$
$x = 356.25$ miles or $x \approx 356$ miles

(c) $\frac{30 \text{ milliliters}}{100 \text{ pounds}} = \frac{x \text{ milliliters}}{34 \text{ pounds}}$
$x \approx 10$ milliliters

Step 5 **State the answer.** Rounded to the nearest mile, the car can travel 357 miles on a full tank of gas.

Step 6 **Check** your work. The answer, 357 miles, is a little more than the estimate of 326 miles, so it is reasonable.

Work Problem 1 at the Side.

Example 2 Solving a Proportion Application

A newspaper report says that 7 out of 10 people surveyed watch the news on TV. At that rate, how many of the 3200 people in town would you expect to watch the news?

Step 1 **Read** the problem. The problem asks how many of the 3200 people in town would be expected to watch TV news.

Step 2 **Work out a plan.** You are comparing people who watch the news to people surveyed. Set up a proportion using the two rates described in the example. Be sure that both rates make the same comparison. "People who watch the news" is mentioned first, so it should be in the numerator of *both* rates.

Step 3 **Estimate** a reasonable answer. To estimate the answer, notice that 7 out of 10 people is more than half the people, but less than all the people. Half of 3200 people is $3200 \div 2 = 1600$, so our estimate is between 1600 and 3200 people.

Step 4 **Solve** the problem. Solve for the unknown number in the proportion.

People who watch news → 7 ; x ← People who watch news
Total group (people surveyed) → 10 ; 3200 ← Total group (people in town)

$$\frac{7}{10} = \frac{x}{3200}$$

$$10 \cdot x = 7 \cdot 3200 \quad \text{Show that cross products are equivalent.}$$

$$10 \cdot x = 22{,}400$$

$$\frac{\overset{1}{\cancel{10}} \cdot x}{\underset{1}{\cancel{10}}} = \frac{22{,}400}{10} \quad \text{Divide each side by 10.}$$

$$x = 2240$$

Step 5 **State the answer.** You would expect 2240 people in town to watch the news on TV.

Step 6 **Check** your work. The answer, 2240 people, is between 1600 and 3200, as called for in the estimate.

Continued on Next Page

CAUTION

Always check that your answer is reasonable. If it is not, look at the way your proportion is set up. Be sure you have matching units in the numerators and matching units in the denominators.

For example, suppose you had set up the last proportion *incorrectly* as shown here.

$$\frac{7}{10} = \frac{3200}{x} \leftarrow \text{Incorrect setup}$$

$$7 \cdot x = 10 \cdot 3200$$

$$\frac{\overset{1}{\cancel{7}} \cdot x}{\underset{1}{\cancel{7}}} = \frac{32{,}000}{7}$$

$$x \approx 4571 \text{ people} \leftarrow \text{Unreasonable answer}$$

This answer is *unreasonable* because there are only 3200 people in the town; it is *not* possible for 4571 people to watch the news.

Work Problem 2 at the Side.

2 Solve each problem to find a reasonable answer. Then flip one side of your proportion to see what answer you get with an *incorrect* setup. Explain why the second answer is *unreasonable.*

(a) A survey showed that 2 out of 3 people would like to lose weight. At this rate, how many people in a group of 150 want to lose weight?

(b) In one state, 3 out of 5 college students receive financial aid. At this rate, how many of the 4500 students at Central Community College receive financial aid?

(c) An advertisement says that 9 out of 10 dentists recommend sugarless gum. If the ad is true, how many of the 60 dentists in our city would recommend sugarless gum?

ANSWERS

2. (a) 100 people (reasonable); incorrect setup gives 225 people (only 150 people in the group).

(b) 2700 students (reasonable); incorrect setup gives 7500 students (only 4500 students at the college).

(c) 54 dentists (reasonable); incorrect setup gives about 67 dentists (only 60 dentists in the city).

Focus on Real-Data Applications

Feeding Hummingbirds

A recipe can be used to make as much of a mixture as you might need as long as the ingredients are kept proportional. Use the recipe for a homemade mixture of sugar water for hummingbird feeders to answer these problems.

1. What is the ratio of sugar to water in the recipe?

What is the ratio of water to sugar in the recipe?

2. Complete each table.

Sugar	Water
1 cup	4 cups
	5 cups
	6 cups
	7 cups
2 cups	8 cups

Sugar	Water
1 cup	4 cups
	3 cups
	2 cups
	1 cup

3. How much water would you need

(a) if you wanted to use 3 cups of sugar?

(b) if you wanted to use 4 cups of sugar?

(c) if you wanted to use $\frac{1}{3}$ cup of sugar?

4. One cup of sugar weighs about 200 grams and one cup of water weighs about 235 grams. If you mix 1 cup of sugar and 4 cups of water, what is the approximate weight of the resulting mixture?

5. The article says that the nectar from wildflowers visited by hummingbirds has an average sugar concentration of 21 percent. That represents a ratio of 21:100. If you wanted to mix a sugar solution in the same proportion as the wildflower nectar, how much water should you use for 1 cup of sugar? What proportion did you set up to solve this problem?

Feeding Hummingbirds

After getting a hummingbird feeder, the next step is to fill it! You have two choices at this point: you can either buy one of the commercial mixtures or you can make your own solution:

> **Recipe for Homemade Mixture:**
> **1 part sugar (not honey)**
> **4 parts water**
> **Boil for 1 to 2 minutes. Cool.**
> **Store extra in refrigerator.**

The concentration of the sugar is important. A 1 to 4 ratio of sugar to water is recommended because it approximates the ratio of sugar to water found in the nectar of many hummingbird flowers. A recent study of 21 native California wildflowers visited by hummingbirds showed that their nectar had an average sugar concentration of 21 percent. This is sweet enough to attract the hummers without being too sweet. If you increase the concentration of sugar, it may be harder for the birds to digest; if you decrease the concentration, they may lose interest.

Boiling the solution helps retard fermentation. Sugar-and-water solutions are subject to rapid spoiling, especially in hot weather.

Source: The Hummingbird Book.

6. As you change the amounts of water and sugar, should you change the length of time that you boil the mixture? Explain your answer.

7. Will the length of time it takes to get the water hot enough to start boiling change? Explain your answer.

5.5 Exercises

FOR EXTRA HELP

Student's Solutions Manual | MyMathLab.com | InterAct Math Tutorial Software | AW Math Tutor Center | www.mathxl.com | Digital Video Tutor CD 3 Videotape 9

Set up and solve a proportion for each application problem. See Example 1.

1. Caroline can sketch four cartoon strips in five hours. How long will it take her to sketch 18 strips?

2. The Cosmic Toads recorded eight songs on their first CD in 26 hours. How long will it take them to record 14 songs for their second CD?

3. Sixty newspapers cost \$27. Find the cost of 16 newspapers.

4. Twenty-two guitar lessons cost \$396. Find the cost of 12 lessons.

5. If three pounds of fescue grass seed cover about 350 square feet of ground, how many pounds are needed for 4900 square feet?

6. Anna earns \$1242.08 in 14 days. How much does she earn in 260 days?

7. Tom makes \$455.75 in 5 days. How much does he make in 3 days?

8. If 5 ounces of a medicine must be mixed with 8 ounces of water, how many ounces of medicine would be mixed with 20 ounces of water?

9. The bag of rice noodles shown below makes 7 servings. (*Source:* Everfresh Foods.) At that rate, how many ounces of noodles do you need for 12 servings, to the nearest ounce?

10. This can of sweet potatoes is enough for 4 servings. (*Source:* Moody Dunbar, Inc.) How many ounces are needed for 9 servings, to the nearest ounce?

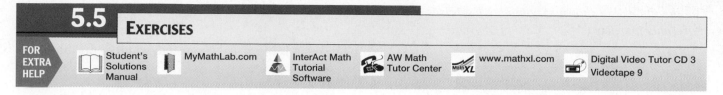

11. Three quarts of a latex enamel paint will cover about 270 square feet of wall surface. (*Source:* Thompson and Formby, Inc.) How many quarts will you need to cover 350 square feet of wall surface in your kitchen and 100 square feet of wall surface in your bathroom?

12. One gallon of clear gloss wood finish covers about 550 square feet of surface. (*Source:* The Flecto Company.) If you need to apply three coats of finish to 400 square feet of surface, how many gallons do you need, to the nearest tenth?

Use the floor plan shown to complete Exercises 13–16. On the plan, one inch represents four feet.

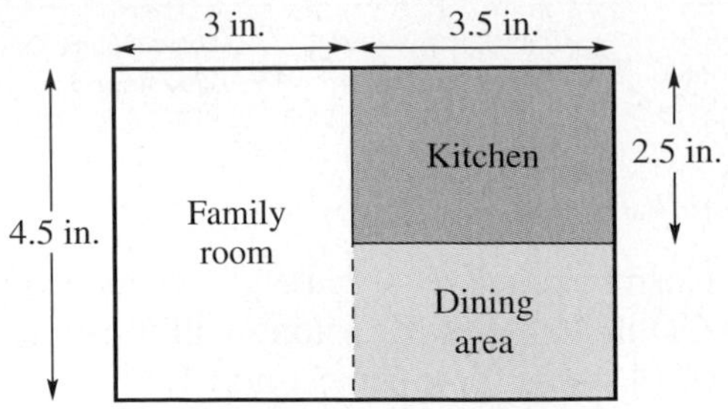

13. What is the actual length and width of the kitchen?

14. What is the actual length and width of the family room?

15. What is the actual length and width of the dining area?

16. What is the actual length and width of the entire floor plan?

17. Vince Carter scored 18 points during the 28 minutes that he played in the Olympic basketball game between the United States and Lithuania. (*Source:* Associated Press.) At that rate, how many points would he make if he played the entire game (40 minutes), to the nearest whole point?

18. In the U.S.–France gold medal Olympic basketball game, Alonzo Mourning played 26 minutes and made 7 rebounds. (*Source:* Associated Press.) How many rebounds would you expect him to make in 40 minutes (a complete Olympic game), to the nearest whole number?

Set up a proportion to solve each problem. Check to see whether your answer is reasonable. Then flip one side of your proportion to see what answer you get with an incorrect setup. Explain why the second answer is unreasonable. See Example 2.

19. About 7 out of 10 people entering a community college need to take a refresher math course. If there are 2950 entering students, how many will probably need refresher math? (*Source:* Minneapolis Community and Technical College.)

20. In a survey, only 3 out of 100 people like their eggs poached. At that rate, how many of the 60 customers who ordered eggs at Soon-Won's restaurant this morning asked to have them poached? Round to the nearest whole person.

21. Nearly 4 out of 5 people choose vanilla as their favorite ice cream flavor. (*Source:* Baskin-Robbins.) If 238 people attend an ice cream social, how many would you expect to choose vanilla? Round to the nearest whole person.

22. In a test of 200 sewing machines, only one had a defect. At that rate, how many of the 5600 machines shipped from the factory have defects?

23. About 9 out of 10 U.S. households have TV remote controls, according to a survey. There were 102,500,000 U.S. households in 1998. If the survey is accurate, how many U.S. households have TV remote controls? (*Source:* Magnavox, U.S. Bureau of the Census.)

24. In a survey, 3 out of 100 dog owners washed their pets by having the dogs go into the shower with them. If the survey is accurate, how many of the 31,200,000 dog owners in the United States use this method? (*Source:* Teledyne Water Pik, American Veterinary Medical Association.)

Set up and solve a proportion for each problem.

25. The stock market report says that 5 stocks went up for every 6 stocks that went down. If 750 stocks went down yesterday, how many went up?

26. The human body contains 90 pounds of water for every 100 pounds of body weight. How many pounds of water are in a child who weighs 80 pounds?

27. The ratio of the length of an airplane wing to its width is 8 to 1. If the length of a wing is 32.5 meters, how wide must it be? Round to the nearest hundredth.

28. The Rosebud School District wants a student-to-teacher ratio of 19 to 1. How many teachers are needed for 1850 students? Round to the nearest whole number.

29. The number of calories you burn depends on your weight. A 150-pound person burns 222 calories during 30 minutes of tennis. How many calories would a 210-pound person burn, to the nearest whole number? (*Source: Wellness Encyclopedia.*)

30. (Complete Exercise 29 first.) A 150-pound person burns 189 calories during 45 minutes of grocery shopping. How many calories would a 115-pound person burn, to the nearest whole number? (*Source: Wellness Encyclopedia.*)

31. At 3 P.M., Coretta's shadow is 1.05 meters long. Her height is 1.68 meters. At the same time, a tree's shadow is 6.58 meters long. How tall is the tree? Round to the nearest hundredth.

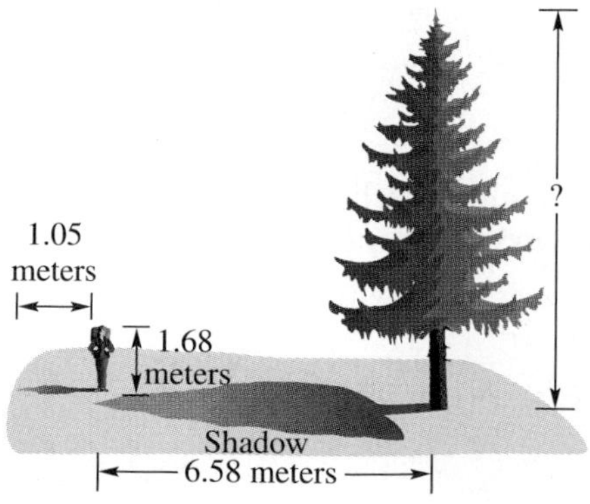

32. Refer to Exercise 31. Later in the day, Coretta's shadow was 2.95 meters long. How long a shadow did the tree have at that time? Round to the nearest hundredth.

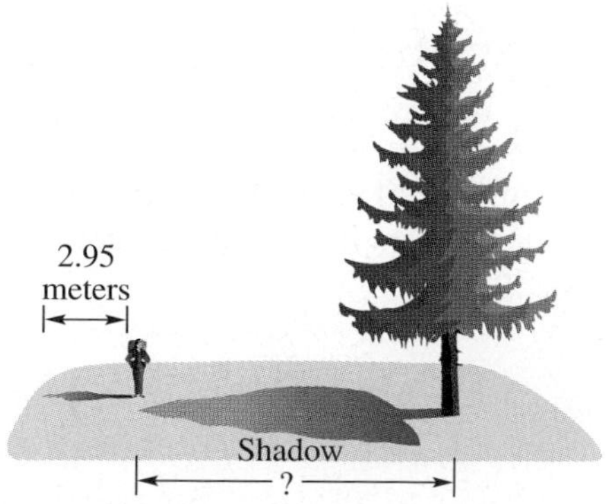

33. Can you set up a proportion to solve this problem? Explain why or why not. Jim is 25 years old and weighs 180 pounds. How much will he weigh when he is 50 years old?

34. Write your own application problem that can be solved by setting up a proportion. Also show the proportion and the steps needed to solve your problem.

35. A survey of college students shows that 4 out of 5 drink coffee. Of the students who drink coffee, 1 out of 8 adds cream to it. How many of the 38,000 students at the University of Minnesota would be expected to use cream in their coffee?

36. About 9 out of 10 adults think it is a good idea to exercise regularly. But of the ones who think it is a good idea, only 1 in 6 actually exercise at least three times a week. At this rate, how many of the 300 employees in our company exercise regularly?

37. The nutrition information on a bran cereal box says that a $\frac{1}{3}$-cup serving provides 80 calories and 8 grams of dietary fiber. (*Source:* Kraft Foods, Inc.) At that rate, how many calories and grams of fiber are in a $\frac{1}{2}$-cup serving?

38. A $\frac{2}{3}$-cup serving of penne pasta has 210 calories and 2 grams of dietary fiber. (*Source:* Borden Foods.) How many calories and grams of fiber would be in a 1-cup serving?

RELATING CONCEPTS (Exercises 39–42) FOR INDIVIDUAL OR GROUP WORK

A box of instant mashed potatoes has the list of ingredients shown in the table. Use this information to ***work Exercises 39–42 in order.***

Ingredient	For 12 Servings
Water	$3\frac{1}{2}$ cups
Margarine	6 Tbsp
Milk	$1\frac{1}{2}$ cups
Potato flakes	4 cups

Source: General Mills.

39. Find the amount of each ingredient for 6 servings. Show *two* different methods for finding the amounts. One method should use proportions.

40. Find the amount of each ingredient for 18 servings. Show *two* different methods for finding the amounts, one using proportions and one using your answers from Exercise 39.

41. Find the amount of each ingredient for 3 servings, using your answers from either Exercise 39 or Exercise 40.

42. Find the amount of each ingredient for 9 servings, using your answers from either Exercise 40 or Exercise 41.

Chapter 5

SUMMARY

KEY TERMS

5.1	**ratio**	A ratio compares two quantities having the same type of units. For example, the ratio of 6 apples to 11 apples is written in fraction form as $\frac{6}{11}$. The common units (apples) divide out.
5.2	**rate**	A rate compares two measurements with different types of units. Examples are 96 dollars for 8 hours, or 450 miles on 18 gallons.
	unit rate	A unit rate has 1 in the denominator.
	cost per unit	Cost per unit is a rate that tells how much you pay for one item or one unit. The lowest cost per unit is the best buy.
5.3	**proportion**	A proportion states that two ratios or rates are equivalent.
	cross products	Cross multiply to get the cross products of a proportion. If the cross products are equal, the proportion is true.

TEST YOUR WORD POWER

See how well you have learned the vocabulary in this chapter. Answers follow the Quick Review.

1. A **ratio**
   (a) can be written only as a fraction
   (b) compares two quantities that have the same type of units
   (c) compares two quantities that have different types of units
   (d) is the reciprocal of a rate.

2. A **rate**
   (a) can be written only as a decimal
   (b) compares two quantities that have the same type of units
   (c) compares two quantities that have different types of units
   (d) is the reciprocal of a ratio.

3. A **unit rate**
   (a) has a numerator of 1
   (b) has a denominator of 1
   (c) is found by cross multiplying
   (d) is usually written in fraction form.

4. **Cost per unit** is
   (a) the best buy
   (b) a ratio written in lowest terms
   (c) found by comparing cross products
   (d) the price of one item or one unit.

5. A **proportion**
   (a) shows that two ratios or rates are equivalent
   (b) contains only whole numbers or decimals
   (c) always has one unknown number
   (d) states that two improper fractions are equivalent.

6. **Cross products** are
   (a) used only with ratios, not with rates
   (b) equal when a proportion is false
   (c) used to find the best buy
   (d) equal when a proportion is true.

QUICK REVIEW

Concepts	*Examples*
5.1 Writing a Ratio A ratio compares two quantities. A ratio is usually written as a fraction with the number that is mentioned first in the numerator. The common units divide out and are not written in the answer. Check that the fraction is in lowest terms.	Write this ratio as a fraction in lowest terms. 60 ounces of medicine **to** 160 ounces of medicine $\frac{60 \text{ ounces}}{160 \text{ ounces}} = \frac{60 \div \mathbf{20}}{160 \div \mathbf{20}} = \frac{3}{8}$ ←Lowest terms ↑ Divide out common units.

Concepts	*Examples*
5.1 Using Mixed Numbers in a Ratio If a ratio has mixed numbers, change the mixed numbers to improper fractions. Rewrite the problem in horizontal format using the "÷" symbol for division. Finally, multiply by the reciprocal of the divisor.	Write as a ratio of whole numbers in lowest terms. $2\frac{1}{2}$ to $3\frac{3}{4}$ $\dfrac{2\frac{1}{2}}{3\frac{3}{4}} = \dfrac{\frac{5}{2}}{\frac{15}{4}}$ Reciprocal $= \frac{5}{2} \div \frac{15}{4} = \frac{5}{2} \bullet \frac{4}{15}$ $= \frac{\cancel{5}^{1}}{\cancel{2}_{1}} \bullet \frac{\cancel{4}^{2}}{\cancel{15}_{3}} = \frac{2}{3}$ ← Ratio in lowest terms
5.1 Using Measurements in Ratios When a ratio compares measurements, both measurements must be in the *same* units. It is usually easier to compare the measurements using the smaller unit, for example, inches instead of feet.	Write as a ratio in lowest terms. 8 in. to 6 ft Compare using the smaller unit, inches. Because 1 ft has 12 in., 6 ft is 6 • **12 in.** = 72 in. The ratio is shown below. $\frac{8\,\cancel{\text{in.}}}{72\,\cancel{\text{in.}}} = \frac{8 \div 8}{72 \div 8} = \frac{1}{9}$ ↑ Divide out common units.
5.2 Writing Rates A rate compares two measurements with different types of units. The units do *not* divide out, so you must write them as part of the rate.	Write the rate as a fraction in lowest terms. 475 miles in 10 hours $\frac{475 \text{ miles} \div 5}{10 \text{ hours} \div 5} = \frac{95 \text{ miles}}{2 \text{ hours}}$ ← Must write units: miles and hours
5.2 Finding a Unit Rate A unit rate has 1 in the denominator. To find the unit rate, divide the numerator by the denominator. Write unit rates using the word **per** or a / mark.	Write as a unit rate: \$1278 in 9 days. $\frac{\$1278}{9 \text{ days}}$ ← The fraction bar indicates division. $9\overline{)1278}$ = 142 so $\frac{\$1278 \div 9}{9 \text{ days} \div 9} = \frac{\$142}{1 \text{ day}}$ Write the answer as \$142 **per** day or \$142/day.

Concepts	*Examples*
5.2 Finding the Best Buy The best buy is the item with the lowest cost per unit. Divide the price by the number of units. Round to thousandths, if necessary. Then compare to find the lowest cost per unit.	Find the best buy on cheese. You have a coupon for 50¢ off on 2 pounds or 75¢ off on 3 pounds. 2 pounds for \$2.75 3 pounds for \$4.15 Find the cost per unit (cost per pound) after subtracting the coupon. 2 pounds cost \$2.75 − \$0.50 = \$2.25. $\frac{\$2.25}{2} = \$1.125$ per pound 3 pounds cost \$4.15 − \$0.75 = \$3.40. $\frac{\$3.40}{3} \approx \$1.133$ per pound The lower cost per pound is \$1.125, so 2 pounds is the better buy.
5.3 Writing Proportions A proportion states that two ratios or rates are equivalent. The proportion "5 is to 6 as 25 is to 30" is written as shown below. $\frac{5}{6} = \frac{25}{30}$	Write as a proportion: 8 is to 40 as 32 is to 160. $\frac{8}{40} = \frac{32}{160}$
To see whether a proportion is true or false, cross multiply one way, then cross multiply the other way. If the two cross products are equal, the proportion is true. If the two cross products are unequal, the proportion is false.	Is this proportion true or false? $\frac{6}{8\frac{1}{2}} = \frac{24}{34}$ Cross multiply. $\frac{\mathbf{6}}{\mathbf{8\frac{1}{2}}} = \frac{\mathbf{24}}{\mathbf{34}}$ $8\frac{1}{2} \cdot 24 = \frac{17}{\cancel{2}_1} \cdot \frac{\cancel{24}^{12}}{1} = \mathbf{204}$ $6 \cdot 34 = \mathbf{204}$ ← Equal The cross products are equal, so the proportion is true.
5.4 Solving Proportions Solve for an unknown number in a proportion by using these steps.	Find the unknown number. $\frac{12}{x} = \frac{6}{8}$ $\frac{12}{x} = \frac{3}{4}$ ← Lowest terms
Step 1 Find the cross products. (If desired, you can rewrite the ratios in lowest terms before finding the cross products.)	$\frac{\mathbf{12}}{\mathbf{x}} = \frac{\mathbf{3}}{\mathbf{4}}$ $x \cdot 3$ $12 \cdot 4$ Find cross products.

(*continued*)

Concepts	*Examples*
Step 2 Show that the cross products are equivalent.	$x \cdot 3 = 12 \cdot 4$ Show that cross products are equivalent. $x \cdot 3 = 48$
Step 3 Divide both products by the number multiplied by x (the number next to x).	$\frac{x \cdot \cancel{3}}{\cancel{3}} = \frac{48}{3}$ Divide each side by 3. $x = 16$
Step 4 Check by writing the solution in the proportion and finding the cross products.	x is 16. → $\frac{12}{16} = \frac{6}{8}$ $16 \cdot 6 = 96$, $12 \cdot 8 = 96$ Equal The cross products are equal, so 16 is the correct solution.

5.5 Solving Application Problems with Proportions

Concepts	*Examples*
Decide what is being compared. Set up and solve a proportion using the two rates described in the problem. Be sure that *both* rates compare things in the *same order.* Use a letter, like x, to represent the unknown number. Use the six problem-solving steps.	If 3 pounds of grass seed cover 450 square feet of lawn, how much seed is needed for 1500 square feet of lawn?
Step 1 **Read** the problem carefully.	The problem asks for the pounds of grass seed needed for 1500 square feet of lawn.
Step 2 **Work out a plan.**	Pounds of seed is compared to square feet of lawn. Set up and solve a proportion using the two given rates. Be sure that pounds of seed is in both numerators and square feet of lawn is in both denominators.
Step 3 **Estimate** a reasonable answer.	Notice that 1500 square feet is about three times as much lawn as 450 square feet. So about three times as much seed will be needed, and $3 \cdot 3$ pounds $= 9$ pounds is a reasonable estimate.
Step 4 **Solve** the problem.	With the proportion set up correctly, solve for the unknown number. Matching units $\frac{3 \text{ pounds}}{450 \text{ square feet}} = \frac{x \text{ pounds}}{1500 \text{ square feet}}$ Matching units Both sides compare pounds to square feet. Ignore the units while finding cross products. $450 \cdot x = 3 \cdot 1500$ Show that cross products are equivalent. $450 \cdot x = 4500$ $\frac{\cancel{450} \cdot x}{\cancel{450}} = \frac{4500}{450}$ Divide each side by 450. $x = 10$
Step 5 **State the answer.**	10 pounds of grass seed are needed.
Step 6 **Check** your work.	The answer, 10 pounds of seed, is close to our estimate of 9 pounds, so it is reasonable.

Answers to Test Your Word Power

1. (b) *Example:* The ratio of 3 miles to 4 miles is $\frac{3}{4}$; the common units (miles) divide out. **2. (c)** *Example:* \$4.50 for 3 pounds is a rate comparing dollars to pounds. **3. (b)** *Example:* $\frac{\$1.79}{1 \text{ pound}}$ is a unit rate. We write it as \$1.79 per pound or \$1.79/pound. **4. (d)** *Example:* \$1.65 per gallon tells the price of one gallon (one unit). **5. (a)** *Example:* $\frac{5}{6} = \frac{25}{30}$ is a proportion because $\frac{5}{6}$ is equivalent to $\frac{25}{30}$. **6. (d)** *Example:* The cross products for $\frac{5}{6} = \frac{25}{30}$ are $6 \cdot 25 = 150$ and $5 \cdot 30 = 150$.

Focus on *Real-Data Applications*

Currency Exchange

When you travel between countries, you will exchange U.S. dollars for the local currency. The exchange rate between currencies changes daily, and you can easily find the updated rates using the Internet or any major newspaper. The table shown below has been extracted from the Bloomberg Currency Calculator Web page.

NORTH AMERICA/CARIBBEAN CURRENCY RATES

			Currency per 1 unit of USD	
Currency	**Symbol**	**Value**	**Net Chg**	**Pct Chg**
Canadian Dollar	CAD	1.4967	+0.0024	+0.1606
Cayman Islands	KYD	0.8282	—	—
Jamaica Dollar	JMD	45.1	+0.1	+0.2222
Mexican Peso	MXN	9.761	−0.014	−0.1432
United States Dollar	USD	1.00		

On February 4, 2001, the currency exchange rate from U.S. dollars to Mexican pesos was given as follows:

$1.00 U.S. was equivalent to 9.761 Mexican pesos

You can set up a proportion to convert dollars to pesos. For example, suppose you want to determine the number of pesos that is equivalent to $50.00.

$$\frac{\$1}{9.761 \text{ pesos}} = \frac{\$50}{x \text{ pesos}} \quad \text{or} \quad \frac{1}{9.761} = \frac{50}{x}$$

$$1 \cdot x = (9.761) \cdot 50$$

$$x = 488.05 \text{ pesos}$$

So $50 buys 488 pesos and 5 centavos.

1. Based on the currency exchange rates for February 4, 2001, find the amount of each local currency that is equivalent to $50 U.S. and find the number of U.S. dollars that is equivalent to 200 units of each local currency. Round your answers to the nearest hundredth.

   (a) $50 = _______ Canadian dollars, and 200 Canadian dollars = _______ U.S. dollars.

   (b) $50 = _______ Cayman Island dollars, and
   200 Cayman Island dollars = _______ U.S. dollars.

   (c) $50 = _______ Jamaican dollars, and 200 Jamaican dollars = _______ U.S. dollars.

2. Set up a proportion to find the number of U.S. dollars that was equivalent to 1 Mexican Peso. 1 Mexican peso was equivalent to $ _______(U.S.).

3. From Problem 2, you should recognize the conversion rate based on 1 Mexican peso as the expression $\frac{1}{9.761}$. What is the mathematical word that describes the relationship between the conversion rates 9.761 and $\frac{1}{9.761}$?

Chapter 5 Review Exercises

[5.1] *Write each ratio as a fraction in lowest terms. Change to the same units when necessary, using the table of measurement comparisons in* ***Section 5.1.*** *Use the information in the graph to answer Exercises 1–3.*

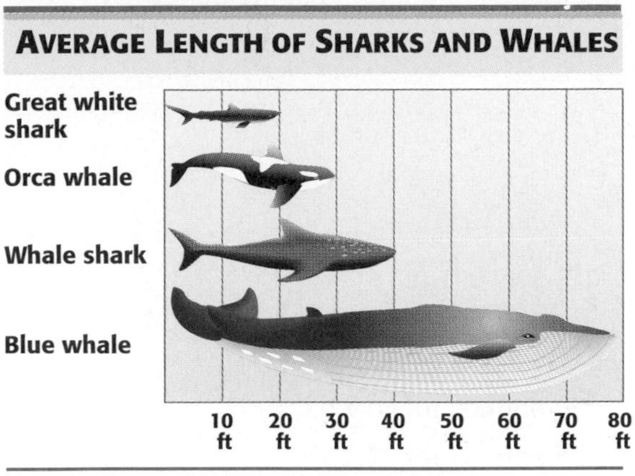

Source: *Grolier Multimedia Encyclopedia.*

1. Ratio of orca whale's length to whale shark's length

2. Ratio of blue whale's length to great white shark's length

3. Ratio of great white shark's length to whale shark's length

4. \$2.50 to \$1.25

5. \$0.30 to \$0.45

6. $1\frac{2}{3}$ cups to $\frac{2}{3}$ cup

7. $2\frac{3}{4}$ miles to $16\frac{1}{2}$ miles

8. 5 hours to 100 minutes

9. 9 in. to 2 ft

10. 1 ton to 1500 pounds

11. 8 hours to 3 days

12. Jake sold \$350 worth of his kachina figures. Ramona sold \$500 worth of her pottery. What is the ratio of her sales to his?

13. Ms. Wei's new car gets 35 miles per gallon. Her old car got 25 miles per gallon. Find the ratio of the new car's mileage to the old car's mileage.

14. This fall, 60 students are taking math and 72 students are taking English. Find the ratio of math students to English students.

[5.2] *Write each rate as a fraction in lowest terms.*

15. \$88 for 8 dozen

16. 96 children in 40 families

17. In his keyboarding class, Patrick can type four pages in 20 minutes. Give his rate in pages per minute and minutes per page.

18. Elena made \$24 in three hours. Give her earnings in dollars per hour and hours per dollar.

Find the best buy.

19. Minced onion

13 ounces for \$2.29

8 ounces for \$1.45

3 ounces for \$0.95

20. Dog food; you have a coupon for \$1 off on 25 pounds or more.

50 pounds for \$19.95

25 pounds for \$10.40

8 pounds for \$3.40

[5.3] *Use either the method of writing in lowest terms or of cross multiplication to decide whether each proportion is true or false.*

21. $\frac{6}{10} = \frac{9}{15}$

22. $\frac{6}{48} = \frac{9}{36}$

23. $\frac{47}{10} = \frac{98}{20}$

24. $\frac{64}{36} = \frac{96}{54}$

25. $\frac{1.5}{2.4} = \frac{2}{3.2}$

26. $\frac{3\frac{1}{2}}{2\frac{1}{3}} = \frac{6}{4}$

[5.4] *Find the unknown number in each proportion. Round answers to the nearest hundredth, if necessary.*

27. $\frac{4}{42} = \frac{150}{x}$

28. $\frac{16}{x} = \frac{12}{15}$

29. $\frac{100}{14} = \frac{x}{56}$

30. $\frac{5}{8} = \frac{x}{20}$

31. $\frac{x}{24} = \frac{11}{18}$

32. $\frac{7}{x} = \frac{18}{21}$

33. $\frac{x}{3.6} = \frac{9.8}{0.7}$

34. $\frac{13.5}{1.7} = \frac{4.5}{x}$

35. $\frac{0.82}{1.89} = \frac{x}{5.7}$

[5.5] *Set up and solve a proportion for each application problem.*

36. The ratio of cats to dogs at the animal shelter is 3 to 5. If there are 45 dogs, how many cats are there?

37. Danielle had 8 hits in 28 times at bat during last week's games. If she continues to hit at the same rate, how many hits will she get in 161 times at bat?

38. If 3.5 pounds of ground beef cost \$9.77, what will 5.6 pounds cost? Round to the nearest cent.

39. About 4 out of 10 students are expected to vote in campus elections. There are 8247 students. How many are expected to vote? Round to the nearest whole number.

40. The scale on Brian's model railroad is 1 in. to 16 ft. One of the scale model boxcars is 4.25 in. long. What is the length of a real boxcar in feet?

41. In the hospital pharmacy, Michiko sees that a certain medicine is to be given at the rate of 3.5 milligrams for every 50 pounds of body weight. How much medicine should be given to a patient who weighs 210 pounds?

42. A 180-pound person burns 284 calories playing basketball for 25 minutes. How many calories would the person burn in 45 minutes, to the nearest whole number? (*Source: Wellness Encyclopedia.*)

43. Marvette makes necklaces to sell at a local gift shop. She made 2 dozen necklaces in $16\frac{1}{2}$ hours. How long will it take her to make 40 necklaces?

Mixed Review Exercises

Find the unknown number in each proportion. Round answers to the nearest hundredth, if necessary.

44. $\frac{x}{45} = \frac{70}{30}$

45. $\frac{x}{52} = \frac{0}{20}$

46. $\frac{64}{10} = \frac{x}{20}$

47. $\frac{15}{x} = \frac{65}{100}$

48. $\frac{7.8}{3.9} = \frac{13}{x}$

49. $\frac{34.1}{x} = \frac{0.77}{2.65}$

Use cross multiplication to decide whether each proportion is true or false. Circle the correct answer.

50. $\frac{55}{18} = \frac{80}{27}$

True False

51. $\frac{5.6}{0.6} = \frac{18}{1.94}$

True False

52. $\frac{\frac{1}{5}}{2} = \frac{1\frac{1}{6}}{11\frac{2}{3}}$

True False

Write each ratio as a fraction in lowest terms. Change to the same units when necessary.

53. 4 dollars to 10 quarters

54. $4\frac{1}{8}$ in. to 10 in.

55. 10 yd to 8 ft

56. \$3.60 to \$0.90

57. 12 eggs to 15 eggs

58. 37 meters to 7 meters

59. 3 pints to 4 quarts

60. 15 minutes to 3 hours

61. $4\frac{1}{2}$ miles to $1\frac{3}{10}$ miles

62. Nearly 7 out of 8 fans buy something to drink at rock concerts. How many of the 28,500 fans at today's concert would be expected to buy a beverage? Round to the nearest hundred fans.

63. Emily spent \$150 on car repairs and \$400 on car insurance. What is the ratio of the amount spent on insurance to the amount spent on repairs?

64. Antonio is choosing among three packages of plastic wrap. Is the best buy 25 ft for \$0.78; 75 ft for \$1.99; or 100 ft for \$2.59? He has a coupon for 50¢ off on either of the larger two packages.

65. On this scale drawing of a backyard patio, 0.5 in. represents 6 ft. If the patio measures 1.25 in. long on the drawing, what will the actual length of the patio be when it is built?

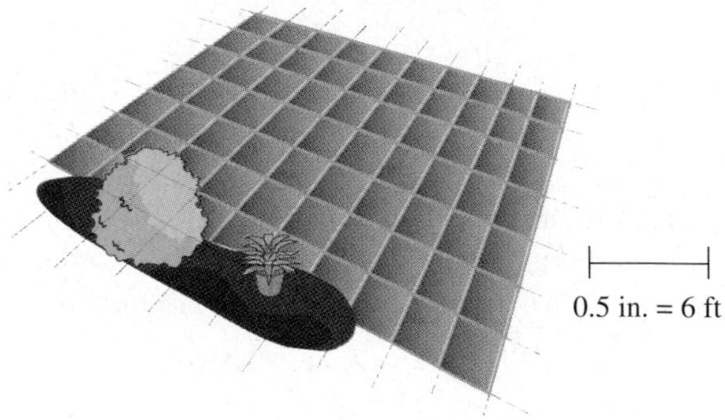

66. A lawn mower uses 0.8 gallon of gas every 3 hours. The gas tank holds 2 gallons. How long can the mower run on a full tank?

67. An antibiotic is to be given at the rate of $1\frac{1}{2}$ teaspoons for every 24 pounds of body weight. How much should be given to an infant who weighs 8 pounds?

68. Charles made 251 points during 169 minutes of playing time last year. If he plays 14 minutes in tonight's game, how many points would you expect him to make? Round to the nearest whole number.

69. Refer to Exercise 67. Explain each step you took in solving the problem. Be sure to tell how you decided which way to set up the proportion and how you checked your answer.

70. A vitamin supplement for cats is to be given at the rate of 1000 milligrams for a 5-pound cat.
(a) How much should be given to a 7-pound cat?

(b) How much should be given to an 8-ounce kitten? (*Source:* St. Jon Pet Care Products.)

Chapter 5 Test

Write each rate or ratio as a fraction in lowest terms. Change to the same units when necessary.

1. 16 fish to 20 fish

1. ______________

2. 300 miles on 15 gallons

2. ______________

3. $15 for 75 minutes

3. ______________

4. The little theater has 320 seats. The auditorium has 1200 seats. Find the ratio of auditorium seats to theater seats.

4. ______________

5. 3 quarts to 60 gallons

5. ______________

6. 3 hours to 40 minutes

6. ______________

7. Find the best buy on spaghetti sauce. You have a coupon for 75¢ off Brand X and a coupon for 25¢ off Brand Y.
28 ounces of Brand X for $3.89
18 ounces of Brand Y for $1.89
13 ounces of Brand Z for $1.29

7. ______________

8. Suppose the ratio of your income last year to your income this year is 3 to 2. Explain what this means. Give an example of the dollars earned last year and this year that fits the 3 to 2 ratio.

8. ______________

9. ______________

Determine whether each proportion is true or false.

9. $\frac{6}{14} = \frac{18}{45}$ **10.** $\frac{8.4}{2.8} = \frac{2.1}{0.7}$

10. ______________

11. ______
12. ______
13. ______
14. ______
15. ______
16. ______
17. ______
18. ______
19. ______
20. ______

Find the unknown number in each proportion. Round the answers to the nearest hundredth, if necessary.

11. $\frac{5}{9} = \frac{x}{45}$

12. $\frac{3}{1} = \frac{8}{x}$

13. $\frac{x}{20} = \frac{6.5}{0.4}$

14. $\frac{2\frac{1}{3}}{x} = \frac{\frac{8}{9}}{4}$

Set up and solve a proportion for each application problem.

15. Pedro types 240 words in five minutes. At that rate, how many words can he type in 12 minutes?

16. Just 0.8 ounce of wildflower seeds is enough for 50 square feet of ground. What weight of seeds is needed for a garden with 225 square feet? (*Source:* White Swan Ltd.)

17. About 2 out of every 15 people are left-handed. How many of the 650 students in our school would you expect to be left-handed? Round to the nearest whole number.

18. A student set up the proportion for Exercise 17 this way and arrived at an answer of 4875.

$$\frac{2}{15} = \frac{650}{x} \qquad \text{Check: } \frac{\mathbf{2}}{\mathbf{15}} = \frac{\mathbf{650}}{\mathbf{4875}} \qquad \begin{array}{l} 15 \cdot 650 = 9750 \\ 2 \cdot 4875 = 9750 \end{array}$$

Because the cross products are equal, the student said the answer is correct. Is the student right? Explain why or why not.

19. A medication is given at the rate of 8.2 grams for every 50 pounds of body weight. How much should be given to a 145-pound person? Round to the nearest tenth.

20. On a scale model, 1 in. represents 8 ft. If a building in the model is 7.5 in. tall, what is the actual height of the building in feet?

Cumulative Review Exercises CHAPTERS 1–5

Name the digit that has the given place value.

1. 216,475,038
thousands
tens
millions
hundred thousands

2. 340.6915
hundredths
ones
ten-thousandths
hundreds

Round each number as indicated.

3. 9903 to the nearest hundred

4. 617.0519 to the nearest tenth

5. \$99.81 to the nearest dollar

6. \$3.0555 to the nearest cent

First use front end rounding to round each number and estimate the answer. Then find the exact answer.

7. *Estimate:* + ____ *Exact:*
$$\begin{array}{r} 28 \\ 5206 \\ +\ 351 \\ \hline \end{array}$$

8. *Estimate:* − ____ *Exact:*
$$\begin{array}{r} 63.1 \\ -\ 5.692 \\ \hline \end{array}$$

9. *Estimate:* × ____ *Exact:*
$$\begin{array}{r} 4716 \\ \times\ 804 \\ \hline \end{array}$$

10. *Estimate:* × ____ *Exact:*
$$\begin{array}{r} 0.982 \\ \times\ 17.8 \\ \hline \end{array}$$

11. *Estimate:* $\overline{)\quad}$ *Exact:* $53\overline{)48{,}071}$

12. *Estimate:* $\overline{)\quad}$ *Exact:* $4.5\overline{)1638}$

13. *Estimate:* ____ • ____ = ____ *Exact:* $1\frac{5}{6} \cdot 3\frac{3}{5}$

14. *Estimate:* ____ ÷ ____ = ____ *Exact:* $5\frac{1}{4} \div \frac{7}{8}$

15. *Estimate:* ____ − ____ = ____ *Exact:* $2\frac{4}{5} - 1\frac{5}{6}$

16. *Estimate:* ____ + ____ = ____ *Exact:* $2\frac{9}{10} + 10\frac{1}{2}$

Add, subtract, multiply, or divide as indicated.

17. 988 + 373,422 + 6

18. 30 − 0.66

19. Write your answer using R for the remainder.

$33\overline{)20{,}157}$

20. (1.9)(0.004)

21. 3020 − 708

22. 0.401 + 62.98 + 5

23. 1.39 ÷ 0.025

24. (6392)(5609)

Use the order of operations to simplify each expression.

25. $36 + 18 \div 6$

26. $8 \div 4 + (10 - 3^2) \cdot 4^2$

27. $88 \div \sqrt{121} \cdot 2^3$

28. $(16.2 - 5.85) - 2.35 \cdot 4$

29. Write 0.0105 in words.

30. Write sixty and seventy-one thousandths in numbers.

Nest boxes for birds are made with different sizes of entry holes. The right size opening allows only certain types of birds to use the nest box and helps prevent entry by predators. Use the information in the table to answer Exercises 31–34. The diameter is the distance across the opening at its widest point.

Eastern bluebird nest box with 1.5 in. opening.

Bird	Diameter of Opening
Eastern bluebird	1.5 in.
Western bluebird	$1\frac{9}{16}$ in.
Chickadee	1.25 in.
Swallow	$1\frac{3}{8}$ in.
Wren	$1\frac{1}{8}$ in.

Source: Duncraft.

31. Write each mixed number measurement in the table as a decimal. Round to the nearest thousandth, if necessary.

32. List the nest box openings in order from smallest to largest.

33. Find the difference in diameter between the largest and smallest openings. Write your answer as a fraction or mixed number in lowest terms.

34. What is the difference in diameter between the openings for an eastern bluebird and a western bluebird? Write your answer as a fraction and as a decimal.

Write each rate or ratio as a fraction in lowest terms. Change to the same units when necessary. Use the circle graph for Exercises 35–37.

Source: National Coffee Association of USA Inc.

35. Write the ratio of those who drink coffee daily to those who drink it occasionally.

36. What is the ratio of those who do not drink coffee to the total number of people in the survey?

37. Find the ratio of all the adults who drink coffee to those who never drink it.

38. Write the ratio of 20 minutes to 4 hours.

39. Find the ratio of 2 ft to 8 in.

40. Find the best buy on instant mashed potatoes. You have a coupon for 50¢ off on either the 36-serving or 48-serving box.

A box that makes 20 servings for \$1.59

A box that makes 36 servings for \$3.24

A box that makes 48 servings for \$4.99

Find the unknown number in each proportion. Round your answers to the nearest hundredth, if necessary.

41. $\frac{9}{12} = \frac{x}{28}$

42. $\frac{7}{12} = \frac{10}{x}$

43. $\frac{x}{\frac{3}{4}} = \frac{2\frac{1}{2}}{\frac{1}{6}}$

44. $\frac{6.7}{x} = \frac{62.8}{9.15}$

Solve each application problem.

45. The college honor society has a goal of collecting 1500 pounds of food to fill Thanksgiving baskets. So far they've collected $\frac{5}{6}$ of their goal. How many more pounds do they need?

46. Tara has a photo that is 10 centimeters wide by 15 centimeters long. If the photo is enlarged to a length of 40 centimeters, find the new width, to the nearest tenth.

47. The distance around Dunning Pond is $1\frac{1}{10}$ miles. Norma ran around the pond four times in the morning and $2\frac{1}{2}$ times in the afternoon. How far did she run in all?

48. Rodney bought 49.8 gallons of gas for his minivan while driving 896.5 miles on a vacation. How many miles per gallon did he get, rounded to the nearest tenth?

49. In a survey, 5 out of 8 apartment residents said they are sometimes bothered by noise from their neighbors. How many of the 224 residents at Harris Towers would you expect to be bothered by noise?

50. The directions on a bottle of plant food call for $\frac{1}{2}$ teaspoon in two quarts of water. How much plant food is needed for five quarts? (*Source:* Schultz Company.)

Use the bar graph of average yearly earnings to answer Exercises 51–56.

AVERAGE YEARLY EARNINGS IN 1998
(Adults age 18 and over)

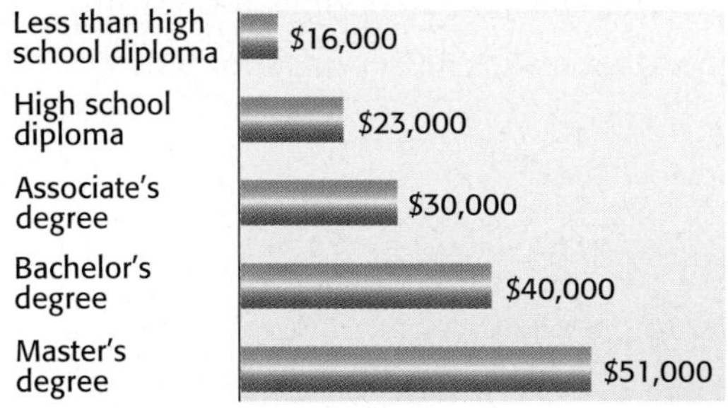

Source: U.S. Bureau of the Census.

51. Write the ratio of bachelor's degree earnings to "less than high school diploma" earnings as a fraction in lowest terms.

52. Explain what the ratio in Exercise 51 is saying about the yearly earnings of the two groups.

53. What is the average monthly salary for a person with a high school diploma, to the nearest dollar?

54. What is the average monthly salary for a person with a master's degree, to the nearest dollar?

55. If an average work year is 2000 hours of work time, find the difference in the hourly wage of a person with an associate's degree and a person without a high school diploma.

56. Compare the lifetime earnings of a person with a high school diploma who works from age 18 to 65 to a person with a bachelor's degree who works from age 22 to 65.

Percent

6

Percents are a large part of our everyday lives. For example, interest rates on automobile loans, home loans, and other installment loans are always given as percents. Sales tax, commission rates, and discounts on sale items are other everyday applications that show the importance of understanding percent. For example, knowing how to calculate the sales tax on any item you purchase allows you to know the true cost of anything you buy. (See Examples 1 and 2, Section 6.6.)

6.1 Basics of Percent

Objectives

1. Learn the meaning of percent.
2. Write percents as decimals.
3. Write decimals as percents.
4. Understand 100%, 200%, and 300%.
5. Use 50%, 10%, and 1%.

1 Write as percents.

(a) In a group of 100 people, 63 are unmarried. What percent are unmarried?

(b) The sales tax is \$5 per \$100. What percent is this?

(c) Out of 100 students, 68 are working part-time. What percent are working part-time?

Answers
1. (a) 63% **(b)** 5% **(c)** 68%

Notice that this figure has one hundred squares of equal size. Eleven of the squares are shaded. The shaded portion is $\frac{11}{100}$, or 0.11, of the total figure.

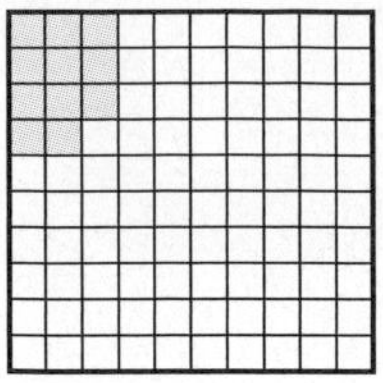

The shaded portion is also 11% of the total, or "eleven parts out of 100 parts." Read **11%** as "eleven percent."

1 Learn the meaning of percent. As we just saw, a percent is a ratio with a denominator of 100.

The Meaning of Percent

Percent means *per one hundred.* The "%" sign is used to show the number of parts out of one hundred parts.

Example 1 Understanding Percent

(a) If 43 out of 100 students are men, then 43 per (out of) 100 or $\frac{43}{100}$ or **43%** of the students are men.

(b) If a person pays a tax of \$7 on every \$100 of purchases, then the tax rate is \$7 per \$100. The ratio is $\frac{7}{100}$ and the percent of tax is **7%.**

Work Problem 1 at the Side.

2 Write percents as decimals. If 8% means 8 parts out of 100 parts or $\frac{8}{100}$, then $p\%$ means p parts out of 100 parts or $\frac{p}{100}$. Because $\frac{p}{100}$ is another way to write the division $p \div 100$, we have

$$p\% = \frac{p}{100} = p \div 100.$$

Writing a Percent as a Decimal

$$\underbrace{p\% = \frac{p}{100}}_{\text{As a fraction}} \quad \text{or} \quad \underbrace{p\% = p \div 100}_{\text{As a decimal}}$$

Example 2 Writing Percents as Decimals

Write each percent as a decimal.

(a) 47%

$$p\% = p \div 100$$
$$47\% = 47 \div 100 = 0.47 \quad \text{Decimal form}$$

Continued on Next Page

(b) 76%	76% = 76 ÷ 100 = 0.76	Decimal form
(c) 28.2%	28.2% = 28.2 ÷ 100 = 0.282	Decimal form
(d) 100%	100% = 100 ÷ 100 = 1.00	Decimal form

CAUTION

In Example 2(d), notice that 100% is 1.00, or 1, which is a whole number. Whenever you have a percent that is 100% or higher, the equivalent decimal number will be a number of 1 or higher.

Work Problem 2 at the Side.

The resulting answers in Example 2 suggest the following procedure for writing a percent as a decimal.

Writing a Percent as a Decimal

Step 1 Drop the percent sign.
Step 2 Divide by 100.

NOTE

Recall from Section 4.6 that a quick way to divide a number by 100 is to move the decimal point two places to the left.

Example 3 Writing Percents as Decimals by Moving the Decimal Point

Write each percent as a decimal by moving the decimal point two places to the left.

(a) 17%

17% = 17.% Decimal point starts at far right side.

0.17 ← Percent sign is dropped. (Step 1)

Decimal point is moved two places to the left. (Step 2)

17% = 0.17

(b) 160%

160% = 160.% = 1.60 or 1.6 Decimal starts at far right side. 1.60 is equivalent to 1.6.

(c) 4.9%

.049% 0 is attached so the decimal point can be moved two places to the left.

4.9% = 0.049

(d) 0.6%

0.6% = 0.006 0s are attached so the decimal point can be moved.

2 Write each percent as a decimal.

(a) 68%

(b) 34%

(c) 58.5%

(d) 175%

ANSWERS
2. **(a)** 0.68 **(b)** 0.34 **(c)** 0.585 **(d)** 1.75

3 Write each percent as a decimal.

(a) 96%

(b) 6%

(c) 24.8%

(d) 0.9%

CAUTION

Look at Example 3(d), where 0.6% is less than 1%. Because 1% is equivalent to 0.01 or $\frac{1}{100}$, any fraction of a percent smaller than 1% is less than 0.01.

Work Problem 3 at the Side.

3 **Write decimals as percents.** You can write any decimal as a percent. For example, the decimal 0.78 is the same as the fraction

$$\frac{78}{100}.$$

This fraction means 78 of 100 parts, or 78%. The following steps give the same result.

Writing a Decimal as a Percent

Step 1 Multiply by 100.

Step 2 Attach a percent sign.

NOTE

A quick way to divide or multiply a number by 100 is to move the decimal point two places to the left or two places to the right, respectively.

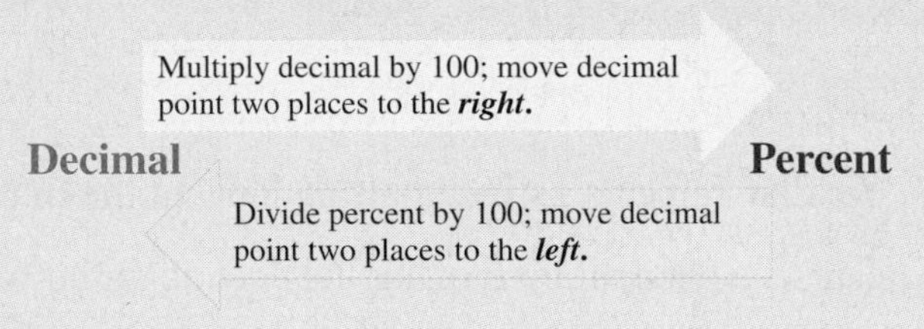

Multiply decimal by 100; move decimal point two places to the ***right.***
Decimal
Percent
Divide percent by 100; move decimal point two places to the ***left.***

Example 4 Writing Decimals as Percents by Moving the Decimal Point

Write each decimal as a percent by moving the decimal point two places to the right.

(a) 0.21

0.21% Percent sign is attached. (Step 1)

Decimal point is moved two places to the right. (Step 2)

0.21 = 21%

Decimal point is not written with whole number percents.

(b) 0.529 = 52.9%

(c) 1.92 = 192%

Continued on Next Page

ANSWERS
3. **(a)** 0.96 **(b)** 0.06 **(c)** 0.248 **(d)** 0.009

(d) 2.5

2.50% 0 is attached so the decimal point can be moved two places to the right.

2.5 = 250%

(e) 3

3. = 3.00% Two 0s are attached.

so 3 = 300%

CAUTION

Look at Examples 4(c), 4(d), and 4(e) where 1.92, 2.5, and 3 are greater than 1. Because the number 1 is equivalent to 100%, all numbers greater than 1 will be greater than 100%.

Work Problem 4 at the Side.

4 Understand 100%, 200%, and 300%. When working with percents, it is helpful to have several reference points. 100%, 200% and 300% are three such helpful reference points.

100% means 100 parts out of 100 parts. That's ***all*** of the parts. If 100% of the 18 people attending last week's meeting attended this week's meeting, then 18 people attended this week (***all*** of them).

If attendance at the meeting this week is 200% of last week's attendance of 18 people, then this week's attendance is 36 people, or *two* times as many people (2 • 18 = 36). Likewise, if attendance is 300% of last week's attendance, then *three* times as many people, or 54 people attended (3 • 18 = 54).

Example 5 Finding 100%, 200%, and 300% of a Number

Fill in the blanks.

(a) 100% of 82 people is ______.
100% is ***all*** of the people. So, 100% of 82 people is <u>82 people</u>.

(b) 200% of \$63 is ______.
200% is twice (2 times) as much money. So, 200% of \$63 is <u>\$126</u>.

(c) 300% of 32 employees is ______.
300% is 3 times as many employees. So, 300% of 32 employees is <u>96 employees</u>.

Work Problem 5 at the Side.

5 Use 50%, 10%, and 1%. 50% means 50 parts out of 100 parts, which is ***half*** of the parts ($\frac{50}{100} = \frac{1}{2}$). 50% of \$18 is \$9 (***half*** of the money).

When using 10%, we have 10 parts out of 100 parts, which is $\frac{1}{10}$ of the parts ($\frac{10}{100} = \frac{1}{10}$). To find 10% or $\frac{1}{10}$ of a number, we move the decimal point one place to the left. 10% of \$285 is \$28.50 (\$28 5. = \$28.50).

To find 1% of a number ($\frac{1}{100}$), we move the decimal point two places to the left. 1% of \$198 is \$1.98 (\$1 98. = \$1.98).

4 Write each decimal as a percent.

(a) 0.95 **(b)** 0.18

(c) 0.09 **(d)** 0.617

(e) 0.834 **(f)** 5.34

(g) 2.8 **(h)** 4

5 Fill in the blanks.

(a) 100% of \$7.80 is ______.

(b) 100% of 1850 workers is ______.

(c) 200% of 24 photographs is ______.

(d) 300% of 780 miles is ______.

ANSWERS

4. **(a)** 95% **(b)** 18% **(c)** 9% **(d)** 61.7% **(e)** 83.4% **(f)** 534% **(g)** 280% **(h)** 400%

5. **(a)** \$7.80 **(b)** 1850 workers **(c)** 48 photographs **(d)** 2340 miles

6 Fill in the blanks.

(a) 50% of 100 windows is ______.

(b) 50% of 64 e-mails is ______.

(c) 10% of 3850 members is ______.

(d) 1% of 240 ft is ______.

Answers
6. **(a)** 50 windows **(b)** 32 e-mails **(c)** 385 members **(d)** 2.4 ft

Example 6 **Finding 50%, 10%, and 1% of a Number**

Fill in the blanks.

(a) 50% of 2420 hours is ______.

50% is half of the hours. So, 50% of 2420 hours is <u>1210 hours</u>.

(b) 10% of 280 pages is ______.

10% is $\frac{1}{10}$ of the pages. Move the decimal point one place to the left. So, 10% of 280. pages is <u>28 pages</u>.

(c) 1% of $540 is ______.

1% is $\frac{1}{100}$ of the money. Move the decimal point two places to the left. So, 1% of $540. is <u>$5.40</u>.

Work Problem 6 at the Side.

6.1 EXERCISES

FOR EXTRA HELP: Student's Solutions Manual · MyMathLab.com · InterAct Math Tutorial Software · AW Math Tutor Center · www.mathxl.com · Digital Video Tutor CD 3 Videotape 10

Write each percent as a decimal. See Examples 2 and 3.

1. 15% **2.** 41% **3.** 60% **4.** 40%

5. 25% **6.** 35% **7.** 140% **8.** 250%

9. 5.5% **10.** 6.7% **11.** 100% **12.** 600%

13. 0.5% **14.** 0.25% **15.** 0.35% **16.** 0.75%

Write each decimal as a percent. See Example 4.

17. 0.8 **18.** 0.4 **19.** 0.58 **20.** 0.25

21. 0.01 **22.** 0.07 **23.** 0.125 **24.** 0.875

25. 0.375 **26.** 0.625 **27.** 2 **28.** 5

29. 3.7 **30.** 2.2 **31.** 0.0312 **32.** 0.0625

33. 4.162 **34.** 8.715 **35.** 0.0028 **36.** 0.0064

37. Fractions, decimals, and percents are all used to describe a part of something. The use of percents is much more common than fractions and decimals. Why do you suppose this is true?

38. List five uses of percent that are or will be part of your life. Consider the activities of working, shopping, saving, and planning for the future.

Write each percent as a decimal and each decimal as a percent. See Examples 2–4.

39. In Lincoln, 45% of the refuse is recycled.

40. At College of DuPage, 82% of the students work part-time.

41. At Bett's Boutique, 18% of the items sold are returned.

42. There was a 43.2% voter turnout at the election.

43. The property tax rate in Alpine County is 0.035.

44. A church building fund has 0.49 of the money needed.

45. The number of people successfully completing CPR training this session is 2 times that of the last session.

46. The number of newspaper subscribers was 4 times as great as last quarter.

47. Only 0.005 of the total population has this genetic defect.

48. The return rate of defective keyboards is 0.0075 of total output.

49. The patient's blood pressure was 153.6% of normal.

50. Success with the diet was 248.7% greater than anticipated.

Fill in the blanks. Remember that 100% is all of something, 200% is two times as many, and 300% is three times as many. See Example 5.

51. There are 20 children in the preschool class. 100% of the children are served breakfast and lunch. How many children are served both meals?

____

52. The company owns 345 vans. 100% of the vans are painted white with blue lettering. How many vans are painted white with blue lettering?

____

53. Last year we had 210 employees. This year we have 200% of that number. How many employees do we have this year? ______________

54. This week's expenses are 200% of last week's $380. This week's expenses are ______________.

55. If we need 300% of the 90 chairs that we needed for last week's meeting, we will need ______________.

56. Jim's new car gets 300% of the 12 miles per gallon that his old car got. His new car gets ______________.

Fill in the blanks. Remember that 50% is half of something; 10% is found by moving the decimal point one place to the left, and 1% is found by moving the decimal point two places to the left. See Example 6.

57. John owes $285 for tuition. Financial aid will pay 50% of the cost. Financial aid will pay ______________.

58. The Animal Humane Society took in 20,000 animals last year. About 50% of them were dogs. The number of dogs taken in was ______________.

59. Only 10% of 8200 commuters are carpooling to work. How many commuters carpool? ______________

60. Sarah knows that she will not be able to sell 10% of the 240 dozen plants in her greenhouse. The number of unsold plants will be ______________.

61. The naturalist said that 1% of the 2600 plants in the park are poisonous. How many plants are poisonous? ______________

62. Of the 4800 accidents, only 1% was caused by mechanical failure. How many accidents were caused by mechanical failure? ______________

63. **(a)** Describe a shortcut method of finding 100% of a number.

(b) Show an example using your shortcut.

64. **(a)** Describe a shortcut method of finding 50% of a number.

(b) Show an example using your shortcut.

65. **(a)** Describe a shortcut method of finding 200% of a number.

(b) Show an example using your shortcut.

66. **(a)** Describe a shortcut method of finding 300% of a number.

(b) Show an example using your shortcut.

67. **(a)** Describe a shortcut method of finding 10% of a number.

(b) Show an example using your shortcut.

68. **(a)** Describe a shortcut method of finding 1% of a number.

(b) Show an example using your shortcut.

College students were asked to rank the most important issues they would like presidential candidates to address. The bar graph shows the ranking of these issues and the percent of students selecting each issue. Use this graph to answer Exercises 69–72. Write each answer as a percent and as a decimal.

69. What portion of the students selected education as a top issue?

70. What portion of the students selected environmental issues as a top issue?

71. **(a)** What was the second most important issue?

(b) Write the portion of the students who selected that issue.

72. **(a)** What was the third most important issue?

(b) Write the portion of the students who selected that issue?

EDUCATION IS TOPS

Education ranks as the most important issue that college students would like presidential candidates to address.

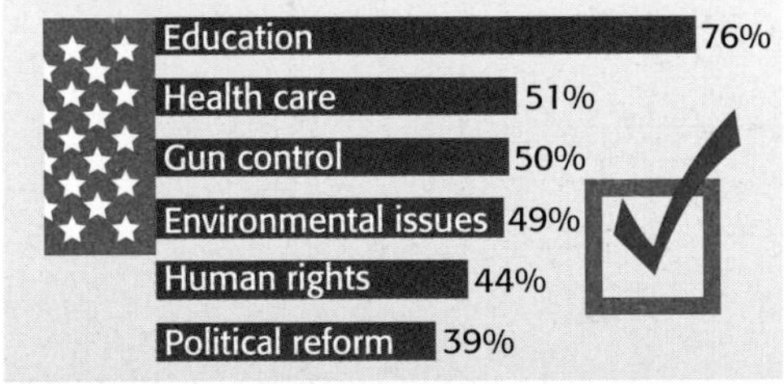

Source: Greenfield Online, YouthStream: Pulsefinder On-Campus Market Study.

In the United States, 12.7% of the population is 65 years old or older. The bar graph shows the countries with the highest percent of people 65 years old or older. Use this graph to answer Exercises 73–76. Write each answer as a percent and as a decimal.

73. What portion of the population of Spain is 65 or older?

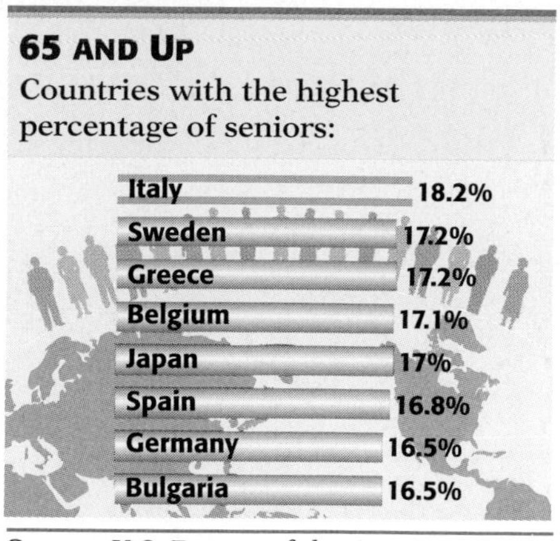

Source: U.S. Bureau of the Census International Programs Center.

74. What portion of the population of Greece is 65 or older?

75. (a) Which two countries have the lowest portion of the population 65 or older?

(b) What portion is this?

76. (a) Which country has the highest portion of the population 65 or older?

(b) What portion is this?

Write a percent for both the shaded and unshaded part of each figure.

77.

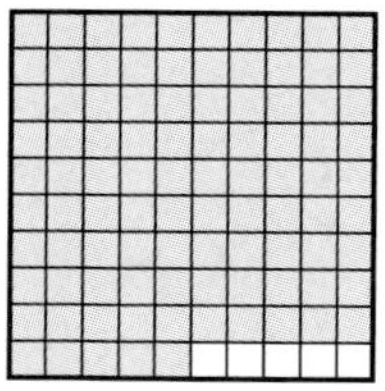

78.

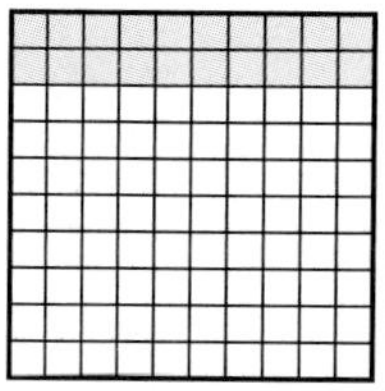

79.

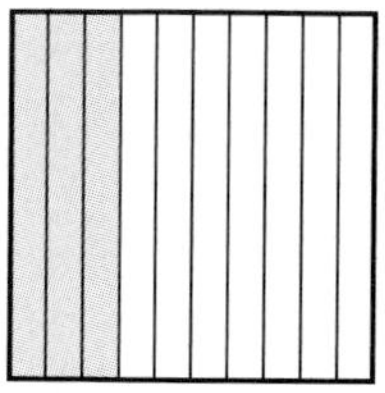

80.

81.

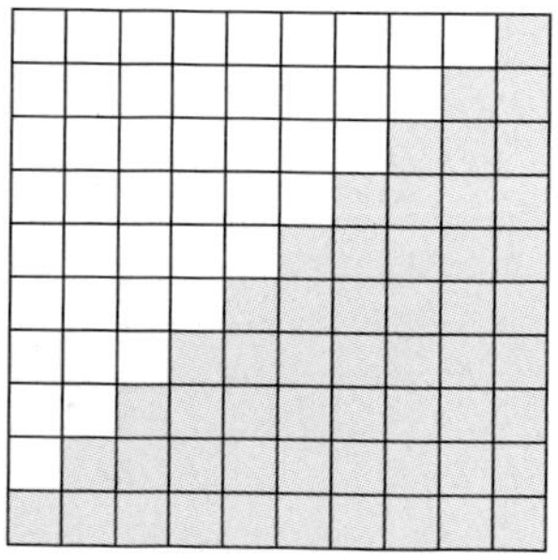

82.

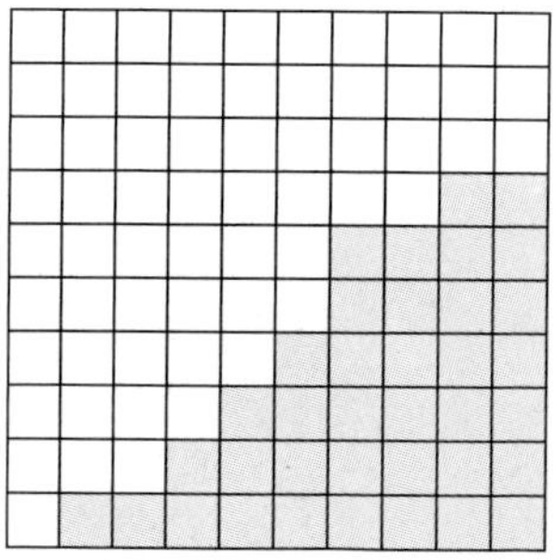

6.2 PERCENTS AND FRACTIONS

OBJECTIVES

1. Write percents as fractions.
2. Write fractions as percents.
3. Use the table of percent equivalents.

1 Write percents as fractions. Percents can be written as fractions by using what we learned in the previous section.

Writing a Percent as a Fraction

$$p\% = \frac{p}{100}, \quad \text{as a fraction.}$$

Example 1 Writing Percents as Fractions

Write each percent as a fraction or mixed number in lowest terms.

(a) 25%

As we saw in the last section, 25% can be written as a decimal.

$$25\% = 25 \div 100 = 0.25 \quad \text{Percent sign dropped}$$

Because 0.25 means 25 hundredths,

$$0.25 = \frac{25 \div 25}{100 \div 25} = \frac{1}{4}. \quad \text{Lowest terms}$$

It is not necessary, however, to write 25% as a decimal first. Just write

$$25\% = \frac{25}{100} \quad \text{25 per 100}$$

$$= \frac{1}{4}. \quad \text{Lowest terms}$$

(b) 76%

Write 76% as $\frac{76}{100}$.

The percent becomes the numerator.

The *denominator* is always 100 because percent means *parts per 100.*

Write $\frac{76}{100}$ in lowest terms.

$$\frac{76 \div 4}{100 \div 4} = \frac{19}{25} \quad \text{Lowest terms}$$

(c) 150%

$$150\% = \frac{150}{100} = \frac{3}{2} = 1\frac{1}{2} \quad \text{Mixed number}$$

NOTE

Remember that percent means ***per 100.***

Work Problem 1 at the Side.

1 Write each percent as a fraction or mixed number in lowest terms.

(a) 50%

(b) 35%

(c) 52%

(d) 23%

(e) 125%

(f) 250%

ANSWERS

1. (a) $\frac{1}{2}$ (b) $\frac{7}{20}$ (c) $\frac{13}{25}$ (d) $\frac{23}{100}$ (e) $1\frac{1}{4}$ (f) $2\frac{1}{2}$

2 Write each percent as a fraction in lowest terms.

(a) 37.5%

(b) 62.5%

(c) 4.5%

(d) $66\frac{2}{3}\%$

(e) $10\frac{1}{3}\%$

(f) $87\frac{1}{2}\%$

ANSWERS

2. **(a)** $\frac{3}{8}$ **(b)** $\frac{5}{8}$ **(c)** $\frac{9}{200}$ **(d)** $\frac{2}{3}$ **(e)** $\frac{31}{300}$ **(f)** $\frac{7}{8}$

The next example shows how to write decimal and fraction percents as fractions.

Example 2 **Writing Decimal or Fraction Percents as Fractions**

Write each percent as a fraction in lowest terms.

(a) 15.5%
Write 15.5 over 100.

$$15.5\% = \frac{15.5}{100}$$

To get a whole number in the numerator, multiply the numerator and denominator by 10. (Recall that multiplying by $\frac{10}{10}$ is the same as multiplying by 1.)

$$\frac{15.5}{100} = \frac{15.5 \cdot \mathbf{10}}{100 \cdot \mathbf{10}} = \frac{155}{1000}$$

Write in lowest terms.

$$\frac{155 \div \mathbf{5}}{1000 \div \mathbf{5}} = \frac{31}{200}$$

(b) $33\frac{1}{3}\%$
Write $33\frac{1}{3}$ over 100.

$$33\frac{1}{3}\% = \frac{33\frac{1}{3}}{100}$$

When we have a mixed number in the numerator, we must write the mixed number as an improper fraction.

Here, we write $33\frac{1}{3}$ as the improper fraction $\frac{100}{3}$. Now,

$$\frac{33\frac{1}{3}}{100} = \frac{\frac{100}{3}}{100}.$$

We rewrite the division problem in a horizontal form. Then we multiply by the reciprocal of the divisor.

Multiply by the reciprocal.

$$\frac{\frac{100}{3}}{100} = \frac{100}{3} \div 100 = \frac{100}{3} \div \frac{100}{1} = \frac{\overset{1}{\cancel{100}}}{3} \cdot \frac{1}{\underset{1}{\cancel{100}}} = \frac{1}{3}$$

NOTE

In Example 2(a) we could have changed 15.5% to $15\frac{1}{2}\%$ and then written it as the improper fraction $\frac{31}{2}$ over 100, as was done in part (b). It is usually easier not to change decimals to fractions but to leave decimal percents as they are.

Work Problem 2 at the Side.

2 **Write fractions as percents.** We will use the formula given at the beginning of this section to write fractions as percents.

$$p\% = \frac{p}{100}$$

Example 3 Writing Fractions as Percents

Write each fraction as a percent. Round to the nearest tenth if necessary.

(a) $\frac{3}{5}$

Write this fraction as a percent by solving for p in the proportion

$$\frac{3}{5} = \frac{p}{100}$$

Find cross products and show that they are equivalent.

$$5 \cdot p = 3 \cdot 100$$
$$5 \cdot p = 300$$

Divide each side by 5.

$$\frac{\overset{1}{\cancel{5}} \cdot p}{\underset{1}{\cancel{5}}} = \frac{300}{5}$$

$$p = 60$$

This result means that $\frac{3}{5} = \frac{60}{100}$ or 60%.

NOTE

Solving proportions can be reviewed in **Section 5.4.**

(b) $\frac{7}{8}$

Write a proportion.

$$\frac{7}{8} = \frac{p}{100}$$

$8 \cdot p = 7 \cdot 100$ Show that cross products are equivalent.

$8 \cdot p = 700$

$\frac{\overset{1}{\cancel{8}} \cdot p}{\underset{1}{\cancel{8}}} = \frac{700}{8}$ Divide each side by 8.

$p = 87.5$

Finally, $\frac{7}{8} = 87.5\%$.

NOTE

If you think of $\frac{700}{8}$ as an improper fraction, changing it to a mixed number gives an answer of $87\frac{1}{2}$. So $\frac{7}{8} = 87.5\%$ or $87\frac{1}{2}\%$.

Continued on Next Page

3 Write as percents. Round to the nearest tenth if necessary.

(a) $\frac{1}{2}$

(b) $\frac{7}{10}$

(c) $\frac{6}{25}$

(d) $\frac{5}{8}$

(e) $\frac{1}{6}$

(f) $\frac{2}{9}$

(c) $\frac{5}{6}$

Start with a proportion.

$$\frac{5}{6} = \frac{p}{100}$$

$6 \cdot p = 5 \cdot 100$ Show that cross products are equivalent.

$6 \cdot p = 500$

$\frac{\overset{1}{\cancel{6}} \cdot p}{\underset{1}{\cancel{6}}} = \frac{500}{6}$ Divide each side by 6.

$p = 83.\overline{3}$ The answer is a repeating decimal.

$p \approx 83.3$ Round to the nearest tenth.

Solving this proportion shows

$$\frac{5}{6} = 83.\overline{3}\% \approx 83.3\% \quad \text{(rounded).}$$

NOTE

You can treat $\frac{500}{6}$ as an improper fraction to get an exact answer of $83\frac{1}{3}\%$.

Work Problem 3 at the Side.

3 **Use the table of percent equivalents.** The more you work with common fractions and mixed numbers and their decimal and percent equivalents, the more familiar you will become with them.

The table on the next two pages shows common and not so common fractions and mixed numbers and their decimal and percent equivalents.

Example 4 Using the Table of Percent Equivalents

Read the following from the table.

(a) $\frac{1}{12}$ as a percent

Find $\frac{1}{12}$ in the "fraction" column. The percent is 8.3% (rounded) or $8\frac{1}{3}\%$ (exact).

(b) 0.375 as a fraction

Look in the "decimal" column for 0.375. The fraction is $\frac{3}{8}$.

(c) $\frac{13}{16}$ as a percent

Find $\frac{13}{16}$ in the "fraction" column. The percent is 81.25% or $81\frac{1}{4}\%$.

Calculator Tip Example 4(c) could be solved on the calculator as

13 ÷ 16 = 0.8125 × 100 = 81.25.

Multiplying by 100 changes the decimal to percent.

ANSWERS

3. **(a)** 50% **(b)** 70% **(c)** 24% **(d)** 62.5%

(e) 16.7% (rounded); $16\frac{2}{3}\%$ (exact)

(f) 22.2% (rounded); $22\frac{2}{9}\%$ (exact)

NOTE

When a fraction like $\frac{1}{12}$ is changed to a decimal, it is a repeating decimal that goes on forever, 0.083333 . . . . In the table these decimals are rounded to the nearest thousandth. When the decimal is then changed to a percent, it will be to the nearest tenth of a percent. Decimals that do not repeat are usually not rounded.

Work Problem 4 at the Side.

Percent, Decimal, and Fraction Equivalents

Percent *(rounded to tenths when necessary)*	Decimal	Fraction
1%	0.01	$\frac{1}{100}$
2%	0.02	$\frac{1}{50}$
4%	0.04	$\frac{1}{25}$
5%	0.05	$\frac{1}{20}$
6.25% or $6\frac{1}{4}$%	0.0625	$\frac{1}{16}$
8.3% (rounded) or $8\frac{1}{3}$% (exact)	$0.08\overline{3}$ rounds to 0.083	$\frac{1}{12}$
10%	0.1	$\frac{1}{10}$
12.5% or $12\frac{1}{2}$%	0.125	$\frac{1}{8}$
16.7% (rounded) or $16\frac{2}{3}$% (exact)	$0.1\overline{6}$ rounds to 0.167	$\frac{1}{6}$
18.75% or $18\frac{3}{4}$%	0.1875	$\frac{3}{16}$
20%	0.2	$\frac{1}{5}$
25%	0.25	$\frac{1}{4}$
30%	0.3	$\frac{3}{10}$
31.25% or $31\frac{1}{4}$%	0.3125	$\frac{5}{16}$
33.3% (rounded) or $33\frac{1}{3}$% (exact)	$0.\overline{3}$ rounds to 0.333	$\frac{1}{3}$
37.5% or $37\frac{1}{2}$%	0.375	$\frac{3}{8}$
40%	0.4	$\frac{2}{5}$
43.75% or $43\frac{3}{4}$%	0.4375	$\frac{7}{16}$

(continued)

4 Read the following fractions, mixed numbers, decimals, and percents from the table on this or the next page. If you already know the answer or can solve the answer quickly, don't use the table.

(a) $\frac{3}{4}$ as a percent

(b) 10% as a fraction

(c) $0.\overline{6}$ as a fraction

(d) $37\frac{1}{2}$% as a fraction

(e) $\frac{7}{8}$ as a percent

(f) $\frac{1}{2}$ as a percent

(g) $33\frac{1}{3}$% as a fraction

(h) $1\frac{3}{4}$ as a percent

ANSWERS

4. (a) 75% **(b)** $\frac{1}{10}$ **(c)** $\frac{2}{3}$ **(d)** $\frac{3}{8}$ **(e)** 87.5% **(f)** 50% **(g)** $\frac{1}{3}$ **(h)** 175%

Percent, Decimal, and Fraction Equivalents (continued)

Percent *(rounded to tenths when necessary)*	**Decimal**	**Fraction**
50%	0.5	$\frac{1}{2}$
56.25% or $56\frac{1}{4}\%$	0.5625	$\frac{9}{16}$
60%	0.6	$\frac{3}{5}$
62.5% or $62\frac{1}{2}\%$	0.625	$\frac{5}{8}$
66.7% (rounded) or $66\frac{2}{3}\%$ (exact)	$0.\overline{6}$ rounds to 0.667	$\frac{2}{3}$
68.75% or $68\frac{3}{4}\%$	0.6875	$\frac{11}{16}$
70%	0.7	$\frac{7}{10}$
75%	0.75	$\frac{3}{4}$
80%	0.8	$\frac{4}{5}$
81.25% or $81\frac{1}{4}\%$	0.8125	$\frac{13}{16}$
83.3% (rounded) or $83\frac{1}{3}\%$ (exact)	$0.8\overline{3}$ rounds to 0.833	$\frac{5}{6}$
87.5% or $87\frac{1}{2}\%$	0.875	$\frac{7}{8}$
90%	0.9	$\frac{9}{10}$
93.75% or $93\frac{3}{4}\%$	0.9375	$\frac{15}{16}$
100%	1.0	1
110%	1.1	$1\frac{1}{10}$
125%	1.25	$1\frac{1}{4}$
133.3% (rounded) or $133\frac{1}{3}\%$ (exact)	$1.\overline{3}$ rounds to 1.333	$1\frac{1}{3}$
150%	1.5	$1\frac{1}{2}$
166.7% (rounded) or $166\frac{2}{3}\%$ (exact)	$1.\overline{6}$ rounds to 1.667	$1\frac{2}{3}$
175%	1.75	$1\frac{3}{4}$
200%	2.0	2

6.2 EXERCISES

FOR EXTRA HELP: Student's Solutions Manual; MyMathLab.com; InterAct Math Tutorial Software; AW Math Tutor Center; www.mathxl.com; Digital Video Tutor CD 4 Videotape 10

Write each percent as a fraction or mixed number in lowest terms. See Examples 1 and 2.

1. 25% **2.** 60% **3.** 75% **4.** 80%

5. 85% **6.** 45% **7.** 62.5% **8.** 87.5%

9. 6.25% **10.** 43.75% **11.** $16\frac{2}{3}\%$ **12.** $83\frac{1}{3}\%$

13. $6\frac{2}{3}\%$ **14.** $46\frac{2}{3}\%$ **15.** 0.5% **16.** 0.8%

17. 120% **18.** 140% **19.** 375% **20.** 225%

Write each fraction as a percent. Round percents to the nearest tenth if necessary. See Example 3.

21. $\frac{1}{2}$ **22.** $\frac{4}{10}$ **23.** $\frac{4}{5}$ **24.** $\frac{3}{10}$

25. $\frac{1}{4}$ **26.** $\frac{3}{4}$ **27.** $\frac{37}{100}$ **28.** $\frac{63}{100}$

29. $\frac{5}{8}$ **30.** $\frac{1}{8}$ **31.** $\frac{7}{8}$ **32.** $\frac{3}{8}$

33. $\frac{9}{25}$ **34.** $\frac{15}{25}$ **35.** $\frac{23}{50}$ **36.** $\frac{18}{50}$

37. $\frac{3}{20}$ **38.** $\frac{9}{20}$ **39.** $\frac{5}{6}$ **40.** $\frac{1}{6}$

41. $\frac{5}{9}$ **42.** $\frac{7}{9}$ **43.** $\frac{1}{7}$ **44.** $\frac{5}{7}$

Complete the chart. Round decimals to the nearest thousandth and percents to the nearest tenth if necessary. See Examples 3 and 4.

	Fraction	Decimal	Percent
45.	______	0.5	______
46.	$\frac{3}{50}$	______	______
47.	______	______	87.5%
48.	$\frac{3}{4}$	______	______
49.	______	0.8	______
50.	______	______	60%
51.	$\frac{1}{6}$	______	______
52.	$\frac{1}{3}$	______	______
53.	______	0.25	______

Fraction	Decimal	Percent
54. ________	________	37.5%
55. ________	________	12.5%
56. $\frac{5}{8}$	________	________
57. $\frac{2}{3}$	________	________
58. ________	0.833 (rounded)	________
59. $\frac{2}{5}$	________	________
60. $\frac{3}{10}$	________	________
61. $\frac{8}{100}$	________	________
62. ________	________	100%
63. $\frac{1}{200}$	________	________
64. $\frac{1}{400}$	________	________

	Fraction	Decimal	Percent
65.	________	2.5	________
66.	________	1.7	________
67.	$3\frac{1}{4}$	________	________
68.	$2\frac{4}{5}$	________	________

69. Select a decimal percent and write it as a fraction. Select a fraction and write it as a percent. Write an explanation of each step of your work.

70. Prepare a table showing fraction, decimal, and percent equivalents for five common fractions and mixed numbers of your choice.

In the following application problems, write the answer as a fraction in lowest terms, as a decimal, and as a percent.

71. Only 13 adults out of every 100 adults consume the recommended 1000 milligrams of calcium daily. What portion consumes the recommended daily amount? (*Source:* Market Facts for Milk Mustache.)

72. About $\frac{1}{3}$ of all books purchased last year were for children. Of these children's books, 27 of every 100 purchased included a coloring activity. What portion of the children's books included a coloring activity? (*Source:* Consumer Research Study on Book Publishing, the American Booksellers, and the Book Industry Study Group.)

73. Many pet owners say they have used the Internet to find pet information. Of 500 people who used the Internet for this purpose, 90 said they used it when buying a pet. What portion used the Internet when buying a pet? (*Source:* American Animal Hospital Association.)

74. In a survey on how people learn to parent, 360 parents out of 800 said they were most influenced by relatives, friends, and spouses. What portion learned to parent this way? (*Source:* Bama Research.)

75. Of the 15 people at the office, 9 are single parents. What portion are single parents?

76. A pizza shop has 25 employees. Of these, 14 are students. What portion are students?

77. An insurance office has 80 employees. If 64 of the employees have cellular phones, what portion of the employees do *not* have cellular phones?

78. A zoo has 125 animals, including 25 that are endangered species. What portion is *not* endangered?

79. An antibiotic is used to treat 380 people. If 342 people do not have side effects from the antibiotic, find the portion that do have side effects.

80. An apple grower's cooperative has 250 members. If 100 of the growers use a certain insecticide, find the portion that do *not* use the insecticide.

The circle graph shows how many of the 4200 students at Metro-Community College use various types of transportation. Use this graph to answer Exercises 81–84, giving the answer as a fraction, as a decimal, and as a percent.

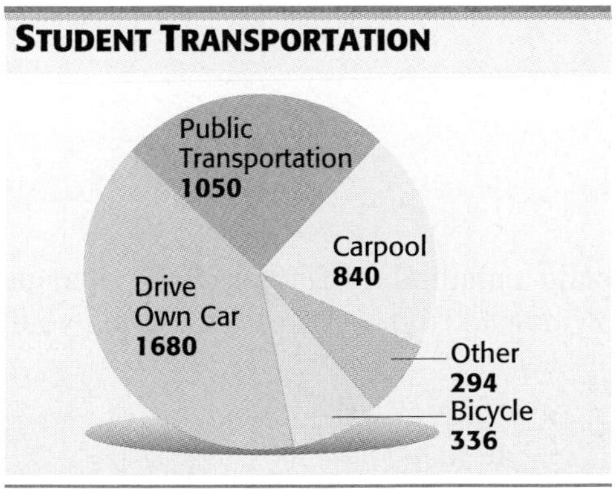

81. What portion of the students use public transportation?

82. What portion of the students use a bicycle?

83. Find the portion that drive their own cars.

84. Find the portion that carpool.

Relating Concepts (Exercises 85–94) For Individual or Group Work

Understanding the basics of percent is an important topic in mathematics.
Work Exercises 85–94 in order.

85. 100% of a number means all of the parts or ______ parts out of ______ parts. 200% means two times as many parts and 300% means three times as many parts.

86. Fill in the blanks.

(a) 100% of 765 workers is __________.

(b) 200% of 48 letters is __________.

(c) 300% of 271 videos is __________.

87. 50% of a number is ______ parts out of ______ parts, which is __________ of the parts.

88. 10% of a number is ______ parts out of ______ parts and is found by moving the decimal point ______ place to the ______.

89. 1% of a number is ______ part out of ______ parts and is found by moving the decimal point ______ places to the ______.

90. Fill in the blanks.

(a) 50% of 1050 homes is __________.

(b) 10% of 370 printers is __________.

(c) 1% of $10,500 sales is __________.

Using the methods for finding 1%, 10%, 50%, 100%, 200%, *and* 300%, *answer Exercises 91–94.*

91. Devise a method for finding 15% of a number. Use your method to find 15% of 160.

92. Devise a method for finding 150% of a number. Use your method to find 150% of 160.

93. Explain a method for finding 90% of a number. Show how to find 90% of $450 using your method.

94. Explain a method for finding 210% of a number. Show how to find 210% of $800 using your method.

6.3 Using the Percent Proportion and Identifying the Components in a Percent Problem

Objectives

1. Learn the percent proportion.
2. Solve for an unknown value in a proportion.
3. Identify the percent.
4. Identify the whole.
5. Identify the part.

There are two ways to solve percent problems. One method uses proportions and is discussed in this and the next section. The other method uses the percent equation and is explained in **Section 6.5.**

1 Learn the percent proportion. We have seen that a statement of two equivalent ratios is called a proportion.

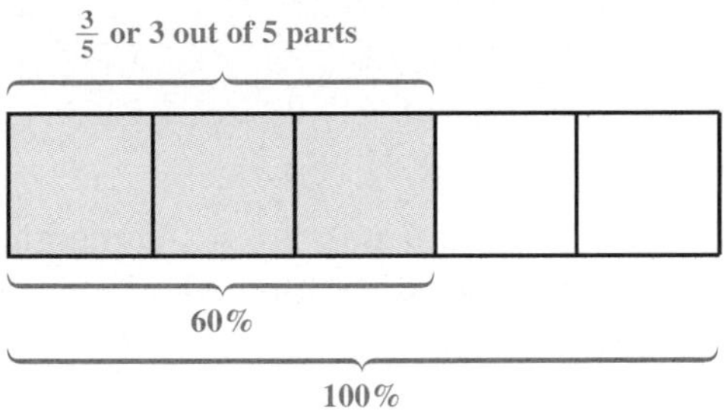

For example, the fraction $\frac{3}{5}$ is the same as the ratio 3 to 5, and 60% is the same as the ratio 60 to 100. As the figure shows, these two ratios are equivalent and make a proportion.

Work Problem 1 at the Side.

The **percent proportion** can be used to solve percent problems.

Percent Proportion

Part is to *whole* as percent is to 100.

$$\frac{\textbf{part}}{\textbf{whole}} = \frac{\textbf{percent}}{\textbf{100}} \quad \leftarrow \text{Always 100 because percent means per 100}$$

In the figure at the top of the page, the **whole** is 5 (the entire quantity), the **part** is 3 (the part of the whole), and the **percent** is 60. Write the percent proportion as follows.

$$\frac{\text{part}}{\text{whole}} \rightarrow \frac{3}{5} = \frac{60}{100} \leftarrow \frac{\text{percent}}{100}$$

2 Solve for an unknown value in a proportion. As shown in **Section 5.4,** if any two of the three values (part, whole, or percent) in the percent proportion are known, the third can be found by solving the proportion.

Example 1 Using the Percent Proportion

Use the percent proportion and solve for the unknown value. Let x represent the unknown value.

(a) part = 12, percent = 25; find the whole.

$$\frac{\text{part}}{\text{whole}} = \frac{\text{percent}}{100} \quad \text{Percent proportion}$$

$$\text{Part} \rightarrow \frac{\mathbf{12}}{\mathbf{x}} = \frac{\mathbf{25}}{\mathbf{100}} \leftarrow \text{Percent} \quad \text{or} \quad \frac{12}{x} = \frac{1}{4} \quad \text{Lowest terms}$$
Whole (unknown) → x

Continued on Next Page

1 As a review of proportions, use the method of cross products to decide whether each proportion is *true* or *false.*

(a) $\frac{1}{2} = \frac{25}{50}$

(b) $\frac{3}{5} = \frac{75}{125}$

(c) $\frac{7}{8} = \frac{180}{200}$

(d) $\frac{32}{53} = \frac{160}{265}$

(e) $\frac{112}{41} = \frac{332}{123}$

Answers

1. **(a)** true **(b)** true **(c)** false **(d)** true **(e)** false

2 Use the percent proportion $\left(\frac{\text{part}}{\text{whole}} = \frac{\text{percent}}{100}\right)$ and solve for the unknown value.

(a) part = 12, percent = 16

(b) part = 30, whole = 120

(c) whole = 210, percent = 20

(d) whole = 4000, percent = 32

(e) part = 74, whole = 185

Find the cross products to solve this proportion.

$$\frac{12}{x} = \frac{1}{4} \qquad x \cdot 1 \qquad 12 \cdot 4$$

Show that the cross products are equivalent.

$$x \cdot 1 = 12 \cdot 4$$
$$x = 48$$

The whole is 48.

CAUTION

You cannot divide out common factors from a numerator and denominator in a proportion. This can be done ***only*** when the fractions are being multiplied.

(b) part = 30, whole = 50; find the percent.
Use the percent proportion.

Percent (unknown) ↓

$$\text{Part} \rightarrow \frac{30}{50} = \frac{x}{100} \leftarrow \text{Whole} \qquad \text{Percent proportion}$$

$$\frac{3}{5} = \frac{x}{100} \qquad \text{Lowest terms}$$

$$5 \cdot x = 3 \cdot 100 \qquad \text{Cross products}$$

$$5 \cdot x = 300$$

$$\frac{\overset{1}{\cancel{5}} \cdot x}{\underset{1}{\cancel{5}}} = \frac{300}{5} \qquad \text{Divide each side by 5.}$$

$$x = 60$$

The percent is 60, written as 60%.

(c) whole = 150, percent = 18; find the part.

Percent ↓

$$\text{Part (unknown)} \rightarrow \frac{x}{150} = \frac{18}{100} \leftarrow \text{Whole} \quad \text{or} \quad \frac{x}{150} = \frac{9}{50} \qquad \text{Lowest terms}$$

$$x \cdot 50 = 150 \cdot 9 \qquad \text{Cross products}$$

$$x \cdot 50 = 1350$$

$$\frac{x \cdot \overset{1}{\cancel{50}}}{\underset{1}{\cancel{50}}} = \frac{1350}{50} \qquad \text{Divide each side by 50.}$$

$$x = 27$$

The part is 27.

Work Problem 2 at the Side.

ANSWERS
2. **(a)** whole = 75 **(b)** percent = 25 (so, the percent is 25%) **(c)** part = 42 **(d)** part = 1280 **(e)** percent = 40 (so, the percent is 40%)

As a help in solving percent problems, keep in mind this basic idea.

Percent Problems

All percent problems involve a comparison between a part of something and the whole.

Solving these problems requires identifying the three components of a percent proportion: part, whole, and percent.

3 **Identify the percent.** Look for the percent first. It is the easiest to identify.

Percent

The **percent** is the ratio of a part to a whole, with 100 as the denominator. In a problem, the percent appears with the word *percent* or with the symbol "%" after it.

Example 2 **Finding the Percent in Percent Problems**

Find the percent in the following.

(a) 32% of the 900 men were retired.
↓
Percent

The percent is 32. The number 32 appears with the symbol %.

(b) $150 is 25 percent of what number?
↓
Percent

The percent is 25 because 25 appears with the word *percent.*

(c) What percent of the 350 women will go?
↓
Percent (unknown)

The word *percent* has no number with it, so the percent is the unknown part of the problem.

Work Problem 3 at the Side.

4 **Identify the whole.** Next, look for the whole.

Whole

The **whole** is the entire quantity. In a percent problem, the whole often appears after the word **of**.

3 Identify the percent.

(a) Of the $1800, 18% will be spent on window coverings.

(b) Of the 620 children, 10% will have birthdays this month.

(c) Find the amount of sales tax by multiplying $590 and $6\frac{1}{2}$ percent.

(d) 105 is 3% of what number?

(e) What percent of the 380 guests will return this year?

ANSWERS

3. **(a)** 18 **(b)** 10 **(c)** $6\frac{1}{2}$ **(d)** 3
(e) The percent is unknown.

4 Identify the whole.

(a) Of the \$1800, 18% will be spent on window coverings.

(b) Of the 620 children, 10% will have birthdays this month.

(c) Find the amount of sales tax by multiplying sales of \$590 and $6\frac{1}{2}$ percent.

(d) \$105 is 3% of what number?

(e) What percent of the 380 guests will return this year?

5 Identify the part.

(a) Of the \$1800, 18%, or \$324 will be spent on window coverings.

(b) Of the 620 children, 10%, or 62 children, will have birthdays this month.

(c) Find the sales tax by multiplying \$590 and $6\frac{1}{2}$ percent.

(d) \$105 is 3% of what number?

(e) 80% of the 380 guests will return this year.

ANSWERS

4. **(a)** \$1800 **(b)** 620 **(c)** \$590 **(d)** what number (an unknown) **(e)** 380

5. **(a)** \$324 **(b)** 62 **(c)** unknown **(d)** \$105 **(e)** unknown

Example 3 Finding the Whole in Percent Problems

Identify the whole in the following.

(a) 32% **of** the 900 men were too large for the imported car.

900 → Whole

The whole is 900. The number 900 appears after the word *of*.

(b) \$150 is 25 percent **of** what number?

what number → Whole The whole is the unknown part of the problem.

(c) 85% **of** 7000 is what number?

7000 → Whole

Work Problem 4 at the Side.

5 **Identify the part.** Finally, look for the part.

Part

The **part** is the portion being compared with the whole.

NOTE

If you have trouble identifying the part, find the whole and percent first. The remaining number is the part.

Example 4 Finding the Part in Percent Problems

Identify the part in the following.

(a) 54% **of** 700 students is 378 students.
First find the percent and the whole.

54% **of** 700 students is 378 students.

54% → Percent; with % sign
700 → Whole; follows "of"

The remaining number is the part.

54% **of** 700 students is 378 students.

54% → Percent
700 → Whole
378 → Part

The part is 378.

(b) \$150 is 25% **of** what number?

25% → Percent
what number → Whole (unknown)

\$150 is the remaining number, so the part is \$150.

(c) 85% **of** \$7000 is what number?

85% → Percent
\$7000 → Whole
what number → Part (unknown)

Work Problem 5 at the Side.

6.3 EXERCISES

FOR EXTRA HELP: Student's Solutions Manual | MyMathLab.com | InterAct Math Tutorial Software | AW Math Tutor Center | www.mathxl.com | Digital Video Tutor CD 4 Videotape 10

Find the unknown value in the percent proportion $\frac{\text{part}}{\text{whole}} = \frac{\text{percent}}{100}$. *Round to the nearest tenth if necessary. If the answer is a percent, be sure to include a percent sign (%). See Example 1.*

1. part = 5, percent = 10

2. part = 20, percent = 25

3. part = 30, percent = 20

4. part = 25, percent = 25

5. part = 28, percent = 40

6. part = 11, percent = 5

7. part = 15, whole = 60

8. part = 105, whole = 35

9. part = 36, whole = 24

10. part = 1.5, whole = 4.5

11. part = 9.25, whole = 27.75

12. part = 12.8, whole = 9.6

13. whole = 52, percent = 50

14. whole = 160, percent = 35

15. whole = 72, percent = 30

16. whole = 115, percent = 38

17. whole = 94.4, part = 25

18. whole = 89.6, part = 50

Solve each problem. If the answer is a percent, be sure to include a percent sign (%). See Examples 2–4.

19. Find the whole if the part is 46 and the percent is 40.

20. The percent is 45 and the whole is 160. Find the part.

21. The whole is 5000 and the part is 20. Find the percent.

22. Suppose the part is 15 and the whole is 2500. Find the percent.

23. Find the percent if the whole is 4300 and the part is $107\frac{1}{2}$.

24. What is the part, if the percent is $12\frac{3}{4}$ and the whole is 5600?

25. The whole is 6480 and the part is 19.44. Find the percent.

26. Suppose the part is 281.25 and the percent is $1\frac{1}{4}$. Find the whole.

Identify the percent, whole, and part in the following. Do not try to solve for any unknowns. See Examples 2–4.

	Percent	Whole	Part
27. 10% of how many bicycles is 60 bicycles?	________	________	________
28. 58% of how many preschoolers is 203 preschoolers?	________	________	________
29. 75% of \$800 is \$600.	________	________	________
30. 93% of \$1500 is \$1395.	________	________	________
31. What is 25% of \$970?	________	________	________
32. What is 61% of 830 homes?	________	________	________
33. 12 injections is 20% of what number of injections?	________	________	________
34. 92 servings is 26% of what number of servings?	________	________	________
35. 34 trophies is 50% of 68 trophies.	________	________	________
36. 410 pallets is $33\frac{1}{3}$% of 1230 pallets.	________	________	________
37. What percent of \$296 is \$177.60?	________	________	________
38. What percent of \$120.80 is \$30.20?	________	________	________
39. 54.34 is 3.25% of what number?	________	________	________
40. 16.74 is 11.9% of what number?	________	________	________
41. 0.68% of \$487 is what amount?	________	________	________
42. What amount is 6.21% of \$704.35?	________	________	________

43. Identify the three components in a percent problem. In your own words, write one sentence telling how you will identify each of these three components.

44. Write one short sentence or statement using numbers and words. The statement should include a percent, a whole, and a part. Identify each of these three components.

Find the percent, whole, and part in each application problem. Do not try to solve for any unknowns.

45. In a tree-planting project, 640 of the 810 trees planted were still living 1 year later. What percent of the trees planted were still living?

46. Ivory Soap is $99\frac{44}{100}$% pure. If a bar of Ivory Soap weighs 9 ounces, how many ounces are pure? (*Source:* Procter & Gamble.)

47. Of the 142 people attending a movie theater, 86 bought buttered popcorn. What percent bought buttered popcorn?

48. On her first check from the Pizza Hut Restaurant, 15% is withheld from Maria's total earnings of \$225. What amount is withheld?

49. Of the lunch and dinner customers at the Heston Grill, 23% prefer a fat-free salad dressing. If the total number of customers is 610, find the number who prefer fat-free dressing.

50. There are 680 1-gigahertz computer chips in a secured storage area designed to hold 2000 computer chips. What percent of the storage area is filled?

51. Of the total candy bars contained in a vending machine, 240 bars have been sold. If 25% of the bars have been sold, find the total number of candy bars that were in the machine.

52. There have been 36 cups of coffee served from a banquet-sized coffee pot. If this is 30% of the capacity of the pot, find the capacity of the pot.

53. In a recent survey of 480 adults, 55% said that they would prefer to have their wedding at a religious site. How many said they would prefer the religious site? (*Source:* National Family Opinion Research.)

54. Sue Ann needs 64 credits to graduate. If she has completed 48 of the credits needed, what percent of the credits has she already completed?

55. In a poll of 822 people, 49.5% said that they get their news from television. Find the number of people who said they get their news from television. (*Source: Brills Content.*)

56. The sales tax on a new car is $820. If the sales tax rate is 5%, find the price of the car before the sales tax is added.

57. A medical clinic found that 16.8% of the patients were late for their appointments. The number of patients who were late was 504. Find the total number of patients.

58. The state troopers tested 924 cars for safety. There were 231 cars that failed the safety test for one or more reasons. Find the percent of cars that failed the test.

6.4 Using Proportions to Solve Percent Problems

This is the percent proportion.

$$\frac{\text{part}}{\text{whole}} = \frac{\text{percent}}{100}$$

As discussed in Section 6.3, if any two of the three values are known, the third can be found by solving the percent proportion.

1 Use the percent proportion to find the part. The first example shows how to use the percent proportion to find the part.

Example 1 Finding the Part with the Percent Proportion

Find 15% of \$160.

Here the percent is 15 and the whole is 160. (Recall that the whole often comes after the word *of*.) Now find the part. Let x represent the unknown part.

$$\frac{\text{part}}{\text{whole}} = \frac{\text{percent}}{100} \quad \text{so} \quad \frac{x}{160} = \frac{15}{100} \quad \text{or} \quad \frac{x}{160} = \frac{3}{20} \qquad \text{Lowest terms}$$

Find the cross products in the proportion.

$$x \cdot 20 = 160 \cdot 3 \qquad \text{Cross products}$$

$$x \cdot 20 = 480$$

$$\frac{x \cdot \overset{1}{\cancel{20}}}{\underset{1}{\cancel{20}}} = \frac{480}{20} \qquad \text{Divide each side by 20.}$$

$$x = 24 \qquad \text{Part}$$

15% of \$160 is **\$24**.

Work Problem 1 at the Side.

Just as with the application problems given earlier, the word ***of*** is an indicator word meaning ***multiply.*** For example:

15% of 160
↓
15% • 160.

Because of this, there is another way to find the part.

Finding the Part Using Multiplication

To find the part:

Step 1 Identify the percent. Write the percent as a decimal.

Step 2 Multiply this decimal by the whole.

Objectives

1. Use the percent proportion to find the part.
2. Find the whole using the percent proportion.
3. Find the percent using the percent proportion.

1 Use the percent proportion to find the part.

(a) 10% of 1250 sailboats

(b) 15% of \$3220

(c) 7% of 2700 miles

(d) 39% of 1220 meters

Answers

1. (a) 125 sailboats **(b)** \$483 **(c)** 189 miles **(d)** 475.8 meters

2 Use multiplication to find the part.

(a) 55% of 10,000 injections

(b) 16% of 120 miles

(c) 135% of 60 dosages

(d) 0.5% of $238

Example 2 **Finding the Part by Using Multiplication**

Use multiplication to find the part.

(a) Find 42% of 830 yards.

Step 1 Here, the percent is 42. Write 42% as the decimal 0.42.

Step 2 Multiply 0.42 and the whole, which is 830.

$$\text{part} = 0.42 \cdot 830$$
$$= 348.6 \text{ yd}$$

It is a good idea to estimate the answer, to make sure no mistakes were made with decimal points. Round 42% to 40% or 0.4, and round 830 as 800. Next, 40% of 800 is

$$0.4 \cdot 800 = 320, \leftarrow \text{Estimate.}$$

so 348.6 is a reasonable answer.

(b) Find 25% of 1680 cars.

Identify the percent as 25. Write 25% in decimal form as 0.25. Now, multiply 0.25 and 1680.

$$\text{part} = 0.25 \cdot 1680 = 420 \text{ cars} \quad \text{Multiply.}$$

You can use a shortcut to estimate the answer. Since 25% means 25 parts out of 100 parts, this is the same as $\frac{1}{4}$ of the whole ($\frac{25}{100} = \frac{1}{4}$). Do you see a shortcut here? You can find $\frac{1}{4}$ of a number by dividing the number by 4. So, this shortcut gives us the exact answer, $1680 \div 4 = 420$.

(c) Find 140% of 60 miles.

In this problem, the percent is 140. Write 140% as the decimal 1.40. Next, multiply 1.40 and 60.

$$\text{part} = 1.40 \cdot 60 = 84 \text{ miles} \quad \text{Multiply.}$$

You can estimate the answer by realizing that 140% is close to 150% (which is $1\frac{1}{2}$) and $1\frac{1}{2}$ times 60 is 90. So, 84 miles is a reasonable answer.

(d) Find 0.4% of 50 kilometers.

$$\text{part} = 0.004 \cdot 50 = 0.2 \text{ kilometers} \quad \text{Multiply.}$$

↑ Write 0.4% as a decimal.

Estimate the answer. 0.4% is less than 1%.

$$1\% \text{ of } 50 \text{ miles} = 50. = 0.5 \text{ miles}$$

So our answer should be *less than* 0.5 miles, and 0.2 miles fits this requirement.

Work Problem 2 at the Side.

Example 3 **Solving for the Part in an Application Problem**

Raley's Markets has 850 employees. Of these employees, 28% are students. How many of the employees are students?

Continued on Next Page

ANSWERS
2. **(a)** 5500 injections **(b)** 19.2 miles **(c)** 81 dosages **(d)** $1.19

Step 1 **Read** the problem. The problem asks us to find the number of employees who are students.

Step 2 **Work out a plan.** Look for the word *of* as an indicator word for multiplication.

28% **of** the employees are students.

↑ Indicator word

The total number of employees is 850, so the whole is 850. The percent is 28. To find the number of students, find the part.

Step 3 **Estimate** a reasonable answer. You can estimate the answer by rounding 28% to 25% and 850 to 900. Remember 25% is 25 parts out of 100, which is equivalent to $\frac{1}{4}$. So divide 900 by 4.

$900 \div 4 = 225$ students ← Estimate

Step 4 **Solve** the problem.

part $= 0.28 \cdot 850 = 238$ Multiply.

↑ Write 28% as a decimal.

Step 5 **State the answer.** Raley's Markets has 238 student employees.

Step 6 **Check.** The answer, 238 students, is close to our estimate of 225 students.

Work Problem 3 at the Side.

Calculator Tip If you are using a calculator, you could solve Example 3 like this.

0.28 (×) 850 (=) 238

Or, you can use this alternate approach on calculators with a % key.

850 (×) 28 (%) (=) 238

2 Fiind the whole using the percent proportion. The next example shows how to use the percent proportion to find the whole.

NOTE

Remember, the whole is the entire quantity.

Example 4 Finding the Whole with the Percent Proportion

(a) 8 tables is 4% of what number of tables?

Here the percent is 4, the whole is unknown, and the part is 8. Use the percent proportion to find the whole. Let x represent the unknown whole.

$$\frac{\text{part}}{\text{whole}} = \frac{\text{percent}}{100} \quad \text{so} \quad \frac{8}{x} = \frac{4}{100} \quad \text{or} \quad \frac{8}{x} = \frac{1}{25} \quad \text{Lowest terms}$$

Find the cross products.

$$x \cdot 1 = 8 \cdot 25$$
$$x = 200$$

8 tables is 4% of **200 tables**.

Continued on Next Page

3 Use the six problem-solving steps to solve each problem.

(a) One day on Jacob's mail route there were 2920 pieces of mail. If 45% of those were advertising pieces, find the number of advertising pieces.

(b) There are 8550 students attending a college. If 28% of the students drink coffee, find the number of coffee drinkers.

ANSWERS

3. (a) 1314 advertising pieces

(b) 2394 coffee drinkers

4 Use the percent proportion to find the unknown whole.

(a) 150 tickets is 25% of what number of tickets?

(b) 28 antiques is 35% of what number of antiques?

(c) 387 customers is 36% of what number of customers?

(d) 292.5 miles is 37.5% of what number of miles?

(b) 135 tourists is 15% of what number of tourists?
The percent is 15 and part is 135, so

$$\text{Part} \rightarrow \frac{135}{x} = \frac{15}{100} \leftarrow \text{Percent} \quad (\text{Whole (unknown)} \rightarrow x)$$

$$\frac{135}{x} = \frac{3}{20} \quad \text{Lowest terms}$$

$$x \cdot 3 = 135 \cdot 20 \quad \text{Cross products}$$

$$x \cdot 3 = 2700$$

$$\frac{x \cdot \cancel{3}^{1}}{\cancel{3}_{1}} = \frac{2700}{3} \quad \text{Divide each side by 3.}$$

$$x = 900.$$

135 tourists is 15% of **900 tourists**.

Work Problem 4 at the Side.

Example 5 Applying the Percent Proportion

At Newark Salt Works, 78 employees are absent because of illness. If this is 5% of the total number of employees, how many employees does the company have?

Step 1 **Read** the problem. The problem asks for the total number of employees.

Step 2 **Work out a plan.** From the information in the problem, the percent is 5 and the part of the total number of employees is 78. The total number of employees or entire quantity, which is the whole, is the unknown.

Step 3 **Estimate** a reasonable answer. Round the number of employees from 78 to 80. Then, 5% is equivalent to the fraction $\frac{1}{20}$. Since 80 is $\frac{1}{20}$ of the total number of employees,

$$80 \cdot 20 = 1600 \text{ employees} \leftarrow \text{Estimate}$$

Step 4 **Solve** the problem. Use the percent proportion to find the whole (the total number of employees).

$$\text{Part} \rightarrow \frac{78}{x} = \frac{5}{100} \leftarrow \text{Percent} \quad (\text{Whole (unknown)} \rightarrow x)$$

$$\frac{78}{x} = \frac{1}{20} \quad \text{Lowest terms}$$

Find the cross products.

$$x \cdot 1 = 78 \cdot 20$$

$$x = 1560$$

Step 5 **State the answer.** The company has **1560 employees**.

Step 6 **Check.** The answer, 1560 employees, is close to our estimate of 1600 employees.

ANSWERS
4. **(a)** 600 tickets **(b)** 80 antiques
(c) 1075 customers **(d)** 780 miles

NOTE

To estimate the answer to Example 5, the 5% was changed to its fraction equivalent $\frac{1}{20}$. Because 80 (rounded) is $\frac{1}{20}$ of the total employees, 80 was multiplied by 20 to get 1600, the estimated answer.

Work Problem 5 at the Side.

3 Find the percent using the percent proportion. Finally, if the part and the whole are known, the percent proportion can be used to find the percent.

Example 6 Using the Percent Proportion to Find the Percent

(a) 13 roofs is what percent of 52 roofs?
The whole is 52 (follows *of*) and the part is 13. Next, find the percent.

$$\frac{\text{part}}{\text{whole}} = \frac{\text{percent}}{100}$$

$$\text{Part} \rightarrow \frac{13}{52} = \frac{x}{100} \leftarrow \text{Percent (unknown)} \quad (\text{Whole} \rightarrow 52)$$

$$\frac{1}{4} = \frac{x}{100} \quad \text{Lowest terms}$$

Find the cross products.

$$4 \cdot x = 1 \cdot 100$$

$$\frac{\overset{1}{\cancel{4}} \cdot x}{\underset{1}{\cancel{4}}} = \frac{100}{4} \quad \text{Divide each side by 4.}$$

$$x = 25$$

13 roofs is **25%** of 52 roofs.

(b) What percent of \$500 is \$100?
The whole is 500 (follows *of*) and the part is 100, so

$$\frac{100}{500} = \frac{x}{100} \leftarrow \text{Percent (unknown)}$$

$$\frac{1}{5} = \frac{x}{100} \quad \text{Lowest terms}$$

$$5 \cdot x = 1 \cdot 100 \quad \text{Cross products}$$

$$5 \cdot x = 100$$

$$\frac{\overset{1}{\cancel{5}} \cdot x}{\underset{1}{\cancel{5}}} = \frac{100}{5} \quad \text{Divide each side by 5.}$$

$$x = 20$$

20% of \$500 is \$100.

5 Use the six problem-solving steps and the percent proportion to solve each problem.

(a) A freeze resulted in a loss of 52% of an avocado crop. If the loss was 182 tons, find the total number of tons in the crop.

(b) A metal alloy contains 450 pounds of zinc, which is 8% of the alloy. Find the total weight of the alloy.

ANSWERS
5. **(a)** 350 tons **(b)** 5625 pounds

6 Use the percent proportion to solve each problem.

(a) \$21 is what percent of \$105?

(b) What percent of 320 Internet companies is 48 Internet companies?

(c) What percent of 2280 court trials is 1026 trials?

(d) 432 snowboarders is what percent of 108 snowboarders?

7 Solve each problem.

(a) The bid price on an auction item is \$289 while the minimum acceptable price is \$425. What percent of the minimum acceptable price is the bid price?

(b) A late-model domestic car gets 38 miles per gallon on the highway and 32.3 miles per gallon around town. What percent of the highway mileage does the car get around town?

ANSWERS

6. **(a)** 20% **(b)** 15% **(c)** 45% **(d)** 400%

7. **(a)** 68% **(b)** 85%

CAUTION

When finding the percent, be sure to label your answer with the percent symbol (%).

Work Problem 6 at the Side.

Example 7 Applying the Percent Proportion

A roof is expected to last 20 years before needing replacement. If the roof is now 15 years old, what percent of the roof's life has been used?

Step 1 **Read** the problem. The problem asks for the percent of the roof's life that is already used.

Step 2 **Work out a plan.** The expected life of the roof is the entire quantity or whole, which is 20. The part of the roof's life that is already used is 15. Use the percent proportion to find the percent of the roof's life used.

Step 3 **Estimate** a reasonable answer. Since the roof is 15 years old, it is $\frac{15}{20}$ or $\frac{3}{4}$ used. Remember that $\frac{3}{4}$ is equivalent to 75%, our estimate.

Step 4 **Solve** the problem. Let x represent the unknown percent.

$$\begin{matrix}\text{Part} \rightarrow \\ \text{Whole} \rightarrow\end{matrix} \frac{15}{20} = \frac{x}{100} \quad \text{or} \quad \frac{3}{4} = \frac{x}{100} \quad \text{Lowest terms}$$

Find the cross products.

$$4 \cdot x = 3 \cdot 100$$

$$4 \cdot x = 300$$

$$\frac{\overset{1}{\cancel{4}} \cdot x}{\underset{1}{\cancel{4}}} = \frac{300}{4} \quad \text{Divide each side by 4.}$$

$$x = 75$$

Step 5 **State the answer.** 75% of the roof's life has been used.

Step 6 **Check.** The answer, 75%, matches our estimate of 75%.

Work Problem 7 at the Side.

Example 8 Applying the Percent Proportion

Rainfall this year was 33 in., while normal rainfall is only 30 in. What percent of normal rainfall is this year's rainfall?

Step 1 **Read** the problem. The problem asks us to find what percent this year's rainfall is of normal rainfall.

Step 2 **Work out a plan.** The normal rainfall is the whole, which is 30. This year's rainfall is all of normal rainfall and more, or 33 (part = 33). You need to find the percent that this year's rainfall is of normal rainfall.

Continued on Next Page

Step 3 **Estimate** a reasonable answer. The increase in rainfall is 3 inches and the whole is 30 inches. The increase is $\frac{3}{30} = \frac{1}{10}$ or 10%. The whole is 100%, so 100% + 10% = 110%, our estimate.

Step 4 **Solve** the problem. Let x represent the unknown percent.

$$\frac{33}{30} = \frac{x}{100} \quad \text{or} \quad \frac{11}{10} = \frac{x}{100} \quad \text{Lowest terms}$$

Find the cross products.

$$10 \cdot x = 11 \cdot 100$$

$$10 \cdot x = 1100$$

$$\frac{\overset{1}{\cancel{10}} \cdot x}{\underset{1}{\cancel{10}}} = \frac{1100}{10}$$

$$x = 110$$

Step 5 **State the answer.** This year's rainfall is 110% of normal rainfall.

Step 6 **Check.** The answer, 110%, matches our estimate of 110%.

Work Problem 8 at the Side.

8 Solve each problem.

(a) The number of students who usually take this class is 300. If 450 students are taking this class now, find the percent of the usual number who are now taking the class.

(b) The service department set a goal of 360 service calls this week. If they made 504 service calls, find the percent of their goal that they completed.

ANSWERS

8. **(a)** 150% **(b)** 140%

Focus on Real-Data Applications

Educational Tax Incentives

The government sponsors tax incentive programs to make education more affordable. To qualify for the programs, you have to have an adjusted gross income below a certain level (typically $40,000). You can find specific information at the Internal Revenue Service Web site: www.irs.ustreas.gov.

- The Hope Scholarship offers 100% of the first $1000 spent for certain expenses, such as tuition and books, during the first year of college, plus 50% of the next $1000 incurred during the second year of college. The scholarship money is payable as a tax refund. The student cannot have completed the first two years of post-secondary education and must meet certain educational goals and workload criteria.
- Lifetime Learning Credits are based on qualified expenses, including tuition and books, and equal 20% of the first $5000 in expenses. It is not based on a student's workload and is not limited to only two years.
- Only one of the credits can be claimed for each student.

Suppose you are paying your own educational costs, and your adjusted gross income meets the guidelines to qualify for the Hope Scholarship or Lifetime Learning Credits. Your goals are to earn an Associate of Arts degree from a community college and then transfer to a state university to complete a Bachelor's degree. Tuition costs for resident students at North Harris Montgomery Community College District (NHMCCD) in Texas are used as an example of educational expenses. (Expenses vary among schools, and you can easily find that information in the college's catalog or Internet site.)

Residents of NHMCCD pay $12 registration fee for each semester enrolled plus $30 per semester hour tuition and fees. Assume that you must study a total of 15 semester hours in developmental work in mathematics, reading, and writing, and to complete an Associate of Arts degree you must study 60 additional semester hours. You decide to limit your course load to 15 credit hours each semester. Assume that one course is 3 semester hours, and you will have to purchase books at an approximate cost of $75 per course.

1. How many semesters and how many courses will it take you to finish the requirements for an Associate of Arts degree?
2. What is the total cost to complete the Associate of Arts degree for **(a)** books and **(b)** tuition and fees?
3. Calculate the total costs for tuition, fees, and books during the first two years (four semesters). What is the maximum tax incentive payable under the Hope Scholarship during **(a)** the first year and **(b)** the second year?
4. Assume that you are returning to school and do not qualify for the Hope Scholarship. What is the maximum tax incentive payable under the Lifetime Learning Credits during **(a)** the first year and **(b)** the second year?
5. How much additional tax incentive would be payable under the Lifetime Learning Credits for the remaining coursework to complete the Associate of Arts degree? How much additional tax incentive would be payable under the Lifetime Learning Credits to complete a Bachelor's degree?

6.4 EXERCISES

FOR EXTRA HELP: Student's Solutions Manual | MyMathLab.com | InterAct Math Tutorial Software | AW Math Tutor Center | MathXL www.mathxl.com | Digital Video Tutor CD 4 Videotape 10

Find the part using the multiplication shortcut. See Example 2.

1. 20% of 80 guests

2. 25% of 3500 salespeople

3. 45% of 4080 military personnel

4. 12% of 3650 Web sites

5. 4% of 120 ft

6. 9% of $150

7. 150% of 210 files

8. 130% of 60 trees

9. 52.5% of 1560 trucks

10. 38.2% of 4250 loads

11. 2% of $164

12. 6% of $434

13. 225% of 680 tables

14. 135% of 800 commuters

15. 17.5% of 1040 homes

16. 46.1% of 843 kilograms

17. 0.9% of $2400

18. 0.3% of $1400

Find the whole using the percent proportion. See Example 4.

19. 80 e-mails is 25% of what number of e-mails?

20. 32 medical exams is 5% of what number of medical exams?

21. 30% of what number of hay bales is 48 hay bales?

22. 55% of what number of experiments is 209 experiments?

23. 495 successful students is 90% of what number of students?

24. 84 letters is 28% of what number of letters?

25. 1496 graduates is 110% of what number of graduates?

26. 154 bicycles is 140% of what number of bicycles?

27. $12\frac{1}{2}\%$ of what number is 350?
$\left(\textit{Hint: } 12\frac{1}{2}\% = 12.5\%\right)$

28. $5\frac{1}{2}\%$ of what number is 176?

Find the percent using the percent proportion. Round your answers to the nearest tenth if necessary. See Example 6.

29. 18 bean burritos is what percent of 36 bean burritos?

30. 62 rooms is what percent of 248 rooms?

31. 26 desks is what percent of 50 desks?

32. 650 liters is what percent of 1000 liters?

33. 32 patients is what percent of 400 patients?

34. 7 bridges is what percent of 350 bridges?

35. 9 rolls is what percent of 600 rolls?

36. 60 cartons is what percent of 2400 cartons?

37. What percent of \$344 is \$64?

38. What percent of \$398 is \$14?

39. What percent of 250 tires is 23 tires?

40. What percent of 105 employees is 54 employees?

41. A student turned in the following answers on a test. You can tell that two of the answers are incorrect without even working the problems. Find the incorrect answers and explain how you identified them (without actually solving the problems).

50% of $84 is <u>$42</u>.

150% of $30 is <u>$20</u>.

25% of $16 is <u>$32</u>.

100% of $217 is <u>$217</u>.

42. Write a percent problem on any topic you choose. Be sure to include only two of the three components so that you can solve for the third component. Identify each component of the problem and then solve it.

Solve each application problem. Round percent answers to the nearest tenth if necessary. See Examples 3, 5, 7, and 8.

43. Aimee Toit, who works part-time, earns $240 per week and has 22% of this amount withheld for taxes, social security, and Medicare. Find the amount withheld.

44. Kirsten Speed needs 124 credits to graduate. If she has already completed 75% of the necessary credits, find the number of credits completed.

45. The guided-missile destroyer USS *Sullivans* has a 335-person crew of which 13% are female. Find the number of female crew members. Round to the nearest whole number. (*Source:* U.S. Navy.)

46. A recent survey of 377 top executives found that 62% of them were upbeat about global growth for the next three years. Find the number of executives that were upbeat about global growth. Round to the nearest whole number.

47. This year, there are 550 scholarship applicants. If 40% of the applicants will receive a scholarship, find the number of students who will receive a scholarship.

48. A U.S. Food and Drug Administration (FDA) biologist found that canned tuna is "relatively clean." Extraneous matter was found in 5% of the 1600 cans of tuna tested. How many cans of tuna contained extraneous matter?

49. On average, a family of two adults and two children pays $95 per night for lodging and $104 a day for food while on vacation this year. If this increases by 3% over the next year, find the total cost per day for lodging and meals for such a family after the increase.

50. The sales of Tropicana pure orange juice in China were only $100 million this year. Market researchers estimate that sales will increase by 35% next year. Find the amount of orange juice sales estimated for next year. (*Source:* Seagram's Tropicana Beverage Group.)

51. A recent study examined 48,000 military jobs, such as Army attack helicopter pilot or Navy gunner's mate. It was found that only 960 of these jobs are filled by women. What percent of these jobs are filled by women? (*Source:* Rand's National Defense Research Institute.)

52. There are more than 55,000 words in Webster's Dictionary, but most educated people can identify only 20,000 of these words. What percent of the words in the dictionary can these people identify?

53. The size of Cuba's labor force is 3.8 million people. If 61% of the labor force is male, find **(a)** the percent who are female and **(b)** the number of workers who are male.

54. In the United States there are 132 million people in the labor force. If 54% of the labor force is male, find **(a)** the percent of the labor force who are female and **(b)** the number of workers who are male.

55. This month's sales goal for Easy Writer Pen Company is 2,380,000 ballpoint pens. If sales of 2,618,000 pens have been made, what percent of the goal has been reached?

56. The number of apartment unit vacancies was predicted to be 2112, while the actual number of vacancies was 2640. The actual number was what percent of the predicted number of vacancies?

57. Americans who are 65 years old or older make up 12.7% of the total population. If there are 35.7 million Americans in this group, find the total U.S. population. (Round to the nearest tenth of a million.)

58. About 61% of the 43,000,000 people who receive Social Security benefits are paid with a direct deposit to their bank. How many of the people receiving benefits are paid with a direct deposit?

The graph (pictograph) shows the percent of chicken noodle soup sold during the cold-and-flu season. Use this information to answer Exercises 59–62.

59. Which of the cold-and-flu season months had the lowest sales of chicken noodle soup?

60. What percent of the chicken noodle soup sales take place in the *non-flu* season months?

61. Find the number of cans of soup sold in the highest sales month.

62. How many more cans of soup were sold in October than in November?

The circle graph shows the sales at fast-food hamburger chains as a percent of total fast-food hamburger sales. Use this information to answer Exercises 63–66.

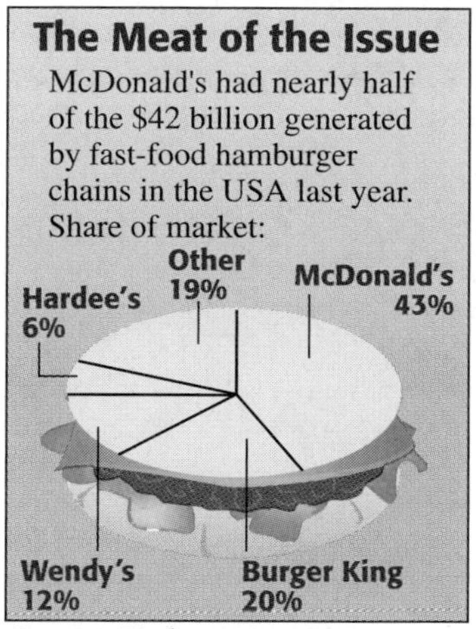

The Meat of the Issue
McDonald's had nearly half of the $42 billion generated by fast-food hamburger chains in the USA last year.
Share of market:
Other 19%
McDonald's 43%
Hardee's 6%
Wendy's 12%
Burger King 20%

Source: Technomic.

63. Which of the hamburger chains had the lowest sales?

64. What percent of the total hamburger sales were made by the two companies having the lowest sales?

65. Find the total annual sales for McDonald's.

66. Find the total annual sales for Burger King.

67. A collection agency, specializing in collecting past-due child support, charges $25 as an application fee plus 20% of the amount collected. What is the total charge for collecting $3100 in past-due child support?

68. Raw steel production by the nation's steel mills decreased by 2.5% from last week. The decrease amounted to 50,475 tons. Find the steel production last week.

69. Marketing Intelligence Service says that there were 15,401 new products introduced last year. If 86% of the products introduced last year failed to reach their business objectives, find the number of products that were successful. (Round to the nearest whole number.)

70. A family of four with a monthly income of $2900 spends 90% of its earnings and saves the balance. Find **(a)** the monthly savings and **(b)** the annual savings of this family.

Relating Concepts (Exercises 71–77) For Individual or Group Work

Knowing and using the percent proportion is useful when solving percent problems.
Work Exercises 71–77 in order.

71. In the percent proportion, part is to ______ as percent is to ______.

72. All percent problems involve a comparison between a part of something and the ______.

Use this Ramen Noodles label of nutrition facts to answer Exercises 73–77.
Read the label very carefully. *Round to the nearest tenth of a percent.*

Nutrition Facts	Amount/serving	%DV*	Amount/serving	%DV*
Serving Size 1/2 Pkg. (15 oz/42.5 g)	**Total Fat 8g**	12%	**Total Carbohydrates 27g**	9%
Servings Per Package 2	Saturated Fat 4g	20%	Dietary Fiber Less Than 1g	3%
Calories 190	**Cholesterol** 0mg	0%	Sugars Less Than 1g	
Calories from Fat 70	**Sodium** 670mg	28%	**Protein** 4g	
	Vitamin A 0% • Vitamin C 0% • Calcium 2% • Iron 4%			
*Percent Daily Values (DV) are based on a 2,000 calorie diet.	**Calories Per Gram** Fat 9 • Carbohydrates 4 • Protein 4			

73. How many calories per serving are from total carbohydrates?

74. How many grams of total carbohydrates are needed to meet the "% DV" (percent daily value)?

75. Find the number of grams of total fat that are needed to meet the "% DV" (percent daily value). Round to the nearest gram.

76. Will a person eating two packages of Ramen Noodles in one day exceed their recommended daily value of sodium? Explain your answer.

77. How many packages of Ramen Noodles must be eaten in a day to meet the recommended daily value of fiber? Would this be possible? Would this result in good nutrition? (Round to the nearest whole package.)

6.5 Using the Percent Equation

Objectives

1. Use the percent equation to find the part.
2. Find the whole using the percent equation.
3. Find the percent using the percent equation.

In the last section you were shown how to use a proportion to solve percent problems. In this section we show another way to solve these problems by using the **percent equation.** The percent equation is just a rearrangement of the percent proportion.

Percent Equation

$$\text{part} = \text{percent} \cdot \text{whole}$$

Be sure to write the percent as a decimal before using the equation.

When using the percent proportion, we did *not* have to write the percent as a decimal because of the 100 in the denominator of the proportion. However, because there is no 100 in the percent *equation,* we *must* first write the percent as a decimal by dividing by 100.

Some of the examples solved earlier will be reworked by using the percent equation. If you want to, you can look back at **Section 6.4** to see how some of these same problems were solved using proportions. This will give you a comparison of the two methods.

1 Use the percent equation to find the part. The first example shows how to find the part.

Example 1 Finding the Part

(a) Find 15% of \$160.

Write 15% as the decimal 0.15. The whole, which comes after the word *of,* is 160. Next, use the percent equation. Let x represent the unknown part.

$$\text{part} = \text{percent} \cdot \text{whole}$$
$$x = 0.15 \cdot 160$$

Multiply 0.15 and 160 to get

$$x = 24.$$

15% of \$160 is **\$24**.

(b) Find 110% of 80 cases.

Write 110% as the decimal 1.10. The whole is 80. Let x represent the unknown part.

$$\text{part} = \text{percent} \cdot \text{whole}$$
$$x = 1.10 \cdot 80$$
$$x = 88$$

110% of 80 cases is **88** cases.

(c) Find 0.4% of 250 patients.

Write 0.4% as the decimal 0.004. The whole is 250. Let x represent the unknown part.

$$\text{part} = \text{percent} \cdot \text{whole}$$
$$x = 0.004 \cdot 250$$
$$x = 1$$

0.4% of 250 patients is **1 patient**.

Continued on Next Page

1 Use the percent equation to find the part.

(a) 14% of 550 magazine titles

(b) 23% of 840 gallons

(c) 120% of $220

(d) 135% of $1080

(e) 0.5% of 1200 test tubes

(f) 0.25% of 1600 lab tests

ANSWERS
1. **(a)** 77 magazine titles **(b)** 193.2 gallons **(c)** $264 **(d)** $1458 **(e)** 6 test tubes **(f)** 4 lab tests

Estimate the answer. 0.4% is approximately 0.5 or $\frac{1}{2}$ of 1%. Because 1% is $\frac{1}{100}$, 1% of 250 is

$$250 \div 100 = 2.5$$

Since 2.5 is equal to 1%, half of 2.5 is equal to 1.25 (2.5 ÷ 2 = 1.25). So, 1 patient is a reasonable answer.

CAUTION

When using the percent equation, the percent must always be *changed to a decimal* before multiplying.

Work Problem 1 at the Side.

2 Find the whole using the percent equation. The next example shows how to use the percent equation to find the whole.

NOTE

Remember that the word *of* is an indicator word for *multiply.*

Example 2 Solving for the Whole

(a) 8 tables is 4% of what number of tables?

The part is 8 and the percent is 4% or the decimal 0.04. The whole is unknown.

8 is 4% **of** what number? ← Indicator word ("of")

Next, use the percent equation

$$\textbf{part} = \textbf{percent} \cdot \textbf{whole}$$

$$8 = 0.04 \cdot x \quad \text{Let } x \text{ represent the unknown whole.}$$

$$\frac{8}{0.04} = \frac{\overset{1}{\cancel{0.04}} \cdot x}{\underset{1}{\cancel{0.04}}} \quad \text{Divide each side by 0.04.}$$

$$200 = x \leftarrow \text{Whole}$$

8 tables is 4% of **200 tables.**

(b) 135 tourists is 15% of what number of tourists?

Write 15% as 0.15. The part is 135. Next, use the percent equation to find the whole.

$$\textbf{part} = \textbf{percent} \cdot \textbf{whole}$$

$$135 = 0.15 \cdot x \quad \text{Let } x \text{ represent the unknown whole.}$$

$$\frac{135}{0.15} = \frac{\overset{1}{\cancel{0.15}} \cdot x}{\underset{1}{\cancel{0.15}}} \quad \text{Divide each side by 0.15.}$$

$$900 = x \leftarrow \text{Whole}$$

135 tourists is 15% of **900 tourists.**

Continued on Next Page

(c) $8\frac{1}{2}\%$ of what number is 102?

Write $8\frac{1}{2}\%$ as 8.5%, or the decimal 0.085. The part is 102. Use the percent equation.

$$\text{part} = \text{percent} \cdot \text{whole}$$

$$102 = 0.085 \cdot x \quad \text{Let } x \text{ represent the unknown whole.}$$

$$\frac{102}{0.085} = \frac{\overset{1}{\cancel{0.085}} \cdot x}{\underset{1}{\cancel{0.085}}} \quad \text{Divide each side by 0.085.}$$

$$1200 = x \longleftarrow \text{Whole}$$

102 is $8\frac{1}{2}\%$ of **1200**.

Estimate the answer. Notice that $8\frac{1}{2}\%$ is close to 10%. If 102 is 10% of a number, then the number is 10 times 102, or 1020. So the answer, 1200, is reasonable.

CAUTION

In Example 2(c) the $8\frac{1}{2}\%$ was changed to 8.5%, which is the decimal form of $8\frac{1}{2}\%$ ($8\frac{1}{2} = 8.5$). The percent sign still remained in 8.5%. Then 8.5% was changed to the decimal 0.085 before dividing.

Work Problem 2 at the Side.

3 Find the percent using the percent equation. The final example shows how to use the percent equation to find the percent.

Example 3 Finding the Percent

(a) 13 roofs is what percent of 52 roofs?

Because 52 follows *of*, the whole is 52. The part is 13, and the percent is unknown. Use the percent equation.

$$\text{part} = \text{percent} \cdot \text{whole}$$

$$13 = x \cdot 52 \quad \text{Let } x \text{ represent the unknown percent.}$$

$$\frac{13}{52} = \frac{x \cdot \overset{1}{\cancel{52}}}{\underset{1}{\cancel{52}}} \quad \text{Divide each side by 52.}$$

$$0.25 = x$$

$$0.25 \text{ is } 25\% \quad \text{Write the decimal as a percent.}$$

13 roofs is **25%** of 52 roofs.

The equation can also be set up by using *of* as an indicator word for multiplication, and *is* as an indicator word for "is equal to."

$$\text{13 is what percent of 52?}$$

$$\downarrow \ \downarrow \qquad \downarrow \qquad \downarrow \ \downarrow$$

$$13 = \quad x \quad \cdot 52$$

$$13 = x \cdot 52 \quad \text{Same equation as above}$$

Continued on Next Page

2 Find the whole using the percent equation.

(a) 36 dance contestants is 18% of what number of dance contestants?

(b) 67.5 containers is 27% of what number of containers?

(c) 666 inoculations is 45% of what number of inoculations?

(d) $5\frac{1}{2}\%$ of what number of policies is 66 policies?

ANSWERS

2. **(a)** 200 dance contestants **(b)** 250 containers **(c)** 1480 inoculations **(d)** 1200 policies

3 Find the percent using the percent equation.

(a) What percent of 70 keyboards is 14 keyboards?

(b) 34 post office boxes is what percent of 85 post office boxes?

(c) What percent of 920 invitations is 1288 invitations?

(d) 9 world class runners is what percent of 1125 runners?

(b) What percent of \$500 is \$100?

The whole is 500 and the part is 100. Let x represent the unknown percent.

part = percent • whole

$100 = x \cdot 500$ Let x represent the unknown percent.

$\frac{100}{500} = \frac{x \cdot \cancel{500}^{1}}{\cancel{500}_{1}}$ Divide each side by 500.

$0.20 = x$

0.20 is 20% ⟵ Write the decimal as a percent.

20% of \$500 is \$100.

(c) What percent of \$300 is \$390?

The whole is 300 and the part is 390. Let x represent the unknown percent.

part = percent • whole

$390 = x \cdot 300$ Let x represent the unknown percent.

$\frac{390}{300} = \frac{x \cdot \cancel{300}^{1}}{\cancel{300}_{1}}$ Divide each side by 300.

$1.3 = x$

1.3 is 130% ⟵ Write the decimal as a percent.

130% of \$300 is \$390.

(d) 6 ladders is what percent of 1200 ladders?

Since 1200 follows *of*, the whole is 1200. The part is 6. Let x represent the unknown percent.

part = percent • whole

$6 = x \cdot 1200$ Let x represent the unknown percent.

$\frac{6}{1200} = \frac{x \cdot \cancel{1200}^{1}}{\cancel{1200}_{1}}$ Divide each side by 1200.

$0.005 = x$

0.005 is 0.5% ⟵ Write the decimal as a percent.

6 ladders is **0.5%** of 1200 ladders.

You can estimate the answer because 1% of 1200 ladders is found by moving the decimal point two places to the left in 1200, resulting in 12. Since 6 ladders is half of 12 ladders, our answer should be $\frac{1}{2}$ of 1% or 0.5%. Our answer is reasonable.

CAUTION

When you use the percent equation to solve for an unknown percent, the answer will always be in decimal form. Notice that in Example 3(a), (b), (c), and (d) the decimal answer had to be changed to a percent by multiplying by 100 and attaching the percent sign. The answers became: (a) 0.25 = 25%; (b) 0.20 = 20%; (c) 1.3 = 130%; and (d) 0.005 = 0.5%.

Work Problem 3 at the Side.

ANSWERS

3. **(a)** 20% **(b)** 40% **(c)** 140% **(d)** 0.8%

6.5 EXERCISES

FOR EXTRA HELP: Student's Solutions Manual · MyMathLab.com · InterAct Math Tutorial Software · AW Math Tutor Center · www.mathxl.com · Digital Video Tutor CD 4 Videotape 11

Find the part using the percent equation. See Example 1.

1. 25% of 540 hamburgers

2. 19% of 700 pages

3. 45% of 1500 garments

4. 75% of 360 dosages

5. 32% of 260 quarts

6. 44% of 430 liters

7. 140% of 500 tablets

8. 145% of 580 donors

9. 12.4% of 8300 meters

10. 26.4% of 4700 miles

11. 0.8% of \$520

12. 0.3% of \$480

Find the whole using the percent equation. See Example 2.

13. 24 patients is 15% of what number of patients?

14. 32 classrooms is 20% of what number of classrooms?

15. 40% of what number of salads is 130 salads?

16. 75% of what number of wrenches is 675 wrenches?

17. 476 circuits is 70% of what number of circuits?

18. 270 lab tests is 45% of what number of lab tests?

19. $12\frac{1}{2}\%$ of what number of people is 135 people?

20. $18\frac{1}{2}\%$ of what number of circuit breakers is 370 circuit breakers?

21. $1\frac{1}{4}\%$ of what number of gallons is 3.75 gallons?

22. $2\frac{1}{4}\%$ of what number of files is 9 files?

Find the percent using the percent equation. See Example 3.

23. 35 rail cars is what percent of 70 rail cars?

24. 90 mail carriers is what percent of 225 mail carriers?

25. 114 tuxedos is what percent of 150 tuxedos?

26. 75 offices is what percent of 125 offices?

27. What percent of \$264 is \$330?

28. What percent of \$480 is \$696?

29. What percent of 160 liters is 2.4 liters?

30. What percent of 600 meters is 7.5 meters?

31. 170 cartons is what percent of 68 cartons?

32. 612 orders is what percent of 425 orders?

33. When using the percent equation, the percent must always be changed to a decimal before doing any calculations. Show and explain how to change a fraction percent to a decimal. Use $2\frac{1}{2}\%$ in your explanation.

34. Suppose a problem on your homework assignment was, "Find $\frac{1}{2}\%$ of \$1300." Your classmates got answers of \$0.65, \$6.50, \$65, and \$650. Which answer is correct? How and why are they getting all of these answers? Explain.

Solve each application problem.

35. A study of office workers found that 27% would like more storage space. If there are 14 million office workers, how many want more storage space? (*Source:* Steelcase Workplace Index.)

36. Most shampoos contain 75% to 90% water. If a 16-ounce bottle of shampoo contains 78% water, find the number of ounces of water in the bottle. Round to the nearest tenth of an ounce.

37. The household lubricant WD-40 is used in 79% of U.S. homes. If there are 104.2 million U.S. homes, find the number of homes in which WD-40 is used. Round to the nearest tenth of a million.

38. Spam and Spam Lite together accounted for 62.2% of the $148 million canned lunch meat sales over a 52-week period (the entire year). Find the total annual sales of these products. Round to the nearest hundredth of a million. (*Source:* Hormel Foods Corporation.)

39. In a recent survey of 1100 employers, it was found that 84% offer only one health plan to employees. How many of these employers offer only one health plan?

40. For a tour of the eastern United States, a travel agent promised a trip of 3300 miles. Exactly 35% of the trip was by air. How many miles would be traveled by air?

41. In the United States, 15 of the 50 states limit the blood alcohol level for drivers to 0.08%. The remaining states limit the level to 0.10%. (*Source: Wall Street Journal.*)

(a) What percent of the states have a blood alcohol limit of 0.08%?

(b) What percent have a limit of 0.10%?

42. Among the 50 companies receiving the greatest number of U.S. patents last year, 18 were Japanese companies. (*Source: Wall Street Journal.*)

(a) What percent of the companies were Japanese companies?

(b) What percent of the companies were not Japanese companies?

43. In a test by *Consumer Reports,* 6 of the 123 cans of tuna analyzed contained more than the 30-microgram intake limit of mercury. What percent of the cans contained this level of mercury? Round to the nearest tenth of a percent.

44. General Nutrition Center now has 3200 stores and plans to add 450 more stores. Find the percent of additional stores that they have planned.

The graph shows the average time spent preparing weekday dinners. Assume that 5400 people were surveyed to gather this data. Use this graph to answer Exercises 45–48.

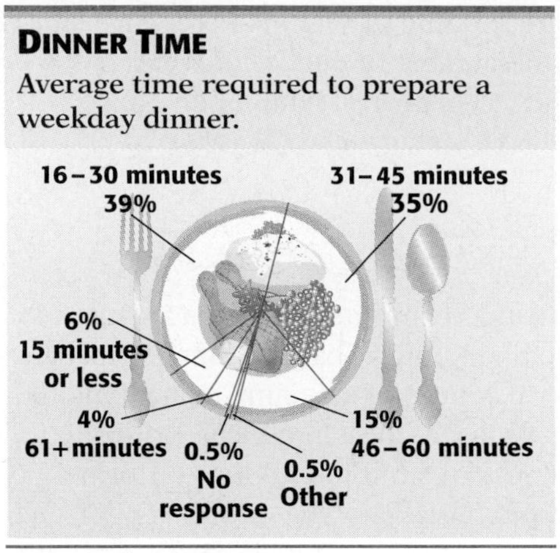

Source: The NPD Group.

45. Find the number of people who said they spend "16–30 minutes" preparing weekday dinners.

46. What total number of people answered "No response" and "Other"?

47. How many people said they spend over 30 minutes preparing weekday dinners?

48. Find the number of people who said they spend 30 minutes or less preparing weekday dinners. Round to the nearest whole number.

49. Chemical Banking Corporation made \$338 million worth of mortgage loans to minorities last year. If this represented 18.6% of all their mortgages, find the total value of all mortgages that they made last year. (Round to the nearest tenth of a million.)

50. Julie Ward has 8.5% of her earnings deposited into the credit union. If this amounts to \$131.75 per month, find her annual earnings.

51. The Chevy Camaro was introduced in 1967. Sales that year were 220,917, which was 46.2% of the number of Ford Mustangs sold in the same year. Find the number of Mustangs sold in the same year. Round to the nearest whole number.

52. A report by the U.S. Bureau of Labor Statistics states that the number of unemployed people is 8.9 million or 7.1% of all workers. What is the size of the total workforce? Round to the nearest tenth of a million.

53. J & K Mustang has increased the sale of auto parts by $32\frac{1}{2}\%$ over last year. If the sale of parts last year amounted to \$385,200, find the volume of sales this year.

54. An ad for steel-belted radial tires promises 15% better mileage. If mileage has been 25.6 miles per gallon in the past, what mileage could be expected after these tires are installed? (Round to the nearest tenth of a mile.)

55. A Polaris Vac-Sweep 380 is priced at \$524 with an allowed trade-in of \$125 on an old unit. If sales tax of $7\frac{3}{4}\%$ is charged on the price of the new Polaris unit before the trade-in, find the total cost to the customer after receiving the trade-in. (*Hint:* Trade-in is subtracted last.)

56. A Palm IIIx^{e} Connected Organizer originally priced at \$332 is marked down 25%. Find the price of the organizer after the markdown.

6.6 SOLVING APPLICATION PROBLEMS WITH PERCENT

OBJECTIVES

1. Find sales tax.
2. Find commissions.
3. Find the discount and sale price.
4. Find the percent of change.

Percent has many applications in our daily lives. This section discusses percent as it applies to sales tax, commissions, discounts, and the percent of change (increase and decrease).

1 Find sales tax. States, counties, and cities often collect taxes on sales to customers. The **sales tax** is a percent of the total sale. The following formula for finding sales tax is based on the percent equation.

Sales Tax Formula

$$\underbrace{\text{amount of sales tax}}_{\text{Part}} = \underbrace{\text{rate of tax}}_{\text{Percent}} \cdot \underbrace{\text{cost of item}}_{\text{Whole}}$$

Example 1 Solving for Sales Tax

Office Max sold a Xerox color copier/printer for \$299. If the sales tax rate is 5%, how much tax was paid? What was the total cost of the copier/printer?

Step 1 **Read** the problem. The problem asks for the total cost of the copier/printer including the sales tax.

Step 2 **Work out a plan.** Use the sales tax formula to find the amount of sales tax. Write the tax rate (5%) as a decimal (0.05). The cost of the item is \$299. Use *a* for the *amount* of tax. Add the sales tax to the cost of the item.

Step 3 **Estimate** a reasonable answer. Round \$299 to \$300. Recall that 5% is equivalent to $\frac{1}{20}$, so divide \$300 by 20.

$$\$300 \div 20 = \$15 \text{ tax}$$

The total estimated cost is \$300 + \$15 = \$315. ← Estimate

Step 4 **Solve** the problem.

$$\underbrace{\text{amount of sales tax}}_{\text{Part}} = \underbrace{\text{rate of tax}}_{\text{Percent}} \cdot \underbrace{\text{cost of item}}_{\text{Whole}}$$

$$a = 5\% \cdot \$299$$

$$a = 0.05 \cdot \$299$$

$$a = \$14.95 \quad \text{Sales tax}$$

The tax paid on the color copier/printer was \$14.95. The customer would pay a total cost of \$299 + \$14.95 = \$313.95.

Step 5 **State the answer.** The total cost of the copier/printer is \$313.95.

Step 6 **Check.** The answer, \$313.95, is close to our estimate of \$315.

Work Problem 1 at the Side.

1 Suppose the sales tax rate in your state is 6%. Find the amount of the tax and the total you would pay for each item.

(a) \$29 Little League bat

(b) \$119 digital answering machine

(c) \$1287 sofa and chair

(d) \$19,300 pickup truck

ANSWERS
1. (a) \$1.74; \$30.74 (b) \$7.14; \$126.14 (c) \$77.22; \$1364.22 (d) \$1158; \$20,458

2 Find the rate of sales tax.

(a) The tax on a $190 table is $15.20.

(b) The tax on a $24 sweatshirt is $1.56.

(c) The tax on $14,820 worth of light fixtures is $592.80.

ANSWERS

2. **(a)** 8% **(b)** 6.5% or $6\frac{1}{2}\%$ **(c)** 4%

Example 2 **Finding the Sales Tax Rate**

The sales tax on a $14,800 Honda Civic is $962. Find the rate of the sales tax.

Step 1 **Read** the problem. This problem asks us to find the sales tax rate.

Step 2 **Work out a plan.** Use the sales tax formula.

sales tax = rate of tax • cost of item

Solve for the rate of tax, which is the percent. The cost of the Honda Civic (the whole) is $14,800, and the amount of sales tax (the part) is $962. Use *r* for the *rate* of tax (the percent).

Step 3 **Estimate** a reasonable answer. Round $14,800 to $15,000 and round $962 to $1000. The sales tax is $\frac{1000}{15,000}$ or $\frac{1}{15}$ of the cost of the car. So divide 1 by 15 to find the percent (rate) of sales tax.

$$\frac{1}{15} = .0\overline{6} = 7\% \text{ Rounded estimate}$$

Step 4 **Solve** the problem.

sales tax = rate of tax • cost of item

$$\$962 = r \cdot \$14,800$$

$$\frac{962}{14,800} = \frac{\overset{1}{\cancel{14,800}} \cdot r}{\underset{1}{\cancel{14,800}}} \quad \text{Divide each side by 14,800.}$$

$$0.065 = r$$

0.065 is 6.5% Write the decimal as a percent.

Step 5 **State the answer.** The sales tax rate is 6.5% or $6\frac{1}{2}\%$.

Step 6 **Check.** The answer, $6\frac{1}{2}\%$, is close to our estimate of 7%.

NOTE

You can use the sales tax formula to find the amount of sales tax, the cost of an item, or the rate of sales tax (the percent).

Work Problem 2 at the Side.

2 **Find commissions.** Many salespeople are paid by *commission* rather than an hourly wage. If you are paid by **commission,** you are paid a certain percent of the total sales dollars. The following formula for finding the commission is based on the percent equation.

Commission Formula

Part = Percent • Whole

amount of commission = rate of commission • amount of sales

Example 3 Determining the Amount of Commission

Scott Samuels had pharmaceutical sales of $42,500 last month. If his commission rate is 9%, find the amount of his commission.

Step 1 **Read** the problem. The problem asks for the amount of commission that Scott earned.

Step 2 **Work out a plan.** Use the commission formula. Write the rate of commission (9%) as a decimal (0.09). The amount of sales ($42,500) is the whole. Use c for the *amount* of commission.

Step 3 **Estimate** a reasonable answer. Round the commission rate of 9% to 10%. Round the amount of sales from $42,500 to $40,000. Since 10% is equivalent to $\frac{1}{10}$, divide $40,000 by 10 to find the amount of commission.

$$\$40{,}000 \div 10 = \$4000 \text{ Our estimate}$$

Step 4 **Solve** the problem.

amount of commission = rate of commission • amount of sales

$$c = 9\% \cdot \$42{,}500$$
$$c = 0.09 \cdot \$42{,}500$$
$$c = \$3825 \quad \text{Amount of commission}$$

Step 5 **State the answer.** Samuels earned a commission of $3825 for selling the pharmaceuticals.

Step 6 **Check.** The answer, $3825, is close to our estimate of $4000.

Work Problem 3 at the Side.

3 Find the amount of commission.

(a) Jill Buteo sells dental equipment at a commission rate of 12% and has sales for the month of $28,750.

(b) Last month Pam Prentice sold $71,800 worth of furniture with a commission rate of 4%.

Example 4 Finding the Rate of Commission

Chris Knudson earned a commission of $510 for selling $17,000 worth of shipping supplies. Find the rate of commission.

Step 1 **Read** the problem. In this problem, we must find the rate (percent) of commission.

Step 2 **Work out a plan.** You could use the commission formula. Another approach is to use the percent proportion. The *whole* is $17,000, the *part* is $510, and the *percent* is unknown. (The rate of commission is the percent.)

Step 3 **Estimate** a reasonable answer. Round the commission, $510, to $500. The commission in fraction form is $\frac{\$500}{\$20{,}000}$, which divides out to $\frac{5}{200}$. Changing $\frac{5}{200}$ to a percent gives $2\frac{1}{2}\%$ (rounded), our estimate.

Step 4 **Solve** the problem.

$$\frac{\text{part}}{\text{whole}} = \frac{x}{100} \quad \leftarrow \text{Percent (unknown)}$$
$$\frac{510}{17{,}000} = \frac{x}{100}$$
$$17{,}000 \cdot x = 510 \cdot 100 \quad \text{Cross multiply.}$$
$$\frac{\overset{1}{\cancel{17{,}000}} \cdot x}{\underset{1}{\cancel{17{,}000}}} = \frac{51{,}000}{17{,}000} \quad \text{Divide each side by 17,000.}$$
$$x = 3$$

Continued on Next Page

ANSWERS
3. (a) $3450 (b) $2872

4 Find the rate of commission.

(a) A commission of $450 is earned on one sale of computer products worth $22,500.

(b) Jamal Story earns $2898 for selling office furniture worth $32,200.

5 Find the amount of the discount and the sale price.

(a) An Easy-Boy leather recliner originally priced at $950 is offered at a 42% discount.

(b) WAL-MART has women's sweater sets on sale at 35% off. One sweater set was originally priced at $30.

Step 5 **State the answer.** The rate of commission is 3%.

Step 6 **Check.** The answer, 3%, is the same as our estimate of $2\frac{1}{2}$%.

Work Problem 4 at the Side.

3 **Find the discount and sale price.** Most of us prefer buying things when they are on sale. A store will reduce prices, or **discount,** to attract additional customers. Use the following formula to find the discount and the sale price.

Discount Formula and Sale Price Formula

amount of discount = rate (or percent) of discount • original price

sale price = original price − amount of discount

Example 5 **Finding a Sale Price**

The Oak Mill Furniture Store has a home entertainment center with an original price of $840 on sale at 15% off. Find the sale price of the entertainment center.

Step 1 **Read** the problem. This problem asks for the price of an entertainment center after a discount of 15%.

Step 2 **Work out a plan.** This problem is solved in two steps. First, find the amount of the discount, that is, the amount that will be "taken off" (subtracted), by multiplying the original price ($840) by the rate of the discount (15%). The second step is to subtract the amount of discount from the original price. This gives you the sale price, which is what you will actually pay for the entertainment center.

Step 3 **Estimate** a reasonable answer. Round the original price from $840 to $800, and the rate of discount from 15% to 20%. Since 20% is equivalent to $\frac{1}{5}$, the discount is $800 ÷ 5 = $160. The sale price is $800 − $160 = $640, our estimate.

Step 4 **Solve** the problem. First find the amount of the discount.

amount of discount = rate of discount • original price

a = 0.15 • $840   Write 15% as a decimal.

a = $126   Amount of discount

Now find the sale price of the entertainment center by subtracting the amount of the discount ($126) from the original price.

sale price = original price − amount of discount

= $840 − $126

= $714   Sale price

Step 5 **State the answer.** The sale price of the entertainment center is $714.

Step 6 **Check.** The answer, $714, is close to our estimate of $640.

Work Problem 5 at the Side.

ANSWERS

4. **(a)** 2% **(b)** 9%

5. **(a)** $399; $551 **(b)** $10.50; $19.50

Calculator Tip In Example 5, you can use a calculator to find the amount of discount and subtract the discount from the original price directly.

840 (−) .15 (×) 840 (=) 714

Original price (840); Amount of discount (.15 × 840); Sale price (714)

A scientific calculator observes the order of operations, so it will automatically do the multiplication before the subtraction.

4 Find the percent of change. We are often interested in looking at increases or decreases in sales, production, population, and many other areas. This type of problem involves finding the *percent of change*. Use the following steps to find the **percent of increase.**

Finding the Percent of Increase

Step 1 Use subtraction to find the amount of increase.

Step 2 Use the percent proportion to find the percent of increase.

$$\frac{\textbf{amount of increase (part)}}{\textbf{original value (whole)}} = \frac{\textbf{percent}}{\textbf{100}}$$

Example 6 Finding the Percent of Increase

Attendance at county parks climbed from 18,300 last month to 56,730 this month. Find the percent of increase.

Step 1 **Read** the problem. The problem asks for the percent of increase.

Step 2 **Work out a plan.** Subtract the attendance last month (18,300) from the attendance this month (56,730) to find the amount of increase in attendance. Next, use the percent proportion. The whole is 18,300 (last month's original attendance), the part is 38,430 (amount of increase in attendance), the percent is unknown.

Step 3 **Estimate** a reasonable answer. Round 18,300 to 20,000 and 56,730 to 60,000. The amount of increase is 60,000 − 20,000 = 40,000. Since 40,000 (the increase) is *twice* as large as the original amount, the percent of increase is 200%, our estimate.

Step 4 **Solve** the problem.

$$56{,}730 - 18{,}300 = 38{,}430 \quad \text{Amount of increase in attendance}$$

$$\frac{38{,}430}{18{,}300} = \frac{x}{100} \quad \text{Percent proportion}$$

Solve this proportion to find that $x = 210$.

Step 5 **State the answer.** The percent of increase is 210%.

Step 6 **Check.** The answer, 210%, is close to our estimate of 200%.

Work Problem 6 at the Side.

6 Find the percent of increase.

(a) A manufacturer of snowboards increased production from 14,100 units last year to 19,035 this year.

(b) The number of flu cases rose from 496 cases last week to 620 this week.

ANSWERS
6. (a) 35% **(b)** 25%

7 Find the percent of decrease.

(a) The number of new trainees fell from 760 last month to 570 this month.

(b) The number of workers applying for unemployment fell from 4850 last month to 3977 this month.

Use the following steps to find the **percent of decrease.**

Finding the Percent of Decrease

Step 1 Use subtraction to find the amount of decrease.

Step 2 Use the percent proportion to find the percent of decrease.

$$\frac{\textbf{amount of decrease (part)}}{\textbf{original value (whole)}} = \frac{\textbf{percent}}{\textbf{100}}$$

Example 7 Finding the Percent of Decrease

The number of production employees this week fell to 1406 people from 1480 people last week. Find the percent of decrease.

Step 1 **Read** the problem. The problem asks for the percent of decrease.

Step 2 **Work out a plan.** Subtract the number of employees this week (1406) from the number of employees last week (1480) to find the amount of decrease. Next, use the percent proportion. The whole is 1480 (last week's original number of employees), the part is 74 (decrease in employees), and the percent is unknown.

Step 3 **Estimate** a reasonable answer. Estimate the answer by rounding 1406 to 1400 and 1480 to 1500. The decrease is $1500 - 1400 = 100$. Since 100 is $\frac{1}{15}$ of 1500 our estimate is $1 \div 15 \approx .07$ or 7%.

Step 4 **Solve** the problem.

$$1480 - 1406 = 74 \quad \text{Decrease in number of employees}$$

$$\frac{74}{1480} = \frac{x}{100} \quad \text{Percent proportion}$$

Solve this proportion to find that $x = 5$.

Step 5 **State the answer.** The percent of decrease is 5%.

Step 6 **Check.** The answer, 5%, is close to our estimate of 7%.

CAUTION

When solving for percent of increase or decrease, the ***whole is always the original value*** or ***value before the change occurred.*** The part is the change in values, that is, how much something went up or went down.

Work Problem 7 at the Side.

ANSWERS
7. (a) 25% **(b)** 18%

6.6 Exercises

FOR EXTRA HELP: Student's Solutions Manual | MyMathLab.com | InterAct Math Tutorial Software | AW Math Tutor Center | www.mathxl.com | Digital Video Tutor CD 4 Videotape 11

Find the amount of sales tax or the tax rate and the total cost (amount of sale + amount of tax = total cost). Round money answers to the nearest cent if necessary. See Examples 1 and 2.

	Amount of Sale	Tax Rate	Amount of Tax	Total Cost
1.	$8	5%	________	________
2.	$15	4%	________	________
3.	$425	________	$12.75	________
4.	$322	________	$19.32	________
5.	$284	________	$14.20	________
6.	$84	________	$5.88	________
7.	$12,229	$5\frac{1}{2}\%$	________	________
8.	$11,789	$7\frac{1}{2}\%$	________	________

Find the commission earned or the rate of commission. Round money answers to the nearest cent if necessary. See Examples 3 and 4.

	Sales	Rate of Commission	Commission
9.	$400	10%	________
10.	$325	4%	________
11.	$3000	________	$600
12.	$7800	________	$1170
13.	$6183.50	3%	________
14.	$4416.70	7%	________
15.	$73,500	9%	________
16.	$55,800	6%	________

Find the amount or rate of discount and the sale price after the discount. Round money answers to the nearest cent if necessary. See Example 5.

	Original Price	Rate of Discount	Amount of Discount	Sale Price
17.	$199.99	10%	______	______
18.	$29.95	15%	______	______
19.	$180	______	$54	______
20.	$38	______	$9.50	______
21.	$17.50	25%	______	______
22.	$76	60%	______	______
23.	$58.40	15%	______	______
24.	$99.80	30%	______	______

25. You are trying to decide between Company A paying a 10% commission and Company B paying an 8% commission. For which company would you prefer to work? Are there considerations other than commission rate that would be important to you? What would they be?

26. Give four examples of where you might use the percent of increase or the percent of decrease in your own personal activities. Think in terms of work, school, home, hobbies, and sports. Write an increase or a decrease problem about one of these four examples, then show how to solve it.

Solve each application problem. Round money answers to the nearest cent and rates to the nearest tenth of a percent if necessary. See Examples 1–7.

27. If the sales tax rate is 4% and today's sales at the Dollar Store are $1453, find the amount of sales tax collected.

28. An Exer-Cycle Machine sells for $590 plus 7% sales tax. Find the amount of sales tax.

29. Diamonds at Discounts sells a diamond engagement ring at 40% off the regular price. Find the sale price of a $\frac{1}{2}$-carat diamond ring normally priced at $1950.

30. During a January inventory sale, Heather Hall purchased a new car at 12% below sticker price. If the sticker price was $18,350, what was her cost?

31. An Anderson wood frame French door is priced at $1980 with a sales tax of $99. Find the rate of sales tax.

32. Textbooks for three classes cost $165 plus sales tax of $11.55. Find the sales tax rate.

33. Harley-Davidson, the only major U.S.-based motorcycle manufacturer, says that it expects to build 145,000 motorcycles this year, up from 131,000 last year. Find the percent of increase in production. (*Source:* Harley-Davidson, Inc.)

34. Americans are eating more fish. This year the average American will eat 15.5 pounds compared to only 12.5 pounds per year a decade ago. Find the percent of increase. (*Source: Consumer Reports.*)

35. The number of industrial accidents this month fell to 989 accidents from 1276 accidents last month. Find the percent of decrease.

36. The average number of hours worked in manufacturing jobs last week fell from 41.1 to 40.9. Find the percent of decrease.

37. A "60% off sale" begins today at Wanda's Women's Wear. What is the sale price of women's wool coats normally priced at $335?

38. What is the sale price of a $769 Kenmore washer/dryer set with a discount of 25%?

The weekly sales record of the top four sales people at the Family Shoe Store is shown in the table. Use this information to answer Exercises 39–42.

Employee	**Sales**	**Rate of Commission**	**Commission**
McKee, J.	$9480	3%	$284.40
Brown, D.	$10,730	3%	$321.90
Poznick, M.	$8840		$353.60
Washington, R.	$11,522		$576.10

39. Find the commission for McKee.

40. Find the commission for Brown.

41. What is the rate of commission for Poznick?

42. What is the rate of commission for Washington?

43. An 8-millimeter camcorder normally priced at $590 is on sale at 18% off. Find the discount and the sale price.

44. A Dodge Durango is offered at 17% off the manufacturer's suggested retail price. Find the discount and the sale price of this Durango, originally priced at $28,700.

45. The price per share of Toys Я Us stock fell from $35.50 to $33.50. Find the percent of decrease in price.

46. In the past five years, the cost of generating electricity from the sun has been brought down from 24 cents per kilowatt hour to 8 cents (less than the newest nuclear power plants). Find the percent of decrease.

47. College students are offered a 6% discount on a dictionary that sells for $18.50. If the sales tax is 6%, find the cost of the dictionary including the sales tax.

48. A personal computer and monitor priced at $698 is marked down 8%. If the sales tax is also 8%, find the cost of the computer and monitor including sales tax.

49. A real estate agent sells a house for $129,605. A sales commission of 6% is charged. The agent gets 55% of this commission. How much money does the agent get?

50. The local real estate agents' association collects a fee of 2% on all money received by its members. The members charge 6% of the selling price of a property as their fee. How much does the association get, if its members sell property worth a total of $8,680,000?

51. What is the total price of a ski boat with an original price of $10,214, if it is sold at a 15% discount? The sales tax rate is $3\frac{3}{4}$%.

52. A commercial security alarm system originally priced at $10,800 is discounted 22%. Find the total price of the system if the sales tax rate is $7\frac{1}{4}$%.

RELATING CONCEPTS (Exercises 53–58) FOR INDIVIDUAL OR GROUP WORK

Knowing how to use the percent equation is important when solving application problems involving sales tax. ***Work Exercises 53–58 in order.***

53. The percent equation is

part = __________ • __________

54. The formula used to find sales tax is an application of the percent equation. The sales tax formula is

sales tax = ________________ • __________

In the United States there are certain items on which an excise tax is charged in addition to a sales tax. A table of federal excise taxes is shown here. Use this table to answer Exercises 55–58. (Excise tax is calculated on the amount of the sale before sales tax is added.) Round answers to the nearest cent.

FEDERAL EXCISE TAXES

Product or Service	Rate	Product or Service	Rate
Telephone service	3%	Tires (by weight)	
Teletypewriter service	3%	Under 40 pounds	No tax
Air transportation	7.5%	40–69 pounds	15¢/pound over 40 pounds
International air travel	$12.20/person	70–89 pounds	$4.50 plus 30¢/pound over 70 pounds
Air freight	6.25%		
Coal		90 pounds and more	$10.50 plus 50¢/pound over 90 pounds
Underground (lower amount)	$1.10/ton or 4.4%		
Surface (lower amount)	55¢/ton or 4.4%	Truck and trailer, chassis and bodies	12%
Fishing rods	10%	Inland waterways fuel	24.4¢/gallon
Bows and arrows	12.4%	Ship passenger tax	$3/passenger
Gasoline	18.4¢/gallon	Luxury cars (amount over $36,000)	6%
Diesel fuel	24.4¢/gallon		
Aviation fuel	21.9¢/gallon		

Source: Publication 510, I.R.S., Excise Taxes for 1999.

55. Some archery equipment (bows and arrows) are priced at $123. Use the federal excise tax table and a sales tax rate of $6\frac{1}{2}$% to find the cost of the equipment, including both taxes. (Round to the nearest cent.)

56. Refer to Exercise 55. Calculate the two taxes separately and then add them together. Now, add the two tax rates together and then find the tax. Are your answers the same? Why or why not? (*Hint:* Recall the commutative and associative properties of multiplication.)

57. The price of an international airline ticket is $1248. Use the federal excise tax table and a sales tax rate of $7\frac{3}{4}$% to find the total cost of one ticket. (Excise tax is not charged on sales tax.)

58. Refer to Exercise 57. Can the federal excise tax be added to the sales tax rate to find the total tax? Why or why not?

6.7 Simple Interest

Objectives

1. Find the simple interest on a loan.
2. Find the total amount due on a loan.

When we open a savings account, we are actually lending money to the bank or credit union. It will in turn lend this money to individuals and businesses. These people then become borrowers. The bank or credit union pays a fee to the savings account holders and charges a higher fee to its borrowers. These fees are called *interest*.

Interest is a fee paid or a charge made for lending or borrowing money. The amount of money borrowed is called the **principal.** The charge for interest is often given as a percent, called the interest rate or **rate of interest.** The rate of interest is assumed to be *per year,* unless stated otherwise.

1 Find the simple interest on a loan. In most cases, interest on a loan is computed on the *original principal* and is called **simple interest.** We use the following **interest formula** to find simple interest.

Formula for Simple Interest

interest = principal • rate • time

The formula is usually written in letters.

$$I = p \cdot r \cdot t$$

NOTE

Simple interest is used for most short-term business loans, most real estate loans, and many automobile and consumer loans.

Example 1 Finding Interest for a Year

Find the interest on $5000 at 6% for 1 year.

The amount borrowed, or principal (p), is $5000. The interest rate (r) is 6%, which is 0.06 as a decimal, and the time of the loan (t) is 1 year. Use the formula.

$$I = p \cdot r \cdot t$$
$$I = 5000 \cdot (0.06) \cdot 1$$
$$I = 300$$

The interest is $300.

Work Problem 1 at the Side.

Example 2 Finding Interest for More Than a Year

Find the interest on $4200 at 8% for three and a half years.

The principal (p) is $4200. The rate ($r$) is 8%, or 0.08 as a decimal, and the time (t) is $3\frac{1}{2}$ or 3.5 years. Use the formula.

$$I = p \cdot r \cdot t$$
$$I = 4200 \cdot (0.08) \cdot (3.5)$$
$$I = 1176$$

The interest is $1176.

Work Problem 2 at the Side.

1 Find the interest.

(a) $1000 at 5% for 1 year

(b) $3650 at 4% for 1 year

2 Find the interest.

(a) $820 at 6% for $3\frac{1}{2}$ years

(b) $4850 at 8% for $2\frac{1}{2}$ years

(c) $16,800 at 5% for $2\frac{3}{4}$ years

Answers

1. (a) $50 **(b)** $146

2. (a) $172.20 **(b)** $970 **(c)** $2310

3 Find the interest.

(a) \$1800 at 8% for 4 months

(b) \$28,000 at $9\frac{1}{2}$% for 3 months

4 Find the total amount due on each loan.

(a) \$3800 at $6\frac{1}{2}$% for 6 months

(b) \$12,400 at 7% for 5 years

(c) \$2400 at 11% for $2\frac{3}{4}$ years

ANSWERS
3. **(a)** \$48 **(b)** \$665
4. **(a)** \$3923.50 **(b)** \$16,740 **(c)** \$3126

CAUTION

It is best to change fractions of percents or fractions of years into their decimal form. In Example 2, the $3\frac{1}{2}$ years becomes 3.5 years.

Interest rates are given *per year.* For loan periods of less than one year, be careful to express time as a fraction of a year.

If time is given in months, for example, use a denominator of 12, because there are 12 months in a year. A loan for 9 months would be for $\frac{9}{12}$ of a year.

Example 3 **Finding Interest for Less Than 1 Year**

Find the interest on \$840 at $8\frac{1}{2}$% for 9 months.

The principal is \$840. The rate is $8\frac{1}{2}$% or 0.085, and the time is $\frac{9}{12}$ of a year. Use the formula $I = p \cdot r \cdot t$.

$$I = \underbrace{840 \cdot (0.085)} \cdot \frac{9}{12} \qquad 9 \text{ months} = \tfrac{9}{12} \text{ of a year.}$$

$$= \quad 71.4 \quad \cdot \frac{3}{4} \qquad \tfrac{9}{12} \text{ in lowest terms is } \tfrac{3}{4}.$$

$$= \frac{(71.4) \cdot 3}{4}$$

$$= \frac{214.2}{4} = 53.55$$

The interest is \$53.55.

Calculator Tip The calculator solution to Example 3 uses chain calculations.

840 (×) 0.085 (×) 9 (÷) 12 (=) 53.55

Work Problem 3 at the Side.

2 **Find the total amount due on a loan.** When a loan is repaid, the interest is added to the original principal to find the total amount due.

Formula for Amount Due

amount due = principal + interest

Example 4 **Calculating the Total Amount Due**

A loan of \$3240 has been made at 12% for 3 months. Find the total amount due.

First find the interest, then add the principal and the interest to find the total amount due.

$$I = 3240 \cdot (0.12) \cdot \frac{3}{12} \qquad 3 \text{ months} = \tfrac{3}{12} \text{ of a year.}$$

$$I = \$97.20$$

The interest is \$97.20.

$$\textbf{amount due} = \text{principal} + \text{interest}$$
$$= \$3240 + \$97.20 = \$3337.20$$

The total amount due is \$3337.20.

Work Problem 4 at the Side.

6.7 EXERCISES

FOR EXTRA HELP: Student's Solutions Manual | MyMathLab.com | InterAct Math Tutorial Software | AW Math Tutor Center | www.mathxl.com | Digital Video Tutor CD 4 Videotape 11

Find the interest. See Examples 1 and 2.

	Principal	Rate	Time in Years	Interest
1.	$100	6%	1	________
2.	$200	3%	1	________
3.	$700	5%	3	________
4.	$900	4%	4	________
5.	$240	4%	3	________
6.	$190	3%	2	________
7.	$2300	$8\frac{1}{2}\%$	$2\frac{1}{2}$	________
8.	$4700	$5\frac{1}{2}\%$	$1\frac{1}{2}$	________
9.	$10,800	$7\frac{1}{2}\%$	$2\frac{3}{4}$	________
10.	$12,400	$6\frac{1}{2}\%$	$3\frac{3}{4}$	________

Find the interest. Round to the nearest cent if necessary. See Example 3.

	Principal	Rate	Time in Months	Interest
11.	$400	5%	6	____________
12.	$600	7%	5	____________
13.	$820	6%	12	____________

	Principal	Rate	Time in Months	Interest
14.	$780	8%	24	________
15.	$940	3%	18	________
16.	$178	4%	12	________
17.	$1225	$5\frac{1}{2}\%$	3	________
18.	$2660	$7\frac{1}{2}\%$	3	________
19.	$15,300	$7\frac{1}{4}\%$	7	________
20.	$13,700	$3\frac{3}{4}\%$	11	________

Find the total amount due on the following loans. Round to the nearest cent if necessary. See Example 4.

	Principal	Rate	Time	Total Amount Due
21.	$200	5%	1 year	________
22.	$400	4%	6 months	________
23.	$740	6%	9 months	________
24.	$1180	3%	2 years	________

	Principal	Rate	Time	Total Amount Due
25.	\$1800	9%	18 months	________
26.	\$9000	6%	7 months	________
27.	\$3250	10%	6 months	________
28.	\$7600	5%	1 year	________
29.	\$16,850	$7\frac{1}{2}\%$	9 months	________
30.	\$19,450	$5\frac{1}{2}\%$	6 months	________

31. The amount of interest paid on savings accounts and charged on loans can vary from one institution to another. However, when the amount of interest is calculated, three factors are used in the calculation. Name these three factors and describe them in your own words.

32. Interest rates are usually given as a rate per year (annual rate). Explain what must be done when time is given in months. Write your own problem where time is given in months and then show how to solve it.

Solve each application problem. Round to the nearest cent if necessary.

33. Reann Chang deposits \$825 at 5% for 1 year. How much interest will she earn?

34. The Jidobu family invests \$18,000 at 9% for 6 months. What amount of interest will the family earn?

35. The Bank of Boston loans \$150,000 to a business at 10% for 18 months. How much interest will the bank earn?

36. Pat Martin, a retiree, deposits \$80,000 at 7% for 3 years. How much interest will he earn?

37. A student borrows $1200 at 8% for 5 months to pay for books and tuition. Find the total amount due.

38. Jill borrows $2750 from her dad for a used car. The loan will be paid back with 8% interest at the end of 9 months. Find the total amount due.

39. Ellen Pate deposits $1568 in a credit union account for 9 months. If the credit union pays $5\frac{1}{2}$% interest, find the amount of interest she will earn.

40. Silvo DiLoreto, owner of Sunset Realtors, borrows $27,000 to update his office computer system. If the loan is for 24 months at $7\frac{1}{4}$%, find the amount of interest he will owe.

41. An investment fund pays $7\frac{1}{4}$% interest. If Beverly Habecker deposits $8800 in her account for $\frac{1}{4}$ year, find the amount of interest she will earn.

42. Ms. Henderson owes $1900 in taxes. She is charged a penalty of $12\frac{1}{4}$% annual interest and pays the taxes and penalty after 6 months. Find the total amount she must pay.

43. A gift shop owner invests his profits of $11,500 at $8\frac{3}{4}$% interest for $\frac{3}{4}$ year. Find the total amount in his account at the end of this time.

44. A pawn shop owner lends $35,400 to another business for $\frac{1}{2}$ year at an interest rate of 14.9%. How much interest will be earned on the loan?

45. The owners of Clear Lake Marina bought six canoes to be used as rentals at a cost of $550 per canoe. If they borrowed 60% of the total cost for 9 months at $12\frac{1}{2}$% interest, find the amount due.

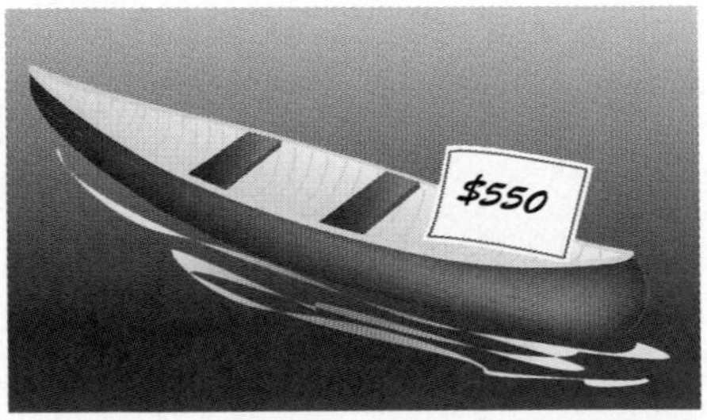

46. The owners of Delta Trucking purchased four diesel-powered tractors for cross-country hauling at a cost of $87,500 per tractor. If they borrowed 80% of the total purchase price for $1\frac{1}{2}$ years at $11\frac{1}{2}$% interest, find the total amount due.

Summary Exercises on PERCENT

Write each percent as a decimal and each decimal as a percent.

1. 6.25% **2.** 380% **3.** 0.375 **4.** 0.006

Write each percent as a fraction or mixed number in lowest terms and each fraction as a percent.

5. 87.5% **6.** 160% **7.** $\frac{5}{8}$ **8.** $\frac{1}{125}$

Find the part, whole, or percent as indicated. Round percent answers to the nearest tenth if necessary.

9. 6% of $780 is what amount?

10. 70 rolls of film is 14% of how many rolls of film?

11. What percent of 320 policies is 176 policies?

12. 0.8% of 3500 screening exams

13. 1016.4 acres is 280% of what number of acres?

14. What percent of 658 circuits is 18 circuits?

Find the amount of sales tax or the tax rate and the total cost. Round to the nearest cent.

	Amount of Sale	Tax Rate	Amount of Tax	Total cost
15.	$176	$5\frac{1}{2}\%$	________	________
16.	$926	________	$64.82	________

Find the commission earned or the rate of commission.

	Sales	Rate of Commission	Commission
17.	$3765	8%	________
18.	$22,482	____________	$1461.33

Find the amount or rate of discount and the amount paid after the discount. Round to the nearest cent.

	Original Price	Rate of Discount	Amount of Discount	Sale Price
19.	$49.95	18%	________	________
20.	$684	________	$239.40	________

Find the interest and the total amount due on the following simple interest loans.

	Principal	Rate	Time	Interest	Total Amount Due
21.	$1000	6%	1 year	________	________
22.	$2380	$7\frac{1}{2}\%$	3 years	________	________
23.	$1470	4%	9 months	________	________
24.	$6820	$6\frac{1}{2}\%$	18 months	________	________

Solve each application problem. Round percent answers to the nearest tenth of a percent if necessary.

25. The number of Harley-Davidson motorcycles sold in 1999 was 136,446, up from 1998 sales of 111,800 motorcycles. Find the percent of increase. (*Source:* Motorcycle Industry Council.)

26. The number of ski lift tickets sold this week was 3419, down from last week's sales of 3820 tickets. Find the percent of decrease.

6.8 Compound Interest

Objectives

1. Understand compound interest.
2. Understand compound amount.
3. Find the compound amount.
4. Use a compound interest table.
5. Find the compound amount and the amount of compound interest.

The interest we studied in **Section 6.7** was *simple interest* (interest only on the original principal). A common type of interest used with savings accounts and most investments is **compound interest,** or interest paid on past interest as well as on the principal.

1 Understand compound interest. Suppose that you make a single deposit of $1000 in a savings account that earns 5% per year. What will happen to your savings over 3 years? At the end of the first year, 1 year's interest on the original deposit is found. Use the simple interest formula.

Interest = principal • rate • time

Year 1 $1000 • (0.05) • 1 = $50
Add the interest to the $1000 to find the amount in your account at the end of the first year. $1000 + $50 = $1050
The interest for the second year is found on $1050; that is, the interest is **compounded.**

Year 2 $1050 • (0.05) • 1 = $52.50
Add this interest to the $1050 to find the amount in your account at the end of the second year. $1050 + $52.50 = $1102.50
The interest for the third year is found on $1102.50.

Year 3 $1102.50 • (0.05) • 1 ≈ $55.13
Add this interest to the $1102.50. $1102.50 + $55.13 = $1157.63

At the end of 3 years, you will have $1157.63 in your savings account. The $1157.63 that you have in your account is called the **compound amount.**

If you had earned only *simple* interest for 3 years, your interest would be as follows.

$$\begin{aligned} I &= \$1000 \cdot (0.05) \cdot 3 \\ &= \$150 \end{aligned}$$

At the end of 3 years, you would have $1000 + $150 = $1150 in your account. Compounding the interest increased your earnings by $7.63 ($1157.63 − $1150).

With *compound* interest, the interest earned during the second year is greater than that earned during the first year, and the interest earned during the third year is greater than that earned during the second year. This happens because the interest earned each year is *added* to the principal, and the new total is used to find the amount of interest in the next year.

Compound Interest

Interest paid on principal plus past interest is called **compound interest.**

2 Understand compound amount. Find the compound amount as follows.

1 Find the compound amount given the following deposits. Round to the nearest cent if necessary.

(a) $500 at 4% for 2 years

(b) $2000 at 7% for 3 years

Example 1 Finding the Compound Amount

Nancy Wegener deposits $3400 in an account that pays 6% interest compounded annually for 4 years. Find the compound amount. Round to the nearest cent when necessary.

Year	Interest	Compound Amount
1	$3400 • (0.06) • 1 = $204	
	$3400 + $204 =	$3604
2	$3604 • (0.06) • 1 = $216.24	
	$3604 + $216.24 =	$3820.24
3	$3820.24 • (0.06) • 1 ≈ $229.21	
	$3820.24 + $229.21 =	$4049.45
4	$4049.45 • (0.06) • 1 ≈ $242.97	
	$4049.45 + $242.97 =	$4292.42

The compound amount is $4292.42.

Work Problem 1 at the Side.

3 **Find the compound amount.** A more efficient way of finding the compound amount is to add the interest rate to 100% and then multiply by the original deposit. Notice that in Example 1, at the end of the first year, you will have $3400 (100% of the original deposit) plus 6% (of the original deposit) or 106% (100% + 6% = 106%).

2 Find the compound amount by multiplying the original deposit by 100% plus the compound interest rate. Round to the nearest cent if necessary.

(a) $1800 at 4% for 3 years

1800 · (1.04) · (1.04) · (1.04) = ______

(b) $900 at 3% for 2 years

(c) $2500 at 5% for 4 years

Example 2 Finding the Compound Amount

Find the compound amount in Example 1 using multiplication.

$$\$3400 \cdot \underbrace{\overset{\text{Year 1}}{(1.06)} \cdot \overset{\text{Year 2}}{(1.06)} \cdot \overset{\text{Year 3}}{(1.06)} \cdot \overset{\text{Year 4}}{(1.06)}}_{100\% + 6\% + 106\% = 1.06} \approx \$4292.42$$

↑ Original deposit (\$3400); ↑ Compound amount (\$4292.42)

Our answer, $4292.42 is the same as Example 1.

Work Problem 2 at the Side.

NOTE

By adding the compound interest rate to 100%, we can then multiply by the original deposit. This will give us the compound amount at the end of each compound interest period.

Calculator Tip If you use a calculator for Example 2 you can use the y^x key (exponent key).

3400 (×) 1.06 (y^x) 4 (=) 4292.42 (rounded)

The 4 following the y^x key represents the number of compound interest periods.

ANSWERS
1. (a) $540.80 (b) $2450.09 (rounded)
2. (a) $2024.76 (rounded) (b) $954.81
(c) $3038.77 (rounded)

4 Use a compound interest table. The calculation of compound interest can be quite tedious. For this reason, compound interest tables have been developed.

Suppose you deposit $1 in a savings account today that earns 4% compounded annually and you allow the deposit to remain for 3 years. The diagram below shows the compound amount at the end of each of the 3 years.

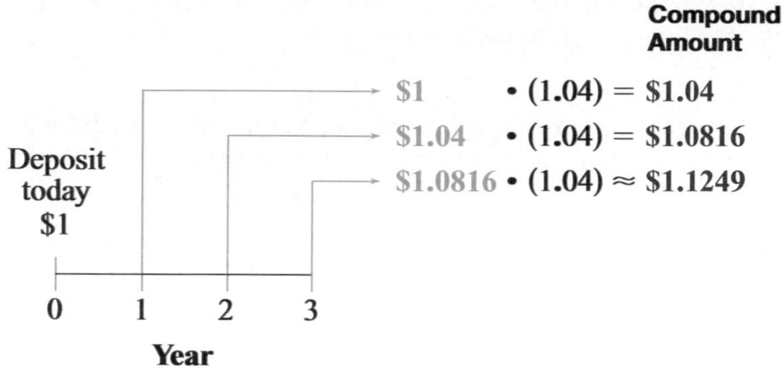

Using the compound amounts for $1, a table can be formed. Look at the table below and find the column headed 4%. The first three numbers for years 1, 2, and 3 are the same as those we have calculated for $1 at 4% for 3 years. This table, giving the compound amounts on a $1 deposit for given lengths of time and interest rates, can be used for finding the compound amount on any amount of deposit.

COMPOUND INTEREST

Time Periods	3.00%	3.50%	4.00%	4.50%	5.00%	5.50%	6.00%	8.00%
1	1.0300	1.0350	1.0400	1.0450	1.0500	1.0550	1.0600	1.0800
2	1.0609	1.0712	1.0816	1.0920	1.1025	1.1130	1.1236	1.1664
3	1.0927	1.1087	**1.1249**	1.1412	1.1576	1.1742	1.1910	1.2597
4	1.1255	1.1475	1.1699	1.1925	1.2155	1.2388	1.2625	1.3605
5	1.1593	1.1877	1.2167	1.2462	1.2763	1.3070	1.3382	1.4693
6	1.1941	1.2293	1.2653	1.3023	1.3401	1.3788	1.4185	1.5869
7	1.2299	1.2723	1.3159	1.3609	1.4071	1.4547	1.5036	1.7138
8	1.2668	1.3168	1.3686	1.4221	1.4775	1.5347	1.5938	1.8509
9	1.3048	1.3629	1.4233	1.4861	1.5513	1.6191	1.6895	1.9990
10	1.3439	1.4106	1.4802	1.5530	**1.6289**	1.7081	1.7908	2.1589
11	1.3842	1.4600	1.5395	1.6229	1.7103	1.8021	1.8983	2.3316
12	1.4258	1.5111	1.6010	1.6959	1.7959	1.9012	2.0122	2.5182

Example 3 Using a Compound Interest Table

Find each compound amount. Round answers to the nearest cent.

(a) $1 is deposited at a 5% interest rate for 10 years.

Look down the column headed 5%, and across to row 10 (because 10 years = 10 time periods). At the intersection of the column and row, read the compound amount, 1.6289, which can be rounded to $1.63.

(b) $1 is deposited at $5\frac{1}{2}$% for 6 years.

The intersection of the $5\frac{1}{2}$% (5.50%) column and row 6 shows 1.3788 as the compound amount. Round this to $1.38.

Work Problem 3 at the Side.

3 Find the compound amount using the compound interest table. Round to the nearest cent.

(a) $1 at 3% for 6 years

(b) $1 at 6% for 8 years

(c) $1 at $4\frac{1}{2}$% for 12 years

ANSWERS

3. (a) $1.19 (rounded) **(b)** $1.59 (rounded) **(c)** $1.70 (rounded)

4 Find the compound amount and the interest.

(a) \$4000 at 8% for 10 years

(b) \$12,600 at $5\frac{1}{2}\%$ for 8 years

(c) \$32,700 at 6% for 12 years

ANSWERS
4. **(a)** \$8635.60; \$4635.60
(b) \$19,337.22; \$6737.22
(c) \$65,798.94; \$33,098.94

5 **Find the compound amount and the amount of compound interest.** Find the compound amount and interest as follows.

Finding the Compound Amount and the Interest

Compound Amount
Find the compound amount for any amount of principal by multiplying the principal by the compound amount for \$1.

Interest
Find the amount of interest earned on a deposit by subtracting the amount originally deposited from the compound amount.

Example 4 Finding Compound Amount and Interest

Find the compound amount and the interest.

(a) \$1000 at $5\frac{1}{2}\%$ interest for 12 years
Look in the table for $5\frac{1}{2}\%$ (5.50%) and 12 periods; find the number 1.9012 but do *not* round it. Multiply this number and the principal of \$1000.

$$\$1000 \cdot (1.9012) = \$1901.20$$

The account will contain \$1901.20 after 12 years.
Find the amount of interest by subtracting the original deposit from the compound amount.

Compound amount → \$1901.20 − Original amount → \$1000 = Amount of interest → \$901.20

$$\$1901.20 - \$1000 = \$901.20$$

(b) \$6400 at 8% for 7 years
Look in the table for 8% and 7 periods, finding 1.7138. Multiply.

$$\$6400 \cdot (1.7138) = \$10{,}968.32 \quad \text{Compound amount}$$

Subtract the original deposit from the compound amount.

$$\$10{,}968.32 - \$6400 = \$4568.32 \quad \text{Interest}$$

A total of \$4568.32 in interest was earned.

Work Problem 4 at the Side.

6.8 EXERCISES

FOR EXTRA HELP: Student's Solutions Manual | MyMathLab.com | InterAct Math Tutorial Software | AW Math Tutor Center | www.mathxl.com | Digital Video Tutor CD 4 Videotape 11

Find the compound amount given the following deposits. Calculate the interest each year, then add it to the previous year's amount. See Example 1.

1. \$500 at 4% for 2 years

2. \$1000 at 5% for 3 years

3. \$1800 at 3% for 3 years

4. \$2000 at 8% for 3 years

5. \$3500 at 7% for 4 years

6. \$5500 at 6% for 4 years

Find each compound amount by multiplying the original amount deposited by 100% plus the compound rate in the following. See Example 2. Round to the nearest cent if necessary.

7. \$1000 at 5% for 2 years

8. \$500 at 4% for 3 years

9. \$1400 at 6% for 5 years

10. \$1800 at 8% for 4 years

11. \$1180 at 7% for 8 years

12. \$12,800 at 6% for 7 years

13. \$10,940 at 8% for 6 years

14. \$15,710 at 10% for 8 years

Use the table on page 439 to find the compound amount and the interest. Interest is compounded annually. Round to the nearest cent if necessary. See Example 3.

15. \$1000 at 4% for 5 years

16. \$10,000 at 3% for 4 years

17. \$8000 at 6% for 10 years

18. \$7800 at 5% for 8 years

19. \$8428.17 at $4\frac{1}{2}$% for 6 years

20. \$10,472.88 at $5\frac{1}{2}$% for 12 years

21. Write a definition for compound interest. Describe in your own words what compound interest means to you.

22. What is the difference between the compound amount and compound interest?

Use the table on page 439 to solve each application problem. Round to the nearest cent if necessary. See Examples 3 and 4.

23. Jane Chavez deposited $8450 in an account that pays 6% interest compounded annually. Find the amount she will have (compound amount) at the end of 8 years.

24. Rob Diamond borrowed $22,500 from his uncle to open Bookkeeping and More. He will repay the loan at the end of 5 years at 8% interest compounded annually. Find the amount he will repay.

25. Al Granard lends $38,000 to the owner of Rick's Limousine Service. He will be repaid at the end of 9 years at 6% interest compounded annually. Find **(a)** the total amount that he should be repaid and **(b)** the amount of interest earned.

26. Sadie Simms has $28,500 in an Individual Retirement Account (IRA) that pays 5% interest compounded annually. Find **(a)** the total amount she will have at the end of 5 years and **(b)** the amount of interest earned.

27. Jennifer Barrister deposits $30,000 at 6% interest compounded annually. Two years after she makes the first deposit, she adds another $40,000, also at 6% compounded annually.

(a) What total amount will she have 5 years after her first deposit?

(b) What amount of interest will she have earned?

28. Kara Ivee invests $25,000 at 8% interest compounded annually. Three years after she makes the first deposit, she adds another $25,000, also at 8% compounded annually.

(a) What total amount will she have 5 years after her first deposit?

(b) What amount of interest will she have earned?

RELATING CONCEPTS (Exercises 29–34) FOR INDIVIDUAL OR GROUP WORK

Knowing how to solve interest problems is important to business people and consumers alike.
Work Exercises 29–34 in order.

29. Simple interest calculation is used for most short-term business loans, most real estate loans, and many automobile and consumer loans. The formula for simple interest is

Interest = ________ • ________ • ________

or $I =$ ________ • ________ • ________.

30. When a loan is repaid, the interest is added to the original principal. The formula for amount due is

amount due = ________ + ________

31. Compound interest is paid on most savings accounts and many other types of investments. Compound interest is interest calculated on ________ plus past ________.

32. The compound amount is the total amount in an account at the end of a period of time. Compound amount is the original ________ + compound ________.

33. Dick Gonsalves has two choices. He can invest \$4350 at 6% simple interest for 6 years, or he can invest the same amount at 6% compounded annually for 6 years.

(a) Find the difference in the amount of interest earned in these two accounts.

(b) If the length of time is doubled from 6 to 12 years, will the difference in the amount of interest earned also double?

(c) Use your own examples to determine that what you found in part (b) is true with other interest rates and other lengths of time.

34. One account is opened with \$20,000 at 8% simple interest for 12 years. Another account is opened with \$16,000 at 8% compounded annually for 12 years.

(a) At the end of the 12 years, which account has the higher balance and by how much?

(b) What does this tell you about compound interest? Why?

Chapter 6

SUMMARY

KEY TERMS

6.1	**percent**	Percent means per one hundred. A percent is a ratio with a denominator of 100.
6.3	**percent proportion**	The proportion $\frac{\text{part}}{\text{whole}} = \frac{\text{percent}}{100}$ is used to solve percent problems.
	whole	The whole in a percent problem is the entire quantity, the total, or the base.
	part	The part in a percent problem is the portion being compared with the whole.
6.5	**percent equation**	The percent equation is part = percent • whole. It is another way to solve percent problems.
6.6	**sales tax**	Sales tax is a percent of the total sales charged as a tax.
	commission	Commission is a percent of the dollar value of total sales paid to a salesperson.
	discount	Discount is often expressed as a percent of the original price; it is then deducted from the original price, resulting in the sale price.
	percent of increase or decrease	Percent of increase or decrease is the amount of increase or decrease expressed as a percent of the original amount.
6.7	**interest**	Interest is a fee paid or a charge for lending or borrowing money.
	interest formula	The interest formula is used to calculate interest. It is interest = principal • rate • time or $I = p \cdot r \cdot t$.
	simple interest	Interest that is computed on the original principal is simple interest.
	principal	Principal is the amount of money on which interest is earned.
	rate of interest	Often referred to as "rate," it is the charge for interest and is given as a percent.
6.8	**compound interest**	Compound interest is interest paid on past interest as well as on principal.
	compound amount	The total amount in an account including compound interest and the original principal.
	compounding	Interest that is compounded once each year is compounded annually.

NEW FORMULAS

To write percents as decimals: $p\% = p \div 100$

To write percents as fractions: $p\% = \frac{p}{100}$

Percent proportion: $\frac{\text{part}}{\text{whole}} = \frac{\text{percent}}{100}$

Percent equation: part = percent • whole

Amount of sales tax: amount of sales tax = rate of tax • cost of item

Amount of commission: amount of commission = rate of commission • amount of sales

Amount of discount: amount of discount = rate of discount • original price

Sale price: sale price = original price − amount of discount

Percent of increase: $\frac{\text{percent}}{100} = \frac{\text{amount of increase}}{\text{original value}}$

Percent of decrease: $\frac{\text{percent}}{100} = \frac{\text{amount of decrease}}{\text{original value}}$

Interest: $I = p \cdot r \cdot t$ (principal • rate • time)

Amount due: amount due = principal + interest

Test Your Word Power

See how well you have learned the vocabulary in this chapter. Answers follow the Quick Review.

1. To write a **percent as a decimal,** you drop the percent sign
   (a) after finding the decimal point
   (b) after removing the decimal point
   (c) after moving the decimal point two places to the right
   (d) after moving the decimal point two places to the left.

2. To write a **decimal as a percent,** you add the percent sign
   (a) after finding the decimal point
   (b) after removing the decimal point
   (c) after moving the decimal point two places to the right
   (d) after moving the decimal point two places to the left.

3. **Percent** means
   (a) the same as interest
   (b) per every ten
   (c) per one thousand
   (d) per one hundred.

4. When you use $\frac{\textbf{part}}{\textbf{whole}} = \frac{\textbf{percent}}{\textbf{100}}$ to solve percent problems, you are using the
   (a) simple interest formula
   (b) percent proportion
   (c) percent equation
   (d) percent sales tax formula.

5. The **percent equation** is
   (a) part = percent • whole
   (b) $I = p \cdot r \cdot t$
   (c) $p\% = \frac{p}{100}$
   (d) amount due = principal + interest.

6. In a **percent of increase or decrease** problem, the increase or decrease is a percent of
   (a) the largest amount
   (b) the original amount
   (c) the new or most recent amount
   (d) all the amounts.

7. In the formula $\boldsymbol{I = p \cdot r \cdot t}$, the $\boldsymbol{p}$ stands for
   (a) proportion
   (b) product
   (c) principal
   (d) percent.

8. The term **rate** in an interest problem represents the
   (a) whole
   (b) percent
   (c) part
   (d) amount of interest.

Quick Review

Concepts	*Examples*
6.1 Basics of Percent **Writing a Percent as a Decimal** To write a percent as a decimal, move the decimal point two places to the left and drop the % sign. **Writing a Decimal as a Percent** To write a decimal as a percent, move the decimal point two places to the right and attach a % sign.	50% (.50%) = 0.50 or just 0.5 3% (.03%) = 0.03 0.75 (0.75) = 75% 3.6 (3.60) = 360%
6.2 Writing a Fraction as a Percent Use a proportion and solve for p to change a fraction to percent.	$\frac{2}{5} = \frac{p}{100}$ Proportion $5 \cdot p = 2 \cdot 100$ Cross products $5 \cdot p = 200$ $\frac{\overset{1}{\cancel{5}} \cdot p}{\underset{1}{\cancel{5}}} = \frac{200}{5}$ Divide each side by 5. $p = 40$ $\frac{2}{5} = 40\%$ Attach % sign.

Concepts	*Examples*
***6.3** Learning the Percent Proportion* Part is to whole as percent is to 100. $\frac{\text{part}}{\text{whole}} = \frac{\text{percent}}{100}$	Use the percent proportion to solve for the unknown value. part = 30, whole = 50; find the percent. Percent (unknown): x $\frac{30}{50} = \frac{x}{100}$ Percent proportion $\frac{3}{5} = \frac{x}{100}$ Lowest terms $5 \cdot x = 3 \cdot 100$ Cross products $5 \cdot x = 300$ $\frac{\overset{1}{\cancel{5}} \cdot x}{\underset{1}{\cancel{5}}} = \frac{300}{5}$ Divide each side by 5. $x = 60$ The percent is 60, which is 60%.
***6.3** Identifying Percent, Whole, and Part in a Percent Problem* The percent appears with the word **percent** or with the symbol %. The whole often appears after the word **of**. The whole is the entire quantity or total. The part is the portion of the total. If the percent and the whole are found first, the remaining number is the part.	Find the percent, whole, and part in the following. 10% of the 500 pies is how many pies? (10%: Percent; 500 pies: Whole; how many pies?: Part (unknown)) 20 cats is 5% of what number of cats? (20 cats: Part; 5%: Percent; what number of cats?: Whole (unknown)) What percent of \$220 is \$33? (What percent: Percent (unknown); \$220: Whole; \$33: Part)
***6.4** Applying the Percent Proportion* Read the problem and identify the percent, whole, and part. Use the percent proportion to solve for the unknown quantity (whole).	A tank contains 35% distilled water. If 28 gallons of distilled water are in the tank when it is full, find the capacity of the tank. percent = 35 and part = 28 Use the percent proportion to find the whole. $\frac{\text{part}}{x} = \frac{\text{percent}}{100}$ (Whole (unknown) → x) $\frac{28}{x} = \frac{35}{100}$ $\frac{28}{x} = \frac{7}{20}$ Lowest terms $x \cdot 7 = 560$ Cross products $\frac{x \cdot \overset{1}{\cancel{7}}}{\underset{1}{\cancel{7}}} = \frac{560}{7}$ Divide each side by 7. $x = 80$ The capacity of the tank is 80 gallons.

Concepts	*Examples*
6.5 Using the Percent Equation The percent equation is part = percent • whole. Identify the percent, whole, and part and solve for the unknown quantity. Always write the percent as a decimal before using the equation.	Solve each problem. **(a)** Find 20% of 220 applicants. $$\text{part (unknown)} = \text{percent} \cdot \text{whole}$$ $$x = 0.2 \cdot 220$$ $$x = 44$$ 20% of 220 applicants is 44 applicants. **(b)** 8 balls is 4% of what number of balls? $$\text{part} = \text{percent} \cdot \text{whole (unknown)}$$ $$8 = 0.04 \cdot x$$ $$\frac{8}{0.04} = \frac{\overset{1}{\cancel{0.04}} \cdot x}{\underset{1}{\cancel{0.04}}}$$ $$x = 200$$ 8 balls is 4% of 200 balls. **(c)** \$13 is what percent of \$52? $$\text{part} = \text{percent (unknown)} \cdot \text{whole}$$ $$13 = x \cdot 52$$ $$\frac{13}{52} = \frac{x \cdot \overset{1}{\cancel{52}}}{\underset{1}{\cancel{52}}}$$ $$x = 0.25 = 25\%$$ \$13 is 25% of \$52.
6.6 Solving Application Problems with Proportions To solve for **sales tax,** use the formula **amount of sales tax = rate of tax • cost of item.**	The cost of a 35-inch television is \$699, and the sales tax is 5%. Find the sales tax. **amount of sales tax** = 5% • \$699 = 0.05 • \$699 = \$34.95
To find **commissions,** use the formula **amount of commission = rate of commission • amount of sales.**	The sales are \$92,000 with a commission rate of 3%. Find the commission. **amount of commission** = 3% • \$92,000 = 0.03 • \$92,000 = \$2760
To find the **discount** and the **sale price,** use these formulas **amount of discount = rate of discount • original price** **sale price** = original price − amount of discount.	A gas oven originally priced at \$480 is offered at a 25% discount. Find the amount of the discount and the sale price. **discount** = 0.25 • \$480 = **\$120** **sale price** = \$480 − **\$120** = **\$360**

Concepts	*Examples*
6.6 *Solving Application Problems with Proportions (continued)* To find the **percent of change,** subtract to find the amount of change (increase or decrease), which is the part. The whole is the original value or value before the change.	The number of parking violations rose from 1980 violations to 2277. Find the percent of increase. $2277 - 1980 = 297$ Increase $\frac{297}{1980} = \frac{\textbf{percent}}{100}$ Solve the proportion to find that the percent = 15, so the percent of increase is 15%.
6.7 *Finding Simple Interest* Use the formula $I = p \cdot r \cdot t$. **interest = principal • rate • time** Time (t) is in years. When the time is given in months, use a fraction with 12 in the denominator because there are 12 months in a year.	\$2800 is deposited at 8% for 3 months. Find the amount of interest. $I = p \cdot r \cdot t$ $= \underbrace{2800 \cdot 0.08}_{} \cdot \frac{3}{12}$ $= 224 \cdot \frac{1}{4} = \frac{224 \cdot 1}{4} = \$56$
6.8 *Finding Compound Amount and Compound Interest* There are three methods for finding the compound amount. **1.** Calculate the interest for each compound interest period, then add it back to the principal. **2.** Multiply the original deposit by 100% plus the compound interest rate. **3.** Use the table to find the interest on \$1. Then, multiply the table value by the principal. The compound interest is found with the formula $\textbf{compound interest} = \textbf{compound amount} - \textbf{original deposit}$	Find the compound amount and interest if \$1500 is deposited at 5% interest for 3 years. (table below) 2. $\$1500 \cdot \underbrace{(1.05) \cdot (1.05) \cdot (1.05)}_{} \approx \$1736.44$ ↑ Original deposit; ↑ Compound amount $100\% + 5\% = 105\% = 1.05$ 3. Locate 5% across the top of the table and 3 periods at the left. The table value is 1.1576. compound amount = \$1500 • (1.1576) = **\$1736.40*** interest = \$1736.40 − \$1500 = **\$236.40**

		Interest	Compound Amount
1.	Year 1	\$1500 • (0.05) • 1 = \$75	\$1500 + \$75 = **\$1575**
	Year 2	\$1575 • (0.05) • 1 = \$78.75	\$1575 + \$78.75 = **\$1653.75**
	Year 3	\$1653.75 • (0.05) ≈ \$82.69	\$1653.75 + \$82.69 = **\$1736.44**

* The difference in the compound amount results from rounding in the table.

Answers to Test Your Word Power

1. (d) *Example:* 50% written as a decimal is 0.50 or 0.5. **2. (c)** *Example:* 0.25 written as a percent is 0.25% or 25%. **3. (d)** *Example:* 8% means 8 per 100. **4. (b)** *Example:* Part = 4, and whole = 25. To find the percent, $\frac{4}{25} = \frac{x}{100}$; $x = 16$ or 16% (percent). **5. (a)** *Example:* Percent = 25, and whole = 300. To find the part, $0.25 \times 300 = 75$ (part). **6. (b)** *Example:* Original = 200, and increase or decrease = 40. To find the percent of increase or decrease, $\frac{40}{200} = 0.2 = 20\%$ (percent of increase or decrease). **7. (c)** *Example:* Principal (p) = \$800, rate (r) = 5%, and time (t) = 1 year. Then, $I = p \cdot r \cdot t$ so $I = \$800 \cdot 0.05 \cdot 1 = \$40$ (interest). **8. (b)** *Example:* Principal (p) = \$1650, rate (r) = 4%, and time $(t) = \frac{1}{2}$ year. To find the interest (I), $I = \$1650 \cdot 0.04 \cdot \frac{1}{2} = \$33$ (interest).

Chapter 6 Review Exercises

[6.1] *Write each percent as a decimal and each decimal as a percent.*

1. 35% **2.** 150% **3.** 99.44% **4.** 0.085%

5. 3.15 **6.** 0.02 **7.** 0.875 **8.** 0.002

[6.2] *Write each percent as a fraction or mixed number in lowest terms and each fraction as a percent.*

9. 15% **10.** 37.5% **11.** 175% **12.** 0.25%

13. $\frac{3}{4}$ **14.** $\frac{5}{8}$ **15.** $3\frac{1}{4}$ **16.** $\frac{1}{200}$

Complete this chart.

Fraction	Decimal	Percent
$\frac{1}{8}$	**17.** ______	**18.** ______
19. ______	0.25	**20.** ______
21. ______	**22.** ______	180%

[6.3] *Find the unknown value in the percent proportion* $\frac{part}{whole} = \frac{percent}{100}$.

23. part = 25, percent = 10

24. whole = 480, percent = 5

Identify each percent, whole, and part. Do not try to solve.

25. 35% of 820 mailboxes is 287 mailboxes.

26. 73 brooms is what percent of 90 brooms?

27. Find 14% of 160 bicycles.

28. 418 curtains is 16% of what number of curtains?

29. A golfer lost three of his eight golf balls. What percent were lost?

30. Only 88% of the door keys cut will operate properly. If there are 1280 keys cut, find the number of keys that will operate properly.

[6.4] *Find the part using the percent proportion or the multiplication shortcut.*

31. 18% of 950 programs

32. 60% of 1450 reference books

33. 0.6% of 5200 acres

34. 0.2% of 1400 kilograms

Find the whole using the percent proportion.

35. 105 crates is 14% of what number of crates?

36. 348 test tubes is 15% of what number of test tubes?

37. 677.6 miles is 140% of what number of miles?

38. 2.5% of what number of cases is 425 cases?

Find the percent using the percent proportion. Round percent answers to the nearest tenth if necessary.

39. 649 tulip bulbs is what percent of 1180 tulip bulbs?

40. What percent of 1620 dinner rolls is 85 dinner rolls?

41. What percent of 380 pairs of socks is 36 pairs?

42. What percent of 650 soup cans is 200 soup cans?

[6.1–6.4] *Solve each application problem. Round percent answers to the nearest tenth if necessary.*

43. Eight years ago there were 112,000 high school math teachers in the United States. Since that time the number of high school math teachers has increased by 25%. How many high school math teachers are there today? (*Source: USA Today.*)

44. Scientists tell us that there are 9600 species of birds and that 1000 of these species are in danger of extinction. What percent of the bird species are in danger of extinction?

[6.5] *Use the percent equation to answer each question.*

45. 32% of $454 is what amount?

46. 155% of 120 trucks is how many trucks?

47. 0.128 ounces is what percent of 32 ounces?

48. 304.5 meters is what percent of 174 meters?

49. 33.6 miles is 28% of what number of miles?

50. $92 is 16% of what number?

[6.6] *Find the amount of sales tax or the tax rate and the total cost. Round to the nearest cent if necessary.*

	Amount of Sale	Tax Rate	Amount of Tax	Total Cost
51.	$630	5%	________	________
52.	$780	________	$58.50	________

Find the commission earned or the rate of commission.

	Sales	Rate of Commission	Commission
53.	$3450	8%	________
54.	$65,300	________	$3265

Find the amount or rate of discount and the amount paid after the discount. Round to the nearest cent if necessary.

	Original Price	Rate of Discount	Amount of Discount	Sale Price
55.	$112.50	30%	________	________
56.	$252	________	$63	________

[6.7] *Find the simple interest due on each loan.*

	Principal	Rate	Time in Years	Interest
57.	$200	4%	1	________
58.	$1080	5%	$1\frac{1}{4}$	________

Find the simple interest paid on each investment.

	Principal	Rate	Time in Months	Interest
59.	$400	7%	3	________
60.	$1560	$6\frac{1}{2}\%$	18	________

Find the total amount due on the following simple interest loans.

	Principal	Rate	Time	Total Amount Due
61.	$750	$5\frac{1}{2}\%$	2 years	________
62.	$1530	6%	9 months	________

[6.8] *Find the compound amount and compound interest in the following. Interest is compounded annually. You may use the table on page 439. Round to the nearest cent if necessary.*

	Principal	Rate	Time in Years	Compound Amount	Compound Interest
63.	$2000	5%	10	________	________
64.	$1870	4%	4	________	________
65.	$3600	8%	3	________	________
66.	$12,500	$5\frac{1}{2}\%$	5	________	________

MIXED REVIEW EXERCISES

Find the unknown value in the percent proportion $\frac{part}{whole} = \frac{percent}{100}$.

67. whole = 80, percent = 15

68. part = 738, percent = 45

Use the percent proportion or equation to answer each question.

69. 12% of 194 meters is how many meters?

70. 327 cars is what percent of 218 cars?

71. 0.6% of $85 is what amount?

72. 396 employees is 20% of what number of employees?

73. 76 chickens is what percent of 190 chickens?

74. 214.484 liters is 43% of what number of liters?

Write each percent as a decimal and each decimal as a percent.

75. 55%

76. 300%

77. 5

78. 4.71

79. 8.6%

80. 0.621

81. 0.375%

82. 0.0006

Write each percent as a fraction in lowest terms and each fraction as a percent.

83. $\frac{3}{4}$

84. 42%

85. 87.5%

86. $\frac{3}{8}$

87. $32\frac{1}{2}\%$

88. $\frac{3}{5}$

89. 0.25%

90. $3\frac{3}{4}$

Solve each application problem. Round percent answers to the nearest tenth and money amounts to the nearest cent if necessary.

91. Jim Bralley invests $16,850 at $10\frac{1}{2}\%$ for 9 months. Find the amount of interest earned.

92. Karl Schmidt borrows $14,750 at 12% for 18 months to buy a small Lionel train collection. Find the total amount due.

93. A Hotpoint refrigerator has a capacity of 11.5 cubic feet in the refrigerator and 5.5 cubic feet in the freezer. What percent of the total capacity is the capacity of the freezer?

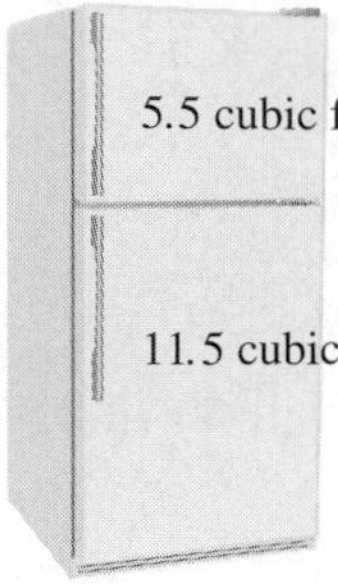

94. Tommy Downs invests the money he inherited from his aunt at 6% compounded annually for 4 years. If the amount of money invested is $12,500, find

(a) the compound amount at the end of 4 years and

(b) the amount of interest that he earned. Do not use the table.

95. Tom Dugally, a real estate agent, sold two properties, one for $125,000 and the other for $290,000. After all of his expenses, he receives a commission of $1\frac{1}{2}\%$ of total sales. Find the commission that he earned.

96. Our mail carrier, Norm, saw his route expand from 481 residential stops to 520 residential stops. Find the percent of increase.

97. A digital camera priced at $598 is marked down 7% to promote the new model. If the sales tax is also 7%, find the cost of the digital camera including sales tax.

98. The DaimlerChrysler Company hopes to sell 35,000 of the new Chrysler PT Cruiser each year in foreign markets. If this amounts to 18.9% of the total annual PT Cruiser sales, find the total predicted annual sales. Round to the nearest whole number. (*Source:* DaimlerChrysler.)

99. Stephen and Heather Hall established a budget allowing 25% for rent, 30% for food, 8% for clothing, 20% for travel and recreation, and the remainder for savings. Stephen takes home $2075 per month, and Heather takes home $32,500 per year. How much money will the couple save in a year?

100. The mileage on a car dropped from 32.8 miles per gallon to 28.5 miles per gallon. Find the percent of decrease.

Chapter 6 Test

Study Skills Workbook
Activity 12

Write each percent as a decimal and each decimal as a percent.

1. 65% **2.** 0.8

3. 1.75 **4.** 0.875

5. 300% **6.** 0.05%

Write each decimal as a fraction in lowest terms.

7. 12.5% **8.** 0.25%

Write each fraction or mixed number as a percent.

9. $\frac{3}{5}$ **10.** $\frac{5}{8}$

11. $2\frac{1}{2}$

Solve each problem.

12. 32 sacks is 4% of what number of sacks?

13. $680 is what percent of $3400?

14. Ann Barnes has saved 65% of the amount needed for a down payment on a home. If she has saved $12,025, find the total down payment needed.

15. The price of a copy machine is $2680 plus sales tax of $6\frac{1}{2}$%. Find the total cost of the copy machine including sales tax.

16. An insurance company pays its salespeople a commission of 8% on all sales. Find the commission earned on insurance sales of $7850.

1. ____________
2. ____________
3. ____________
4. ____________
5. ____________
6. ____________
7. ____________
8. ____________
9. ____________
10. ____________
11. ____________
12. ____________
13. ____________
14. ____________
15. ____________
16. ____________

17. ______________________

17. Attendance at the homecoming game increased from 4320 fans last year to 5616 fans this year. Find the percent of increase.

18. ______________________

18. A problem includes last year's salary, this year's salary, and asks for the percent of increase. Explain how you would identify the part, the whole, and the percent in the problem. Show the percent proportion that you would use.

19. ______________________

19. Write the formula used to find interest. Explain the difference in what to do if the time is expressed in months or in years. Write a problem that involves finding interest for 9 months and another problem that involves finding interest for $2\frac{1}{2}$ years. Use your own numbers for the principal and the rate. Show how to solve your problems.

Find the amount of discount and the sale price. Round answers to the nearest cent.

	Original Price	Rate of Discount
20. ______________________	**20.** \$96	12%
21. ______________________	**21.** \$280	32.5%

Find the simple interest on each loan.

	Principal	Rate	Time
22. ______________________	**22.** \$4200	6%	$1\frac{1}{2}$ years
23. ______________________	**23.** \$6400	5%	9 months

24. ______________________

24. A parent borrows \$5300 to help her son start college. The loan is for 9 months at 9% interest. Find the total amount due on the loan.

25. (a) ______________________

(b) ______________________

25. The River City School PTA Emergency Fund deposited \$4000 at 6% compounded annually. Two years after the first deposit, they add another \$5000, also at 6% compounded annually. Use the compound interest table.

(a) What total amount will they have 4 years after their first deposit? Round to the nearest dollar.

(b) What amount of interest will they have earned?

Cumulative Review Exercises CHAPTERS 1–6

Round the numbers in each problem using front end rounding. Then add, subtract, multiply, or divide the rounded numbers, as indicated, to estimate the answer. Finally, solve for the exact answer.

1. *Estimate:* $+$ ___ *Exact:* $5608 + 94 + 739$

2. *Estimate:* $+$ ___ *Exact:* $0.56 + 49.614 + 8.4$

3. *Estimate:* $-$ ___ *Exact:* $75{,}078 - 46{,}090$

4. *Estimate:* $-$ ___ *Exact:* $7.8 - 3.5029$

5. *Estimate:* $\times$ ___ *Exact:* 6538×708

6. *Estimate:* $\times$ ___ *Exact:* 65.3×8.7

7. *Estimate:* $\overline{)\quad}$ *Exact:* $43\overline{)38{,}786}$

8. *Estimate:* $\overline{)\quad}$ *Exact:* $7.6\overline{)2432}$

9. *Estimate:* $\overline{)\quad}$ *Exact:* $0.8\overline{)6.76}$

Use the order of operations to simplify each expression.

10. $6^2 - 3 \cdot 6$

11. $\sqrt{49} + 5 \cdot 4 - 8$

12. $9 + 6 \div 3 + 7 \cdot 4$

Round each number to the place shown.

13. 4677 to the nearest ten

14. 7,583,281 to the nearest hundred thousand

15. $513.499 to the nearest dollar

16. $362.735 to the nearest cent

Add, subtract, multiply, or divide as indicated. Write answers in lowest terms and as whole or mixed numbers when possible.

17. $\frac{3}{4}+\frac{5}{8}$

18. $\frac{1}{3}+\frac{3}{4}$

19. $\begin{array}{r} 5\frac{3}{4} \\ +\ 7\frac{5}{8} \\ \hline \end{array}$

20. $\frac{7}{8}-\frac{3}{4}$

21. $\begin{array}{r} 8\frac{3}{8} \\ -\ 4\frac{1}{2} \\ \hline \end{array}$

22. $\begin{array}{r} 26\frac{1}{3} \\ -\ 17\frac{4}{5} \\ \hline \end{array}$

23. $\frac{7}{8} \bullet \frac{2}{3}$

24. $7\frac{3}{4} \bullet 3\frac{3}{8}$

25. $36 \bullet \frac{4}{5}$

26. $\frac{5}{8} \div \frac{5}{7}$

27. $12 \div \frac{3}{4}$

28. $2\frac{3}{4} \div 7\frac{1}{2}$

Write < or > to make a true statement.

29. $\frac{5}{8}$ ______ $\frac{2}{3}$

30. $\frac{8}{15}$ ______ $\frac{11}{20}$

31. $\frac{2}{3}$ ______ $\frac{7}{12}$

Simplify each expression. Use the order of operations as needed.

32. $\left(\frac{7}{8}-\frac{1}{2}\right) \bullet \frac{2}{3}$

33. $\frac{7}{8} \div \left(\frac{3}{4}+\frac{1}{8}\right)$

34. $\left(\frac{5}{6}-\frac{5}{12}\right) - \left(\frac{1}{2}\right)^2 \bullet \frac{2}{3}$

Write each fraction as a decimal. Round to the nearest thousandth if necessary.

35. $\frac{3}{5}$

36. $\frac{7}{8}$

37. $\frac{7}{12}$

38. $\frac{11}{20}$

Write each ratio as a fraction in lowest terms. Be sure to make all necessary conversions.

39. 2 hours to 40 minutes

40. If there are 27 people and 36 life jackets, what is the ratio of life jackets to people?

41. $1\frac{5}{8}$ to 13

Use cross multiplication to decide whether each proportion is true *or* false. *Circle the correct answer.*

42. $\frac{5}{8} = \frac{45}{72}$

True False

43. $\frac{64}{144} = \frac{48}{108}$

True False

Find the unknown value in each proportion.

44. $\frac{1}{6} = \frac{x}{36}$

45. $\frac{315}{45} = \frac{21}{x}$

46. $\frac{8}{x} = \frac{72}{144}$

47. $\frac{x}{120} = \frac{7.5}{30}$

Write each percent as a decimal. Write each decimal as a percent.

48. 78%

49. 3%

50. 200%

51. 0.5%

52. 0.87

53. 3.8

54. 0.023

Write each percent as a fraction or mixed number in lowest terms. Write each fraction as a percent.

55. 8%

56. 62.5%

57. 175%

58. $\frac{7}{8}$

59. $\frac{3}{10}$

60. $4\frac{1}{5}$

Solve each percent problem.

61. 35% of 1400 watches is how many watches?

62. $5\frac{1}{2}$% of \$720 is how much?

63. 72 tires is 40% of what number of tires?

64. $4\frac{1}{2}$% of what number of miles is 76.5 miles?

65. What percent of 656 books is 328 books?

66. 72 hours is what percent of 180 hours?

Find the amount of sales tax or the tax rate and the total cost. Round to the nearest cent if necessary.

	Amount of Sale	Tax Rate	Amount of Tax	Total Cost
67.	$108.95	5%	______	______
68.	$460	______	$29.90	______

Find the commission earned or the rate of commission.

	Sales	Rate of Commission	Commission
69.	$12,538	8%	______
70.	$225,300	______	$5632.50

Find the amount or rate of discount and the amount paid after the discount. Round to the nearest cent if necessary.

	Original Price	Rate of Discount	Amount of Discount	Sale Price
71.	$456	45%	______	______
72.	$1085	______	$162.75	______

Find the total amount due on each simple interest loan. Round to the nearest cent if necessary.

	Principal	Rate	Time	Total Amount To Be Repaid
73.	$970	10%	$1\frac{1}{2}$ years	______
74.	$18,350	11%	9 months	______

Set up and solve a proportion for each problem.

75. Carol can test 11 cars for emissions in 4 hours. Find the number of cars that she can test in 12 hours.

76. If 3.5 ounces of weed killer is needed to make 6 gallons of spray, how much weed killer is needed for 102 gallons of spray?

Solve each application problem. Use the table to answer Exercises 77–80. Round answers to the nearest tenth of a percent if necessary.

EXISTING HOME SALES

Region	Last Year	This Year
Northeast	32,000	36,000
Midwest	65,000	66,300
South	82,000	77,500
West	54,000	49,600

77. Find the percent of increase in sales in the northeastern region.

78. Find the percent of increase in sales in the midwestern region.

79. What is the percent of decrease in sales in the southern region?

80. What is the percent of decrease in sales in the western region?

81. Joan Ong has $26,880 invested in her home. If this amount is 25% of her total assets, find her total assets.

82. Pat Ueda deposits $15,000 in a savings account that pays 5.5% interest compounded annually. Find

(a) the amount that he will have in the account at the end of 5 years, and

(b) the amount of interest earned. Do not use the table.

Appendix A
An Introduction to Calculators

A SCIENTIFIC CALCULATORS

OBJECTIVES

1. Learn the basic calculator keys.
2. Understand the (C), (CE), and (ON/C) or (ON/AC) keys.
3. Understand the floating decimal point.
4. Use the (%) key.
5. Use the (x^2) and the (x^3) keys.
6. Use the (y^x) and ($\sqrt{x}$) keys.
7. Use the ($a^b/_c$) key.
8. Solve problems with negative numbers.
9. Use the calculator memory function.
10. Solve chain calculations using the order of operations.
11. Use the parentheses keys.

Calculators are among the more popular inventions of the last three decades. Each year better calculators are developed and costs drop. The first all-transistor desktop calculator was introduced to the market in 1966; it weighed 55 lb, cost \$2500, and was slow. Today, these same calculations are performed quite well on a calculator costing less than \$10. And today's \$200 calculators have more ability to solve problems than some of the early computers.

Many colleges allow students to use calculators in basic mathematics courses. Although you can still purchase a basic four-function calculator, you're probably better off spending \$10 to \$20 on a **scientific calculator.** A scientific calculator will allow you to do a lot more than the basic four-function calculators. A **graphing calculator** allows you to graph functions and data, but it is beyond the scope of this text. The discussion here is confined to the common scientific calculator with the percent key, reciprocal key, exponent key, square root key, memory function, order of operations, and parentheses keys.

NOTE

For explanations of specific calculator models or special function keys, refer to the instruction booklet supplied with your calculator.

1 Learn the basic calculator keys. Most calculators use **algebraic logic.** Some problems can be solved by entering number and function keys in the same order as you would solve problems by hand, but many others require a knowledge of the order of operations when entering the problem. The problem 14 + 28 would be entered as

14 (+) 28 (=)

and 42 would appear as the answer. Enter 387 − 62 as

387 (−) 62 (=)

and 325 appears as the answer. If your calculator does not work problems in this way, check its instruction book to see how to proceed.

2 Understand the (C), (CE), and (ON/C) or (ON/AC) keys. All calculators have a

(C), (ON/C), or (ON/AC)

key. Pressing this key erases everything in the calculator and prepares the calculator to begin a new problem. Some calculators also have a

(CE)

key. Pressing this key erases *only* the number displayed and allows the person using the calculator to correct a mistake without having to start the problem over.

Many calculators combine the (C) key and the (CE) key and use an (ON/C) key. This key turns the calculator on and is also used to erase the calculator display. If the (ON/C) key is pressed after the (=) or one of the operation keys ((+), (−), (×), (÷)), everything in the calculator is erased. If the wrong operation key is pressed, you press the correct key and the error is corrected. For example, 7 (+) (−) 3 (=) 4. Pressing the (−) key cancels out the previous (+) key entry.

CAUTION

Be sure to look at the directions that come with your calculator to see how to clear the memory.

3 **Understand the floating decimal point.** Most calculators have a **floating decimal,** which locates the decimal point in the final result. For example, to find the cost of 55.75 square yards of vinyl floor covering at a cost of $18.99 per square yard, proceed as follows.

55.75 (×) 18.99 (=) 1058.6925

The decimal point is automatically placed in the answer. You should **round** money answers to the nearest cent. Draw a cut-off line after the hundredths place.

Look only at the first digit being cut off.
↓
1058.69|25
↑
Cent position (hundredths)

Because the first digit being cut off is 4 or less, the part you are keeping remains the same. The answer is rounded to $1058.69. If the first digit being cut off had been 5 or more, you would have rounded up by adding 1 to the cent position.

When using a calculator with a floating decimal, enter the decimal point as needed. For example, enter $47 as

47

with no decimal point, but enter 95¢ as

(·) 95

One problem in using a floating decimal is shown by the following example (adding $21.38 and $1.22).

21.38 (+) 1.22 (=) 22.6

The calculator does not show the final 0. You must remember that the problem dealt with money and write the final 0, making the answer $22.60.

4 **Use the (%) key.** The (%) key moves the decimal point two places to the left when pressed following multiplication or division. The problem 8% of $4205 is solved as follows:

4205 (×) 8 (%) (=) 336.4

Because the problem involved money, write the answer as $336.40.

5 Use the x^2 and the x^3 keys. The squaring key, x^2, allows you to square the number in the display (multiply the number by itself). The square of 7 is found as follows.

7 x^2 49

The cubing key, x^3, allows you to find the cube of a number (the number is multiplied by itself three times). To find the cube of 6.8 (that is, 6.8 • 6.8 • 6.8), follow these keystrokes.

6.8 x^3 314.432

6 Use the y^x and $\sqrt{x}$ keys. The product of $3 \times 3 \times 3 \times 3 \times 3$ can be written as

$$3^5$$

(Exponent: 5; Base: 3)

The exponent, 5, shows how many times the base is multiplied by itself (multiply 3 by itself five times). The y^x key raises a base to any desired power. Find 3^5 as

3 y^x 5 = 243.

Since $3^2 = 9$, the number 3 is called the square root of 9. Square roots are written with the symbol $\sqrt{\ }$. Use the $\sqrt{x}$ key to find the square root of 144, $\sqrt{144}$, as follows:

144 $\sqrt{x}$ 12

Find $\sqrt{20}$ as

20 $\sqrt{x}$ 4.472135955,

which may be rounded to the desired position.

7 Use the $a^{b}/_{c}$ key. The $a^{b}/_{c}$ key is used when solving problems containing fractions and mixed numbers.

Solve $\frac{3}{4} + \frac{6}{11}$ as

3 $a^{b}/_{c}$ 4 + 6 $a^{b}/_{c}$ 11 = 1_13⌋44

The answer is $1\frac{13}{44}$.

Solve the mixed number problem $4\frac{7}{8} \div 3\frac{4}{7}$ as

4 $a^{b}/_{c}$ 7 $a^{b}/_{c}$ 8 ÷ 3 $a^{b}/_{c}$ 4 $a^{b}/_{c}$ 7 = 1_73⌋200

The answer is $1\frac{73}{200}$.

NOTE

The calculator automatically shows fractions in lowest terms and as mixed numbers when possible.

8 Solve problems with negative numbers. Negative numbers may be entered by first entering the number and then using the (+/–) key. This changes the number entered to a negative number. For example, solve $-10 + 6 - 8$ as follows.

10 (+/–) (+) 6 (–) 8 (=) –12

9 Use the calculator memory function. Many calculators feature memory keys, which are a sort of electronic scratch paper. These memory keys are used to store intermediate steps in a calculation. On some basic calculators, a key labeled (M) is used to store the numbers in the display, with (MR) used to recall the numbers from memory.

Some calculators have (M+) and (M–) keys. The (M+) key adds the number displayed to the number already in memory. For example, if the memory contains the number 0 at the beginning of a problem, and the calculator display contains the number 29.4, then pushing (M+) will cause 29.4 to be stored in the memory (the result of adding 0 and 29.4). If 57.8 is then entered into the display, pushing (M+) will cause

$$29.4 + 57.8 = 87.2$$

to be stored. If 11.9 is then entered into the display, with (M–) pushed, the memory will contain

$$87.2 - 11.9 = 75.3.$$

The (MR) key is used to recall the number in memory as needed, with (MC) used to clear the memory.

Scientific calculators typically have one or more storage registers in which to store numbers. These memory keys are usually labeled as (STO) for store and (RCL) for recall. For example, you can store 25.6 in register 1 by pressing

25.6 (STO) 1

or you can store it in register 2 by pressing 25.6 (STO) 2 and so forth for other registers. Values are retrieved from a particular memory register by using the (RCL) key followed by the number of the register, for example, (RCL) 2 recalls the contents of memory in register 2.

With a scientific calculator, a number stays in memory until it is replaced by another number or until the memory is cleared. With some calculators, the contents of the memory is saved even when the calculator is turned off.

Here is an example of a problem that uses the memory keys. Suppose an elevator technician wants to find the average weight of a person using an elevator. To do this, she counts the number of people entering an elevator and also measures the weight of each group of people.

Number of People	Total Weight (in pounds)
6	839
8	1184
4	640

First, find the total weight of all three groups and store this in memory register 1.

839 (+) 1184 (+) 640 (=) 2663 (STO) 1

Then, find the total number of people.

6 (+) 8 (+) 4 (=) 18 (STO) 2

Finally, divide the contents of memory register 1 (total weight) by the contents of memory register 2 (18 people).

RCL 1 ÷ RCL 2 = 147.9444444 lb

This answer can be rounded as needed.

10 **Solve chain calculations using the order of operations.** Long calculations involving several different operations are called **chain calculations.** They must be done in a specific sequence called the **order of operations.** The logic of the following order of operations is built into most scientific calculators and can help you work problems without having to store or write down a lot of intermediate steps.

Order of Operations

Step 1 Do all operations inside parentheses or other grouping symbols.
Step 2 Simplify any expressions with exponents and find any square roots.
Step 3 Multiply and divide, proceeding from left to right.
Step 4 Add and subtract, proceeding from left to right.

Your scientific calculator can be used to solve the problem $3 + 7 \times 9\frac{3}{4}$.

3 + 7 × 9 ab/c 3 ab/c 4 = $71\frac{1}{4}$

The calculator automatically multiplies $7 \times 9\frac{3}{4}$ *before* adding 3.

The problem $42.1 \times 5 - 90 \div 4$ is solved as

42.1 × 5 − 90 ÷ 4 = 188.

The calculator automatically multiplies 42.1×5 and divides $90 \div 4$ *before* doing the subtraction.

CAUTION

Scientific calculators keep track of the order of operations for us. All we have to do is enter the problem correctly into the calculator and the calculator does the rest. However, the basic four-function calculator is *not* programmed to observe the order of operations and will calculate correctly *only* if you enter numbers in the proper order.

11 **Use the parentheses keys.** The parentheses keys allow you to group numbers in a complex chain calculation. For example, $\frac{4}{5 + 7}$ can be written as $\frac{4}{(5 + 7)}$, which can be solved as follows.

Left parentheses key

4 ÷ (5 + 7) = 0.333333333

Right parentheses key

Without the parentheses the calculator would have automatically divided 4 by 5 before adding 7, giving an *incorrect* answer of 7.8.

To solve the problem

$$\frac{16 \div 2.5}{39.2 - 29.8 \times 0.6}$$

you must think of the problem as

$$\frac{(16 \div 2.5)}{(39.2 - 29.8 \times 0.6)}.$$

Use parentheses to set off the numerator and the denominator.

(16 ÷ 2.5) ÷ (39.2 − 29.8 × .6) = 0.300187617

Appendix B Inductive and Deductive Reasoning

B Inductive and Deductive Reasoning

Objectives

1. Use inductive reasoning to analyze patterns.
2. Use deductive reasoning to analyze arguments.
3. Use deductive reasoning to solve problems.

1 Use inductive reasoning to analyze patterns. In many scientific experiments, conclusions are drawn from specific outcomes. After many repetitions and similar outcomes, the findings are generalized into statements that appear to be true. When general conclusions are drawn from specific observations, we are using a type of reasoning called **inductive reasoning.** The next several examples illustrate this type of reasoning.

Example 1 Using Inductive Reasoning

Find the next number in the sequence 3, 7, 11, 15, ...

To discover a pattern, calculate the difference between each pair of successive numbers.

$$7 - 3 = 4$$
$$11 - 7 = 4$$
$$15 - 11 = 4$$

As shown, the difference is always 4. Each number is 4 greater than the previous one. Thus, the next number in the pattern is $15 + 4$, or 19.

Work Problem 1 at the Side.

1. Find the next number in the sequence 2, 8, 14, 20, ... Describe the pattern.

Example 2 Using Inductive Reasoning

Find the next number in this sequence.

7, 11, 8, 12, 9, 13, . . .

The pattern in this example can be determined as follows.

$$7 + 4 = 11$$
$$11 - \mathbf{3} = 8$$
$$8 + 4 = 12$$
$$12 - \mathbf{3} = 9$$
$$9 + 4 = 13$$

To get the second number, we add 4 to the first number. To get the third number, we subtract 3 from the second number. To obtain subsequent numbers, we continue the pattern. The next number is $13 - \mathbf{3}$, or 10.

Work Problem 2 at the Side.

2. Find the next number in the sequence 6, 11, 7, 12, 8, 13, ... Describe the pattern.

Answers

1. 26; add 6 each time.
2. 9; add 5, subtract 4.

3 Find the next number in the sequence 2, 6, 18, 54, ... Describe the pattern.

Example 3 Using Inductive Reasoning

Find the next number in the sequence 1, 2, 4, 8, 16, ...

Each number after the first is obtained by multiplying the previous number by 2. So the next number would be $16 \cdot \mathbf{2} = 32$.

Work Problem 3 at the Side.

4 Find the next shape in this sequence.

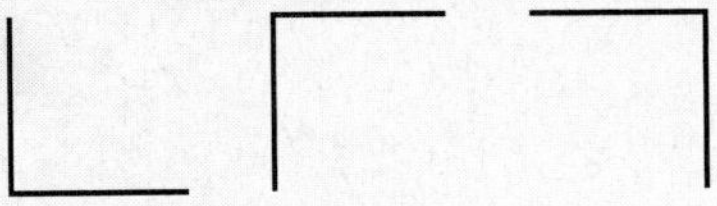

Example 4 Using Inductive Reasoning

(a) Find the next geometric shape in this sequence.

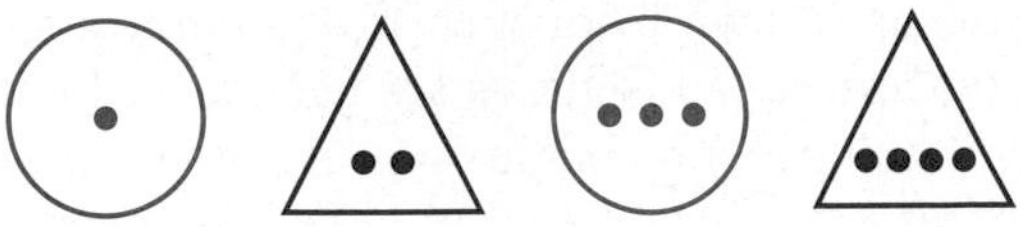

The figures alternate between a circle and a triangle. Also, the number of dots increases by 1 in each subsequent figure. Thus, the next figure should be a circle with five dots inside it.

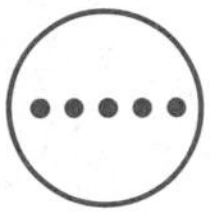

(b) Find the next geometric shape in this sequence.

The first two shapes consist of vertical lines with horizontal lines at the bottom extending first left and then right. The third shape is a vertical line with a horizontal line at the top extending to the left. Therefore, the next shape should be a vertical line with a horizontal line at the top extending to the right.

Work Problem 4 at the Side.

2 **Use deductive reasoning to analyze arguments.** In the previous discussion, specific cases were used to find patterns and predict the next event. There is another type of reasoning called **deductive reasoning,** which moves from general cases to specific conclusions.

Example 5 Using Deductive Reasoning

Does the conclusion follow from the premises in this argument?

All Buicks are automobiles.
All automobiles have horns.
$\therefore$ All Buicks have horns.

In this example, the first two statements are called *premises* and the third statement (below the line) is called a *conclusion.* The symbol $\therefore$ is a mathematical symbol meaning "**therefore**." The entire set of statements is called an *argument.*

Continued on Next Page

ANSWERS
3. 162; multiply by 3.
4.

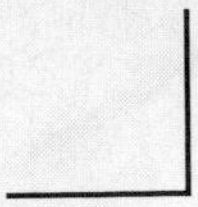

The focus of deductive reasoning is to determine whether the conclusion follows (is valid) from the premises. A set of circles called **Euler circles** is used to analyze the argument.

In Example 5, the statement "All Buicks are automobiles" can be represented by two circles, one for Buicks and one for automobiles. Note that the circle representing Buicks is totally inside the circle representing automobiles.

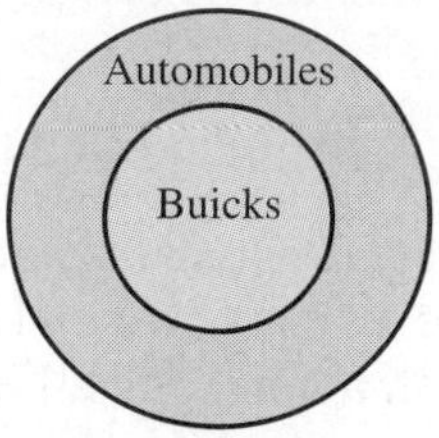

Now, a circle is added to represent the second statement, vehicles with horns. This circle must completely surround the circle representing automobiles.

To analyze the conclusion, notice that the circle representing Buicks is completely inside the circle representing vehicles with horns. It must follow that all Buicks have horns. ***The conclusion is valid.***

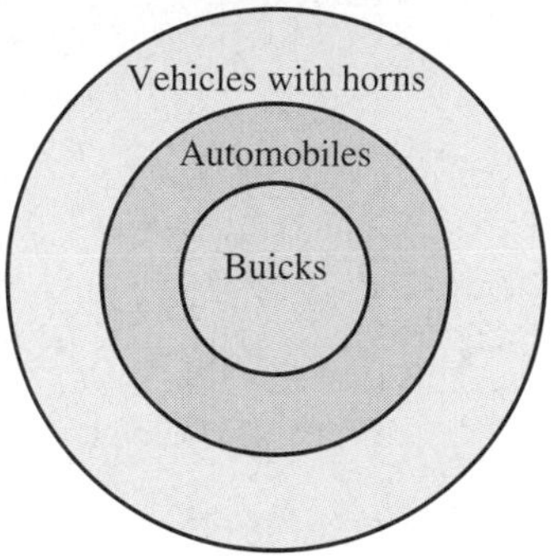

Work Problem ❺ at the Side.

Example 6 **Using Deductive Reasoning**

Does the conclusion follow from the premises in this argument?

All tables are round.
All glasses are round.
∴ All glasses are tables.

Using Euler circles, a circle representing tables is drawn inside a circle representing round objects.

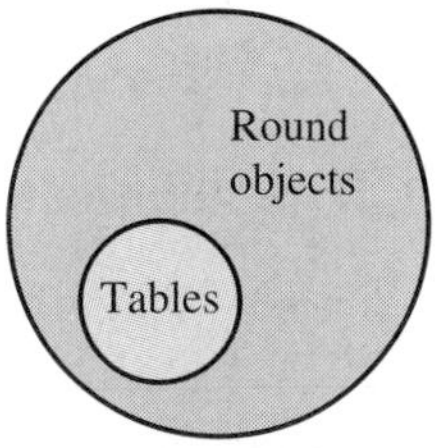

The second statement requires that a circle representing glasses must now be drawn inside the circle representing round objects, but not necessarily inside the circle representing tables. The conclusion does *not* follow from the premises. This means that ***the conclusion is invalid or untrue.***

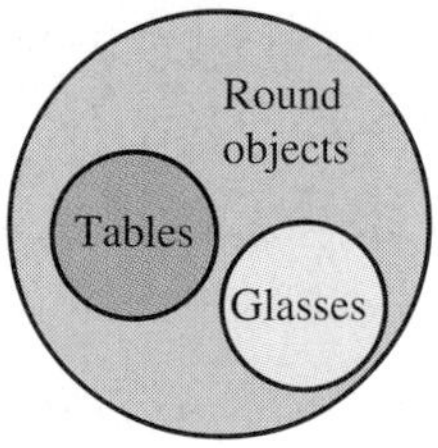

Work Problem ❻ at the Side.

❺ Does the conclusion follow from the premises in the following argument?

All cars have four wheels.
All Fords are cars.
∴ All Fords have four wheels.

❻ Does each conclusion follow from the premises?

(a) All animals are wild.
All cats are animals.
∴ All cats are wild.

(b) All students use math.
All adults use math.
∴ All adults are students.

ANSWERS
5. The conclusion follows from the premises.
6. (a) The conclusion follows from the premises.
(b) The conclusion does *not* follow from the premises.

7 In a college class of 100 students, 35 take both math and history, 50 take history, and 40 take math. How many take neither math nor history?

8 A Chevy, BMW, Cadillac, and Ford are parked side by side. The known facts are:

(a) The Ford is on the right end.

(b) The BMW is next to the Cadillac.

(c) The Chevy is between the Ford and the Cadillac.

Which car is parked on the left end?

3 **Use deductive reasoning to solve problems.** Another type of deductive reasoning problem occurs when a set of facts is given in a problem and a conclusion must be drawn by using these facts.

Example 7 Using Deductive Reasoning

There were 25 students enrolled in a ceramics class. During the class, 10 of the students made a bowl and 8 students made a birdbath. Three students made both a bowl and a birdbath. How many students did not make either a bowl or a birdbath?

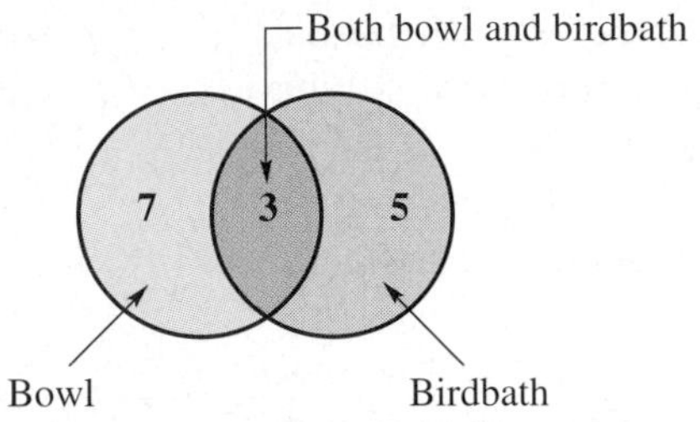

This type of problem is best solved by organizing the data using a drawing called a **Venn diagram.** Two overlapping circles are drawn, with each circle representing one item made by students.

In the region where the circles overlap, write the number of students who made *both* items, namely, 3. In the remaining portion of the birdbath circle, write the number 5, which when added to 3 will give the total number of students who made a birdbath, namely, 8. In a similar manner, write 7 in the remaining portion of the bowl circle, since $7 + 3 = 10$, the total number of students who made a bowl. The total of all three numbers written in the circles is 15. Since there are 25 students in the class, this means $25 - 15$ or 10 students did not make either a birdbath or a bowl.

Work Problem 7 at the Side.

Example 8 Using Deductive Reasoning

Four cars in a race finish first, second, third, and fourth. The following facts are known.

(a) Car A beat Car C.

(b) Car D finished between Cars C and B.

(c) Car C beat Car B.

In which order did the cars finish?

To solve this type of problem, it is helpful to use a line diagram.

1. *Write A before C,* because Car A beat Car C (fact a).

A C

2. *Write B after C,* because Car C beat Car B (fact c).

A C B

3. *Write D between C and B,* because Car D finished between Cars C and B (fact b).

The correct order of finish is shown below.

A C D B

Work Problem 8 at the Side.

Answers
7. 45 students
8. BMW

Appendix B Exercises

FOR EXTRA HELP: Student's Solutions Manual | MyMathLab.com | InterAct Math Tutorial Software | AW Math Tutor Center | www.mathxl.com

Find the next number in each sequence. Describe the pattern in each sequence. See Examples 1–3.

1. 2, 9, 16, 23, 30, . . .

2. 5, 8, 11, 14, 17, . . .

3. 0, 10, 8, 18, 16, . . .

4. 3, 9, 7, 13, 11, . . .

5. 1, 2, 4, 8, . . .

6. 1, 4, 16, 64, . . .

7. 1, 3, 9, 27, 81, . . .

8. 3, 6, 12, 24, 48, . . .

9. 1, 4, 9, 16, 25, . . .

10. 6, 7, 9, 12, 16, . . .

Find the next shape in each sequence. See Example 4.

11.

12.
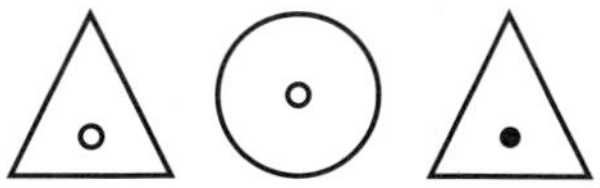

13.

14.
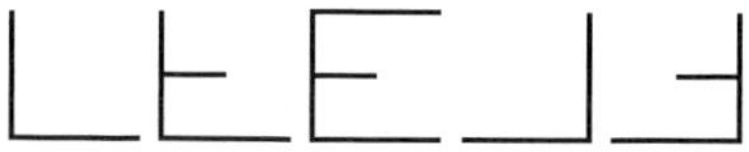

In each argument, state whether or not the conclusion follows from the premises. See Examples 5 and 6.

15. All animals are wild.
All lions are animals.
∴ All lions are wild.

16. All students are hard workers.
All business majors are students.
∴ All business majors are hard workers.

17. All teachers are serious.
All mathematicians are serious.
∴ All mathematicians are teachers.

18. All boys ride bikes.
All Americans ride bikes.
∴ All Americans are boys.

Solve each application problem. See Examples 7 and 8.

19. In a given 30-day period, a man watched television 20 days and his wife watched television 25 days. If they watched television together 18 days, how many days did neither watch television?

20. In a class of 40 students, 21 students take both calculus and physics. If 30 students take calculus and 25 students take physics, how many do not take either calculus or physics?

21. Tom, Dick, Mary, and Joan all work for the same company. One is a secretary, one is a computer operator, one is a receptionist, and one is a mail clerk.

(a) Tom and Joan eat dinner with the computer operator.

(b) Dick and Mary carpool with the secretary.

(c) Mary works on the same floor as the computer operator and the mail clerk.

Who is the computer operator?

22. Four cars—a Ford, a Buick, a Mercedes, and an Audi—are parked in a garage in four spaces.

(a) The Ford is in the last space.

(b) The Buick and Mercedes are next to each other.

(c) The Audi is next to the Ford but not next to the Buick.

Which car is in the first space?

Answers to Selected Exercises

In this section we provide the answers that we think most students will obtain when they work the exercises using the methods explained in the text. If your answer does not look exactly like the one given here, it is not necessarily wrong. In many cases there are equivalent forms of the answer that are correct. For example, if the answer section shows $\frac{3}{4}$ and your answer is 0.75, you have obtained the right answer but written it in a different (yet equivalent) form. Unless the directions specify otherwise, 0.75 is just as valid an answer as $\frac{3}{4}$.

In general, if your answer does not agree with the one given in the text, see whether it can be transformed into the other form. If it can, then it is the correct answer. If you still have doubts, talk with your instructor.

Diagnostic Pretest

(page xxv)

1. 89,023,507 **2.** 4331 **3.** 697 **4.** 89,000 **5.** $29\frac{3}{8}$
6. $2^3 \cdot 7^2$ **7.** $11\frac{1}{4}$ cups **8.** *Estimate:* $6 \cdot 2 = 12$; *Exact:* $12\frac{7}{32}$
9. 120 **10.** $\frac{19}{30}$ **11.** *Estimate:* $8 + 13 = 21$; *Exact:* $21\frac{1}{18}$
12. $\frac{35}{48}$ **13.** \$1.39 **14.** 6.556 (rounded) **15.** 24.25
16. \$8.43 (rounded) **17.** $\frac{5}{24}$ **18.** false **19.** $26\frac{2}{3}$ **20.** \$340
21. 58.2% **22.** 875% **23.** \$135.68 **24.** 7%

Chapter 1

Section 1.1 (page 7)

1. 2; 7 **3.** 1; 0 **5.** 8; 2 **7.** 2; 768; 543 **9.** 60; 0; 502; 109 **11.** Evidence suggests that this is true. It is common to count using fingers. **13.** twenty-three thousand, one hundred fifteen **15.** three hundred forty-six thousand, nine **17.** twenty-five million, seven hundred fifty-six thousand, six hundred sixty-five **19.** 32,526 **21.** 10,000,223 **23.** \$198 **25.** 2,000,000,000 **27.** 532,000 **29.** 800,000,621,020,215 **31.** public transportation; six million, sixty-nine thousand, five hundred eighty-nine **33.** seven million, eight hundred ninety-four thousand, nine hundred eleven

Section 1.2 (page 15)

1. 79 **3.** 89 **5.** 889 **7.** 889 **9.** 7785 **11.** 1578 **13.** 7676 **15.** 78,446 **17.** 8928 **19.** 59,224 **21.** 114 **23.** 155 **25.** 121 **27.** 145 **29.** 102 **31.** 1651 **33.** 1154 **35.** 413 **37.** 1771 **39.** 1410 **41.** 9253 **43.** 11,624 **45.** 17,611 **47.** 15,954 **49.** 10,648 **51.** 15,594 **53.** 11,557 **55.** 12,078 **57.** 4250 **59.** 12,268 **61.** correct **63.** incorrect; should be 769 **65.** correct **67.** incorrect; should be 11,577 **69.** correct **71.** Changing the order in which numbers are added does not change the sum. You can add from bottom to top when checking addition. **73.** 33 mi **75.** 38 mi **77.** \$89 **79.** 699 people **81.** \$16,342 **83.** 550 ft **85.** 72 ft **87.** 9421 **88.** 1249 **89.** 77,762 **90.** 22,267 **91.** 9,994,433 **92.** 3,334,499 **93.** Write the largest digits on the left, using the smaller digits as you move right. **94.** Write the smallest digits on the left, using the larger digits as you move right.

Section 1.3 (page 25)

1. 31 **3.** 33 **5.** 17 **7.** 213 **9.** 101 **11.** 6211 **13.** 3412 **15.** 2111 **17.** 13,160 **19.** 41,110 **21.** correct **23.** incorrect; should be 62 **25.** incorrect; should be 121 **27.** correct **29.** incorrect; should be 7222 **31.** 18 **33.** 45 **35.** 19 **37.** 281 **39.** 519 **41.** 9177 **43.** 7589 **45.** 8859 **47.** 3 **49.** 23 **51.** 19 **53.** 2833 **55.** 7775 **57.** 503 **59.** 156 **61.** 2184 **63.** 5687 **65.** 19,038 **67.** 31,556 **69.** 6584 **71.** correct **73.** correct **75.** correct **77.** correct **79.** Possible answers are 1. $3 + 2 = 5$ could be charged to $5 - 2 = 3$ or $5 - 3 = 2$ 2. $6 - 4 = 2$ could be changed to $2 + 4 = 6$ or $4 + 2 = 6$. **81.** 15 calories **83.** 367 ft **85.** 121 passengers **87.** 1329 students **89.** 9539 flags **91.** \$263 **93.** 758 people **95.** \$57,500 **97.** 48 deliveries **99.** 284 deliveries

Section 1.4 (page 35)

1. 16 **3.** 48 **5.** 0 **7.** 24 **9.** 30 **11.** 0 **13.** Factors may be multiplied in any order to get the same answer. They are the same; you may add or multiply numbers in any order. **15.** 150 **17.** 238 **19.** 3210 **21.** 1872 **23.** 8612 **25.** 10,084 **27.** 20,488 **29.** 258,447 **31.** 150 **33.** 480 **35.** 2220 **37.** 3600 **39.** 3750 **41.** 44,550 **43.** 270,000 **45.** 86,000,000 **47.** 48,500 **49.** 350,000 **51.** 1,940,000 **53.** 540 **55.** 2400 **57.** 3735 **59.** 2378 **61.** 6164 **63.** 15,792 **65.** 21,665 **67.** 15,730 **69.** 82,320 **71.** 183,996 **73.** 2,468,928 **75.** 66,005 **77.** 86,028 **79.** 19,422,180 **81.** 2,278,410 **83.** To multiply by 10, 100, or 1000, just add the number of 0s to the number you are multiplying and that's your answer. **85.** 18,000 cartons **87.** 216 plants **89.** 418 mi **91.** \$255 **93.** \$1560 **95.** \$112,888 **97.** 50,568 **99.** 38,250 trees **101.** 1058 calories **103.** \$4820 **104.** **(a)** 452 **(b)** 452 **105.** commutative **106.** **(a)** 281 **(b)** 281 **107.** associative **108.** **(a)** 15,840 **(b)** 15,840 **109.** commutative **110.** **(a)** 6552 **(b)** 6552 **111.** associative **112.** No. Some examples are 1. $7 - 5 = 2$, but $5 - 7$ does not equal 2 2. $12 - 6 = 6$, but $6 - 12$ does not equal 6 3. $(8 - 2) - 5 = 1$, but $8 - (2 - 5)$ does not equal 1. **113.** No. Some examples are 1. $10 \div 2 = 5$, but $2 \div 10$ does not equal 5 2. $(16 \div 8) \div 2 = 1$, but $16 \div (8 \div 2)$ does not equal 1.

Section 1.5 (page 49)

1. $3\overline{)15}$ with quotient 5; $\frac{15}{3} = 5$ **3.** $9\overline{)45}$ with quotient 5; $45 \div 9 = 5$ **5.** $16 \div 2 = 8$; $\frac{16}{2} = 8$
7. 1 **9.** 7 **11.** undefined **13.** 24 **15.** 0 **17.** undefined **19.** 15 **21.** 8 **23.** 15 **25.** 24 **27.** 304 **29.** 627 R1 **31.** 1522 R5 **33.** 309 **35.** 3005 **37.** 5006 **39.** 811 R1 **41.** 2589 R2 **43.** 7324 R2 **45.** 3157 R2 **47.** 5522 **49.** 12,458 R3

51. 10,253 R5 **53.** 18,377 R6 **55.** correct **57.** incorrect; should be 1908 R1 **59.** incorrect; should be 670 R2 **61.** incorrect; should be 3568 R1 **63.** correct **65.** correct **67.** incorrect; should be 9628 R3 **69.** correct **71.** Multiply the quotient by the divisor and add any remainder. The result should be the dividend. **73.** 233 place settings **75.** 11,200 people each day **77.** $18,200 **79.** 205 acres **81.** $225,000 **83.** $9135 **85.** √ √ √ √ **87.** √ X X X **89.** X X √ X **91.** X √ X X **93.** √ √ X X **95.** X X X X

Section 1.6 (page 59)

1. 22 **3.** 250 **5.** 120 **R**7 **7.** 1308 **R**9 **9.** 7134 **R**12 **11.** 900 **R**100 **13.** 108 **R**4 **15.** 183 **R**22 **17.** 2407 **R**1 **19.** 1146 **R**15 **21.** 3331 **R**82 **23.** 850 **25.** incorrect; should be 101 **R**14 **27.** incorrect; should be 658 **29.** incorrect; should be 62 **31.** When dividing by 10, 100, or 1000, drop the same number of 0s from the dividend to get the quotient. One example is 2500 ÷ 100 = 25. **33.** 18 hr **35.** 56 floor clocks **37.** $108 **39.** 1680 circuit boards **41.** $39 per week **43.** $0 **44.** 0 **45.** undefined **46.** impossible; if you have 6 cookies, it is not possible to divide them among 0 people. **47. (a)** 14 **(b)** 17 **(c)** 38 **48.** Yes. some examples are 18 • 1 = 18; 26 • 1 = 26; 43 • 1 = 43. **49. (a)** 3200 **(b)** 320 **(c)** 32 **50.** Drop the same number of 0s that appear in the divisor. The result is the quotient. With the divisor 10, drop one 0; with 100, drop two 0s; with 1000, drop three 0s.

Section 1.7 (page 69)

1. 410 **3.** 980 **5.** 6800 **7.** 86,800 **9.** 34,500 **11.** 6000 **13.** 16,000 **15.** 78,000 **17.** 8000 **19.** 10,000 **21.** 600,000 **23.** 9,000,000 **25.** 2370; 2400; 2000 **27.** 3370; 3400; 3000 **29.** 5050; 5000; 5000 **31.** 4240; 4200; 4000 **33.** 19,540; 19,500; 20,000 **35.** 26,290; 26,300; 26,000 **37.** 64,500; 64,500; 65,000 **39.** 1. Locate the place to be rounded and underline it. 2. Look only at the next digit to the right. If this digit is 5 or more, increase the underlined digit by 1. 3. Change all digits to the right of the underlined place to 0s. **41.** 90 30 80 90 290; 290 **43.** 90 30 60; 52 **45.** 70 30 2100; 2278 **47.** 900 700 400 800 2800; 2828 **49.** 800 400 400; 387 **51.** 400 400 160,000; 160,448 **53.** 8000 60 700 4000 12,760; 12,605 **55.** 800 400 400; 357 **57.** 900 30 27,000; 27,231 **59.** Perhaps the best explanation is that 3492 is closer to 3500 than 3400, but 3492 is closer to 3000 than to 4000. **61.** 80 million people; 280 million people **63.** 50 yr; 80 yr **65.** 3,025,940,000 pesos; 3,026,000,000 pesos; 3,000,000,000 pesos **67.** $8,490,487,600,000; $8,490,500,000,000; $8,490,000,000,000 **69.** 71,500 **70.** 72,499 **71.** 7500 **72.** 8499 **73.** $330; $500; $550; $730; $950; $1380; $1390; $2550 **74.** $300; $500; $600; $700; $1000; $1000; $1000; $3000 **75. (a)** When using front end rounding, all digits are 0 except the first digit. These numbers are easier to work with when estimating answers. **(b)** When using front end rounding to estimate an answer, the estimated answer can vary greatly from the exact answer.

Section 1.8 (page 75)

1. 2; 4; 16 **3.** 2; 5; 25 **5.** 2; 12; 144 **7.** 2; 15; 225 **9.** 3 **11.** 8 **13.** 10 **15.** 12 **17.** 64; 64 **19.** 400; 400 **21.** 1225; 1225 **23.** 1600; 1600 **25.** 2916; 2916 **27.** A perfect square is the square of a whole number. The number 25 is the square of 5 because 5 • 5 = 25. The number 50 is not a perfect square. There is no whole number that can be squared to get 50. **29.** 18 **31.** 25 **33.** 5 **35.** 20 **37.** 45 **39.** 63 **41.** 118 **43.** 18 **45.** 40 **47.** 102 **49.** 10 **51.** 63 **53.** 33 **55.** 70 **57.** 7 **59.** 17 **61.** 55 **63.** 108 **65.** 28 **67.** 9 **69.** 21 **71.** 16 **73.** 9 **75.** 3 **77.** 7 **79.** 20 **81.** 14 **83.** 25 **85.** 16 **87.** 23 **89.** 233

Section 1.9 (page 81)

1. 15 thousand **3.** United States **5.** 3 thousand **7.** 9 people **9. (a)** Saw ad **(b)** 25 people **11.** 9 people **13.** 2002; 7000 homes sales **15.** 4500 home sales **17.** Possible answers are 1. shortage of homes for sale 2. lack of qualified buyers 3. poor economy 4. high interest rates on home loans. **19.** (5 + 1) • 8 − 2 **20.** (4 + 2) • (5 + 1) **21.** 36 ÷ (3 • 3) • 4 **22.** 48 ÷ (2 • 2 • 2) + 2 **23. (a)** (100 + 50 + 65 + (50 − 15) + (100 − 65) + 15) • 12 **(b)** 3600 ft

Section 1.10 (page 89)

1. *Estimate:* 80 + 80 + 100 + 40 + 50 = 350 mi; *Exact:* 382 mi **3.** *Estimate:* 200 − 70 = 130 more crimes; *Exact:* 200 − 70 = 130 more crimes **5.** *Estimate:* 200 × 20 = 4000 kits; *Exact:* 5664 kits **7.** *Estimate:* 3000 ÷ 700 ≈ 4 toys; *Exact:* 4 toys **9.** *Estimate:* 8000 − 4000 = 4000 people; *Exact:* 4174 people **11.** *Estimate:* $10 × 5 = $50; *Exact:* $70 **13.** *Estimate:* $40,000 − $30,000 = $10,000; *Exact:* $9700 **15. (a)** *Estimate:* $20,000 + $30,000 = $50,000 Garrett; *Exact:* $20,000 + $40,000 = $60,000 Harcos; *Estimate:* $51,500 Garrett; *Exact:* $55,700 Harcos; Mr. and Mrs. Harcos **(b)** *Estimate:* $60,000 − $50,000 = $10,000; *Exact:* $4200 **17.** *Estimate:* $2000 − $500 − $300 − $300 − $200 − $200 = $500; *Exact:* $193 **19.** *Estimate:* 40,000 × 100 = 4,000,000 ft^2; *Exact:* 6,011,280 ft^2 **21.** *Estimate:* $400 + $600 + $200 + $100 + $100 = $1400; *Exact:* $1375 **23.** *Estimate:* $400 + $600 + $300 + $900 = $2200; $2200 − $1600 = $600; *Exact:* $530 **25.** *Estimate:* ($1000 × 6) + ($900 × 20) = $24,000; *Exact:* $20,961 **27.** Possible answers are Addition: more; total; gain of Subtraction: less; loss of; decreased by Multiplication: twice, of; product Division: divided by; goes into; per Equals: is; are **29.** Estimating the answer can help you avoid careless mistakes like decimal or calculation errors. Examples of reasonable answers in daily life might be a $25 bag of groceries, $20 to fill the gas tank, or $45 for a phone bill. **31.** $165 **33.** 2477 lb **35.** $500 **37.** $375 **39.** 20 seats

Chapter 1 Review Exercises (page 97)

1. 3; 582 **2.** 64; 234 **3.** 105; 724 **4.** 1; 768; 710; 618 **5.** six hundred thirty-five **6.** fifteen thousand, three hundred ten **7.** three hundred nineteen thousand, two hundred fifteen **8.** sixty-two million, five hundred thousand, five **9.** 10,008 **10.** 200,000,455 **11.** 90 **12.** 137 **13.** 5464 **14.** 15,657 **15.** 10,986 **16.** 9845 **17.** 40,602 **18.** 49,855 **19.** 34 **20.** 23 **21.** 189 **22.** 184 **23.** 5755 **24.** 4327 **25.** 224 **26.** 25,866 **27.** 36 **28.** 0 **29.** 32 **30.** 64 **31.** 35 **32.** 56 **33.** 56 **34.** 81 **35.** 48 **36.** 45 **37.** 48 **38.** 8 **39.** 0 **40.** 42 **41.** 48 **42.** 0 **43.** 160 **44.** 368 **45.** 522 **46.** 98 **47.** 4123 **48.** 5467 **49.** 5396 **50.** 45,815 **51.** 16,728 **52.** 32,640 **53.** 465,525 **54.** 174,984 **55.** 675 **56.** 1764 **57.** 1176 **58.** 5100 **59.** 15,576 **60.** 30,184 **61.** 887,169 **62.** 500,856 **63.** $300 **64.** $1064 **65.** $20,352 **66.** $1512 **67.** 14,000 **68.** 23,800 **69.** 206,800 **70.** 318,500 **71.** 128,000,000 **72.** 90,300,000 **73.** 5 **74.** 5 **75.** 6 **76.** 2 **77.** 6 **78.** 4 **79.** 7 **80.** 0 **81.** undefined **82.** 0 **83.** 8 **84.** 9 **85.** 92 **86.** 35 **87.** 4422 **88.** 352 **89.** 150 R4 **90.** 124 R25 **91.** 480 **92.** 14,300 **93.** 21,000 **94.** 70,000 **95.** 3490; 3500; 3000 **96.** 20,070; 20,100; 20,000 **97.** 98,200; 98,200; 98,000 **98.** 352,120; 352,100; 352,000 **99.** 5 **100.** 8 **101.** 12 **102.** 14 **103.** 3; 7; 343 **104.** 6; 3; 729 **105.** 3; 5; 125 **106.** 5; 4; 1024 **107.** 54 **108.** 8 **109.** 9 **110.** 4 **111.** 9 **112.** 6 **113.** 8 parents

114. 5 parents **115.** Keeping bedroom clean **116.** Hanging up wet bath towels **117.** *Estimate:* 80 × \$10 = \$800; *Exact:* \$750 **118.** *Estimate:* 1000 × 60 = 60,000 revolutions; *Exact:* 84,000 revolutions **119.** *Estimate:* 100 × 4 = 400 cups; *Exact:* 380 cups **120.** *Estimate:* 6000 × 30 = 180,000 brackets; *Exact:* 180,000 brackets **121.** *Estimate:* 2000 × 10 = 20,000 hr; *Exact:* 24,000 hr **122.** *Estimate:* 80 × 5 = 400 mi; *Exact:* 400 mi
123. *Estimate:* (\$20 × 20) + (\$10 × 30) = \$700; *Exact:* \$582
124. *Estimate:* (60 × \$20) + (20 × \$7) = \$1340; *Exact:* \$1139
125. *Estimate:* \$400 − \$200 = \$200; *Exact:* \$180
126. *Estimate:* \$900 − \$400 − \$200 = \$300; *Exact:* \$252
127. *Estimate:* 9000 ÷ 200 = 45 lb; *Exact:* 50 lb
128. *Estimate:* 30,000 ÷ 1000 = 30 hr; *Exact:* 33 hr
129. *Estimate:* 30,000 ÷ 600 = 50 acres; *Exact:* 52 acres
130. *Estimate:* 6000 ÷ 200 = 30 homes; *Exact:* 32 homes
131. 280 **132.** 664 **133.** 139 **134.** 588 **135.** 1041
136. 1661 **137.** 32,062 **138.** 24,947 **139.** 3 **140.** 7
141. 93,635 **142.** 83,178 **143.** undefined **144.** 7
145. 6900 **146.** 2305 **147.** 1,079,040 **148.** 103,268
149. 108 **150.** 207 **151.** three hundred seventy-six thousand, eight hundred fifty-three **152.** four hundred eight thousand, six hundred ten **153.** 7500 **154.** 600,000 **155.** 7 **156.** 9
157. \$5940 **158.** \$31,080 **159.** \$2288 **160.** \$15,782
161. 468 cards **162.** 3600 textbooks **163.** \$280
164. \$114,635 **165.** \$1905 **166.** \$12,420

Chapter 1 Test (page 105)

1. six thousand, one hundred six **2.** eighty-five thousand, fifty-five **3.** 426,005 **4.** 8714 **5.** 112,630 **6.** 1053
7. 3084 **8.** 96 **9.** 171,000 **10.** 1710 **11.** 4,450,743
12. 7047 **13.** undefined **14.** 458 R5 **15.** 160
16. 5240 **17.** 68,000 **18.** 48 **19.** 28 **20.** *Estimate:*
\$500 + \$500 + \$500 + 400 − \$800 = \$1100; *Exact:* \$1140
21. *Estimate:* 70,000 ÷ 500 = 140 days; *Exact:* 125 days
22. *Estimate:* \$1000 − \$700 − \$200 − \$70 = \$30;
Exact: \$165 **23.** *Estimate:* (200 × 4) + (200 × 4) =
1600 cameras; *Exact:* 1784 cameras **24.** 1. Locate the place to which you are rounding and underline it. 2. Look only at the next digit to the right. If this digit is a 4 or less, do not change the underlined digit. If the digit is 5 or more, increase the underlined digit by 1. 3. Change all digits to the right of the underlined place to 0s. Each person's rounding example will vary. **25.** 1. Read the problem carefully. 2. Work out a plan. 3. Estimate a reasonable answer. 4. Solve the problem. 5. State the answer. 6. Check your work.

Chapter 2

Section 2.1 (page 111)

1. $\dfrac{3}{4}$; $\dfrac{1}{4}$ **3.** $\dfrac{1}{3}$; $\dfrac{2}{3}$ **5.** $\dfrac{7}{5}$; $\dfrac{3}{5}$ **7.** $\dfrac{2}{11}$ **9.** $\dfrac{8}{25}$ **11.** $\dfrac{49}{134}$ **13.** 3; 8

15. 11; 10 **17.** Proper $\dfrac{1}{3}, \dfrac{5}{8}, \dfrac{7}{16}$ Improper $\dfrac{8}{5}, \dfrac{6}{6}, \dfrac{12}{2}$ **19.** Proper

$\dfrac{3}{4}, \dfrac{9}{11}, \dfrac{7}{15}$ Improper $\dfrac{3}{2}, \dfrac{5}{5}, \dfrac{19}{18}$

21. One possibility is

$\dfrac{3}{4}$ ← Numerator / ← Denominator

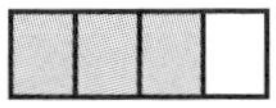

The denominator shows the number of equal parts in the whole and the numerator shows how many of the parts are being considered.
23. 7; 8 **25.** 4; 32

Section 2.2 (page 117)

1. $\dfrac{3}{2}$ **3.** $\dfrac{17}{5}$ **5.** $\dfrac{13}{2}$ **7.** $\dfrac{22}{3}$ **9.** $\dfrac{18}{11}$ **11.** $\dfrac{19}{3}$ **13.** $\dfrac{81}{8}$ **15.** $\dfrac{43}{4}$

17. $\dfrac{27}{8}$ **19.** $\dfrac{41}{6}$ **21.** $\dfrac{54}{11}$ **23.** $\dfrac{131}{4}$ **25.** $\dfrac{233}{13}$ **27.** $\dfrac{269}{15}$

29. $\dfrac{187}{24}$ **31.** $2\dfrac{1}{2}$ **33.** $2\dfrac{1}{4}$ **35.** 8 **37.** $7\dfrac{3}{5}$ **39.** $4\dfrac{7}{8}$ **41.** 9

43. $15\dfrac{3}{4}$ **45.** $5\dfrac{2}{9}$ **47.** $8\dfrac{1}{8}$ **49.** $16\dfrac{4}{5}$ **51.** $21\dfrac{3}{5}$ **53.** $26\dfrac{1}{7}$

55. Multiply the denominator by the whole number and add the numerator. The result becomes the new numerator, which is placed over the original denominator.

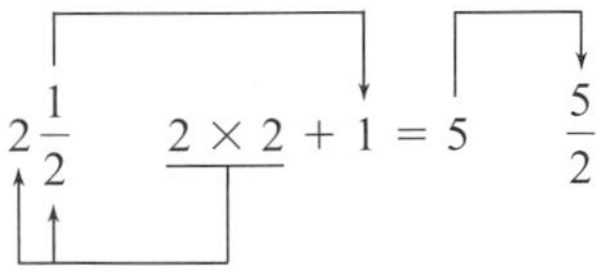

$$2\frac{1}{2} \qquad 2 \times 2 + 1 = 5 \qquad \frac{5}{2}$$

57. $\dfrac{443}{4}$ **59.** $\dfrac{1000}{3}$ **61.** $\dfrac{4179}{8}$ **63.** $104\dfrac{5}{8}$ **65.** 171

67. $122\dfrac{13}{32}$ **69.** $\dfrac{2}{3}, \dfrac{4}{5}, \dfrac{3}{4}, \dfrac{7}{10}$ **70. (a)** numerator; denominator

(b)

(c) less **71.** $\dfrac{5}{5}, \dfrac{10}{3}, \dfrac{6}{5}$ **72. (a)** numerator; denominator

(b)

(c) greater **73.** $\dfrac{5}{3} = 1\dfrac{2}{3}; \dfrac{7}{7} = 1; \dfrac{11}{6} = 1\dfrac{5}{6}$ **74. (a)** improper; greater than or equal to

(b)

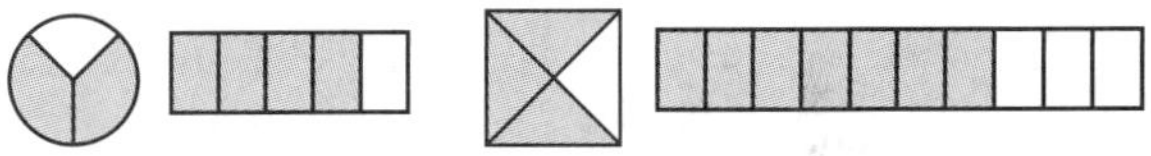

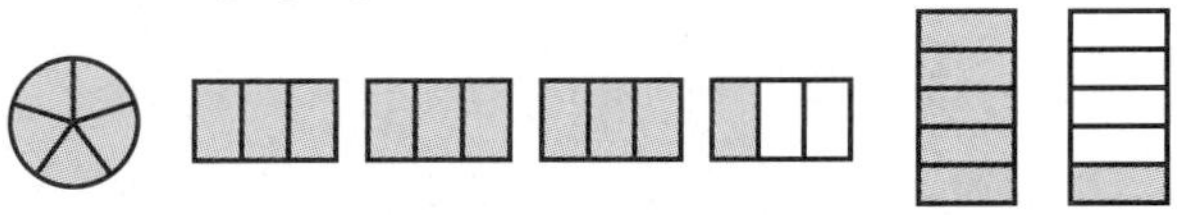

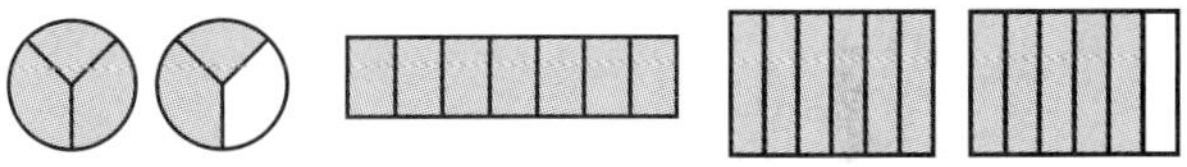

(c) Divide the numerator by the denominator and place the remainder over the denominator.

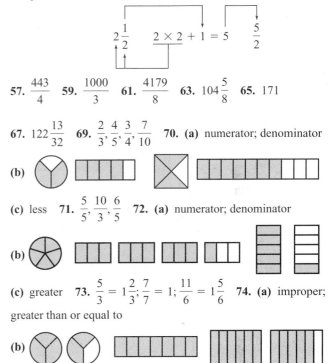

Section 2.3 (page 127)

1. 1, 2, 3, 6 **3.** 1, 2, 4, 8 **5.** 1, 2, 3, 4, 6, 8, 12, 16, 24, 48
7. 1, 2, 3, 4, 6, 9, 12, 18, 36 **9.** 1, 2, 4, 5, 8, 10, 20, 40
11. 1, 2, 4, 8, 16, 32, 64 **13.** composite **15.** prime
17. composite **19.** prime **21.** prime **23.** composite
25. composite **27.** composite **29.** 2 • 5 **31.** 2^2 • 5 **33.** 5^2
35. 2^2 • 3^2 **37.** 2^2 • 17 **39.** 2^3 • 3^2 **41.** 2^2 • 11 **43.** 2^2 • 5^2
45. 5^3 **47.** 2^2 • 3^2 • 5 **49.** 2^6 • 5 **51.** 2^3 • 3^2 • 5
53. A composite number has a factor(s) other than itself or 1. Examples include 4, 6, 8, 9, 10. A prime number is a whole number that has exactly two *different* factors, itself and 1. Examples include 2, 3, 5, 7, 11. The numbers 0 and 1 are neither prime nor composite.
55. All the possible factors of 24 are 1, 2, 3, 4, 6, 8, 12, and 24. This list includes both prime numbers and composite numbers. The prime factors of 24 include only prime numbers. The prime factorization of 24 is 2 • 2 • 2 • 3 = 2^3 • 3 **57.** 2^6 • 5
59. 2^6 • 3 • 5 **61.** 2^3 • 3 • 5 • 13 **63.** 2^2 • 3^2 • 5 • 7
65. 2, 3, 5, 7, 11, 13, 17, 19, 23, 29, 31, 37, 41, 43, 47
66. A prime number is a whole number that is evenly divisible by itself and 1 only. **67.** It is true because any even number is divisible by the number 2 in addition to itself and 1.

68. No. A multiple of a prime number can never be prime because it will always be divisible evenly by the prime number.
69. $2 \cdot 2 \cdot 3 \cdot 5 \cdot 5 \cdot 7$ **70.** $2^2 \cdot 3 \cdot 5^2 \cdot 7$

Section 2.4 (page 133)

1. ✓ ✓ ✓ ✓ **3.** ✓ ✓ X X **5.** ✓ X ✓ ✓ **7.** ✓ ✓ X X
9. $\frac{3}{4}$ **11.** $\frac{2}{3}$ **13.** $\frac{5}{8}$ **15.** $\frac{6}{7}$ **17.** $\frac{9}{10}$ **19.** $\frac{6}{7}$ **21.** $\frac{4}{7}$ **23.** $\frac{1}{50}$
25. $\frac{8}{11}$ **27.** $\frac{5}{9}$

29. $\frac{\overset{1}{\cancel{2}} \cdot \overset{1}{\cancel{3}} \cdot 3}{\underset{1}{\cancel{2}} \cdot 2 \cdot 2 \cdot \underset{1}{\cancel{3}}} = \frac{3}{4}$ **31.** $\frac{\overset{1}{\cancel{3}} \cdot 7}{2 \cdot 2 \cdot 2 \cdot \underset{1}{\cancel{3}}} = \frac{7}{8}$

33. $\frac{\overset{1}{\cancel{2}} \cdot \overset{1}{\cancel{3}} \cdot \overset{1}{\cancel{3}} \cdot \overset{1}{\cancel{3}}}{\underset{1}{\cancel{2}} \cdot 2 \cdot \underset{1}{\cancel{3}} \cdot \underset{1}{\cancel{3}} \cdot \underset{1}{\cancel{3}}} = \frac{1}{2}$ **35.** $\frac{\overset{1}{\cancel{2}} \cdot \overset{1}{\cancel{2}} \cdot \overset{1}{\cancel{3}} \cdot 3}{\underset{1}{\cancel{2}} \cdot \underset{1}{\cancel{2}} \cdot \underset{1}{\cancel{3}}} = 3$

37. $\frac{2 \cdot 2 \cdot 2 \cdot \overset{1}{\cancel{3}} \cdot \overset{1}{\cancel{3}}}{\underset{1}{\cancel{3}} \cdot \underset{1}{\cancel{3}} \cdot 5 \cdot 5} = \frac{8}{25}$ **39.** equivalent **41.** not equivalent
43. not equivalent **45.** equivalent **47.** not equivalent
49. equivalent **51.** A fraction is in lowest terms when the numerator and the denominator have no common factors other than 1. Some examples are $\frac{1}{2}, \frac{3}{8}$, and $\frac{2}{3}$. **53.** $\frac{7}{8}$ **55.** $\frac{2}{1} = 2$

Section 2.5 (page 141)

1. $\frac{3}{8}$ **3.** $\frac{5}{12}$ **5.** $\frac{3}{4}$ **7.** $\frac{1}{4}$ **9.** $\frac{5}{12}$ **11.** $\frac{9}{32}$ **13.** $\frac{2}{5}$ **15.** $\frac{13}{32}$
17. $\frac{21}{128}$ **19.** 5 **21.** 40 **23.** 21 **25.** $13\frac{1}{2}$ **27.** $31\frac{1}{2}$
29. 240 **31.** $189\frac{1}{3}$ **33.** 400 **35.** 810 **37.** $\frac{1}{4}$ mi^2
39. 9 m^2 **41.** $\frac{3}{10}$ in.^2 **43.** Multiply the numerators and multiply the denominators. An example is $\frac{3}{4} \cdot \frac{1}{2} = \frac{3 \cdot 1}{4 \cdot 2} = \frac{3}{8}$.
45. $1\frac{1}{2}$ yd^2 **47.** 1 mi^2 **49.** They are both the same size: $\frac{3}{64}$ mi^2
51. 470,000 vehicles **52.** 457,895 vehicles **53.** 37,500; 35,013 Ford Explorers **54.** 17,143; 15,190 Dodge Rams
55. $\frac{3}{4} \cdot 48{,}000$ (multiple of 4) $= 36{,}000$ Ford Explorers
56. $\frac{3}{7} \cdot 35{,}000$ (multiple of 7) $= 15{,}000$ Dodge Rams

Section 2.6 (page 149)

1. $\frac{1}{2}$ yd^2 **3.** $\frac{8}{9}$ ft^2 **5.** 61 players **7.** \$2568 **9.** 375 students
11. 325 women **13.** 0–24 years; 40 million or 40,000,000 books
15. 250 million or 250,000,000 books **17.** Because everyone is included and fractions are given for *all* age groups, the sum of the fractions must be *1* or *all* of the people. **19.** \$38,000
21. \$7600 **23.** \$2375
25. The correct solution is

$$\frac{9}{10} \times \frac{20}{21} = \frac{\overset{3}{\cancel{9}}}{\underset{1}{\cancel{10}}} \times \frac{\overset{2}{\cancel{20}}}{\underset{7}{\cancel{21}}} = \frac{6}{7}$$

27. \$42 **29.** 9000 votes **31.** $\frac{1}{32}$ of the estate

Section 2.7 (page 159)

1. $\frac{3}{2}$ **3.** $\frac{5}{8}$ **5.** $\frac{6}{5}$ **7.** $\frac{1}{4}$ **9.** $\frac{1}{3}$ **11.** $2\frac{5}{8}$ **13.** $\frac{9}{20}$ **15.** 4
17. 6 **19.** $\frac{13}{16}$ **21.** $\frac{4}{5}$ **23.** 18 **25.** $22\frac{1}{2}$ **27.** $\frac{1}{14}$ **29.** $\frac{2}{9}$ acre
31. 15 times **33.** 88 dispensers **35.** 60 trips **37.** 12 batches
39. You can divide two fractions by using the reciprocal of the second fraction (divisor) and then multiplying. **41.** 76 mi
43. \$120,000 **45.** double, twice, times, product, twice as much
46. goes into, divide, per, quotient, divided by **47.** reciprocal
48. $\frac{4}{3}; \frac{8}{7}; \frac{1}{5}; \frac{19}{12}$ **49.** $3\frac{3}{4}$ in; Add the lengths of all the sides.
50. $\frac{225}{256}$ in.^2; Multiply the length by the width.

Section 2.8 (page 169)

1. *Exact:* $8\frac{1}{8}$; *Estimate:* $3 \cdot 3 = 9$ **3.** *Exact:* $4\frac{1}{2}$; *Estimate:* $2 \cdot 3 = 6$ **5.** *Exact:* 4; *Estimate:* $3 \cdot 1 = 3$ **7.** *Exact:* 50; *Estimate:* $8 \cdot 6 = 48$ **9.** *Exact:* $49\frac{1}{2}$; *Estimate:* $5 \cdot 2 \cdot 5 = 50$
11. *Exact:* 12; *Estimate:* $3 \cdot 2 \cdot 3 = 18$ **13.** *Exact:* $\frac{1}{3}$; *Estimate:* $3 \div 8 = \frac{3}{8}$ **15.** *Exact:* $\frac{5}{6}$; *Estimate:* $3 \div 3 = 1$ **17.** *Exact:* $3\frac{3}{5}$; *Estimate:* $9 \div 3 = 3$ **19.** *Exact:* $\frac{3}{7}$; *Estimate:* $1 \div 2 = \frac{1}{2}$
21. *Exact:* $\frac{3}{10}$; *Estimate:* $2 \div 6 = \frac{1}{3}$ **23.** *Exact:* $\frac{17}{18}$; *Estimate:* $6 \div 6 = 1$ **25. (a)** *Estimate:* $3 \cdot 2 = 6$ cups; *Exact:* 5 cups of Quaker Oats **(b)** *Estimate:* $1 \cdot 2 = 2$ cups; *Exact:* $2\frac{1}{2}$ cups of brown sugar **(c)** *Estimate:* $2 \cdot 2 = 4$ cups; *Exact:* $3\frac{1}{2}$ cups of flour **27. (a)** *Estimate:* $1 \div 2 = \frac{1}{2}$ cup; *Exact:* $\frac{5}{8}$ cup of brown sugar **(b)** *Estimate:* $1 \div 2 = \frac{1}{2}$ cup; *Exact:* $\frac{1}{4}$ cup granulated sugar **(c)** *Estimate:* $2 \div 2 = 1$ cup; *Exact:* $\frac{7}{8}$ cup of flour **29.** *Estimate:* $1314 \div 110 \approx 12$ homes; *Exact:* 12 homes **31.** *Estimate:* $2 \cdot 13 = 26$ oz; *Exact:* $21\frac{7}{8}$ oz
33. The answer should include *Step 1* Change mixed numbers to improper fractions. *Step 2* Multiply the fractions. *Step 3* Write the answer in lowest terms, changing to mixed or whole numbers where possible. **35.** *Estimate:* $25{,}730 \div 10 = 2573$ anchors; *Exact:* 2480 anchors **37.** *Estimate:* $10 \div 1 = 10$ spacers; *Exact:* 13 spacers **39.** *Estimate:* $13 \cdot 28 = 364$; *Exact:* 471 gal
$7 \cdot 16 = +\underline{112}$
476 gal

Chapter 2 Review Exercises (page 177)

1. $\frac{1}{3}$ **2.** $\frac{5}{8}$ **3.** $\frac{2}{4}$ **4.** Proper $\frac{1}{8}, \frac{3}{4}, \frac{2}{3}$; Improper $\frac{4}{3}, \frac{5}{5}$
5. Proper $\frac{15}{16}, \frac{1}{8}$; Improper $\frac{6}{5}, \frac{16}{13}, \frac{5}{3}$ **6.** $\frac{17}{3}$ **7.** $\frac{54}{5}$ **8.** $2\frac{1}{8}$
9. $12\frac{3}{5}$ **10.** 1, 2, 3, 6 **11.** 1, 2, 3, 4, 6, 8, 12, 24
12. 1, 5, 11, 55 **13.** 1, 2, 3, 5, 6, 9, 10, 15, 18, 30, 45, 90 **14.** 3^3
15. $2 \cdot 3 \cdot 5^2$ **16.** $2^3 \cdot 3 \cdot 7$ **17.** 36 **18.** 200 **19.** 1728

20. 2048 **21.** $\frac{3}{4}$ **22.** $\frac{5}{6}$ **23.** $\frac{15}{16}$

24. $\frac{\overset{1}{\cancel{3}}\cdot 5}{2\cdot 2\cdot 3\cdot \underset{1}{\cancel{3}}}$; $\frac{5}{12}$ **25.** $\frac{\overset{1}{\cancel{2}}\cdot\overset{1}{\cancel{2}}\cdot\overset{1}{\cancel{2}}\cdot\overset{1}{\cancel{2}}\cdot\overset{1}{\cancel{2}}\cdot 2\cdot 2\cdot\overset{1}{\cancel{3}}}{\underset{1}{\cancel{2}}\cdot\underset{1}{\cancel{2}}\cdot\underset{1}{\cancel{2}}\cdot\underset{1}{\cancel{2}}\cdot\underset{1}{\cancel{2}}\cdot\underset{1}{\cancel{3}}}$; 4

26. equivalent **27.** not equivalent **28.** $\frac{1}{2}$ **29.** $\frac{3}{16}$ **30.** $\frac{1}{7}$

31. $\frac{4}{21}$ **32.** 15 **33.** 625 **34.** $\frac{2}{3}$ **35.** $\frac{5}{3} = 1\frac{2}{3}$ **36.** $\frac{5}{2} = 2\frac{1}{2}$

37. 2 **38.** 8 **39.** 24 **40.** $\frac{2}{15}$ **41.** $\frac{2}{15}$ **42.** $\frac{4}{13}$ **43.** $\frac{9}{40}$ ft^2

44. $\frac{7}{12}$ in.^{2} **45.** $7\frac{1}{2}$ ft^2 **46.** 36 yd^2 **47.** *Exact:* $6\frac{7}{8}$; *Estimate:* $6 \cdot 1 = 6$ **48.** *Exact:* $21\frac{3}{8}$; *Estimate:* $2 \cdot 7 \cdot 1 = 14$

49. *Exact:* $5\frac{1}{6}$; *Estimate:* $16 \div 3 = 5\frac{1}{3}$ **50.** *Exact:* $\frac{3}{4}$; *Estimate:* $5 \div 6 = \frac{5}{6}$ **51.** 336 boxes **52.** $\frac{2}{15}$ of the estate

53. *Estimate:* $158 \div 4 \approx 40$ pull cords; *Exact:* 36 pull cords
54. *Estimate:* $9 \cdot 38 = \$342$; *Exact:* \$323 **55.** 25 lb **56.** \$510

57. $\frac{5}{32}$ of the budget **58.** $\frac{1}{12}$ ton **59.** $\frac{1}{4}$ **60.** $\frac{2}{5}$ **61.** $31\frac{1}{4}$

62. $28\frac{1}{8}$ **63.** $\frac{1}{10}$ **64.** $\frac{5}{32}$ **65.** 30 **66.** $2\frac{1}{6}$ **67.** $2\frac{1}{3}$

68. $28\frac{3}{5}$ **69.** $\frac{17}{3}$ **70.** $\frac{307}{8}$

71. $\frac{\overset{1}{\cancel{2}}\cdot\overset{1}{\cancel{2}}\cdot 2}{\underset{1}{\cancel{2}}\cdot\underset{1}{\cancel{2}}\cdot 3} = \frac{2}{3}$ **72.** $\frac{\overset{1}{\cancel{2}}\cdot 2\cdot\overset{1}{\cancel{3}}\cdot 3\cdot 3}{\underset{1}{\cancel{2}}\cdot\underset{1}{\cancel{3}}\cdot 5\cdot 7} = \frac{18}{35}$

73. $\frac{2}{3}$ **74.** $\frac{1}{3}$ **75.** $\frac{2}{5}$ **76.** $\frac{1}{3}$ **77.** *Estimate:* $4 \cdot 44 = 176$ oz; *Exact:* $152\frac{4}{9}$ oz **78.** *Estimate:* $7 \cdot 26 = 182$ qt; *Exact:* $184\frac{7}{8}$ qt

79. $\frac{49}{64}$ in.^{2} **80.** $\frac{5}{16}$ yd^2

Chapter 2 Test (page 181)

1. $\frac{5}{6}$ **2.** $\frac{3}{8}$ **3.** $\frac{2}{3}, \frac{6}{7}, \frac{1}{4}, \frac{5}{8}$ **4.** $\frac{20}{3}$ **5.** $23\frac{3}{5}$ **6.** 1, 2, 3, 6, 9, 18

7. $3^2 \cdot 7$ **8.** $2^5 \cdot 3$ **9.** $2^2 \cdot 5^3$ **10.** $\frac{3}{4}$ **11.** $\frac{5}{6}$

12. $\frac{56}{84} = \frac{\overset{1}{\cancel{2}}\cdot\overset{1}{\cancel{2}}\cdot 2\cdot\overset{1}{\cancel{7}}}{\underset{1}{\cancel{2}}\cdot\underset{1}{\cancel{2}}\cdot 3\cdot\underset{1}{\cancel{7}}} = \frac{2}{3}$

13. Multiply fractions by multiplying the numerators and multiplying the denominators. Divide two fractions by using the reciprocal of the second fraction (divisor) and multiplying.

14. $\frac{3}{20}$ **15.** 16 **16.** $\frac{3}{8}$ yd^2 **17.** 3696 drivers **18.** $\frac{9}{10}$

19. $15\frac{3}{4}$ **20.** 200 sports bottles **21.** *Estimate:* $4 \cdot 4 = 16$; *Exact:* $14\frac{7}{16}$ **22.** *Estimate:* $2 \cdot 4 = 8$; *Exact:* $7\frac{17}{18}$ **23.** *Estimate:* $10 \div 2 = 5$; *Exact:* $4\frac{4}{15}$ **24.** *Estimate:* $9 \div 2 = 4\frac{1}{2}$; *Exact:* $5\frac{1}{10}$

25. *Estimate:* $3 \cdot 12 = 36$; *Exact:* $30\frac{5}{8}$ g

Cumulative Review Exercises: Chapters 1–2 (page 183)

1. hundreds 5; tens 7 **2.** millions 8; ten thousands 2 **3.** 166
4. 149,199 **5.** 4452 **6.** 3,221,821 **7.** 476 **8.** 96
9. 2,168,232 **10.** 450,400 **11.** 9 **12.** 7581 **13.** 8471 R2
14. 22 R26 **15.** 5740; 5700; 6000 **16.** 76,270; 76,300; 76,000
17. 3 **18.** 5 **19.** \$992 **20.** \$130 **21.** 47,000 hairs

22. \$285 **23.** $\frac{11}{16}$ ft^2 **24.** 6¢ **25.** proper **26.** improper

27. proper **28.** $\frac{27}{8}$ **29.** $\frac{32}{5}$ **30.** 2 **31.** $12\frac{7}{8}$ **32.** $2^3 \cdot 3^2$

33. $2 \cdot 3^2 \cdot 7$ **34.** $2 \cdot 5^2 \cdot 7$ **35.** 64 **36.** 288 **37.** 640

38. $\frac{7}{8}$ **39.** $\frac{2}{3}$ **40.** $\frac{5}{9}$ **41.** $\frac{3}{8}$ **42.** 12 **43.** 25 **44.** $\frac{24}{25}$

45. $\frac{7}{12}$ **46.** $2\frac{2}{5}$

Chapter 3

Section 3.1 (page 189)

1. $\frac{4}{5}$ **3.** $\frac{5}{6}$ **5.** $\frac{1}{3}$ **7.** $1\frac{1}{5}$ **9.** $\frac{1}{3}$ **11.** $\frac{13}{20}$ **13.** $\frac{11}{15}$ **15.** $1\frac{1}{2}$

17. $\frac{11}{27}$ **19.** $\frac{3}{8}$ **21.** $\frac{6}{11}$ **23.** $\frac{3}{5}$ **25.** $1\frac{1}{7}$ **27.** $\frac{1}{5}$ **29.** $1\frac{1}{6}$

31. $1\frac{1}{10}$ **33.** Three steps to add like fractions are:
1. Add the numerators of the fractions to find the numerator of the sum (the answer). 2. Use the denominator of the fractions as the denominator of the sum. 3. Write the answer in lowest terms.

35. $\frac{3}{4}$ **37.** $\frac{1}{3}$ **39.** $\frac{1}{2}$ acre

Section 3.2 (page 197)

1. 4 **3.** 15 **5.** 36 **7.** 14 **9.** 30 **11.** 100 **13.** 20 **15.** 72

17. 36 **19.** 120 **21.** 180 **23.** 144 **25.** $\frac{8}{24}$ **27.** $\frac{18}{24}$

29. $\frac{20}{24}$ **31.** 2 **33.** 15 **35.** 28 **37.** 55 **39.** 32 **41.** 72

43. 136 **45.** 96 **47.** 27 **49.** It probably depends on how large the numbers are. If the numbers are small, the method using multiples of the largest number seems best. If numbers are larger, or there are more of them, then the factorization method will be better. **51.** 7200 **53.** 10,584 **55.** like; unlike **56.** numerators; denominator; lowest **57.** least; smallest **58.** 40 is the least common multiple. **59.** 72 **60.** 450 **61.** 240 is a common multiple but twice as large as the least common multiple; 120 is the LCM. **62.** The least common multiple can be no smaller than the largest number in a group and the number 1760 is a multiple of 55.

Section 3.3 (page 205)

1. $\frac{5}{6}$ **3.** $\frac{8}{9}$ **5.** $\frac{3}{4}$ **7.** $\frac{39}{40}$ **9.** $\frac{23}{36}$ **11.** $\frac{14}{15}$ **13.** $\frac{29}{36}$ **15.** $\frac{17}{20}$

17. $\frac{23}{30}$ **19.** $\frac{7}{12}$ **21.** $\frac{23}{48}$ **23.** $\frac{3}{8}$ **25.** $\frac{1}{2}$ **27.** $\frac{7}{15}$ **29.** $\frac{1}{6}$

31. $\frac{19}{45}$ **33.** $\frac{3}{40}$ **35.** $\frac{17}{48}$ **37.** $\frac{23}{24}$ yd^3 **39.** $\frac{7}{12}$ acre **41.** $\frac{31}{40}$ in.

43. $\frac{3}{8}$ gal **45.** You cannot add or subtract until all the fractional pieces are the same size. For example, halves are larger than fourths, so you cannot add $\frac{1}{2} + \frac{1}{4}$ until you rewrite $\frac{1}{2}$ as $\frac{2}{4}$.

47. $\frac{1}{4}$ **49.** work and travel; 8 hr **51.** $\frac{3}{16}$ in.

Section 3.4 (page 213)

1. *Estimate:* $6 + 3 = 9$; *Exact:* $8\frac{5}{6}$

3. *Estimate:* $7 + 4 = 11$; *Exact:* $11\frac{1}{2}$

5. *Estimate:* $1 + 4 = 5$; *Exact:* $4\frac{5}{24}$

7. *Estimate:* $25 + 19 = 44$; *Exact:* $43\frac{2}{3}$

9. *Estimate:* $34 + 19 = 53$; *Exact:* $52\frac{1}{10}$

11. *Estimate:* $23 + 15 = 38$; *Exact:* $38\frac{5}{28}$

13. *Estimate:* $11 + 18 + 16 = 45$; *Exact:* $43\frac{5}{6}$

15. *Estimate:* $12 - 10 = 2$; *Exact:* $2\frac{1}{8}$

17. *Estimate:* $13 - 1 = 12$; *Exact:* $11\frac{7}{15}$

19. *Estimate:* $28 - 6 = 22$; *Exact:* $22\frac{7}{30}$

21. *Estimate:* $17 - 7 = 10$; *Exact:* $10\frac{3}{8}$

23. *Estimate:* $19 - 6 = 13$; *Exact:* $12\frac{19}{20}$

25. *Estimate:* $16 - 11 = 5$; *Exact:* $5\frac{7}{8}$

27. $9\frac{3}{8}$ **29.** $11\frac{1}{2}$ **31.** $3\frac{5}{6}$ **33.** $6\frac{11}{12}$ **35.** $8\frac{1}{8}$ **37.** $\frac{5}{6}$

39. $2\frac{7}{8}$ **41.** $2\frac{7}{12}$ **43.** $5\frac{9}{20}$ **45.** $3\frac{16}{21}$

47. Find the least common denominator. Change the fraction parts so that they have the same denominator. Add the fraction parts. Add the whole number parts. Write the answer as a mixed number.

49. *Estimate:* $143 - 29 = 114$ tons; *Exact:* $114\frac{1}{10}$ tons

51. *Estimate:* $4 - 2 = 2$ deaths per 1000; *Exact:* $2\frac{3}{10}$ deaths per 1000

53. *Estimate:* $13 + 9 = 22$ ft; *Exact:* $21\frac{1}{6}$ ft

55. *Estimate:* $3 + 6 + 5 + 3 + 6 = 23$ hr; *Exact:* $22\frac{7}{8}$ hr

57. *Estimate:* $24 + 35 + 24 + 35 = 118$ in.; *Exact:* $116\frac{1}{2}$ in.

59. *Estimate:* $9 - 3 - 3 - 2 = 1$ yd^3; *Exact:* $1\frac{5}{8}$ yd^3

61. *Estimate:* $527 - 108 - 151 - 139 = 129$ ft; *Exact:* 130 ft

63. *Estimate:* $3 + 7 + 2 + 3 + 7 = 22$ tons; *Exact:* $21\frac{7}{12}$ tons

65. $4\frac{11}{16}$ in. **67.** $21\frac{3}{8}$ in. **69.** **(a)** 30 **(b)** 28 **(c)** 25 **(d)** 264

70. least common denominator **71.** **(a)** $\frac{23}{24}$ **(b)** $\frac{8}{15}$ **(c)** $\frac{43}{48}$

(d) $\frac{4}{21}$ **72.** fraction parts **73.** improper

74. **(a)** $4\frac{5}{8} + 3\frac{2}{8} = 7\frac{7}{8}$; $\frac{37}{8} + \frac{26}{8} = \frac{63}{8} = 7\frac{7}{8}$

(b) $11\frac{56}{40} - 8\frac{35}{40} = 3\frac{21}{40}$; $\frac{496}{40} - \frac{355}{40} = \frac{141}{40} = 3\frac{21}{40}$

Section 3.5 (page 225)

1.–12.

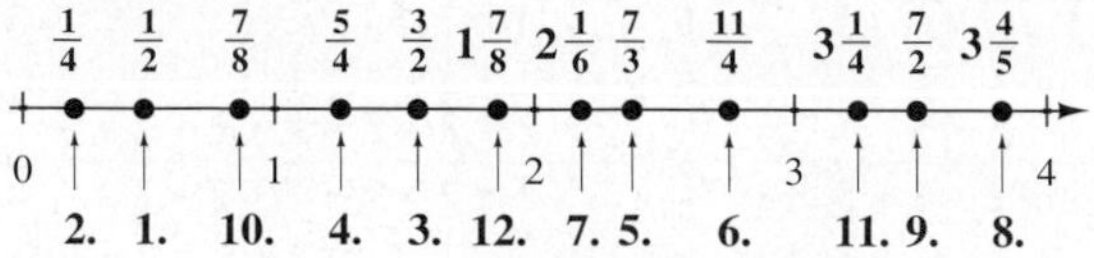

13. $>$ **15.** $<$ **17.** $>$ **19.** $<$ **21.** $>$ **23.** $>$

25. $\frac{1}{4}$ **27.** $\frac{25}{49}$ **29.** $\frac{9}{16}$ **31.** $\frac{64}{125}$ **33.** $\frac{81}{16} = 5\frac{1}{16}$ **35.** $\frac{81}{256}$

37. A number line is a horizontal line with a range of numbers placed on it. The lowest number is on the left and the highest number is on the right. It can be used to compare the size or value of numbers.

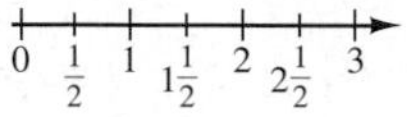

39. 3 **41.** 16 **43.** 1 **45.** $\frac{3}{16}$ **47.** $\frac{4}{9}$ **49.** $\frac{1}{3}$ **51.** $\frac{1}{2}$ **53.** $\frac{3}{8}$

55. $\frac{1}{4}$ **57.** $1\frac{1}{2}$ **59.** $\frac{1}{12}$ **61.** 3 **63.** $\frac{5}{16}$ **65.** $\frac{1}{4}$

67. $\frac{1}{32}$ **69.** $\frac{9}{25}$ is greater. **71.** $<$; $>$

72. **(a)** like; numerators; numerator **(b)** Answers will vary.

73. parentheses; exponents; square; multiply; divide; add; subtract

74. $\frac{2}{45}$ **75.** $\frac{4}{9}$ **76.** $\frac{9}{64}$ **77.** $1\frac{169}{343}$ **78.** $2\frac{113}{256}$ **79.** 2 **80.** $\frac{2}{45}$

Summary Exercises on Fractions (page 229)

1. proper **2.** improper **3.** improper **4.** proper

5. $\frac{4}{5}$ **6.** $\frac{7}{8}$ **7.** $\frac{3}{7}$ **8.** $\frac{23}{47}$ **9.** $\frac{1}{2}$ **10.** $\frac{3}{8}$ **11.** 35 **12.** $\frac{5}{6}$

13. $1\frac{1}{6}$ **14.** 56 **15.** $1\frac{13}{24}$ **16.** $1\frac{13}{16}$ **17.** $2\frac{1}{12}$ **18.** $\frac{1}{12}$

19. $\frac{11}{24}$ **20.** $\frac{2}{15}$ **21.** *Exact:* $7\frac{7}{8}$; *Estimate:* $4 \cdot 2 = 8$

22. *Exact:* $17\frac{15}{32}$; *Estimate:* $5 \cdot 3 = 15$

23. *Exact:* $107\frac{2}{3}$; *Estimate:* $8 \cdot 6 \cdot 2 = 96$

24. *Exact:* $1\frac{1}{6}$; *Estimate:* $4 \div 4 = 1$

25. *Exact:* $3\frac{7}{16}$; *Estimate:* $7 \div 2 = 3\frac{1}{2}$

26. *Exact:* $6\frac{1}{6}$; *Estimate:* $5 \div 1 = 5$

27. *Estimate:* $4 + 4 = 8$; *Exact:* $7\frac{7}{8}$

28. *Estimate:* $22 + 8 = 30$; *Exact:* $30\frac{1}{6}$

29. *Estimate:* $15 + 11 = 26$; *Exact:* $25\frac{4}{15}$

30. *Estimate:* $8 - 3 = 5$; *Exact:* $4\frac{9}{10}$

31. *Estimate:* $14 - 7 = 7$; *Exact:* $6\frac{5}{8}$

32. *Estimate:* $32 - 23 = 9$; *Exact:* $9\frac{1}{4}$ **33.** $\frac{1}{12}$ **34.** $\frac{9}{10}$

35. $\frac{5}{18}$ **36.** 40 **37.** 72 **38.** 84 **39.** 35 **40.** 12 **41.** 55

42. $<$ **43.** $>$ **44.** $>$

Chapter 3 Review Exercises (page 235)

1. $\frac{7}{9}$ **2.** $\frac{5}{7}$ **3.** $\frac{3}{4}$ **4.** $\frac{1}{8}$ **5.** $\frac{1}{5}$ **6.** $\frac{1}{6}$ **7.** $\frac{13}{31}$ **8.** $\frac{1}{3}$ **9.** $\frac{3}{4}$ of her patients **10.** $\frac{1}{4}$ Web page less **11.** 12 **12.** 20 **13.** 60 **14.** 24 **15.** 120 **16.** 180 **17.** 8 **18.** 21 **19.** 10 **20.** 45 **21.** 32 **22.** 20 **23.** $\frac{5}{6}$ **24.** $\frac{7}{8}$ **25.** $\frac{5}{8}$ **26.** $\frac{5}{12}$ **27.** $\frac{13}{24}$ **28.** $\frac{17}{36}$ **29.** $\frac{23}{24}$ of the sack **30.** $\frac{59}{60}$ of the amount needed

31. *Estimate:* 19 + 14 = 33; *Exact:* $32\frac{3}{8}$

32. *Estimate:* 23 + 15 = 38; *Exact:* $38\frac{1}{9}$

33. *Estimate:* 13 + 9 + 10 = 32; *Exact:* $31\frac{43}{80}$

34. *Estimate:* 32 − 15 = 17; *Exact:* $17\frac{1}{12}$

35. *Estimate:* 34 − 16 = 18; *Exact:* $18\frac{1}{3}$

36. *Estimate:* 215 − 136 = 79; *Exact:* $79\frac{7}{16}$

37. $9\frac{1}{10}$ **38.** $10\frac{5}{12}$ **39.** $3\frac{1}{4}$ **40.** $1\frac{2}{3}$ **41.** $5\frac{1}{2}$ **42.** $2\frac{19}{24}$

43. *Estimate:* 15 − 6 − 7 = 2 gal; *Exact:* $2\frac{5}{12}$ gal

44. *Estimate:* 29 + 25 = 54 tons; *Exact:* $53\frac{5}{12}$ tons

45. *Estimate:* 9 + 9 + 7 = 25 lb; *Exact:* $24\frac{23}{24}$ lb

46. *Estimate:* 9 − 2 − 3 = 4 acres; *Exact:* $4\frac{1}{16}$ acres

47.–50.

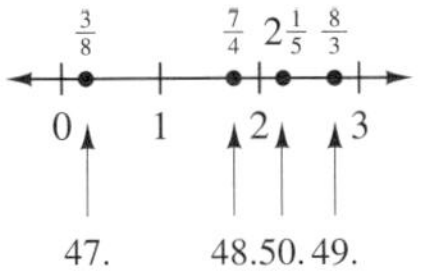

51. < **52.** < **53.** > **54.** > **55.** < **56.** > **57.** < **58.** > **59.** $\frac{1}{4}$ **60.** $\frac{4}{9}$ **61.** $\frac{27}{1000}$ **62.** $\frac{81}{4096}$ **63.** $\frac{2}{3}$ **64.** $6\frac{2}{3}$ **65.** $\frac{4}{9}$ **66.** 1 **67.** $\frac{3}{16}$ **68.** $1\frac{25}{64}$ **69.** $\frac{2}{3}$ **70.** $\frac{1}{8}$ **71.** $\frac{19}{32}$ **72.** $\frac{11}{16}$ **73.** $2\frac{1}{6}$ **74.** $26\frac{1}{4}$ **75.** $5\frac{3}{8}$ **76.** $11\frac{43}{80}$ **77.** $15\frac{5}{12}$ **78.** $\frac{8}{11}$ **79.** $\frac{1}{250}$ **80.** $\frac{1}{2}$ **81.** $\frac{2}{9}$ **82.** $\frac{11}{27}$ **83.** > **84.** < **85.** < **86.** > **87.** 36 **88.** 120 **89.** 126 **90.** 18 **91.** 108 **92.** 60

93. *Estimate:* 93 − 14 − 22 = 57 ft; *Exact:* $56\frac{7}{8}$ ft

94. *Estimate:* 15 − 4 − 9 = 2 gal; *Exact:* $2\frac{7}{8}$ gal

Chapter 3 Test (page 241)

1. $\frac{3}{4}$ **2.** $\frac{1}{2}$ **3.** $\frac{2}{5}$ **4.** $\frac{1}{6}$ **5.** 12 **6.** 30 **7.** 108 **8.** $\frac{5}{8}$ **9.** $\frac{23}{36}$ **10.** $\frac{5}{24}$ **11.** $\frac{1}{40}$

12. *Estimate:* 8 + 5 = 13; *Exact:* $12\frac{1}{2}$

13. *Estimate:* 16 − 12 = 4; *Exact:* $4\frac{11}{15}$

14. *Estimate:* 19 + 9 + 12 = 40; *Exact:* $40\frac{29}{60}$

15. *Estimate:* 24 − 18 = 6; *Exact:* $5\frac{5}{8}$

16. Probably addition and subtraction of fractions is more difficult because you have to find the least common denominator and then change the fractions to the same denominator. **17.** Round mixed numbers to the nearest whole number. Then add, subtract, multiply, or divide to estimate the answer. The estimate may vary from the exact answer but it lets you know if your answer is reasonable.

18. *Estimate:* 6 + 5 + 4 + 7 + 5 = 27 hr; *Exact:* $26\frac{3}{4}$

19. *Estimate:* 148 − 69 − 37 − 6 = 36 gal; *Exact:* $35\frac{7}{8}$ gal

20. > **21.** > **22.** 2 **23.** $\frac{13}{48}$ **24.** $1\frac{3}{4}$ **25.** $1\frac{1}{3}$

Cumulative Review Exercises: Chapters 1–3 (page 243)

1. 8, 1 **2.** 5, 9 **3.** 1440; 1400; 1000 **4.** 59,800; 59,800; 60,000 **5.** *Estimate:* 2000 + 400 + 50,000 + 30,000 = 82,400; *Exact:* 79,779 **6.** *Estimate:* 20,000 − 10,000 = 10,000; *Exact:* 14,3897. *Estimate:* 4000 × 300 = 1,200,000; *Exact:* 1,251,040 **8.** *Estimate:* 100,000 ÷ 40 = 2500; *Exact:* 3211 **9.** 26 **10.** 1,255,609 **11.** 591 **12.** 2,801,695 **13.** 120 **14.** 126 **15.** 160 **16.** 456 **17.** 369,408 **18.** 17,000 **19.** 135 **20.** 2693 R2 **21.** 32 R166 **22.** *Estimate:* 20 + 9 + 5 + 20 + 9 + 5 = 68 ft; *Exact:* 64 ft **23.** *Estimate:* 20 • 10 = 200 ft^2; *Exact:* 252 ft^2 **24.** *Estimate:* 40,000 ÷ 30 ≈ 1333 cartons; *Exact:* 1260 cartons **25.** *Estimate:* 4000 × 60 = 240,000 revolutions; *Exact:* 216,000 revolutions

26. *Estimate*: 2 • 3 = 6 yd^2; *Exact*: $4\frac{2}{3}$ yd^2

27. *Estimate*: 5 • 7 = 35 mi^2; *Exact*: $30\frac{5}{6}$ mi^2

28. *Estimate*: 5 • 4 = 20 cords; *Exact*: $18\frac{3}{8}$ cords

29. *Estimate*: 1537 − 83 = 1454 ft; *Exact*: $1454\frac{3}{8}$ ft

30. $2 \cdot 3^2$ **31.** $2^3 \cdot 5^2$ **32.** $5^2 \cdot 7^2$ **33.** 144 **34.** 108 **35.** 972 **36.** 6 **37.** 9 **38.** 12 **39.** 9 **40.** 44 **41.** $\frac{4}{45}$ **42.** $\frac{9}{10}$ **43.** $1\frac{1}{16}$ **44.** proper **45.** improper **46.** improper **47.** $\frac{5}{12}$ **48.** $\frac{7}{8}$ **49.** $\frac{9}{10}$ **50.** $\frac{1}{2}$ **51.** $\frac{5}{16}$ **52.** $36\frac{3}{4}$ **53.** $1\frac{2}{3}$ **54.** $2\frac{3}{16}$ **55.** $13\frac{1}{2}$ **56.** $\frac{25}{28}$ **57.** $\frac{13}{16}$ **58.** $\frac{7}{36}$ **59.** *Estimate*: 3 + 5 = 8; *Exact*: $7\frac{7}{8}$

60. *Estimate*: 22 + 4 = 26; *Exact*: $26\frac{7}{24}$

61. *Estimate*: 5 − 2 = 3; *Exact*: $2\frac{5}{8}$

62. 24 **63.** 120 **64.** 144 **65.** 36 **66.** 56 **67.** 81 **68.** 60 **69.–72.**

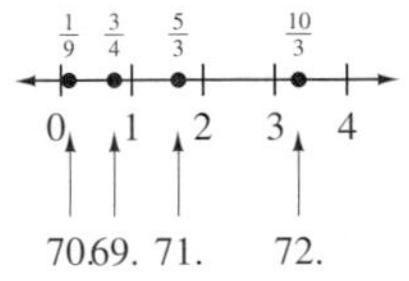

73. < **74.** > **75.** <

Chapter 4

Section 4.1 (page 253)

1. 7; 0; 4 **3.** 5; 1; 8 **5.** 4; 7; 0 **7.** 1; 6; 3 **9.** 1; 8; 9 **11.** 6; 2; 1 **13.** 410.25 **15.** 6.5432 **17.** 5406.045 **19.** $\frac{7}{10}$ **21.** $13\frac{2}{5}$ **23.** $\frac{1}{4}$ **25.** $\frac{33}{50}$ **27.** $10\frac{17}{100}$ **29.** $\frac{3}{50}$ **31.** $\frac{41}{200}$ **33.** $5\frac{1}{500}$ **35.** $\frac{343}{500}$ **37.** five tenths **39.** seventy-eight hundredths **41.** one hundred five thousandths **43.** twelve and four hundredths **45.** one and seventy-five thousandths **47.** 6.7 **49.** 0.32 **51.** 420.008 **53.** 0.0703 **55.** 75.030 **57.** Anne should not say "and" because that denotes a decimal point. **59.** ten thousandths inch; $\frac{10}{1000} = \frac{1}{100}$ **61.** 12 pounds **63.** 3-C **65.** 4-A **67.** One and six hundred two thousandths centimeters **69.** millionths, ten-millionths, hundred-millionths, billionths; these match the words on the left side of the chart with "ths" added. **70.** First place to left of decimal point is ones, so first place to right could be one*ths*, like tens and ten*ths*. But anything that is 1 or more is to the left of the decimal point. **71.** Seventy-two million four hundred thirty-six thousand nine hundred fifty-five hundred-millionths **72.** Six hundred seventy-eight thousand five hundred fifty-four billionths **73.** Eight thousand six and five hundred thousand one millionths **74.** Twenty thousand, sixty and five hundred five millionths **75.** 0.0302040 **76.** 9,876,543,210.100200300

Section 4.2 (page 263)

1. 16.9 **3.** 0.956 **5.** 0.80 **7.** 3.661 **9.** 794.0 **11.** 0.0980 **13.** 49 **15.** 9.09 **17.** 82.0002 **19.** \$0.82 **21.** \$1.22 **23.** \$0.50 **25.** \$48,650 **27.** \$310 **29.** \$849 **31.** \$500 **33.** \$1.00 **35.** \$1000 **37.** (a) 186.0 miles per hour (b) 763.0 miles per hour **39.** (a) 322 miles per hour (b) 163 miles per hour **41.** Rounds to \$0 (zero dollars) because \$0.499 is closer to \$0 than to \$1. **42.** Round amounts less than \$1.00 to nearest cent instead of nearest dollar. **43.** Rounds to \$0.00 (zero cents) because \$0.0015 is closer to \$0.00 than to \$0.01. **44.** Both round to \$0.60. Rounding to nearest thousandth (tenth of a cent) would allow you to identify \$0.597 as less than \$0.601.

Section 4.3 (page 269)

1. 17.48 **3.** 23.013 **5.** 7.763 **7.** 77.006 **9.** 20.104 **11.** 0.109 **13.** 330.86895 **15.** (a) 24.75 in. (b) 3.95 in. **17.** (a) 62.27 in. (b) 0.39 in. **19.** 6 should be written 6.00; sum is 46.22. **21.** *Estimate:* \$20 − 7 = \$13; *Exact:* \$13.16 **23.** *Estimate:* 400 + 1 + 20 = 421; *Exact:* 414.645 **25.** *Estimate:* 9 − 4 = 5; *Exact:* 4.849 **27.** *Estimate:* 60 + 500 + 6 = 566; *Exact:* 608.4363 **29.** 0.275 **31.** 6.507 **33.** 1.81 **35.** 6056.7202 **37.** *Estimate:* \$50 − \$40 = \$10; *Exact:* \$8.91 **39.** *Estimate:* 2 − 2 = 0; *Exact:* 0.019 in. **41.** *Estimate:* 400 − 300 = 100 million people; *Exact:* 112.48 million people **43.** *Estimate:* 11 − 10 = 1 ounce; *Exact:* 0.65 ounce **45.** *Estimate:* 20 + 6 + 20 + 6 = 52 in.; *Exact:* 52.1 in. **47.** *Estimate:* \$29 − \$9 = \$20; *Exact:* \$20.50 **49.** *Estimate:* \$13 + 18 + 6 + 3 + 2 = \$42; *Exact:* \$41.60 **51.** \$1939.36 **53.** \$3.97 **55.** \$598.22 **57.** $b = 1.39$ cm **59.** $q = 23.843$ ft

Section 4.4 (page 275)

1. 0.1344 **3.** 159.10 **5.** 15.5844 **7.** \$34,500.20 **9.** 43.2 **11.** 0.432 **13.** 0.0432 **15.** 0.00432 **17.** 0.0000312 **19.** 0.000006 **21.** Multiplying by 10, decimal point moves one place to the right; by 100, two places to the right; by 1000, three places to the right. **23.** *Estimate:* 40 × 5 = 200; *Exact:* 190.08 **25.** *Estimate:* 40 × 40 = 1600; *Exact:* 1558.2 **27.** *Estimate:* 7 × 5 = 35; *Exact:* 30.038 **29.** *Estimate:* 3 × 7 = 21; *Exact:* 19.24165 **31.** unreasonable; \$189.00 **33.** reasonable **35.** unreasonable; \$3.19 **37.** unreasonable; 9.5 pounds **39.** \$945.87 (rounded) **41.** \$2.45 (rounded) **43.** \$28.82 (rounded) **45.** \$12,271 **47.** \$347.52; \$719.40 **49.** \$76.50 **51.** \$73.45 **53.** \$388.34 **55.** \$4.09 (rounded) **57.** \$129.25 **59.** (a) \$70.05 (b) \$25.80

Section 4.5 (page 285)

1. 3.9 **3.** 0.47 **5.** 400.2 **7.** 36 **9.** 0.06 **11.** 6000 **13.** 25.3 **15.** 516.67 (rounded) **17.** 24.291 (rounded) **19.** 10,082.647 (rounded) **21.** Dividing by 10, decimal point moves one place to the left; by 100, two places to the left; by 1000, three places to the left. **23.** unreasonable; 40 ÷ 8 = 5; Correct answer is 4.725. **25.** reasonable; 50 ÷ 50 = 1 **27.** unreasonable; 300 ÷ 5 = 60; Correct answer is 60.2. **29.** unreasonable; 9 ÷ 1 = 9; Correct answer is 7.44. **31.** \$4.00 (rounded) **33.** \$19.46 **35.** \$0.30 **37.** \$8.92 per hour **39.** 21.2 miles per gallon (rounded) **41.** 8.81 m (rounded) **43.** 0.03 meter **45.** 26.72 meters **47.** 14.25 **49.** 73.4 **51.** 1.205 **53.** 0.334 **55.** \$0.03 (rounded) per can **57.** \$0.04 (rounded) **59.** 100,000 box tops **61.** 2632 box tops (rounded)

Section 4.6 (page 293)

1. 0.5 **3.** 0.75 **5.** 0.3 **7.** 0.9 **9.** 0.6 **11.** 0.875 **13.** 2.25 **15.** 14.7 **17.** 3.625 **19.** 0.333 (rounded) **21.** 0.833 (rounded) **23.** 1.889 (rounded) **25. (a)** A proper fraction is less than 1, so it cannot be equivalent to a mixed number. **(b)** $\frac{5}{9}$ means 5 ÷ 9 or $9\overline{)5}$ so correct answer is 0.556 (rounded). This makes sense because both the fraction and decimal are less than 1. **26. (a)** $2.035 = 2\frac{35}{1000} = 2\frac{7}{200}$, not $2\frac{7}{20}$. **(b)** Adding the whole number part gives 2 + 0.35, which is 2.35, not 2.035. To check, $2.35 = 2\frac{35}{100} = 2\frac{7}{20}$. **27.** Just add the whole number part to 0.375. So $1\frac{3}{8} = 1.375$; $3\frac{3}{8} = 3.375$; $295\frac{3}{8} = 295.375$. **28.** It works only when the fraction part has a one-digit numerator and a denominator of 10, a two-digit numerator and a denominator of 100, and so on. **29.** $\frac{2}{5}$ **31.** $\frac{5}{8}$ **33.** $\frac{7}{20}$ **35.** 0.35 **37.** $\frac{1}{25}$ **39.** $\frac{3}{20}$ **41.** 0.2 **43.** $\frac{9}{100}$ **45.** shorter; 0.72 in. **47.** too much; 0.005 gram **49.** 0.9991 cm, 1.0007 cm **51.** more; 0.05 in. **53.** 0.5399, 0.54, 0.5455 **55.** 5.0079, 5.79, 5.8, 5.804 **57.** 0.6009, 0.609, 0.628, 0.62812 **59.** 2.8902, 3.88, 4.876, 5.8751 **61.** 0.006, 0.043, $\frac{1}{20}$, 0.051 **63.** 0.37, $\frac{3}{8}$, $\frac{2}{5}$, 0.4001 **65.** red box **67.** 0.01 in. **69.** 1.4 in. (rounded) **71.** 0.3 in. (rounded) **73.** 0.4 in. (rounded)

Chapter 4 Review Exercises (page 301)

1. 0; 5 **2.** 0; 6 **3.** 8; 9 **4.** 5; 9 **5.** 7; 6 **6.** $\frac{1}{2}$ **7.** $\frac{3}{4}$ **8.** $4\frac{1}{20}$ **9.** $\frac{7}{8}$ **10.** $\frac{27}{1000}$ **11.** $27\frac{4}{5}$ **12.** eight tenths

13. four hundred and twenty-nine hundredths **14.** twelve and seven thousandths **15.** three hundred six ten-thousandths **16.** 8.3 **17.** 0.205 **18.** 70.0066 **19.** 0.30 **20.** 275.6 **21.** 72.79 **22.** 0.160 **23.** 0.091 **24.** 1.0 **25.** $15.83 **26.** $0.70 **27.** $17,625.79 **28.** $350 **29.** $130 **30.** $100 **31.** $29 **32.** *Estimate:* 6 + 400 + 20 = 426; *Exact:* 444.86 **33.** *Estimate:* 80 + 1 + 100 + 1 + 30 = 212; *Exact:* 233.515 **34.** *Estimate:* 300 − 20 = 280; *Exact:* 290.7 **35.** *Estimate:* 9 − 8 = 1; *Exact:* 1.2684 **36.** *Estimate:* 100 million − 60 million = 40 million; *Exact:* 38.8 million **37.** *Estimate:* $200 + $40 = $240; *Exact:* $260.00 **38.** *Estimate:* $2 + $5 + $20 = $27; $30 − $27 = $3; *Exact:* $4.14 **39.** *Estimate:* 2 + 4 + 5 = 11 kilometers; *Exact:* 11.55 kilometers **40.** *Estimate:* 6 × 4 24; *Exact:* 22.7106 **41.** *Estimate:* 40 × 3 120; *Exact:* 141.57 **42.** 0.0112 **43.** 0.000355 **44.** reasonable; 700 ÷ 10 = 70 **45.** unreasonable; 30 ÷ 3 = 10; Correct answer is 9.5. **46.** 14.467 (rounded) **47.** 1200 **48.** 0.4 **49.** $708 (rounded) **50.** $2.99 (rounded) **51.** 133 shares (rounded) **52.** $3.12 (rounded) **53.** 29.215 **54.** 10.15 **55.** 3.8 **56.** 0.64 **57.** 1.875 **58.** 0.111 (rounded) **59.** 3.6008, 3.68, 3.806 **60.** 0.209, 0.2102, 0.215, 0.22 **61.** $\frac{1}{8}$, $\frac{3}{20}$, 0.159, 0.17 **62.** 404.865 **63.** 254.8 **64.** 3583.261 (rounded) **65.** 29.0898 **66.** 0.03066 **67.** 9.4 **68.** 175.675 **69.** 9.04 **70.** 19.50 **71.** 8.19 **72.** 0.928 **73.** 35 **74.** 0.259 **75.** 0.3 **76.** $3.00 (rounded) **77.** $2.17 (rounded) **78.** $35.96 **79.** $199.71 **80.** $78.50 **81.** **(a)** baked potato with skin **(b)** cup of orange juice **(c)** 0.59 milligram **82.** **(a)** 2.06 milligrams **(b)** more, by 0.06 milligram

Chapter 4 Test (page 305)

1. $18\frac{2}{5}$ **2.** $\frac{3}{40}$ **3.** sixty and seven thousandths **4.** two hundred eight ten-thousandths **5.** 725.6 **6.** 0.630 **7.** $1.49 **8.** $7860 **9.** *Estimate*: 8 + 80 + 40 = 128; *Exact*: 129.2028 **10.** *Estimate*: 80 − 4 = 76; *Exact*: 75.498 **11.** *Estimate*: 6 • 1 = 6; *Exact*: 6.948 **12.** *Estimate*: 20 ÷ 5 = 4; *Exact*: 4.175 **13.** 839.762 **14.** 669.004 **15.** 0.0000483 **16.** 480 **17.** 2.625 **18.** 0.44, $\frac{9}{20}$, 0.4506, 0.451 **19.** 35.49 **20.** $446.87 **21.** 1988, 0.33 meter **22.** $5.35 (rounded) **23.** 2.8 degrees **24.** $4.55 per meter **25.** Answer varies.

Cumulative Review Exercises: Chapters 1–4 (page 307)

1. 5, 9, 2 **2.** 0, 5, 8 **3.** 500,000 **4.** 602.49 **5.** $710 **6.** $0.05 **7.** *Estimate*: 4000 + 600 + 9000 = 13,600; *Exact*: 13,339 **8.** *Estimate*: 4 + 20 + 1 = 25; *Exact*: 20.683 **9.** *Estimate*: 5000 − 2000 = 3000; *Exact*: 3209 **10.** *Estimate*: 50 − 7 = 43; *Exact*: 44.506 **11.** *Estimate*: 3000 × 200 = 600,000; *Exact*: 550,622 **12.** *Estimate*: 7 × 7 = 49; *Exact*: 49.786 **13.** *Estimate*: 100,000 ÷ 50 = 2000; *Exact*: 2690 **14.** *Estimate*: 40 ÷ 8 = 5; *Exact*: 4.5 **15.** *Estimate*: 2 • 4 = 8; *Exact*: $7\frac{1}{8}$ **16.** *Estimate*: 2 ÷ 1 = 2; *Exact*: $2\frac{4}{5}$ **17.** *Estimate*: 2 + 2 = 4; *Exact*: $3\frac{7}{15}$ **18.** *Estimate*: 5 − 2 = 3; *Exact*: $2\frac{5}{8}$ **19.** 9.671 **20.** $1\frac{4}{9}$ **21.** 73,225 **22.** $1\frac{2}{5}$ **23.** 4914 **24.** 93.603 **25.** 404 R3 **26.** $1\frac{17}{24}$ **27.** 233,728 **28.** 0.03264 **29.** 8 **30.** 45 **31.** $\frac{4}{31}$ **32.** $\frac{2}{3}$ **33.** 0.51 (rounded) **34.** 4 **35.** 14 **36.** $\frac{1}{4}$ **37.** 20.81 **38.** 576 **39.** 14 **40.** $2^3 \cdot 5^2$ **41.** forty and thirty-five thousandths **42.** 0.0306 **43.** $\frac{1}{8}$ **44.** $3\frac{2}{25}$ **45.** 2.6 **46.** 0.636 (rounded) **47.** > **48.** 7.005, 7.5, 7.5005, 7.505 **49.** 0.8, 0.8015, $\frac{21}{25}$, $\frac{7}{8}$ **50.** size M (medium) **51.** XXS $1\frac{1}{2}$ in.; XS $\frac{3}{4}$ in.; S $\frac{3}{4}$ in.; M $\frac{7}{8}$ in.; L $\frac{3}{4}$ in. **52.** 21.125 − 20.25 = 0.875 in.; 7 ÷ 8 = 0.875 in. or $\frac{875}{1000} = \frac{875 \div 125}{1000 \div 125} = \frac{7}{8}$ in. **53.** Answers will vary. Many people prefer using decimals because you do not need to find a common denominator or rewrite answers in lowest terms. **54.** *Estimate*: $20 − $8 − $1 = $11; *Exact*: $11.17 **55.** *Estimate*: 50 − 47 = 3 in.; *Exact*: $3\frac{3}{8}$ in. **56.** *Estimate*: $10 × 20 = $200; *Exact*: $191.90 (rounded) **57.** *Estimate*: (8 × 20) + (10 × 30) = 160 + 300 = 460 students; *Exact*: 488 students **58.** *Estimate*: 2 + 4 = 6 yards; *Exact*: $6\frac{5}{24}$ yards **59.** *Estimate*: $30 + $200 − $40 − $20 = $170; *Exact*: $191.50 **60.** *Estimate*: $3 ÷ 3 = $1 per pound; *Exact*: $0.95 per pound (rounded) **61.** *Estimate*: $80,000 ÷ 100 = $800; *Exact*: $729 (rounded) **62.** (a) 9 days (rounded) (b) 15.6 pounds (rounded) **63.** Average weight of food eaten by bee is 0.32 ounce **64.** 0.112 ounce; $\frac{14}{125}$ ounce; 14 ÷ 125 = 0.112

Chapter 5

Section 5.1 (page 317)

1. $\frac{8}{9}$ **3.** $\frac{2}{1}$ **5.** $\frac{1}{3}$ **7.** $\frac{8}{5}$ **9.** $\frac{3}{8}$ **11.** $\frac{9}{7}$ **13.** $\frac{6}{1}$ **15.** $\frac{5}{6}$ **17.** $\frac{8}{5}$ **19.** $\frac{1}{12}$ **21.** $\frac{5}{16}$ **23.** $\frac{4}{1}$ **25.** $\frac{1}{2}$ **27.** $\frac{36}{1}$ **29.** Answers will vary. One possibility is stocking cards of various types in the same ratios as those in the table. **31.** Comparing the violin to piano, guitar, organ, clarinet, and drums gives ratios of $\frac{1}{11}$, $\frac{1}{10}$, $\frac{1}{3}$, $\frac{1}{2}$, and $\frac{2}{3}$, respectively. **33.** Answers will vary. Possibilities include: guitars are less expensive than drums and easier to carry around; more guitar players than drummers are needed in a band. **35.** **(a)** $\frac{6}{1}$ **(b)** $\frac{5}{2}$ **(c)** $\frac{7}{16}$ **37.** $\frac{7}{5}$ **39.** $\frac{6}{1}$ **41.** $\frac{38}{17}$ **43.** $\frac{1}{4}$ **45.** $\frac{34}{35}$ **47.** $\frac{1}{1}$; as long as the sides all have the same length, any measurement you choose will maintain the ratio. **48.** Answers will vary. Some possibilities are $\frac{4}{5} = \frac{8}{10} = \frac{12}{15} = \frac{16}{20} = \frac{20}{25} = \frac{24}{30} = \frac{28}{35}$. **49.** It is not possible. Amelia would have to be older than her mother to have a ratio of 5 to 3. **50.** Answers will vary, but a ratio of 3 to 1 means your income is 3 times your friend's income.

Section 5.2 (page 325)

1. $\frac{5\text{ cups}}{3\text{ people}}$ **3.** $\frac{3\text{ feet}}{7\text{ seconds}}$ **5.** $\frac{1\text{ person}}{2\text{ dresses}}$ **7.** $\frac{5\text{ letters}}{1\text{ minute}}$ **9.** $\frac{\$21}{2\text{ visits}}$ **11.** $\frac{18\text{ miles}}{1\text{ gallon}}$ **13.** \$12 per hour or \$12/hour **15.** 5 eggs per chicken or 5 eggs/chicken **17.** 1.25 pounds/person **19.** \$103.30/day **21.** 325.9 21.0 (rounded) **23.** 338.6 20.9 (rounded) **25.** 4 ounces for \$0.89 **27.** 15 ounces for \$3.15 **29.** 18 ounces for \$1.79 **31.** Answers will vary. For example, you might choose Brand B because you like more chicken, so the cost per chicken chunk may actually be the same or less than Brand A. **33.** 1.75 pounds/week **35.** \$12.26/hour **37.** \$11.50/share **39.** 0.11 seconds/meter; $9.\overline{09}$ or 9.1 meters/second (rounded) **41.** (a) \$0.167 or 16.7¢ (rounded); (b) \$0.125 or 12.5¢; (c) \$0.083 or 8.3¢ (rounded) **43.** One battery for \$1.79; like getting 3 batteries so \$1.79 ÷ 3 ≈ \$0.597 per battery **45.** Brand P with the 50¢ coupon is the best buy. (\$3.39 − \$0.50 = \$2.89 ÷ 16.5 ounces ≈ \$0.175 per ounce) **47.** \$25 ÷ 12 ≈ \$2.08 per month **48.** Round to thousandths to see that Verizon is the best buy at \$0.028 per minute; Qwest ≈ \$0.030; VoiceStream ≈ \$0.062; Sprint ≈ \$0.060 **49.** 1000 min ≈ 16.7 hours; 104 weekend days per year ÷ 12 ≈ 8.7 weekend days per month; 16.7 hours ÷ 8.7 days ≈ 1.9 hours of talking per day. **50.** Verizon ≈ \$1.17; Qwest ≈ \$1.27; VoiceStream ≈ \$0.95; Sprint ≈ \$1.05

Section 5.3 (page 333)

1. $\frac{\$9}{12\text{ cans}} = \frac{\$18}{24\text{ cans}}$ **3.** $\frac{200\text{ adults}}{450\text{ children}} = \frac{4\text{ adults}}{9\text{ children}}$ **5.** $\frac{120}{150} = \frac{8}{10}$ **7.** true **9.** true **11.** false **13.** true **15.** true **17.** false **19.** True **21.** False **23.** False **25.** True **27.** False **29.** True **31.** True **33.** False **35.** False **37.** $\frac{16\text{ hits}}{50\text{ at bats}} = \frac{128\text{ hits}}{400\text{ at bats}}$ $50 \cdot 128 = 6400$, $16 \cdot 400 = 6400$ Cross products are *equal* so the proportion is *true*; they hit equally well.

Section 5.4 (page 339)

1. 4 **3.** 2 **5.** 88 **7.** 91 **9.** 5 **11.** 10 **13.** 24.44 (rounded) **15.** 50.4 **17.** 17.64 (rounded) **19.** 1 **21.** $3\frac{1}{2}$ **23.** 0.2 or $\frac{1}{5}$ **25.** 0.005 or $\frac{1}{200}$

27. Find cross products: 20 ≠ 30, so the proportion is false.

$$\frac{6\frac{2}{3}}{4} = \frac{5}{3} \text{ or } \frac{10}{6} = \frac{5}{3} \text{ or } \frac{10}{4} = \frac{7.5}{3} \text{ or } \frac{10}{4} = \frac{5}{2}$$

28. Find cross products: 192 ≠ 180, so the proportion is false.

$$\frac{6.4}{8} = \frac{24}{30} \text{ or } \frac{6}{7.5} = \frac{24}{30} \text{ or } \frac{6}{8} = \frac{22.5}{30} \text{ or } \frac{6}{8} = \frac{24}{32}$$

Section 5.5 (page 345)

1. 22.5 hours **3.** \$7.20 **5.** 42 pounds **7.** \$273.45 **9.** 10 ounces (rounded) **11.** 5 quarts **13.** 14 ft, 10 ft **15.** 14 ft, 8 ft **17.** 26 points (rounded) **19.** 2065 students (reasonable); about 4214 students with incorrect setup (only 2950 students in the group). **21.** about 190 people (reasonable); about 298 people with incorrect setup (only 238 people attended). **23.** 92,250,000 households (reasonable); about 113,888,889 households with incorrect setup (only 102,500,000 U.S. households). **25.** 625 stocks **27.** 4.06 meters (rounded) **29.** 311 calories (rounded) **31.** 10.53 m (rounded) **33.** You cannot solve this problem using a proportion because the ratio of age to weight is not constant. As Jim's age increases, his weight may decrease, stay the same, or increase. **35.** 3800 students use cream **37.** 120 calories and 12 grams of fiber **39.** $1\frac{3}{4}$ cups water, 3 Tbsp margarine, $\frac{3}{4}$ cup milk, 2 cups flakes **40.** $5\frac{1}{4}$ cups water, 9 Tbsp margarine, $2\frac{1}{4}$ cups milk, 6 cups flakes **41.** $\frac{7}{8}$ cup water, $1\frac{1}{2}$ Tbsp margarine, $\frac{3}{8}$ cup milk, 1 cup flakes **42.** $2\frac{5}{8}$ cups water, $4\frac{1}{2}$ Tbsp margarine, $1\frac{1}{8}$ cups milk, 3 cups flakes

Chapter 5 Review Exercises (page 355)

1. $\frac{3}{4}$ **2.** $\frac{4}{1}$ **3.** $\frac{1}{2}$ **4.** $\frac{2}{1}$ **5.** $\frac{2}{3}$ **6.** $\frac{5}{2}$ **7.** $\frac{1}{6}$ **8.** $\frac{3}{1}$ **9.** $\frac{3}{8}$ **10.** $\frac{4}{3}$ **11.** $\frac{1}{9}$ **12.** $\frac{10}{7}$ **13.** $\frac{7}{5}$ **14.** $\frac{5}{6}$ **15.** $\frac{\$11}{1\text{ dozen}}$ **16.** $\frac{12\text{ children}}{5\text{ families}}$ **17.** 0.2 page/minute or $\frac{1}{5}$ page/minute 5 minutes/page **18.** \$8/hour 0.125 hour/dollar or $\frac{1}{8}$ hour/dollar **19.** 13 ounces for \$2.29 **20.** 25 pounds for \$10.40 − \$1 coupon **21.** true **22.** false **23.** false **24.** true **25.** true **26.** true **27.** 1575 **28.** 20 **29.** 400 **30.** 12.5 **31.** 14.67 (rounded) **32.** 8.17 (rounded) **33.** 50.4 **34.** 0.57 (rounded) **35.** 2.47 (rounded) **36.** 27 cats **37.** 46 hits **38.** \$15.63 (rounded) **39.** 3299 students (rounded) **40.** 68 feet **41.** 14.7 milligrams **42.** 511 calories (rounded) **43.** $27\frac{1}{2}$ hours or 27.5 hours **44.** 105 **45.** 0 **46.** 128 **47.** 23.08 (rounded) **48.** 6.5 **49.** 117.36 (rounded) **50.** False **51.** False **52.** True **53.** $\frac{8}{5}$ **54.** $\frac{33}{80}$ **55.** $\frac{15}{4}$ **56.** $\frac{4}{1}$ **57.** $\frac{4}{5}$ **58.** $\frac{37}{7}$ **59.** $\frac{3}{8}$ **60.** $\frac{1}{12}$ **61.** $\frac{45}{13}$ **62.** 24,900 fans (rounded) **63.** $\frac{8}{3}$ **64.** 75 ft for \$1.99 − \$0.50 coupon **65.** 15 ft **66.** 7.5 hours or $7\frac{1}{2}$ hours **67.** $\frac{1}{2}$ teaspoon or 0.5 teaspoon **68.** 21 points (rounded) **69.** Set up the proportion to compare teaspoons to pounds on both sides. $\frac{1.5\text{ tsp}}{24\text{ pounds}} = \frac{x\text{ tsp}}{8\text{ pounds}}$ Show that cross products are equal.

$24 \cdot x = 1.5 \cdot 8$ Divide each side by 24.

$$\frac{\overset{1}{\cancel{24}} \cdot x}{\underset{1}{\cancel{24}}} = \frac{12}{24} \quad \text{so } x = \frac{1}{2} \text{ tsp or } 0.5 \text{ tsp}$$

70. **(a)** 1400 milligrams **(b)** 100 milligrams

Chapter 5 Test (page 359)

1. $\frac{4}{5}$ **2.** $\frac{20\text{ miles}}{1\text{ gallon}}$ **3.** $\frac{\$1}{5\text{ minutes}}$ **4.** $\frac{15}{4}$ **5.** $\frac{1}{80}$ **6.** $\frac{9}{2}$ **7.** 18 ounces for \$1.89 − \$0.25 coupon **8.** You earned less this year. An example is: $\frac{\text{Last year} \rightarrow \$15{,}000}{\text{This year} \rightarrow \$10{,}000} = \frac{3}{2}$ **9.** false **10.** true **11.** 25 **12.** 2.67 (rounded) **13.** 325 **14.** $10\frac{1}{2}$ **15.** 576 words

16. 3.6 ounces **17.** 87 students (rounded) **18.** No, 4875 cannot be correct because there are only 650 students in the whole school. **19.** 23.8 grams (rounded) **20.** 60 ft

Cumulative Review Exercises: Chapters 1–5 (page 361)

1. 5; 3; 6; 4 **2.** 9; 0; 5; 3 **3.** 9900 **4.** 617.1 **5.** \$100 **6.** \$3.06 **7.** *Estimate:* 30 + 5000 + 400 = 5430; *Exact:* 5585 **8.** *Estimate:* 60 − 6 = 54; *Exact:* 57.408 **9.** *Estimate:* 5000 × 800 = 4,000,000; *Exact:* 3,791,664 **10.** *Estimate:* 1 × 18 = 18; *Exact:* 17.4796 **11.** *Estimate:* 50,000 ÷ 50 = 1000; *Exact:* 907 **12.** *Estimate:* 2000 ÷ 5 = 400; *Exact:* 364 **13.** *Estimate:* 2 • 4 = 8; *Exact:* $6\frac{3}{5}$ **14.** *Estimate:* 5 ÷ 1 = 5; *Exact:* 6 **15.** *Estimate:* 3 − 2 = 1; *Exact:* $\frac{29}{30}$ **16.** *Estimate:* 3 + 11 = 14; *Exact:* $13\frac{2}{5}$ **17.** 374,416 **18.** 29.34 **19.** 610 R27 **20.** 0.0076 **21.** 2312 **22.** 68.381 **23.** 55.6 **24.** 35,852,728 **25.** 39 **26.** 18 **27.** 64 **28.** 0.95 **29.** one hundred five ten-thousandths **30.** 60.071 **31.** $1\frac{9}{16} \approx 1.563$; $1\frac{3}{8} = 1.375$; $1\frac{1}{8} = 1.125$ **32.** $1\frac{1}{8}$ in., 1.25 in., $1\frac{3}{8}$ in., 1.5 in., $1\frac{9}{16}$ in. **33.** $\frac{7}{16}$ in. **34.** $\frac{1}{16}$ in. or 0.063 in. (rounded) **35.** $\frac{11}{5}$ **36.** $\frac{1}{5}$ **37.** $\frac{4}{1}$ **38.** $\frac{1}{12}$ **39.** $\frac{3}{1}$ **40.** 36 servings for \$3.24 − \$0.50 coupon **41.** 21 **42.** 17.14 (rounded) **43.** $11\frac{1}{4}$ **44.** 0.98 (rounded) **45.** 250 pounds **46.** 26.7 centimeters (rounded) **47.** $7\frac{3}{20}$ miles **48.** 18.0 miles per gallon (rounded) **49.** 140 residents **50.** $1\frac{1}{4}$ teaspoons **51.** $\frac{5}{2}$ **52.** Answers will vary. Some possibilities are: people with bachelor's degrees earn $2\frac{1}{2}$ times as much as people without high school diplomas; for every \$5 earned by the first group, only \$2 is earned by the second group. **53.** \$1917 (rounded) **54.** \$4250 **55.** \$7 per hour difference **56.** \$1,081,000 over 47 years for a person with a high school diploma; \$1,720,000 over 43 years for the person with a bachelor's degree, or \$639,000 more.

Chapter 6

Section 6.1 (page 371)

1. 0.15 **3.** 0.60 or 0.6 **5.** 0.25 **7.** 1.40 or 1.4 **9.** 0.055 **11.** 1.00 or 1 **13.** 0.005 **15.** 0.0035 **17.** 80% **19.** 58% **21.** 1% **23.** 12.5% **25.** 37.5% **27.** 200% **29.** 370% **31.** 3.12% **33.** 416.2% **35.** 0.28% **37.** Answers will vary. Some possibilities are: No common denominators are needed with percents. The denominator is always 100 with percent, which makes comparisons easier to understand. **39.** 0.45 **41.** 0.18 **43.** 3.5% **45.** 200% **47.** 0.5% **49.** 1.536 **51.** 20 children **53.** 420 employees **55.** 270 chairs **57.** \$142.50 **59.** 820 commuters **61.** 26 plants **63.** **(a)** Since 100% means 100 parts out of 100 parts, 100% is all of the number. **(b)** Answers will vary. **65.** **(a)** Since 200% is two times a number, find 200% of the number by multiplying the number by 2 (double it). **(b)** Answers will vary. **67.** **(a)** Since 10% means 10 parts out of 100 parts or $\frac{1}{10}$, the shortcut for finding 10% of a number is to move the decimal point in the number one place to the left. **(b)** Answers will vary. **69.** 76%; 0.76 **71.** **(a)** health care **(b)** 51%; 0.51 **73.** 16.8%; 0.168 **75.** **(a)** Germany; Bulgaria **(b)** 16.5%; 0.165 **77.** 95%; 5% **79.** 30%; 70% **81.** 55%; 45%

Section 6.2 (page 383)

1. $\frac{1}{4}$ **3.** $\frac{3}{4}$ **5.** $\frac{17}{20}$ **7.** $\frac{5}{8}$ **9.** $\frac{1}{16}$ **11.** $\frac{1}{6}$ **13.** $\frac{1}{15}$ **15.** $\frac{1}{200}$ **17.** $1\frac{1}{5}$ **19.** $3\frac{3}{4}$ **21.** 50% **23.** 80% **25.** 25% **27.** 37% **29.** 62.5% **31.** 87.5% **33.** 36% **35.** 46% **37.** 15% **39.** 83.3% (rounded) **41.** 55.6% (rounded) **43.** 14.3% (rounded) **45.** $\frac{1}{2}$; 50% **47.** $\frac{7}{8}$; 0.875 **49.** $\frac{4}{5}$; 80% **51.** 0.167 (rounded) 16.7% (rounded) **53.** $\frac{1}{4}$; 25% **55.** $\frac{1}{8}$; 0.125 **57.** 0.667 (rounded); 66.7% (rounded) **59.** 0.4; 40% **61.** 0.08; 8% **63.** 0.005; 0.5% **65.** $2\frac{1}{2}$; 250% **67.** 3.25; 325% **69.** There are many possible answers. Examples 2 and 3 show the steps that students should include in their answers. **71.** $\frac{13}{100}$; 0.13; 13% **73.** $\frac{9}{50}$; 0.18; 18% **75.** $\frac{3}{5}$; 0.6; 60% **77.** $\frac{1}{5}$; 0.20; 20% **79.** $\frac{1}{10}$; 0.1; 10% **81.** $\frac{1}{4}$; 0.25; 25% **83.** $\frac{2}{5}$; 0.4; 40% **85.** 100; 100 **86.** **(a)** 765 workers **(b)** 96 letters **(c)** 813 videos **87.** 50; 100; half or $\frac{1}{2}$ **88.** 10; 100; 1; left **89.** 1; 100; 2; left **90.** **(a)** 525 homes **(b)** 37 printers **(c)** \$105

91. Find 10% of 160, then add $\frac{1}{2}$ of 10%.

10%		5%		15%
↓		↓		↓
16	+	8	=	24

92. Find 100% of 160, then add 50% of 160.

100%		50%		150%
↓		↓		↓
160	+	80	=	240

93. From 100% of \$450, subtract 10% of \$450.

100%		10%		90%
↓		↓		↓
\$450	−	\$45	=	\$405

94. To 100% of \$800, add 100% of \$800, and then add 10% of \$800.

100%		100%		10%		210%
↓		↓		↓		↓
\$800	+	\$800	+	\$80	=	\$1680

Section 6.3 (page 393)

1. 50 **3.** 150 **5.** 70 **7.** 25% **9.** 150% **11.** 33.3% (rounded) **13.** 26 **15.** 21.6 **17.** 26.5% (rounded) **19.** 115 **21.** 0.4% **23.** 2.5% **25.** 0.3% **27.** 10; unknown; 60 **29.** 75; \$800; \$600 **31.** 25; \$970; unknown **33.** 20; unknown; 12 **35.** 50; 68; 34 **37.** unknown; \$296; \$177.60 **39.** 3.25; unknown; 54.34 **41.** 0.68; \$487; unknown **43.** percent—the ratio of the part to the whole. It appears with the word *percent* or "%" after it. whole—the entire quantity. Often appears after the word *of*. part—the part being compared with the whole. **45.** percent is unknown; 810; 640 **47.** percent is unknown; 142; 86 **49.** 23; 610; part is unknown **51.** 25; whole is unknown; 240 **53.** 55; 480; part is unknown **55.** 49.5; 822; part is unknown **57.** 16.8; whole is unknown; 504

Section 6.4 (page 405)

1. 16 guests **3.** 1836 military personnel **5.** 4.8 ft **7.** 315 files **9.** 819 trucks **11.** \$3.28 **13.** 1530 tables **15.** 182 homes **17.** \$21.60 **19.** 320 e-mails **21.** 160 hay bales **23.** 550 students **25.** 1360 graduates **27.** 2800 **29.** 50% **31.** 52% **33.** 8% **35.** 1.5% **37.** 18.6% (rounded) **39.** 9.2% **41.** 150% of \$30 cannot be less than \$30 because 150% is greater than 1 (100%). The answer must be greater than \$30. 25% of \$16 cannot be greater than \$16 because 25% is less than 1 (100%). The answer must be less than \$16. **43.** \$52.80 **45.** 44 females **47.** 220 students **49.** \$204.97 **51.** 2% **53.** **(a)** 39% female **(b)** 2.3 million male workers (rounded) **55.** 110% **57.** 281.1 million **59.** March **61.** 52.5 million cans **63.** Hardee's **65.** \$18.06 billion **67.** \$645 **69.** 2156 products **71.** whole; 100 **72.** whole **73.** 108 calories **74.** 300 grams **75.** 67 grams **76.** Yes, since they would eat 28% × 4 servings = 112% of the daily value. **77.** 17 packages (rounded). It may be possible but would result in a diet that is high in total fat, saturated fat, sodium, and total carbohydrates.

Section 6.5 (page 415)

1. 135 hamburgers **3.** 675 garments **5.** 83.2 quarts **7.** 700 tablets **9.** 1029.2 meters **11.** \$4.16 **13.** 160 patients **15.** 325 salads **17.** 680 circuits **19.** 1080 people **21.** 300 gallons **23.** 50% **25.** 76% **27.** 125% **29.** 1.5% **31.** 250% **33.** You must first change the fraction in the percent to a decimal, then divide the percent by 100 to change it to a decimal.

$2\frac{1}{2}\% = 2.5\% = 0.025 \leftarrow$ 2.5% as a decimal

↑ ↑

Change $2\frac{1}{2}$ to 2.5.

35. 3.78 million or 3,780,000 office workers **37.** 82.3 million homes **39.** 924 employers **41.** **(a)** 30%; **(b)** 70% **43.** 4.9% **45.** 2106 people **47.** 2916 people **49.** \$1817.2 million **51.** 478,175 Mustangs **53.** \$510,390 **55.** \$439.61

Section 6.6 (page 425)

1. \$0.40; \$8.40 **3.** 3%; \$437.75 **5.** 5%; \$298.20 **7.** \$672.60 (rounded); \$12,901.60 (rounded) **9.** \$40 **11.** 20% **13.** \$185.51 (rounded) **15.** \$6615 **17.** \$20 (rounded); \$179.99 (rounded) **19.** 30%; \$126 **21.** \$4.38; \$13.12 **23.** \$8.76; \$49.64 **25.** On the basis of commission alone you would choose Company A. Other considerations might be: reputation of the company; expense allowances; other fringe benefits; travel; promotion and training, to name a few. **27.** \$58.12 **29.** \$1170 **31.** 5% **33.** 10.7% (rounded) **35.** 22.5% (rounded) **37.** \$134 **39.** \$284.40 **41.** 4% **43.** \$106.20; \$483.80 **45.** 5.6% (rounded) **47.** \$18.43 (rounded) **49.** \$4276.97 (rounded) **51.** \$9007.47 (rounded) **53.** percent; whole **54.** rate (percent) of tax; cost of item **55.** \$146.25 (rounded) **56.** Yes, the same; (12.4% • \$123) + (6.5% • \$123) = \$23.25 (rounded) and (12.4% + 6.5%) • (\$123) = \$23.25 **57.** \$1356.92 **58.** No. In Exercise 57 the excise tax of \$12.20 cannot be added to the sales tax rate of $7\frac{3}{4}\%$ because the excise tax is an amount, not a percent.

Section 6.7 (page 431)

1. \$6 **3.** \$105 **5.** \$28.80 **7.** \$488.75 **9.** \$2227.50 **11.** \$10 **13.** \$49.20 **15.** \$42.30 **17.** \$16.84 (rounded) **19.** \$647.06 (rounded) **21.** \$210 **23.** \$773.30 **25.** \$2043 **27.** \$3412.50 **29.** \$17,797.81 (rounded) **31.** The answer should include: amount of principal—This is the amount of money borrowed or loaned. interest rate—This is the percent used to calculate the interest. time of loan—The length of time that money is loaned or borrowed is an important factor in determining interest. **33.** \$41.25 **35.** \$22,500 **37.** \$1240 **39.** \$64.68 **41.** \$159.50 **43.** \$12,254.69 (rounded) **45.** \$2165.63 (rounded)

Summary Exercises on Percent (page 435)

1. 0.0625 **2.** 3.80 or 3.8 **3.** 37.5% **4.** 0.6% **5.** $\frac{7}{8}$

6. $1\frac{3}{5}$ **7.** 62.5% **8.** 0.8% **9.** \$46.80 **10.** 500 rolls

11. 55% **12.** 28 screening exams **13.** 363 acres

14. 2.7% (rounded) **15.** \$9.68; \$185.68 **16.** 7%; \$990.82

17. \$301.20 **18.** 6.5% or $6\frac{1}{2}\%$

19. \$8.99 (rounded) \$40.96 (rounded) **20.** 35% \$444.60 **21.** \$60 \$1060 **22.** \$535.50 \$2915.50 **23.** \$44.10 \$1514.10 **24.** \$664.95 \$7484.95 **25.** 22.0% increase (rounded) **26.** 10.5% decrease (rounded)

Section 6.8 (page 441)

1. \$540.80 **3.** \$1966.91 (rounded) **5.** \$4587.79 (rounded) **7.** \$1102.50 **9.** \$1873.52 (rounded) **11.** \$2027.46 (rounded) **13.** \$17,360.41 (rounded) **15.** \$1216.70; \$216.70 **17.** \$14,326.40; \$6326.40 **19.** \$10,976.01 (rounded); \$2547.84 (rounded) **21.** Interest paid on past interest as well as on the principal. Many people describe compound interest as "interest on interest." **23.** \$13,467.61 **25.** **(a)** \$64,201 **(b)** \$26,201 **27.** **(a)** \$87,786 **(b)** \$17,786 **29.** principal; rate; time p; r; t **30.** principal; interest **31.** principal; interest **32.** principal; interest **33.** **(a)** \$254.48 **(b)** Yes; it has more than doubled (about five times). **(c)** yes **34.** **(a)** the compound interest account; \$1091.20 more **(b)** Greater amounts of interest are earned with compound interest than simple interest because interest is earned on both principal and past interest.

Chapter 6 Review Exercises (page 451)

1. 0.35 **2.** 1.5 **3.** 0.9944 **4.** 0.00085 **5.** 315% **6.** 2%

7. 87.5% **8.** 0.2% **9.** $\frac{3}{20}$ **10.** $\frac{3}{8}$ **11.** $1\frac{3}{4}$ **12.** $\frac{1}{400}$

13. 75% **14.** 62.5% or $62\frac{1}{2}\%$ **15.** 325% **16.** 0.5%

17. 0.125 **18.** 12.5% **19.** $\frac{1}{4}$ **20.** 25% **21.** $1\frac{4}{5}$ **22.** 1.8

23. 250 **24.** 24 **25.** 35; 820; 287 **26.** unknown; 90; 73 **27.** 14; 160; unknown **28.** 16; unknown; 418 **29.** unknown; 8; 3 **30.** 88; 1280; unknown **31.** 171 programs **32.** 870 reference books **33.** 31.2 acres **34.** 2.8 kilograms **35.** 750 crates **36.** 2320 test tubes **37.** 484 miles **38.** 17,000 cases **39.** 55% **40.** 5.2% (rounded) **41.** 9.5% (rounded) **42.** 30.8% (rounded)

43. 140,000 math teachers **44.** 10.4% (rounded) **45.** \$145.28 **46.** 186 trucks **47.** 0.4% **48.** 175% **49.** 120 miles **50.** \$575 **51.** \$31.50; \$661.50 **52.** $7\frac{1}{2}\%$; \$838.50 **53.** \$276 **54.** 5% **55.** \$33.75; \$78.75 **56.** 25% \$189 **57.** \$8 **58.** \$67.50 **59.** \$7 **60.** \$152.10 **61.** \$832.50 **62.** \$1598.85 **63.** \$3257.80; \$1257.80 **64.** \$2187.71 (rounded); \$317.71 (rounded) **65.** \$4534.92; \$934.92 **66.** \$16,337.50; \$3837.50 **67.** 12 **68.** 1640 **69.** 23.28 meters **70.** 150% **71.** \$0.51 **72.** 1980 employees **73.** 40% **74.** 498.8 liters **75.** 0.55 **76.** 3 **77.** 500% **78.** 471% **79.** 0.086 **80.** 62.1% **81.** 0.00375 **82.** 0.06% **83.** 75% **84.** $\frac{21}{50}$ **85.** $\frac{7}{8}$ **86.** 37.5% or $37\frac{1}{2}\%$ **87.** $\frac{13}{40}$ **88.** 60% **89.** $\frac{1}{400}$ **90.** 375% **91.** \$1326.94 (rounded) **92.** \$17,405 **93.** 32.4% (rounded) **94. (a)** \$15,780.96 (rounded) **(b)** \$3280.96 (rounded) **95.** \$6225 **96.** 8.1% (rounded) **97.** \$595.07 (rounded) **98.** 185,185 annual sales (rounded) **99.** \$9758 **100.** 13.1% (rounded)

Chapter 6 Test (page 457)

1. 0.65 **2.** 80% **3.** 175% **4.** 87.5% or $87\frac{1}{2}\%$ **5.** 3.00 or 3 **6.** 0.0005 **7.** $\frac{1}{8}$ **8.** $\frac{1}{400}$ **9.** 60% **10.** 62.5% or $62\frac{1}{2}\%$ **11.** 250% **12.** 800 sacks **13.** 20% **14.** \$18,500 **15.** \$2854.20 **16.** \$628 **17.** 30%

18. $\frac{\text{amount of increase}}{\text{last year's salary}} = \frac{p}{100}$

A possible answer is: Part is the increase in salary.

this year − last year = increase

Whole is last year's salary. Percent of increase is unknown.

19. $I = p \cdot r \cdot t$

$$I = 1000 \cdot 0.05 \cdot \frac{9}{12} = \$37.50$$

$$I = 1000 \cdot 0.05 \cdot 2.5 = \$125$$

The interest formula is $I = p \cdot r \cdot t$. If time is in months, it is expressed as a fraction with 12 as the denominator. If time is expressed in years, it is placed over 1 or shown as a decimal number.

20. \$11.52; \$84.48 **21.** \$91; \$189 **22.** \$378 **23.** \$240 **24.** \$5657.75 **25. (a)** \$10,668 (rounded) **(b)** \$1668

Cumulative Review Exercises: Chapters 1–6 (page 459)

1. *Estimate:* 6000 + 90 + 700 = 6790; *Exact:* 6441
2. *Estimate:* 1 + 50 + 8 = 59; *Exact:* 58.574
3. *Estimate:* 80,000 − 50,000 = 30,000; *Exact:* 28,988
4. *Estimate:* 8 − 4 = 4; *Exact:* 4.2971
5. *Estimate:* 7000 × 700 = 4,900,000; *Exact:* 4,628,904
6. *Estimate:* 70 × 9 = 630; *Exact:* 568.11
7. *Estimate:* 40,000 ÷ 40 = 1000; *Exact:* 902
8. *Estimate:* 2000 ÷ 8 = 250; *Exact:* 320
9. *Estimate:* 7 ÷ 1 = 7; *Exact:* 8.45 **10.** 18 **11.** 19 **12.** 39 **13.** 4680 **14.** 7,600,000 **15.** \$513 **16.** \$362.74 **17.** $1\frac{3}{8}$ **18.** $1\frac{1}{12}$ **19.** $13\frac{3}{8}$ **20.** $\frac{1}{8}$ **21.** $3\frac{7}{8}$ **22.** $8\frac{8}{15}$ **23.** $\frac{7}{12}$ **24.** $26\frac{5}{32}$ **25.** $28\frac{4}{5}$ **26.** $\frac{7}{8}$ **27.** 16 **28.** $\frac{11}{30}$ **29.** < **30.** < **31.** > **32.** $\frac{1}{4}$ **33.** 1 **34.** $\frac{1}{4}$ **35.** 0.6 **36.** 0.875 **37.** 0.583 (rounded) **38.** 0.55 **39.** $\frac{3}{1}$ **40.** $\frac{4}{3}$ **41.** $\frac{1}{8}$ **42.** True **43.** True **44.** 6 **45.** 3 **46.** 16 **47.** 30 **48.** 0.78 **49.** 0.03 **50.** 2.00 or 2 **51.** 0.005 **52.** 87% **53.** 380% **54.** 2.3% **55.** $\frac{2}{25}$ **56.** $\frac{5}{8}$ **57.** $1\frac{3}{4}$ **58.** 87.5% or $87\frac{1}{2}\%$ **59.** 30% **60.** 420% **61.** 490 watches **62.** \$39.60 **63.** 180 tires **64.** 1700 miles **65.** 50% **66.** 40% **67.** \$5.45 (rounded); \$114.40 **68.** 6.5% or $6\frac{1}{2}\%$; \$489.90 **69.** \$1003.04 **70.** 2.5% or $2\frac{1}{2}\%$ **71.** \$205.20; \$250.80 **72.** 15%; \$922.25 **73.** \$1115.50 **74.** \$19,863.88 (rounded) **75.** 33 cars **76.** 59.5 ounces **77.** 12.5% **78.** 2% **79.** 5.5% (rounded) **80.** 8.1% (rounded) **81.** \$107,520 **82. (a)** \$19,604.40 (rounded) **(b)** \$4604.40 (rounded)

Appendix B

Appendix B Exercises (page A-11)

1. 37; add 7 **3.** 26; add 10, subtract 2 **5.** 16; multiply by 2 **7.** 243; multiply by 3 **9.** 36; add 3, add 5, add 7, etc.; or 1^2, 2^2, 3^2, etc.

11.

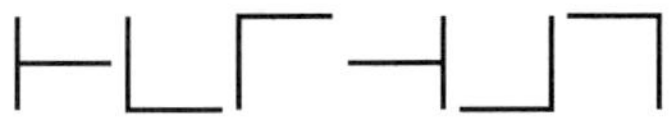

13.

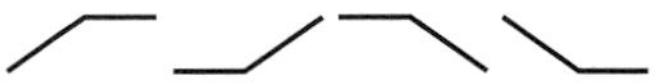

15. Conclusion follows **17.** Conclusion does not follow **19.** 3 days **21.** Dick

Index

Videotape Index

The purpose of this index is to show those exercises from the text that are used in the Real to Reel videotape series.

Section	Exercises	Section	Exercises
1.1	9, 13, 17, 19, 21	4.1	35
1.2	37, 49	4.2	9
1.3	9, 19, 55, 69, 85	4.3A	23
1.4	21, 41, 74, 87	4.5	3, 17
1.5	15, 17, 37, 45	4.6	13, 61
1.6	23	5.1	43
1.9	1, 5, 15	5.2	25
1.10	7, 17, 19	5.3	21
2.1	19	5.4	5, 7, 9, 11, 13, 15
2.2	27	5.5	11, 17
2.3	43	6.1	57, 59
2.4	9, 15, 23, 41, 49	6.2	9, 77
2.5	11, 43	6.3	27, 31, 35, 37, 41
2.7	37	6.4	1, 5, 7, 43, 55, 57
2.8	3, 29	6.5	5, 17, 25
3.2	5, 17	6.6	11, 23, 33
3.3	31	6.7	43
3.4	53	6.8	11
3.5	55		